Wetland Ecosystems

Wetland Ecosystems

Editor: Ewan Lewis

Callisto Reference
www.callistoreference.com

Callisto Reference,
118-35 Queens Blvd., Suite 400,
Forest Hills, NY 11375, USA

Visit us on the World Wide Web at:
www.callistoreference.com

ISBN: 978-1-63239-832-1 (Hardback)

The publisher's policy is to use permanent paper from mills that operate a sustainable forestry policy. Furthermore, the publisher ensures that the text paper and cover boards used have met acceptable environmental accreditation standards.

Printed in China.

Cataloging-in-publication Data

Wetland ecosystems / edited by Ewan Lewis.
p. cm.
Includes bibliographical references and index.
ISBN 978-1-63239-832-1
1. Wetland ecology. 2. Wetland conservation. 3. Biodiversity conservation. 4. Biotic communities. I. Lewis, Ewan.
QH541.5.M3 W48 2017
577.68--dc23

Table of Contents

Preface

Wetlands are areas that are saturated with water for most periods of a season. It becomes a distinct ecosystem due to the presence of hydric soil and unique vegetation that is found there. Wetlands naturally act as barriers against floods as well as provide water sources to animals and humans. Over usage and despoiling of wetlands, which otherwise take longer time to replenish and sustain itself, have led to newer research that focuses on preservation and management of wetlands. This book brings forth some of the most innovative concepts elucidates the unexplored aspects of wetland ecosystems. It strives to provide a fair idea about this discipline and to help better understanding within this field. Scientists and students actively engaged in this field will find this book full of crucial and unexplored concepts.

The purpose of the book is to provide a glimpse into the dynamics and to present opinions and studies of some of the scientists engaged in the development of new ideas in the field from very different standpoints. This book will prove useful to students and researchers owing to its high content quality.

At the end, I would like to appreciate all the efforts made by the authors in completing their chapters professionally. I express my deepest gratitude to all of them for contributing to this book by sharing their valuable works. A special thanks to my family and friends for their constant support in this journey.

Editor

Aboveground Allometric Models for Freeze-Affected Black Mangroves (*Avicennia germinans*): Equations for a Climate Sensitive Mangrove-Marsh Ecotone

Michael J. Osland[1]*, Richard H. Day[1], Jack C. Larriviere[2], Andrew S. From[2]

1 U.S. Geological Survey, National Wetlands Research Center, Lafayette, Louisiana, United States of America, 2 Five Rivers Services, LLC, U.S. Geological Survey, National Wetlands Research Center, Lafayette, Louisiana, United States of America

Abstract

Across the globe, species distributions are changing in response to climate change and land use change. In parts of the southeastern United States, climate change is expected to result in the poleward range expansion of black mangroves (*Avicennia germinans*) at the expense of some salt marsh vegetation. The morphology of *A. germinans* at its northern range limit is more shrub-like than in tropical climes in part due to the aboveground structural damage and vigorous multi-stem regrowth triggered by extreme winter temperatures. In this study, we developed aboveground allometric equations for freeze-affected black mangroves which can be used to quantify: (1) total aboveground biomass; (2) leaf biomass; (3) stem plus branch biomass; and (4) leaf area. Plant volume (i.e., a combination of crown area and plant height) was selected as the optimal predictor of the four response variables. We expect that our simple measurements and equations can be adapted for use in other mangrove ecosystems located in abiotic settings that result in mangrove individuals with dwarf or shrub-like morphologies including oligotrophic and arid environments. Many important ecological functions and services are affected by changes in coastal wetland plant community structure and productivity including carbon storage, nutrient cycling, coastal protection, recreation, fish and avian habitat, and ecosystem response to sea level rise and extreme climatic events. Coastal scientists in the southeastern United States can use the identified allometric equations, in combination with easily obtained and non-destructive plant volume measurements, to better quantify and monitor ecological change within the dynamic, climate sensitive, and highly-productive mangrove-marsh ecotone.

Editor: Ben Bond-Lamberty, DOE Pacific Northwest National Laboratory, United States of America

Funding: This project was supported by the Department of Interior Southeast Climate Science Center (doi.gov/csc/southeast/index.cfm). The funders had no role in study design, data collection and analysis, decision to publish, or preparation of the manuscript.

Competing Interests: JCL & ASF are employees of Fiver Rivers Services, LLC, who provide technical support via a contract with the USGS. Via this contract with the USGS, Five Rivers Services personnel "provide technical support in the following areas: biological, ecological, and hydrologic research; environmental chemistry; geography; river studies; geographic information systems and spatial analysis; information technology, and administrative services." More information about Five Rivers Services and this contract with USGS can be found at this link: http://www.fiveriversservices.com/2011/04/26/five-rivers-services-awarded-us-geological-survey-contract/. There are no patents, products in development or marketed products to declare. This does not alter the authors' adherence to all the PLOS ONE policies on sharing data and materials.

* Email: mosland@usgs.gov

Introduction

In response to climate change and land use change, species distributions are changing across the globe [1–3]. Some of these distributional changes are resulting in comparatively large ecological transformations, especially where traditional ecosystem boundaries (i.e., ecotones) are migrating [4–6]. In areas with shifting ecotones, ecologists and environmental managers seek information that will enable them to better understand the ecological implications of ecosystem transformations.

In coastal wetland ecosystems, abrupt ecotones are common and occur across multiple abiotic gradients (e.g., inundation, salinity, and macroclimatic gradients) [7–10]. Since tidal wetland ecotones involve foundation plant species, sensu [11–14], with divergent growth forms and functionality (e.g., graminoid plants, microbial mats, succulent plants, and woody mangrove plants), the ecological implications of shifting ecotones in coastal wetlands can be substantial.

In the southeastern United States and other tropical-to-temperate climatic transition zones, the mangrove-salt marsh ecotone is dynamic, visually striking, and highly productive [15,16]. Mangroves are woody foundation plant species that are sensitive to extreme winter temperatures and dominant in warmer climates [17,18]. In contrast, salt marsh graminoids and salt marsh succulent plant foundation species are dominant in cooler coastal reaches [8,19] where extreme winter temperature events lead to mangrove mortality and/or limit mangrove forest development, reproduction, and dispersal [10,20,21]. At the poleward mangrove-marsh ecotone, mangrove abundance and coverage is winter temperature-sensitive in that it increases and decreases in response to the absence or presence of extreme winter events, respectively [22–24]. In response to changing climatic conditions, a decrease in the frequency, duration, and/or intensity of extreme winter temperatures is expected which would facilitate poleward mangrove range expansion at the expense of salt marsh vegetation [10]. In the southeastern U.S., the northward expansion of mangroves is expected to occur in Texas, Louisiana, and parts of Florida.

The ecological implications of mangrove expansion are poorly understood. Mangroves and salt marsh ecosystems provide many important ecosystem goods and services including fish and wildlife habitat, carbon storage, nutrient and sediment retention, coastal protection, and maintenance of coastal food webs and fisheries [25]. Although salt marsh and mangrove forest ecosystems are both highly valuable and have many ecosystem goods and services in common, there are structural and functional differences between these two ecosystems as well as differences in the provision of some ecosystem goods and services. Like many other ecotones, structural and functional complexity is higher at the mangrove-marsh ecotone than in adjacent graminoid or mangrove-dominated ecosystems. As a result, it is likely that ecological resilience and the supply of some ecosystem goods and services within the mangrove-marsh ecotone is equivalent to or even greater than in adjacent marsh or mangrove-dominated areas. However, our understanding of the ecosystem goods and services that would be affected by mangrove expansion and replacement of salt marsh is limited, and coastal wetland ecologists seek information and tools that will enable them to better quantify ecological changes associated with mangrove expansion at the expense of salt marshes [26–31]. Here, we provide equations that can be used to quantify changes in important aspects of aboveground mangrove structure near and within the dynamic mangrove-marsh ecotone.

Aboveground biomass is an important and fundamental metric for quantifying change in coastal wetland ecosystems. Many important coastal wetland ecological functions and services are affected by changes in aboveground biomass, structure, and productivity, including resilience to sea level rise [32–34], carbon storage [18,35–37], coastal protection [18,38,39], recreation, and fish and avian habitat [40,41]. Allometric equations provide an approach for estimating plant biomass and structural attributes via easily-obtained and non-destructive measurements. Most allometric models for determining mangrove aboveground biomass have been developed in tropical climates in the absence of extreme freeze events [42]. Within the freeze-dependent mangrove-marsh ecotone, traditional metrics for quantifying mangrove forest aboveground biomass are not appropriate because the morphology and architecture of freeze-affected mangrove trees is different than tropical mangrove trees. In the absence of freeze events and where freshwater and nutrient resources are abundant, tropical mangrove plants can develop into comparatively tall and straight trees. In contrast, at the poleward range limit of mangroves, severe freeze events can either lead to total mangrove mortality or cause aboveground structural damage that is followed by vigorous basal resprouting (Figure 1). As a result, mangroves in freeze-prone areas are often shorter, wider, multi-stemmed, and more shrub-like relative to their tropical counterparts growing in resource-rich environments, but see [43]. Tree-focused allometric models for quantifying aboveground biomass (i.e., dbh and/or height-based models) [42,44–49] are not suitable for freeze-affected and shrubby mangrove individuals.

Our objective was to develop aboveground allometric models for freeze-affected black mangroves (*Avicennia germinans*). In the southeastern United States, *A. germinans* is the most cold-tolerant mangrove species and the most abundant mangrove species at the northern range limit of mangrove forests. Based upon studies from other shrub-dominated ecosystems [50–53] and models for stunted mangroves in New Zealand [54], southern Florida (USA) [55,56], and Iran [57], we expected that crown area and height measurements could be used to develop allometric models for quantifying total aboveground biomass, leaf area, and leaf biomass for freeze-stunted *A. germinans* individuals.

Figure 1. Photos of freeze-affected black mangroves (*Avicennia germinans*) in Louisiana (USA) near their northern range limit. The upper two photos show the shrub-like morphology. The middle left photo shows the size of the smallest individuals included in the analyses. The middle right photo shows leaf damage from an extreme winter temperature event in January, 2014. The lower two photos show the high stem density of freeze-affected individuals.

Materials and Methods

Study Area

Allometric equations were developed using mangroves from two tidal wetland sites near Port Fourchon, Louisiana (USA; 29°8′52″N, −90°14′38″W and 29°6′58″N, −90°11′26″W; Figure 2), which is within the Mississippi River Deltaic Plain. Tidal wetlands in this area are abundant and dominated primarily by *Spartina alterniflora* (smooth cordgrass) and/or *A. germinans*. For more information about the study area, see [26,29,58–62]. Port Fourchon is near the northern limit of *A. germinans* along the Gulf of Mexico coast and is periodically exposed to extreme winter events that can damage or kill *A. germinans* (see photos in Figure 1), reducing areal coverage and limiting structural development. Historical accounts of *A. germinans* in the region document freeze-induced mortality during the 1960's [63] and 1980's [24]. The aerial extent and abundance of *A. germinans* has expanded and contracted in response to extreme winter events [22,24,63,64].

Although most of the *A. germinans* stands in the Port Fourchon vicinity were less than 3 m in height at the time of this study, some individuals were between 4.5 and 5 m in height and some stands were in the 3–4 m height range (RHD and MJO, personal observation). This study was conducted, with permission, on private lands owned by The Louisiana Land and Exploration Company LLC (LL&E) and managed by the Conoco Phillips Company/LL&E office in Houma, Louisiana.

Allometric Measurements

In February and May 2013, we selected a total of 56 *A. germinans* individuals for allometric equation development. We selected 40 individuals in February and 16 individuals in May. For each individual, we measured the height and the basal diameter at 30 cm above the soil surface. We also measured crown diameter in two perpendicular directions, across the widest crown section (CD_1) and at the widest section perpendicular to the first section (CD_2). The two crown diameter measurements were used to calculate crown area via the equation for an ellipse: crown area $= [(CD_1)/2)*(CD_2)/2)]*\pi$. When an individual had multiple stems, height was measured only on the tallest stem and basal diameter was measured only on the thickest stem. Whereas height was measured to the base of the highest leaf, crown diameter measurements were made to the edge of the most horizontally-distant leaf tips. Crown area and height were integrated into a single estimate of volume via the following equation: volume = crown area*height. Once *in situ* measurements were recorded, we cut each individual at the soil surface and transported the samples back to the laboratories of the U.S. Geological Survey's National Wetlands Research Center in Lafayette, Louisiana.

In the laboratory, the February and May samples were processed in a slightly different fashion since the May samples were collected explicitly for the development of leaf area and leaf biomass allometric equations. Individuals collected in February were dried to a constant mass at 60°C to determine total aboveground biomass (i.e., leaves were not separated from stems and branches prior to drying and weighing). In contrast, the May samples were separated into leaves and non-leaves (i.e., stem+ branch) prior to drying, and the area of fresh leaves was determined using a Li-3000C (Licor, Inc., Lincoln, NE, USA). The separated leaf and non-leaf samples were then dried to a constant mass at 60°C to determine leaf biomass, stem+branch biomass, and total aboveground biomass. The data for this study can be found in Dataset S1.

Data Analyses

Linear regression models were developed for the following response variables: (1) total aboveground biomass; (2) leaf biomass; (3) stem plus branch biomass; and (4) leaf area. The independent variables (i.e., crown area, height, volume, and basal diameter) were evaluated alone and in combination via simple and multiple regression analyses. Response and independent variables were natural log transformed prior to analysis. For each equation, we calculated a correction factor (CF) sensu Sprugel [65], where $CF = \exp(SEE^2/2)$ and where SEE is the Standard Error of the Estimate. These correction factors are to be applied to the back-transformed estimates of the response variables in order to correct for a small underestimation associated with using log transformed data in the regression analyses [50,66]. For example, for an allometric equation of the form $\ln(y) = a+b * \ln(x)$, y should be calculated as follows: $y = CF * (\exp[a+b * \ln(x)])$. Data analyses were conducted in SAS Version 9.3 (SAS Institute, Cary, NC, USA) and Sigma Plot Version 12.5 (Systat Software, Inc., San Jose, CA, USA).

Results

We measured *A. germinans* individuals within a height range of 31–157 cm, a crown diameter range of 4–180 cm, and an aboveground biomass range of 1–2,452 g (Table 1). For the smallest plants, the measurements of crown diameter were essentially the lengths of two pairs of leaves on a single un-

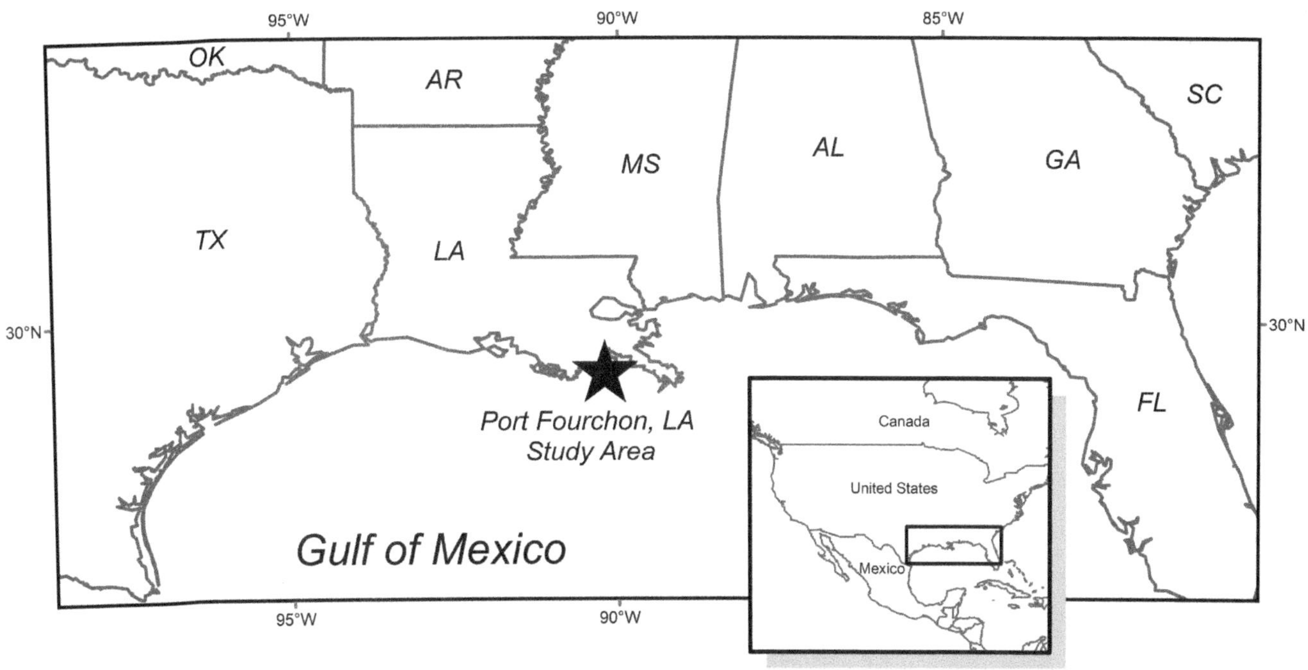

Figure 2. Map highlighting the mangrove-marsh ecotone where this study was conducted (Port Fourchon, Louisiana [USA]).

branched stem. In contrast, the largest plants contained multiple stems and branches extending to almost 1 m from a central base (Figure 1). We developed and compared multiple allometric equations using various combinations of predictor and response variables resulting in the selection of those shown in Table 2 and Table S1. The range limits presented in Table 1 identify boundaries for the application of the developed equations and should be used to avoid unwarranted extrapolation. Plant volume was selected as the best predictor of all of the response variables (i.e., leaf area, leaf biomass, stem plus branch biomass, and total above ground biomass) (Table 2, Figure 3, Table S1). Additional models based upon other combinations of measurements were also significant (Table S1). These additional models may be useful for researchers that want to determine biomass using alternative combinations of variables.

Discussion

In this study, our primary objective was to develop aboveground allometric equations for freeze-affected *A. germinans* individuals near and within the poleward mangrove-marsh ecotone in the southeastern United States. Allometric equations enable scientists to estimate plant community structural attributes via easily-obtained and non-destructive measurements. Although above-ground allometric models have been developed for mangrove species in different abiotic settings [42], most of these models are for trees growing in tropical wet climates. At the poleward range limit of mangrove forests, extreme winter temperature events affect the morphology of mangrove plants and result in plants that sometimes resemble shrubs (i.e., short, wide, and multi-stemmed individuals) rather than trees. To our knowledge, our study is the first to develop allometric models for freeze-affected and shrub-like *A. germinans* individuals at their northern range limit.

Mangrove shrub or dwarf morphologies can result from various abiotic conditions, including drought stress, freeze stress, salinity stress, nutrient limitation, and hydrologic isolation [44,67–71]. Shrub-based aboveground allometric models have been developed for some mangroves in some of these abiotic conditions. In New Zealand, near the southern latitudinal limit of mangroves and in an area with shrubby mangroves less than 1 m tall, Woodroffe [54] developed allometric equations that predict *A. marina* biomass using mean canopy width measurements. In an arid high-latitude mangrove forest in Iran, Parvaresh et al. [57] developed allometric equations that predict *A. marina* biomass from measurements of crown diameter. In a south Florida (USA) dwarf mangrove forest located in an irregularly-inundated basin setting, Ross et al. [55] developed allometric equations that predict *A. germinans*, *Rhizophora mangle*, and *Laguncularia racemosa* biomass from measurements of stem basal diameter and crown volume. In another south Florida dwarf mangrove forest, Coronado-Molina et al. [56] developed allometric equations that predict *Rhizophora mangle* biomass from measurements of height, crown diameter, and the number of prop roots. Collectively, these four studies along with our results show that crown area-focused measurements, sometimes in combination with other measurements (e.g., plant height, number of prop roots), can be used to develop allometric equations for estimating aboveground biomass of mangrove individuals with dwarf or shrub morphologies.

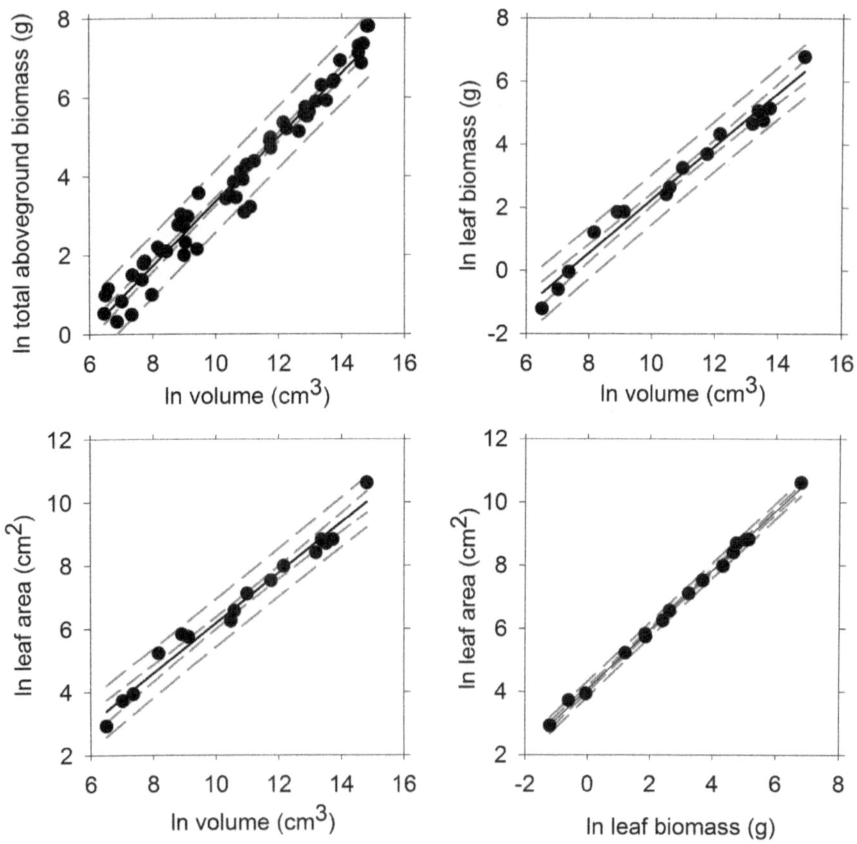

Figure 3. Aboveground allometric relationships for freeze-affected black mangrove (*Avicennia germinans*) individuals. The short and long dashed lines show the 95% confidence and prediction bands, respectively. Note the natural log scale on both axes.

Table 1. Sample size and measurement range of the variables used to develop allometric models for freeze-affected black mangroves (*Avicennia germinans*).

Variable	Sample Size	Minimum	Maximum
Height (cm)	56	31	157
Basal Diameter at 30 cm (mm)	56	1	31
Crown Diameter (cm)	56	4	180
Crown Area (cm^2)	56	19	23,629
Volume (cm^3)	56	641	2,811,792
Total Aboveground Biomass (g)	56	1	2,452
Leaf Biomass (g)	16	0.3	881
Stem and Branch Biomass (g)	16	2	1,556
Leaf Area (cm^2)	16	19	40,941

How do we envision that these equations will be used? In response to future changing winter temperature regimes and accelerated sea level rise, black mangroves are expected to migrate both poleward [10] and landward [72] along the northern Gulf of Mexico coast, in many cases at the expense of salt marshes, tidal freshwater systems, and/or upland forests. At many of these ecotones, ecologists will be challenged to characterize and monitor the structural changes that are already occurring or are expected to occur in the future. To the north and south of these ecotones (i.e., in salt marshes and large mangrove forests, respectively), methods for characterizing aboveground biomass and vegetation structure are well-established. Vegetation methods used in salt marshes are quite different than those used in mangrove forests. For example, whereas ground-layer measurements and vegetation samples (e.g., biomass clip plots and stem measurements from plants less than 2 m in height) are typically taken in small plots (~1-m^2) within marshes, tree-focused measurements (e.g., dbh and/or tree heights) are typically taken in large plots (~100-m^2) within tropical mangrove forests. Vegetation measurements in the freeze-affected and structurally-complex mangrove-marsh ecotone require a hybrid sampling approach that includes measurements tailored for the stunted morphology of freeze-affected mangroves (e.g., crown area and plant height measurements) as well as the multiple other vegetation strata that could be present ranging from graminoid and succulent marsh plants to large (i.e., 2–10 m tall) mangrove forest trees. See Osland et al. [36] for an example of the multiple plot sizes and stratum-specific measurements required to sample marshes and mangroves within a single study design. Sampling designs within mangrove-marsh ecotones often need to be customized to fit the range of plant species and morphologies present. Where conditions allow multi-stemmed shrubby mangroves to grow into large multi-stemmed trees, traditional tree-focused metrics can be applied to each individual stem contained within a tree [43].

Our findings and allometric equations provide a foundation for characterizing aboveground structural changes of freeze-affected black mangroves within mangrove-marsh ecotones. In combination with a robust sampling design, these equations can be used to quantify the temporal and spatial changes in plant community structure in response to the presence or lack of extreme climatic events including freeze events, drought, and hurricanes, as well as slower abiotic changes associated with sea level rise. Plant community structural information can be used to investigate the effects of ecosystem transformations and ecotone migration upon important ecosystem functions and services (e.g., sediment trapping, soil elevation change, nutrient cycling, fish and wildlife abundances, wave attenuation, carbon storage). When paired with carbon concentration data and information regarding belowground processes and soil carbon accumulation, our equations can provide a foundation for quantifying the aboveground carbon present in freeze-affected black mangrove forests which is information that can contribute to estimates of carbon sequestration and inform best practices for carbon management and climate change mitigation [35,37]. The biomass equations can also be used to evaluate plant community development and functional equivalency after wetland restoration and/or management. The leaf area equations may be incorporated into ecophysiological and/or biogeochemical models, for example [60,73].

In summary, we developed aboveground allometric equations for freeze-affected black mangroves (*A. germinans*) which can be used to quantify the following: (1) total aboveground biomass; (2) leaf biomass; (3) stem plus branch biomass; and (4) leaf area. Plant

Table 2. Selected allometric equations for freeze-affected black mangrove (*Avicennia germinans*) individuals.

Response (y)	Predictor (x)	*a* (SE)	*b* (SE)	Adj-R^2	RMSE	CF	d.f.
Total Aboveground Biomass	Volume	−4.8045 (0.2250)	0.8157 (0.0204)	0.97	0.39	1.0781	56
Leaf Biomass	Volume	−6.2219 (0.3834)	0.8468 (0.0348)	0.98	0.36	1.0654	16
Stem plus Branch Biomass	Volume	−4.5075 (0.2971)	0.7682 (0.0269)	0.98	0.28	1.0388	16
Leaf Area	Volume	−1.8036 (0.3681)	0.7981 (0.0334)	0.97	0.34	1.0602	16
Leaf Area	Leaf Biomass	4.0626 (0.0437)	0.9418 (0.0121)	0.99	0.11	1.0057	16

These are all equations of the following form: $\ln(y) = a + b*\ln(x)$. CF is the correction factor sensu Sprugel [65]. Additional equations can be found in Table S1.

volume (i.e., a combination of crown area and plant height), was selected as the optimal predictor of these response variables. Coastal wetland scientists in the southeastern United States can use these equations to better quantify and monitor ecological changes within the dynamic and climate sensitive mangrove-marsh ecotone.

Supporting Information

Table S1 Additional allometric regression equations for freeze-affected black mangrove (*Avicennia germinans*) individuals.

Dataset S1 The data used to develop allometric equations for freeze-affected black mangrove (*Avicennia germinans*) individuals. An excel file that has two spreadsheets: (1) a data spreadsheet (spreadsheet entitled "data"); and (2) a spreadsheet that includes a description of column names and units contained within the data spreadsheet (spreadsheet entitled "legend").

Acknowledgments

We thank Erik Yando for assistance in the field, Ken Krauss for leaf area measurement recommendations, and the ConocoPhillips Company/Louisiana Land and Exploration Company LLC for permission to conduct research on their land. The contents of the manuscript are solely the responsibility of the authors and do not necessarily represent the views of the Southeast Climate Science Center or the USGS. Any use of trade, product, or firm names is for descriptive purposes only and does not imply endorsement by the U.S. Government.

Author Contributions

Conceived and designed the experiments: MJO RHD JCL ASF. Performed the experiments: MJO RHD JCL ASF. Analyzed the data: MJO RHD. Wrote the paper: MJO RHD JCL ASF.

References

1. Parmesan C, Yohe G (2003) A globally coherent fingerprint of climate change impacts across natural systems. Nature 421: 37–42.
2. Millenium Ecosystem Assessment (2005) Ecosystems and human well-being: Synthesis. Washington, D.C.: Island Press. 137 p.
3. Chen IC, Hill JK, Ohlemüller R, Roy DB, Thomas CD (2011) Rapid range shifts of species associated with high levels of climate warming. Science 333: 1024–1026.
4. Allen CD, Breshears DD (1998) Drought-induced shift of a forest–woodland ecotone: rapid landscape response to climate variation. Proceedings of the National Academy of Sciences 95: 14839–14842.
5. Neilson RP (1993) Transient ecotone response to climatic change: some conceptual and modelling approaches. Ecological Applications: 385–395.
6. Risser PG (1995) The status of the science examining ecotones: A dynamic aspect of landscape is the area of steep gradients between more homogeneous vegetation associations. BioScience 45: 318–325.
7. Jiang J, Gao D, DeAngelis DL (2012) Towards a theory of ecotone resilience: Coastal vegetation on a salinity gradient. Theoretical Population Biology 82: 29–37.
8. Pennings SC, Bertness MD (2001) Salt marsh communities. In: Bertness MD, Gaines SD, Hay M, editors. Marine Community Ecology. Sunderland: Sinauer Associates. pp. 289–316.
9. Marani M, D'Alpaos A, Lanzoni S, Carniello L, Rinaldo A (2010) The importance of being coupled: Stable states and catastrophic shifts in tidal biomorphodynamics. Journal of Geophysical Research: Earth Surface 115: F04004.
10. Osland MJ, Enwright N, Day RH, Doyle TW (2013) Winter climate change and coastal wetland foundation species: salt marshes versus mangrove forests in the southeastern U.S. Global Change Biology 19: 1482–1494.
11. Ellison AM, Bank MS, Clinton BD, Colburn EA, Elliott K, et al. (2005) Loss of foundation species: consequences for the structure and dynamics of forested ecosystems. Frontiers in Ecology and the Environment 3: 479–486.
12. Angelini C, Altieri AH, Silliman BR, Bertness MD (2011) Interactions among foundation species and their consequences for community organization, biodiversity, and conservation. BioScience 61: 782–789.
13. Bruno JF, Bertness MD (2001) Habitat modification and facilitation in benthic marine communities. In: Bertness MD, Gaines SD, Hay J, editors. Marine Community Ecology. Sunderland: Sinauer Associates. pp. 201–218.
14. Dayton PK (1972) Toward an understanding of community resilience and the potential effects of enrichments to the benthos at McMurdo Sound, Antartica. In: Parker BC, editor. Proceedings of the Colloquium on Conservation Problems in Antartica. Lawrence: Allen Press. pp. 81–96.
15. McKee K, Rogers K, Saintilan N (2012) Response of salt marsh and mangrove wetlands to changes in atmospheric CO_2, climate, and sea level. In: Middleton BA, editor. Global Change and the Function and Distribution of Wetlands: Global Change Ecology and Wetlands. Dordrecht: Springer. pp. 63–96.
16. Saintilan N, Wilson NC, Rogers K, Rajkaran A, Krauss KW (2014) Mangrove expansion and salt marsh decline at mangrove poleward limits. Global Change Biology 20: 147–157.
17. Tomlinson PB (1986) The Botany of Mangroves. New York: Cambridge University Press. 413 p.
18. Alongi DM (2009) The Energetics of Mangrove Forests. New York: Springer. 216 p.
19. Adam P (1990) Saltmarsh Ecology. Cambridge: Cambridge University Press.
20. Cavanaugh KC, Kellner JR, Forde AJ, Gruner DS, Parker JD, et al. (2014) Poleward expansion of mangroves is a threshold response to decreased frequency of extreme cold events. Proceedings of the National Academy of Sciences 111: 723–727.
21. Pickens CN, Hester MW (2011) Temperature tolerance of early life history stages of black mangrove *Avicennia germinans*: implications for range expansion. Estuaries and Coasts 34: 824–830.
22. Giri C, Long J, Tieszen L (2011) Mapping and monitoring Louisiana's mangroves in the aftermath of the 2010 Gulf of Mexico Oil Spill. Journal of Coastal Research 27: 1059–1064.
23. Ross MS, Ruiz PL, Sah JP, Hanan EJ (2009) Chilling damage in a changing climate in coastal landscapes of the subtropical zone: a case study from south Florida. Global Change Biology 15: 1817–1832.
24. Woodroffe CD, Grindrod J (1991) Mangrove biogeography: the role of Quaternary environmental and sea-level change. Journal of Biogeography 18: 479–492.
25. Barbier EB, Hacker SD, Kennedy C, Koch EW, Stier AC, et al. (2011) The value of estuarine and coastal ecosystem services. Ecological Monographs 81: 169–193.
26. Perry CL, Mendelssohn IA (2009) Ecosystem effects of expanding populations of *Avicennia germinans* in a Louisiana salt marsh. Wetlands 29: 396–406.
27. Comeaux RS, Allison MA, Bianchi TS (2012) Mangrove expansion in the Gulf of Mexico with climate change: Implications for wetland health and resistance to rising sea levels. Estuarine, Coastal and Shelf Science 96: 81–95.
28. Chavez-Ramirez F, Wehtje W (2012) Potential impact of climate change scenarios on whooping crane life history. Wetlands 32: 11–20.
29. Henry KM, Twilley RR (2013) Soil development in a coastal Louisiana wetland during a climate-induced vegetation shift from salt marsh to mangrove. Journal of Coastal Research 29: 1273–1283.
30. Lewis DB, Brown JA, Jimenez KL (2014) Effects of flooding and warming on soil organic matter mineralization in *Avicennia germinans* mangrove forests and *Juncus roemerianus* salt marshes. Estuarine, Coastal and Shelf Science 139: 11–19.
31. Bianchi TS, Allison MA, Zhao J, Li X, Comeaux RS, et al. (2013) Historical reconstruction of mangrove expansion in the Gulf of Mexico: linking climate change with carbon sequestration in coastal wetlands. Estuarine, Coastal and Shelf Science 119: 7–16.
32. Krauss KW, McKee KL, Lovelock CE, Cahoon DR, Saintilan N, et al. (2014) How mangrove forests adjust to rising sea level. New Phytologist 202: 19–34.
33. Kirwan ML, Megonigal JP (2013) Tidal wetland stability in the face of human impacts and sea-level rise. Nature 504: 53–60.
34. Morris JT, Sundareshwar PV, Nietch CT, Kjerfve B, Cahoon D (2002) Responses of coastal wetlands to rising sea level. Ecology 83: 2869–2877.
35. Duarte CM, Losada IJ, Hendriks IE, Mazarrasa I, Marbà N (2013) The role of coastal plant communities for climate change mitigation and adaptation. Nature Climate Change 3: 961–968.
36. Osland MJ, Spivak AC, Nestlerode JA, Lessmann JM, Almario AE, et al. (2012) Ecosystem development after mangrove wetland creation: plant-soil change across a 20-year chronosequence. Ecosystems 15: 848–866.
37. Mcleod E, Chmura GL, Bouillon S, Salm R, Björk M, et al. (2011) A blueprint for blue carbon: toward an improved understanding of the role of vegetated coastal habitats in sequestering CO_2. Frontiers in Ecology and the Environment 9: 552–560.
38. Gedan KB, Kirwan ML, Wolanski E, Barbier EB, Silliman BR (2011) The present and future role of coastal wetland vegetation in protecting shorelines: answering recent challenges to the paradigm. Climatic Change 106: 7–29.
39. Feagin RA, Lozada-Bernard SM, Ravens TM, Möller I, Yeager KM, et al. (2009) Does vegetation prevent wave erosion of salt marsh edges? Proceedings of the National Academy of Sciences 106: 10109–10113.
40. Odum WE, McIvor CC, Smith TJ III (1982) The ecology of mangroves of south Florida: a community profile: U.S. Fish and Wildlife Service, Office of Biological Services, Washington, D.C. FWS/OBS-81/24. 144 p.

41. Caudill MC (2005) Nekton utilization of black mangrove (*Avicennia germinans*) and smooth cordgrass (*Spartina alterniflora*) sites in southwestern Caminada Bay, Louisiana. M.S. Thesis. Louisiana State University, Department of Oceanography and Coastal Sciences, Baton Rouge, LA.
42. Komiyama A, Ong JE, Poungparn S (2008) Allometry, biomass, and productivity of mangrove forests: A review. Aquatic Botany 89: 128–137.
43. Clough BF, Dixon P, Dalhaus O (1997) Allometric relationships for estimating biomass in multi-stemmed mangrove trees. Australian Journal of Botany 45: 1023–1031.
44. Cintrón G, Schaeffer Novelli Y (1984) Methods for studying mangrove structure. In: Snedacker SC, Snedaker JG, editors. The mangrove ecosystem: research methods. Paris: UNESCO. pp. 91–113.
45. Smith TJ III, Whelan KR (2006) Development of allometric relations for three mangrove species in South Florida for use in the Greater Everglades Ecosystem restoration. Wetlands Ecology and Management 14: 409–419.
46. Day JW, Conner WH, Ley-Lou F, Day RH, Navarro AM (1987) The productivity and composition of mangrove forests, Laguna de Términos, Mexico. Aquatic Botany 27: 267–284.
47. Fromard F, Puig H, Mougin E, Marty G, Betoulle JL, et al. (1998) Structure, above-ground biomass and dynamics of mangrove ecosystems: new data from French Guiana. Oecologia 115: 39–53.
48. Clough BF, Scott K (1989) Allometric relationships for estimating above-ground biomass in six mangrove species. Forest Ecology and Management 27: 117–127.
49. Clough B, Dixon P, Dalhaus O (1997) Allometric relationships for estimating biomass in multi-stemmed mangrove trees. Australian Journal of Botany 45: 1023–1031.
50. Cleary MB, Pendall E, Ewers BE (2008) Testing sagebrush allometric relationships across three fire chronosequences in Wyoming, USA. Journal of Arid Environments 72: 285–301.
51. Uresk DW, Gilbert RO, Rickard WH (1977) Sampling big sagebrush for phytomass. Journal of Range Management 30: 311–314.
52. Northup BK, Zitzer SF, Archer S, McMurtry CR, Boutton TW (2005) Above-ground biomass and carbon and nitrogen content of woody species in a subtropical thornscrub parkland. Journal of Arid Environments 62: 23–43.
53. McClaran MP, McMurtry CR, Archer SR (2013) A tool for estimating impacts of woody encroachment in arid grasslands: allometric equations for biomass, carbon and nitrogen content in *Prosopis velutina*. Journal of Arid Environments 88: 39–42.
54. Woodroffe CD (1985) Studies of a mangrove basin, Tuff Crater, New Zealand: I. Mangrove biomass and production of detritus. Estuarine, Coastal and Shelf Science 20: 265–280.
55. Ross MS, Ruiz PL, Telesnicki GJ, Meeder JF (2001) Estimating above-ground biomass and production in mangrove communities of Biscayne National Park, Florida (USA). Wetlands Ecology and Management 9: 27–37.
56. Coronado-Molina C, Day JW, Reyes E, Perez BC (2004) Standing crop and aboveground biomass partitioning of a dwarf mangrove forest in Taylor River Slough, Florida. Wetlands Ecology and Management 12: 157–164.
57. Parvaresh H, Parvaresh E, Zahedi G (2012) Establishing allometric relationship using crown diameter for the estimation of above-ground biomass of grey mangrove, *Avicennia Marina* (Forsk) Vierh in mangrove forests of Sirik, Iran. Journal of Basic and Applied Scientific Research 2: 1763–1769.
58. Patterson S, McKee KL, Mendelssohn IA (1997) Effects of tidal inundation and predation on *Avicennia germinans* seedling establishment and survival in a sub-tropical mangal/salt marsh community. Mangroves and Salt Marshes 1: 103–111.
59. Stagg CL, Mendelssohn IA (2011) Controls on resilience and stability in a sediment-subsidized salt marsh. Ecological Applications 21: 1731–1744.
60. Krauss KW, McKee KL, Hester MW (2014) Water use characteristics of black mangrove (*Avicennia germinans*) communities along an ecotone with marsh at a northern geographical limit. Ecohydrology 7: 354–365.
61. Alleman LK, Hester MW (2011) Refinement of the fundamental niche of black mangrove (*Avicennia germinans*) seedlings in Louisiana: applications for restoration. Wetlands Ecology and Management 19: 47–60.
62. McKee KL, Rooth JE (2008) Where temperate meets tropical: multi-factorial effects of elevated CO_2, nitrogen enrichment, and competition on a mangrove-salt marsh community. Global Change Biology 14: 971–984.
63. West RC (1977) Tidal salt-marsh and mangal formations of Middle and South America. In: Chapman VJ, editor. Wet coastal ecosystems. Amsterdam: Elsevier. pp. 193–213.
64. Michot TC, Day RH, Wells CJ (2010) Increase in black mangrove abundance in coastal Louisiana. Lousiana Natural Resource News. Newsletter of the Lousiana Association of Professional Biologists, January 2010. 4–5.
65. Sprugel D (1983) Correcting for bias in log-transformed allometric equations. Ecology 64: 209–210.
66. Beauchamp JJ, Olson JS (1973) Corrections for bias in regression estimates after logarithmic transformation. Ecology 54: 1403–1407.
67. Lugo AE, Snedaker SC (1974) The ecology of mangroves. Annual Review of Ecology and Systematics 5: 39–64.
68. Pool DJ, Snedaker SC, Lugo AE (1977) Structure of mangrove forests in Florida, Puerto Rico, Mexico, and Costa Rica. Biotropica 9: 195–212.
69. Feller IC, McKee KL, Whigham DF, O'Neill JP (2003) Nitrogen vs. phosphorus limitation across an ecotonal gradient in a mangrove forest. Biogeochemistry 62: 145–175.
70. Castañeda-Moya E, Twilley RR, Rivera-Monroy VH (2013) Allocation of biomass and net primary productivity of mangrove forests along environmental gradients in the Florida Coastal Everglades, USA. Forest Ecology and Management 307: 226–241.
71. Cintrón G, Lugo AE, Pool DJ, Morris G (1978) Mangroves of arid environments in Puerto Rico and adjacent islands. Biotropica 10: 110–121.
72. Doyle TW, Krauss KW, Conner WH, From AS (2010) Predicting the retreat and migration of tidal forests along the northern Gulf of Mexico under sea-level rise. Forest Ecology and Management 259: 770–777.
73. Lovelock CE, Ball MC, Choat B, Engelbrecht BM, Holbrook NM, et al. (2006) Linking physiological processes with mangrove forest structure: phosphorus deficiency limits canopy development, hydraulic conductivity and photosynthetic carbon gain in dwarf Rhizophora mangle. Plant, cell & environment 29: 793–802.

2

Physical Stress, Not Biotic Interactions, Preclude an Invasive Grass from Establishing in Forb-Dominated Salt Marshes

Qiang He[1], Baoshan Cui[1]*, Yuan An[2]

1 State Key Laboratory of Water Environmental Simulation, School of Environment, Beijing Normal University, Beijing, China, **2** School of Agriculture and Biology, MOA Key Laboratory of Urban Agriculture (South), Shanghai Jiao Tong University, Shanghai, China

Abstract

Background: Biological invasions have become the focus of considerable concern and ecological research, yet the relative importance of abiotic and biotic factors in controlling the invasibility of habitats to exotic species is not well understood. *Spartina* species are highly invasive plants in coastal wetlands; however, studies on the factors that control the success or failure of *Spartina* invasions across multiple habitat types are rare and inconclusive.

Methodology and Principal Findings: We examined the roles of physical stress and plant interactions in mediating the establishment of the smooth cordgrass, *Spartina alterniflora*, in a variety of coastal habitats in northern China. Field transplant experiments showed that cordgrass can invade mudflats and low estuarine marshes with low salinity and frequent flooding, but cannot survive in salt marshes and high estuarine marshes with hypersaline soils and infrequent flooding. The dominant native plant *Suaeda salsa* had neither competitive nor facilitative effects on cordgrass. A common garden experiment revealed that cordgrass performed significantly better when flooded every other day than when flooded weekly. These results suggest that physical stress rather than plant interactions limits cordgrass invasions in northern China.

Conclusions and Significance: We conclude that *Spartina* invasions are likely to be constrained to tidal flats and low estuarine marshes in the Yellow River Delta. Due to harsh physical conditions, salt marshes and high estuarine marshes are unlikely to be invaded. These findings have implications for understanding *Spartina* invasions in northern China and on other coasts with similar biotic and abiotic environments.

Editor: Justin Wright, Duke University, United States of America

Funding: Funding was provided by the National Natural Science Foundation of China (U0833002; 41071330) (http://www.nsfc.gov.cn/); China National Funds for Distinguished Young Scientists (51125035) (http://www.nsfc.gov.cn/); and the Fundamental Research Funds for the Central Universities (2009SD-24)(http://www.moe.edu.cn/). The funders had no role in study design, data collection and analysis, decision to publish, or preparation of the manuscript.

Competing Interests: The authors have declared that no competing interests exist.

* E-mail: cuibs@bnu.edu.cn

Introduction

Biological invasions have recently become the focus of considerable concern and ecological research [1]. Although the ecological consequences of biological invasions have been well documented [2,3], the mechanisms that control invasions and the factors that mediate the susceptibility or resistance of habitats to exotic species invasions, i.e., invasibility [2,4], remain unclear [5].

The invasibility of habitats to a specific exotic species after propagules arrive is potentially determined by several abiotic and biotic factors [2,4,6]. Many studies have investigated the role of either the physical environment or properties of native communities in determining the invasibility of habitats, yet studies investigating the relative importance of both these factors across habitat types are only a few [6,7,8,9]. Here, we examine the roles of physical stress and plant interactions in mediating the establishment of smooth cordgrass, *Spartina alterniflora*, across estuarine and non-estuarine habitats in northern China.

Estuarine and coastal systems are among the most heavily invaded systems in the world [10]. *S. alterniflora* and several other species in the widely invasive genus *Spartina* have successfully invaded coastal wetlands in China and across the globe [11,12,13]. These invasions can dramatically impact the structure and function of invaded wetlands [12,14]. The morphology, physiology (e.g. [15,16,17,18,19,20]) and invasiveness (e.g. [12,21,22,23,24,25,26]) of *Spartina* species have all been well documented. However, the invasibility of habitats to *Spartina*, which is of particular importance for predicting future *Spartina* invasions [6,8,25,26,27], needs to be further examined.

Spartina alterniflora (hereafter referred to as cordgrass), introduced to China in 1979 [22,28] was suggested to be capable of invading both high salinity coastal marshes and low salinity estuaries [22]. Previous studies focused on the impact of cordgrass invasions on the structure and function of native communities (see [28,29] for reviews). A previous study [8] used pot experiments to investigate the interactive effects of plant competition and environmental gradients on cordgrass invasions in the Yangtze estuary in southern China, but such experiments were seldom done in the field, particularly in northern China, which differs from southern China in both the physical environment and community structure.

In contrast to southern coastal marshes dominated by perennial grass species, most northern salt marshes are dominated by the annual succulent *Suaeda salsa*, a highly stress-tolerant but competitively inferior species [28,30,31].

In this study, we focus on the invasibility of habitats to cordgrass in the Yellow River Delta in northern China by testing the hypothesis that cordgrass is prevented from invading hypersaline marshes by harsh physical conditions rather than interspecific competition. The habitats examined occur in different zones in both estuarine and non-estuarine systems. We first describe the physical environment and vegetation in these habitats, and then examine how cordgrass performance is mediated by physical stress and plant interactions in different habitats using transplant and common garden experiments.

Materials and Methods

Study sites

Field work was conducted in the Yellow River Delta National Nature Reserve, northern China (Figure 1). We got permits from the managers of the reserve to enter any parts of the reserve, do field survey and manipulative experiments, and collect soil and plant samples. No specific permissions were required for the study sites/activities, and no endangered or protected species were involved in the present study. An estuarine site and a non-estuarine salt marsh site in the reserve were used. The estuarine site is a low salinity system with significant freshwater input, particularly during the summer. In contrast, the salt marsh site is a high salinity system dominated by seawater input. Both sites experience irregular, semidiurnal tides and a warm-temperate climate. Substrate salinity, one of the most limiting physical stressors in coastal habitats, increases with elevation at both sites [31,32].

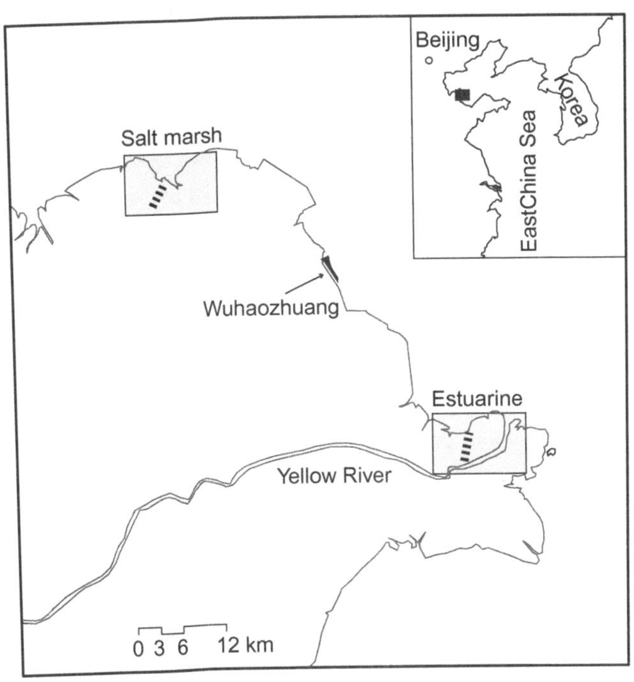

Figure 1. Map showing study sites in the Yellow River Delta, northern China. Cordgrass was artificially planted at Wuhaozhuang in the 1980s, and has invaded the estuarine site. Cordgrass was collected at Wuhaozhuang and transplanted into different zones (along the dotted lines) at the salt marsh and estuarine sites.

At each site, three zones were defined based on elevation and vegetation: mudflat, low marsh and high marsh. At the estuarine site, cordgrass has recently colonized the mudflats and is in the early stages of invasion [32]. *Suaeda salsa*, an annual halophyte, dominates the low and high marsh zones. Salt pans and *Tamarix chinensis*, a shrubby halophyte, are also common at upper elevations, while patches of *Phragmites australis* are found in the low marsh zone. Other plants observed at the estuarine site include *Salicornia europaea*, *Scirpus* spp. and *Juncus* spp. [32].

Cordgrass has not invaded the mudflat zone at the salt marsh site, which remains unvegetated (Qiang He, personal observation). The low and high marsh zones at the salt marsh site are similarly dominated by *S. salsa* and sparse *T. chinensis* at upper elevations, but overall vegetation cover is much lower and hypersaline salt pans without vegetation are prevalent [31]. Few other plant species occur there.

Quantification of physical stresses and vegetation

To quantify physical stresses at the two sites, we randomly collected eight soil samples (5.05 cm diameter×5 cm depth) in each of the three zones at each site in July 2010. At the salt marsh site where the cordgrass transplant experiment was initiated in 2009, soil samples were also collected in July 2009. As our previous studies [31,32] have documented seasonal variation in physical stresses at these sites, we simply quantified relative differences among zones and sites, and did not sample on multiple dates.

Soil cores were oven-dried and weighed to determine soil moisture. Dried soil was mixed with a known volume of deionized water. The salinity of the supernatant was measured after 24 hours using an electronic meter, and the original soil pore water salinity was calculated based on the initial water content of the core [33]. We estimated flooding frequency for each zone based on long-term field observations at each site (see [31,32]).

Vegetation in the low and high marsh zones at each site was quantified in August 2010. We randomly located 15 quadrats (1×1 m) in each habitat at each site and visually estimated percentage vegetation covers (mostly *S. salsa* plants).

Differences in abiotic factors and vegetation cover among zones and sites in 2010 were analyzed using generalized linear models, as these data did not meet the assumptions of ANOVA. Generalized linear models are a widely accepted alternative test when data cannot be transformed to meet the assumptions of ANOVA [34]. We used Gamma distribution with log link for soil salinity and moisture, and Poisson distribution with log link for vegetation cover data. Differences in soil salinity and moisture between 2009 and 2010 at the salt marsh site were examined using one-way ANOVAs, or non-parametric Wilcoxon tests if the data did not meet the assumptions of ANOVA after transformation.

Field transplant experiments

To determine the invasibility of various habitats to cordgrass in the Yellow River Delta, we transplanted cordgrass replicates into the mudflat, low marsh and high marsh zones at the salt marsh site in May 2009. As the mudflat and the high marsh zones had little vegetation, the experimental design was not fully factorial with neighboring vegetation. We excavated a number of substrate blocks (16×16×20 cm) containing emerging cordgrass tillers at Wuhaozhuang where cordgrass was introduced in the 1980s (Figure 1). These substrate blocks were immediately transplanted into the mudflat ($n=15$), low marsh ($n=16$) and high marsh ($n=12$) zones at the salt marsh site. We mainly aimed to compare cordgrass performance among different zones at this site, and did not transplant cordgrass back to Wuhaozhuang as controls. The transplants were watered with fresh water every other day for a

Table 1. Physical stresses and vegetation cover in different zones at the salt marsh and estuarine sites.

	Salt marsh site			Estuarine site			P		
	High marsh	Low marsh	Mudflat	High marsh	Low marsh	Mudflat	Zone	Site	Site*Zone
Salinity (g/kg)									
2009	160.6(17.1)	109.0(3.1)	46.9(2.3)	-	-	-	-	-	-
2010	211.29(12.8)	76.64(3.4)	30.68(1.5)	164.68(9.7)	45.1(0.6)	29.7(0.8)	<0.01	<0.01	<0.01
Moisture (%)									
2009	18.9(0.3)	23.1(0.2)	22.3(0.2)	-	-	-	-	-	-
2010	17.4(0.7)	22.6(0.3)	23.4(0.4)	19.0(0.6)	24.5(0.5)	22.2(0.4)	<0.01	0.30	0.02
Flooding (%)	5–20	30–50	70–90	5–15	40–60	80–90	-	-	-
Vegetation cover (%)	3.87(1.5)	10.47(2.0)	-	83.67(1.6)	38.67(2.7)	-	0.17	<0.01	<0.01

Data are means ± 1SE. Sample sizes are 8 for salinity, moisture, and 15 for vegetation cover. Flooding frequency (shown as a range) was estimated based on previous studies and field observations.
Statistical analyses were conducted using generalized linear models. Gamma regressions with log link were used for salinity and moisture while a Poisson regression with log link was used for vegetation cover. -, no data. Flooding indicates the percentage of days flooded in a year.

week to reduce transplant shock. Half of the transplants in the low marsh were assigned to neighbor-removal treatments. Neighbors surrounding the transplants were removed by clipping a 25 cm-radius border at the soil surface every other week as needed. Neighbors adjacent to the remaining transplants were unmanipulated. Neighbor removal treatments were maintained semi-monthly as needed throughout the growing season. We did not conduct neighbor removal treatments for the high marsh transplants, as native plants were very sparse in the high marsh with only a few scattered *S. salsa* and stunted *T. chinensis*. Transplants were maintained for two growing seasons. The number of stems, inflorescences and maximum plant height of each transplant were determined at the end of the growing season in September 2009 and 2010. Above-ground biomass was harvested in September 2010, oven-dried at 60°C for 48 hours, and weighed.

We conducted a second transplant experiment at the low salinity estuarine site in 2010 to investigate additional habitats with varying abiotic stress. Substrate blocks (same size and origin as 2009) with emerging cordgrass tillers were excavated and transplanted into the mudflat ($n = 8$), low marsh ($n = 16$) and high marsh ($n = 16$) zones. These transplants were watered for the first week, and half of the transplants in the low and high marsh zones were assigned and maintained as neighbor-removal treatments as described above. Neighbors adjacent to the remaining transplants were unmanipulated. In September 2010, the number of stems, inflorescences and maximum plant height of each transplant were determined, above-ground biomass harvested, oven-dried and weighed.

The effect of zone on cordgrass performance was analyzed using non-parametric multiple comparisons (Steel test, [35]), as high mortality of cordgrass transplants made the data highly skewed. The effect of native vegetation was examined using one-way ANOVAs or non-parametric Wilcoxon tests, depending on whether the assumptions of ANOVA were met.

Common garden experiment

To examine the role of flooding on cordgrass performance in the Yellow River Delta salt marshes, we conducted a common garden experiment. In May 2010, substrate blocks (16×16×20 cm) with emerging cordgrass tillers were excavated at Wuhaozhuang and transplanted into a common garden at our field station. Two cordgrass blocks were planted in 0.9×0.9 m tanks ($n = 12$) filled with soil collected from a salt marsh near the common garden. The soils were mixed thoroughly to ensure homogeneity among plots prior to treatments. The soil in each plot was approximately 30 cm deep. We did not plant *S. salsa* in these artificial plots, as our 2009 field experiment suggested that *S. salsa* had no effect on cordgrass performance. Plots were watered with seawater every other day for the first two weeks to reduce transplant shock. We then randomly assigned half of the plots to one of two different flooding regimes: every other day and weekly, to simulate the natural flooding of the low and high marsh zones, respectively. Seawater collected from the nearest tidal channel (filled with seawater at high tide) was used for flooding treatments. Flooding treatments were done by pouring ~ 25 L seawater into each plot in late afternoons and letting it drain through a 6 cm hole on one side of the plot. In late September 2010, we determined the performance of cordgrass transplants in each plot by counting the number of stems and measuring maximum plant height. Above-ground biomass in each plot was harvested, oven-dried and weighed. The effect of flooding on cordgrass performance was examined using one-way ANOVAs or non-parametric Wilcoxon tests, depending on whether the assumptions of ANOVA were met.

Results

Physical stresses and vegetation

Physical stresses and vegetation varied between zones and sites (Table 1). For both the salt marsh and estuarine sites, soil salinity was lowest in the mudflat, intermediate in the low marsh and highest in the high marsh (Table 1). Soil salinity in the low and high marsh zones was much higher at the salt marsh site than at the estuarine site, but soil salinity in the mudflat did not differ between sites in 2010. Soil moisture was much higher in the low marsh and the mudflat than in the high marsh at both sites (Table 1). Flooding frequency decreased from the mudflat to the high marsh at both sites (Table 1). Although the overall pattern of soil salinity across zones at the salt marsh site was the same between 2009 and 2010, there were some significant differences (Table 1). Soil salinity was higher in the low marsh and the mudflat but much lower in the high marsh in 2009 than in 2010 ($P<0.05$). Vegetation cover (mostly *S. salsa*) was significantly higher in the high marsh than in the low marsh at the estuarine site (Table 1).

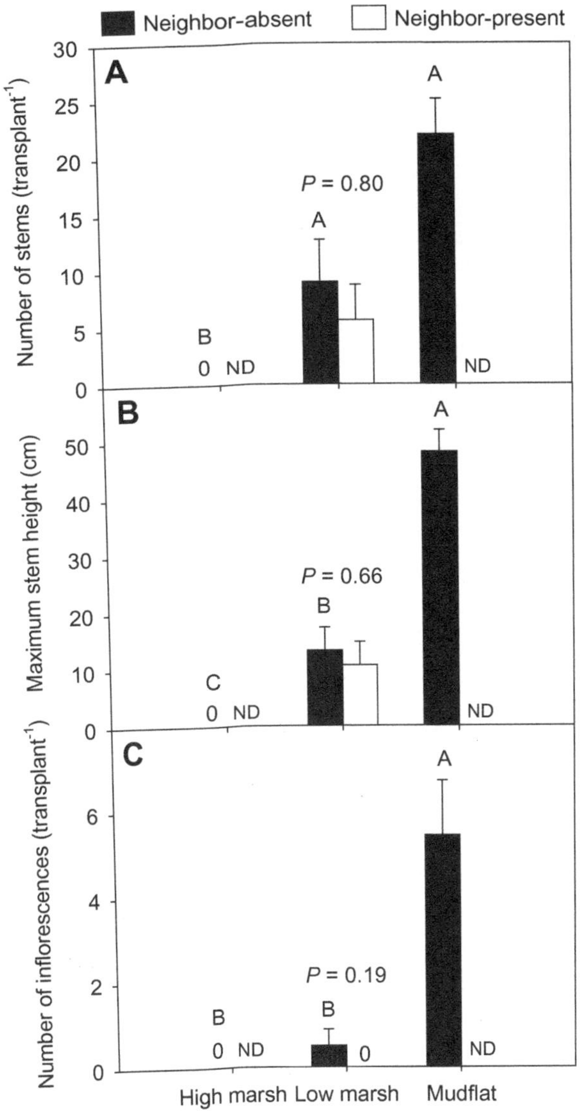

Figure 2. Cordgrass performance in different zones at the salt marsh site in 2009. (A) number of stems (B) maximum plant height and (C) number of inflorescences. Data are means + 1SE. ND indicates no data. Bars sharing a capital letter were not significantly different from one another (nonparametric multiple comparisons, Steel-test). *P*-values from ANOVAs or Wilcoxon tests investigating the effect of neighbors are indicated above the bar group for low marsh transplants.

However, vegetation cover was very low in both the low and high marsh zones at the salt marsh site (Table 1).

Field transplant experiments

At the salt marsh site, cordgrass transplants grew well in the mudflat in September 2009 (after one growing season), but their growth decreased with increasing elevation (Figure 2). The number of stems, maximum plant height and number of inflorescences were all highest in the mudflat. In the low marsh, a few cordgrass transplants survived, but their growth and reproduction were much lower than those in the mudflat (Figure 2b and c). None of the cordgrass transplants survived in the high marsh. In September 2010, after two growing seasons, however, most cordgrass transplants in all of the three habitats died. The number of stems, biomass, number of inflorescences and maximum plant height for the few surviving transplants were all very low, and did not differ among zones and neighbor treatments (Table S1).

At the estuarine site, cordgrass transplants also performed best in the mudflat and worst in the high marsh (Figure 3). Number of stems, maximum plant height, number of inflorescences and biomass of cordgrass transplants were all highest in the mudflat. In the low marsh, most cordgrass transplants also survived and performed relatively well, although maximum plant height and biomass were significantly lower in the low marsh than in the mudflat (Figure 3b and d). Most of the cordgrass transplants died in the high marsh, regardless of neighbor treatments. None of the fitness parameters differed significantly between neighbor-removal and neighbor-present treatments in the low or high marsh at the estuarine site (Figure 3).

Common garden experiment

Cordgrass transplants grew much better when flooded every other day than when flooded weekly (Figure 4). Cordgrass stem number, maximum plant height and biomass in the every other day flooding treatment were much higher than those in the weekly flooding treatment (Figure 4).

Discussion

Our results reveal that cordgrass transplants generally perform well in mudflats and low estuarine marshes with low salinities and frequent flooding. They cannot establish, however, in high estuarine marshes and salt marshes where soils are hypersaline and seldom flooded. We found neither a positive nor negative effect of native vegetation on cordgrass, even in estuarine marshes where neighbor effects were expected (due to abundant native vegetation and the well performance of the transplants). These results suggest that physical stress rather than competition from native plants are likely to control cordgrass invasions in the Yellow River Delta.

Despite the well-documented ability of *Spartina* species to tolerate salt and flooding stress [18,20,36], we found that hypersaline habitats, particularly at upper elevations, prevented the establishment, growth and sexual reproduction of cordgrass transplants. This result is consistent with a previous study which suggests that salinity is a critical physical stressor that restricts *S. anglica* invasions in the Pacific Northwest, USA [6]. The performance of *Spartina* species is generally believed to decrease with increasing salinity [18,19,21,37]. The results from our field and common garden experiments suggest that frequent flooding favors the establishment and growth of cordgrass, results that are consistent with previous studies reporting increased biomass of cordgrass under frequent flooding regimes [6,8,36,38]. This may be due to infrequent flooding leading to elevated salt stress by soil heating and evaporation of pore water [6,25,26,39,40]. Although we did not quantify salinity differences between weekly and every other day flooding treatments in the common garden experiment, the accumulation of salt on the soil surface in weekly flooding treatments was visually striking, suggesting that increased salt stress was responsible for the lower performance of cordgrass there.

We found no evidence that competition with native species mediates the performance of cordgrass transplants in the Yellow River Delta, despite the paradigm that lower competitive ability restricts cordgrass to lower zones in its native ranges [33,36,41]. At the salt marsh site, the lack of positive or negative interactions between native vegetation and the transplanted cordgrass can be due to the low abundance of native vegetation. In the low estuarine marsh where both cordgrass and native vegetation grow

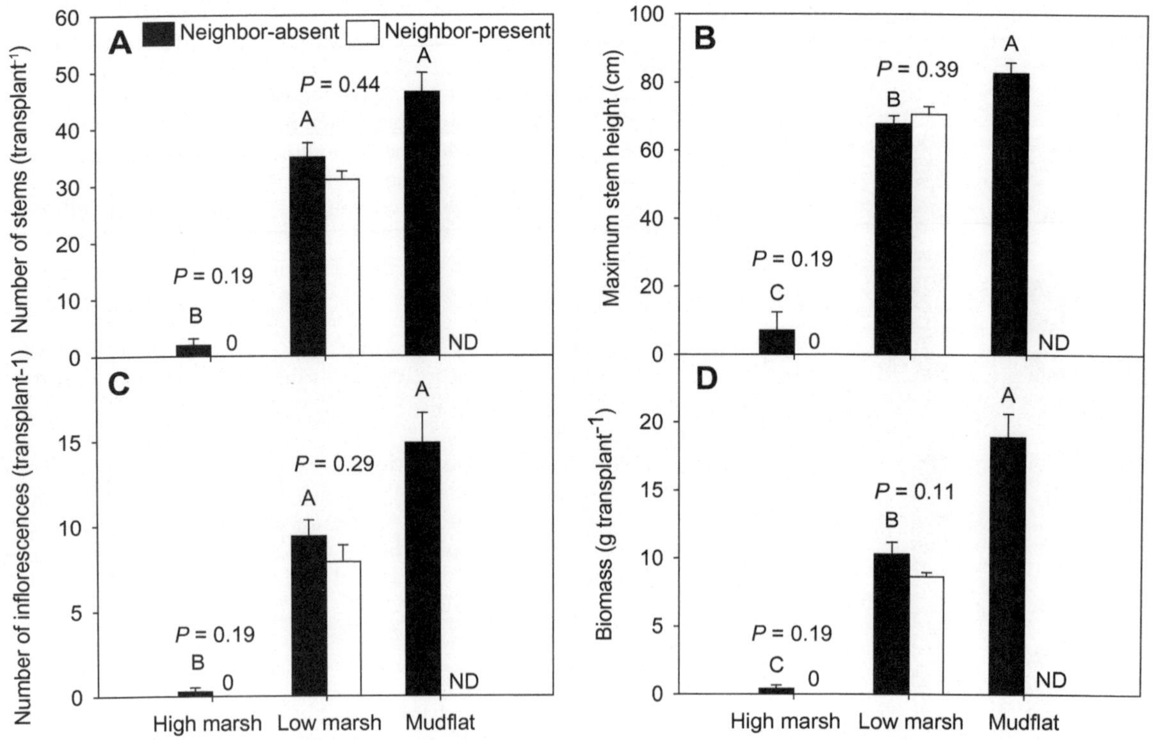

Figure 3. Cordgrass performance in different zones at the estuarine site in 2010. (A) number of stems (B) maximum plant height (C) number of inflorescences and (D) biomass. Data are means + 1SE. ND indicates no data. Bars sharing a capital letter were not significantly different from one another (nonparametric multiple comparisons, Steel-test). *P*-values from ANOVAs or Wilcoxon tests investigating the effect of neighbors are indicated above each bar group for low marsh and high marsh transplants.

well, however, the lack of a neighbor effect should be a result of no interactions, instead of a result of having no neighbors; in the high estuarine marsh where the native vegetation grow well while cordgrass performs very poorly, no neighbor effect can be a result of the lack of cordgrass. Because the estuarine marshes are dominated by stunted *S. salsa*, its ability to reduce soil salinity by shading is minor [32,42], and as a result, *S. salsa* has little positive effect on cordgrass growth and persistence. In a previous study, Dethier and Hacker (2005) [6] also found that species interactions had no effect on *S. anglica* performance. In addition, although it

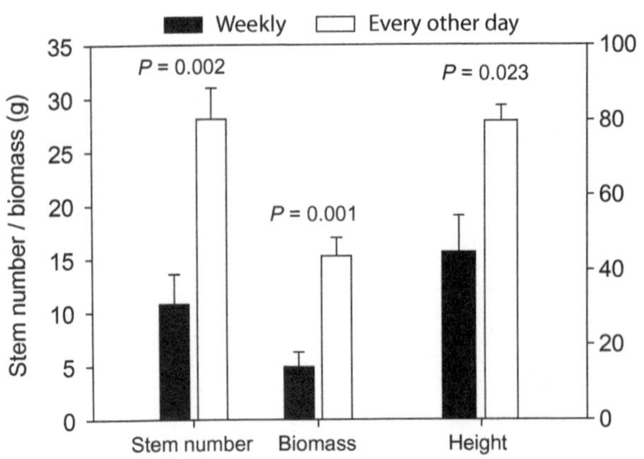

Figure 4. Cordgrass performance in different flooding treatments in the common garden experiment. (A) number of stems and maximum plant height and (B) biomass. Data are means + 1SE ($n = 6$). *P*-values from ANOVAs or Wilcoxon tests are indicated.

has been suggested that the outcome of interactions between native *P. australis* and cordgrass can change along environmental gradients and that native plants can outcompete cordgrass under some conditions [8], we found no evidence that *S. salsa* can outcompete cordgrass at any of the habitats we examined.

Two limitations of our experiments should be noted. First, we focused on how the establishment and reproduction of transplanted cordgrass ramets with rhizomes varied among different habitats, as the documented large-scale invasions of this species in China have been shown to rely mainly on human-aided transplanting [28], although after plantation this species expands at a local range by both sexual and vegetative reproduction [43,44]. Nevertheless, as *Spartina* species can be more susceptive to physical stress and species interactions at the germination or seedling establishment stage than it is at the clone stage [6,37], their dynamics, particularly in hypersaline conditions where seedling recruitment and growth are often limited, should be mainly controlled by vegetative growth [37,45]. Second, in the field experiments, we did not establish transplant controls at Wuhaozhuang where cordgrass were excavated. Transplanting may affect the vigor of species. But as the physical conditions (tides and non-estuarine freshwater input) at Wuhaozhuang were similar to those in the mudflats at the salt marsh site and cordgrass transplants performed well there, transplanting artifacts should not comprise our results.

In conclusion, our results indicate that cordgrass invasions in the Yellow River Delta are likely to be limited to tidal flats and low estuarine marshes where flooding is frequent and soil salinity is relatively low. Salt marshes and high estuarine marshes dominated by native *S. salsa* and *T. chinensis* plants are unlikely to be invaded due to harsh physical conditions. These findings have important implications for understanding *Spartina* invasions in northern China [43,46] and on other coasts with similar biotic and abiotic environments [31,33,47].

Acknowledgments

Mark Bertness and Tyler Coverdale helped greatly with the writing of the manuscript, Yanzi Cai and Wenjie He with field/laboratory work, and Yueliang Liu and Juanzhang Lv with access to the study sites.

Author Contributions

Conceived and designed the experiments: QH BC YA. Performed the experiments: QH. Analyzed the data: QH. Wrote the paper: QH BC YA.

References

1. Compagnoni A, Halpern CB (2009) Properties of native plant communities do not determine exotic success during early forest succession. Ecography 32: 449–458.
2. Davis MA, Grime JP, Thompson K (2000) Fluctuating resources in plant communities: a general theory of invasibility. Journal of Ecology 88: 528–534.
3. Shea K, Chesson P (2002) Community ecology theory as a framework for biological invasions. Trends in Ecology & Evolution 17: 170–176.
4. Alpert P, Bone E, Holzapfel C (2000) Invasiveness, invasibility and the role of environmental stress in the spread of non-native plants. Perspectives in Plant Ecology, Evolution and Systematics 3: 52–66.
5. Eschtruth AK, Battles JJ (2009) Assessing the relative importance of disturbance, herbivory, diversity, and propagule pressure in exotic plant invasion. Ecological Monographs 79: 265–280.
6. Dethier MN, Hacker SD (2005) Physical factors vs. biotic resistance in controlling the invasion of an estuarine marsh grass. Ecological Applications 15: 1273–1283.
7. D'Antonio CM (1993) Mechanisms controlling invasion of coastal plant communities by the alien succulent *Carpobrotus edulis*. Ecology 74: 83–95.
8. Wang Q, Wang C, Zhao B, Ma Z, Luo Y, et al. (2006) Effects of growing conditions on the growth of and interactions between salt marsh plants: implications for invasibility of habitats. Biological Invasions 8: 1547–1560.
9. Cheng BS, Hovel KA (2010) Biotic resistance to invasion along an estuarine gradient. Oecologia 164: 1049–1059.
10. Grosholz E (2002) Ecological and evolutionary consequences of coastal invasions. Trends in Ecology & Evolution 17: 22–27.
11. Ayres DR, Smith DL, Zaremba K, Klohr S, Strong DR (2004) Spread of exotic cordgrasses and hybrids (*Spartina* sp.) in the tidal marshes of San Francisco Bay, California, USA. Biological Invasions 6: 221–231.
12. Strong DR, Ayres DR (2009) *Spartina* introductions and consequences in salt marshes: arrive, survive, thrive, and sometimes hybridize. In: Silliman BR, Grosholz ED, Bertness MD, eds. Human Impacts on Salt Marshes: A Global Perspective. Berkeley: University of California Press. pp 3–22.
13. Grosholz ED, Levin LA, Tyler AC, Neira C (2009) Changes in community structure and ecosystem function of *Spartina* invasion of Pacific estuaries. In: Silliman BR, Grosholz ED, Bertness MD, eds. Human Impacts on Salt Marshes: A Global Perspective. Berkeley: University of California Press. pp 23–40.
14. Levin LA, Neira C, Grosholz ED (2006) Invasive cordgrass modifies wetland trophic function. Ecology 87: 419–432.
15. Naidoo G, McKee K, Mendelssohn I (1992) Anatomical and metabolic responses to waterlogging and salinity in *Spartina alterniflora* and *S. patens* (Poaceae). American Journal of Botany 79: 765–770.
16. Hwang YH, Morris JT (1994) Whole-plant gas exchange responses of *Spartina alterniflora* (Poaceae) to a range of constant and transient salinities. American Journal of Botany 81: 659–665.
17. Colmer TD, Teresa W (1996) Interactive effects of salinity, nitrogen and sulphur on the organic solutes in *Spartina alterniflora* leaf blades. Journal of Experimental Botany 47: 369–375.
18. Smart RM, Barko JW (1978) Influence of sediment salinity and nutrients on the physiological ecology of selected salt marsh plants. Estuarine and Coastal Marine Science 7: 487–495.
19. Smart RM, Barko JW (1980) Nitrogen nutrition and salinity tolerance of *Distichlis spicata* and *Spartina alterniflora*. Ecology 61: 630–638.
20. Seneca ED, Broome SW, Woodhouse WW Jr (1985) The influence of duration-of-inundation on development of a man-initiated *Spartina alterniflora* Loisel. marsh in North Carolina. Journal of Experimental Marine Biology and Ecology 94: 259–268.
21. Castillo JM, Rubio-Casal AE, Redondo S, Alvarez-Lopez AA, Luque T, et al. (2005) Short-term responses to salinity of an invasive cordgrass. Biological Invasions 7: 29–35.
22. Jiang LF, Luo YQ, Chen JK, Li B (2009) Ecophysiological characteristics of invasive *Spartina alterniflora* and native species in salt marshes of Yangtze River estuary, China. Estuarine, Coastal and Shelf Science 81: 74–82.
23. Castillo JM, Figueroa E (2009) Effects of abiotic factors on the life span of the invasive cordgrass *Spartina densiflora* and the native *Spartina maritima* at low salt marshes. Aquatic Ecology 43: 51–60.
24. Grosholz E (2010) Avoidance by grazers facilitates spread of an invasive hybrid plant. Ecology Letters 13: 145–153.
25. Hacker SD, Heimer D, Hellquist CE, Reeder TG, Reeves B, et al. (2001) A marine plant (*Spartina anglica*) invades widely varying habitats: potential mechanisms of invasion and control. Biological Invasions 3: 211–217.
26. Hacker SD, Dethier MN (2006) Community modification by a grass invader has differing impacts for marine habitats. Oikos 113: 279–286.
27. Mateos-Naranjo E, Redondo-Gómez S, Luque CJ, Castellanos EM, Davy AJ, et al. (2008) Environmental limitations on recruitment from seed in invasive *Spartina densiflora* on a southern European salt marsh. Estuarine, Coastal and Shelf Science 79: 727–732.
28. An S, Gu B, Zhou C, Wang Z, Deng Z, et al. (2007) *Spartina* invasion in China: implications for invasive species management and future research. Weed Research 47: 183–191.
29. Li B, Liao C, Zhang X, Chen H, Wang Q, et al. (2009) *Spartina alterniflora* invasions in the Yangtze River estuary, China: an overview of current status and ecosystem effects. Ecological Engineering 35: 511–520.
30. Editorial Committee of Wetland Vegetation in China (1999) Wetland Vegetation in China. Beijing: Science Press. pp 553.
31. He Q, Cui B, Cai Y, Deng J, Sun T, et al. (2009) What confines an annual plant to two separate zones along coastal topographic gradients? Hydrobiologia 630: 327–340.
32. Cui BS, He Q, An Y (2011) Community structure and abiotic determinants of salt marsh plant zonation vary across topographic gradients. Estuaries and Coasts 34: 459–469.
33. Pennings SC, GRANT MB, Bertness MD (2005) Plant zonation in low latitude salt marshes: disentangling the roles of flooding, salinity and competition. Journal of Ecology 93: 159–167.
34. McCullagh P, Nelder JA (1989) Generalized linear models. London: Chapman & Hall.
35. R Development Core Team (2008) R: A language and environment for statistical computing. R Foundation for Statistical Computing, Vienna, Austria.
36. Bertness MD (1991) Zonation of *Spartina patens* and *Spartina alterniflora* in a New England salt marsh. Ecology 72: 138–148.
37. Kittelson PM, Boyd MJ (1997) Mechanisms of expansion for an introduced species of cordgrass, *Spartina densiflora*, in Humboldt Bay, California. Estuaries and Coasts 20: 770–778.
38. Morris JT, Sundareshwar P, Nietch CT, Kjerfve B, Cahoon D (2002) Responses of coastal wetlands to rising sea level. Ecology 83: 2869–2877.
39. Bertness MD, Pennings SC (2000) Spatial variation in process and pattern in salt marsh plant communities in eastern North America. In: Weinstein M, Kreeger D, eds. Concepts and Controversies in Tidal Marsh Ecology. New York: Kluwer. pp 39–57.
40. Noe GB, Zedler JB (2001) Variable rainfall limits the germination of upper intertidal marsh plants in southern California. Estuaries and Coasts 24: 30–40.
41. Levine JM, Brewer JS, Bertness MD (1998) Nutrients, competition and plant zonation in a New England salt marsh. Journal of Ecology 86: 285–292.
42. He Q, Cui B, An Y (2011) The importance of facilitation in the zonation of shrubs along a coastal salinity gradient. Journal of Vegetation Science 22: 828–836.
43. Zhang R, Shen Y, Lu L, Yan S, Wang Y, et al. (2004) Formation of *Spartina alterniflora* salt marshes on the coast of Jiangsu Province, China. Ecological Engineering 23: 95–105.
44. Wang C, Tang L, Fei S, Wang J, Gao Y, et al. (2009) Determinants of seed bank dynamics of two dominant helophytes in a tidal salt marsh. Ecological Engineering 35: 800–809.
45. Shumway SW, Bertness MD (1992) Salt stress limitation of seedling recruitment in a salt marsh plant community. Oecologia 92: 490–497.
46. Ding JJ, Wang L, Ji YH, Ding YH, Li RJ (2009) Landscape pattern change of Yancheng coastal wetland in Jiangsu. Wetland Science 7: 202–207.
47. Wang H, Hsieh YP, Harwell MA, Huang W (2007) Modeling soil salinity distribution along topographic gradients in tidal salt marshes in Atlantic and Gulf coastal regions. Ecological Modelling 201: 429–439.

The Influence of Typhoon Khanun on the Return Migration of *Nilaparvata lugens* (Stål) in Eastern China

Gao Hu[1], Fang Lu[1], Ming-Hong Lu[2], Wan-Cai Liu[2], Wei-Gen Xu[3], Xue-Hui Jiang[3], Bao-Ping Zhai[1]*

1 Key Laboratory of Integrated Management of Crop Diseases and Insect Pests (Ministry of Education), College of Plant Protection, Nanjing Agricultural University, Nanjing, China, **2** Division of Pest Forecasting, China National Agro-Tec Extension and Service Center, Beijing, China, **3** Plant Protection Station of Zhejiang Province, Hangzhou, China

Abstract

Migratory insects adapt to and exploit the atmospheric environment to complete their migration and maintain their population. However, little is known about the mechanism of insect migration under the influence of extreme weather conditions such as typhoons. A case study was conducted to investigate the effect of typhoon Khanun, which made landfall in the eastern China in Sept. 2005, on the migration of brown planthopper, *Nilaparvata lugens* (Stål). The migration pathways of *N. lugens* were reconstructed for the period under the influence of the typhoon by calculating trajectories using the MM5, a mesoscale numerical weather prediction model, and migration events were examined in 7 counties of the Yangtze River Delta region with ancillary information. The light trap catches and field observations indicated that the migration peak of *N. lugens* coincided with the period when the typhoon made landfall in this region. The trajectory analyses revealed that most emigrations from this region during this period were hampered or ended in short distances. The sources of the light-trap catches were mainly located the nearby regions of each station (i.e. mostly less than 100 km away, with a few exceeding 200 km but all less than 300 km). This disrupted emigration was very different from the usual *N. lugens* migration which would bring them to Hunan, Jiangxi, and southern Anhui from this region at this time of year. This study revealed that the return migration of *N. lugens* was suppressed by the typhoon Khanun, leading to populations remaining high in the Yangtze River Delta and exacerbating later outbreaks.

Editor: Olle Terenius, Swedish University of Agricultural Sciences, Sweden

Funding: Funding provided by the National '973' Program of China (2010CB126200) and the Youth Sci-Tech Innovation Fund of Nanjing Agricultural University (Y0201100097). The funders had no role in study design, data collection and analysis, decision to publish, or preparation of the manuscript.

Competing Interests: The authors have declared that no competing interests exist.

* E-mail: bpzhai@njau.edu.cn

Introduction

Many insects undertake regular long-distance migrations each year to track seasonal changes in resources and habitats. By spreading their breeding efforts in space and time, migratory insects maintain their populations over a range of environmental conditions [1]. Various features of the atmospheric environment affect insect migration at different spatial scales. In most cases, migratory insects adapt to these features to maximise their flight efficiency and succeed in spatial redistribution [2–4]. Hence, prevailing seasonal airflows are crucial for annual regular migrations [4–7]. As aerial densities of migrating insects may be increased by strong wind convergence, migrants may land in favourable or unfavourable places at large number, either building up rapidly or resulting in high mortality. Therefore, migration is essentially a high-risk strategy [8,9], especially under extreme weather, such as typhoons. Under the influence of such a weather system, usual migration routes are often perturbed and migrants may not reach a suitable habitat. Although there have been several studies of rice planthopper migration under the influence of typhoons (see below), little is known about changes in insect migration process and much less about the changes in density and distribution of migratory population under extreme weathers. Therefore, evaluation of the influence of extreme weather on insect migration is of great significance.

Brown planthopper, *Nilaparvata lugens* (Stål) (Hemiptera: Delphacidae), one of the most serious rice pests in East and Southeast Asia, undertakes an annual return migration in East Asia [10–12]. *N. lugens* cannot overwinter in temperate zones such as mainland China, Japan, and the Korean Peninsula. Instead, an infestation is initiated by windborne spring/summer migrants from the south [13,14]. It begins with a northward migration occurring in late March every year. The distribution of *N. lugens* then further expands northward by progeny of the migrant population and can cover the whole rice-growing region in China, as well as into Japan and the Korean Peninsula. From September onwards the general direction of planthopper migration becomes predominantly southbound. In the end of October, most *N. lugens* populations are only present in the safe overwintering areas, mainly in the Indo-China Peninsula, within which there is a proportion of successfully returned migrants [10,13,15,16]. In the past few years new outbreaks of *N. lugens* have been continuously recorded. For example, the loss of rice yield in China caused by *N. lugens* was approximately 1,880,000 t in 2005 [16–18].

N. lugens is a micro insect and a slow flyer of only 0.3 m/s [19] and its windborne migration is significantly influenced by meteorological factors and phenomena. Take-off usually occurs in a narrow time window at dusk or dawn under calm weather, temperatures greater than 17°C, and wind speeds less than 3.1 m/s [20–23]. Long-distance displacement is induced by a long-lasting warm and humid wind. A low-level jet sustained for a long time over a wide scale is the most favourable current for northward migrations [15,22,24–28]. Landing is often observed under windy, rainy, or even stormy weather when a frontal system passes over regardless of the time of the day. Downdrafts also can terminate a migration, especially during southward migration [22,24,25,29–31].

A typhoon, which can produce substantial winds of 250 kilometres per hour (kph) with gusts of up to 300 kph, as well as intense bursts of rain and ocean surges that cause extensive flooding, may change the *N. lugens* population distribution patterns. A typhoon can last for weeks and cover thousands of miles. Therefore, it could reroute *N. lugens* migrations, and thus cause either sudden infestation or population collapse. For example, *N. lugens* could have unusually migrated from Philippines to Taiwan and southern China, assisted by the circulation of a typhoon [32,33]. In China, about seven typhoons in average make landfall every year. Most typhoon landings occur from July to September, which partially overlaps with the period of *N. lugens* migration (April to October). The influence of typhoon on the migration of *N. lugens* has been studied since the 1970s. In 1976, the synchronous immigration of *N. lugens* in Zhejiang, Jiangsu, and Anhui provinces from Aug. 12 to 24 was thought to be correlated with the inverted trough produced by the thirteenth typhoon [34]. Wang et al. (1982) categorised typhoons into three types according to the location of typhoon landing, namely, Beibu Gulf, South China Sea, and East China Sea. Among these, the East China Sea type was thought to be favourable for the southward or southeastward migration of planthoppers from Jiang-Huai Valley (Central and North Jiangsu, Central and North Anhui) [35]. From 1991 to 2005, 52 typhoons had made landfall in China, 16 of which influenced *N. lugens* migrations [36]. The serious *N. lugens* outbreaks in 2005 have been attributed to the increased number and intensity of typhoons that made landfall in the year [18,37,38]. Nevertheless, knowledge of the influence of typhoon on planthopper migration remains limited and ambiguous, largely due to the complexity of typhoon's structure and the uncertainty of typhoon's movement.

This paper reports a case study on the effect of typhoon Khanun on the return migration of *N. lugens* in eastern China in 2005, in order to improve the understanding of the influence of typhoon on this pest's population distribution. In 2005, the population of *N. lugens* in the Yangtze River Valley increased exponentially in September. 'Hopperburn', which is browning of the leaves or withering of the whole plant, occurred widely in mid-September in the Yangtze River Delta (including Zhejiang and Jiangsu provinces and Shanghai municipality) [18,37–39]. Based on previous researches, *N. lugens* in this region would undertake return migration in this time of year, and the local population would decrease and the usual destination would be Hunan, Jiangxi and southern Anhui provinces [13,16,23,40]. Six typhoons made landfall between Mid July and Mid October 2005 in eastern China. Among of these, the intensity of typhoon Khanun was the strongest. It landed in Zhejiang in mid September when *N. lugens* population level was very high in the Yangtze River Delta [18,37,38,41]. By investigating the influence of typhoon Khanun on the population dynamics and distribution of *N. lugens* in this region, this study aims to improve the understanding of the relationship between insect migration process and the meteorological background.

Materials and Methods

Light trap data and field observation data

Daily light trap data from plant protection stations (PPS) for seven counties in the Yangtze Delta region, including Tiantai, Xiaoshan, Dongyang, Zhuji, Songyang, Fengxian and Yixing, were obtained from Zhejiang province, Shanghai municipality and Jiangsu province in 2005 (Fig. 1). In all but the Fengxian and Yixing PPS, Jiaduo Brand automatic light traps (black-light lamp with the power of 20 W, Jiaduo Brand, Jiaduo Science, Industry and Trade Co. Ltd., Henan Province, China) were used for catching planthoppers. Traditional black-light traps were used in Fengxian and Yixing. These lamps were switched on at 1900 h in the evening and off at 0700 h Beijing Time (same thereafter) in the next morning every day from Apr. 1 to Nov. 15. At the time when this study was done, there had been no comparative study on the efficacy of these two types of light trap for capturing *N. lugens*. Comparisons between light traps at different stations were not made in this study.

Field observations were also collected from these stations as well as other counties in this region. Most field observation data were obtained from China National Agro-Tec Extension and Service Center (NATESC), while others were collected from pest reports published by related PPS and research papers [37–39,42].

Meteorological data and model

MM5, which is a limited-area, nonhydrostatic, terrain-following sigma-coordinate model designed to simulate or predict mesoscale atmospheric circulation, was used in this study. MM5 has several advanced features, including multiple nesting capability, various global model output data input, physical parameterisation choices, and the capability to run on various computer platforms.

The MM5 model system includes TERRAIN, REGRID, LITTLE_R, INTERPF, MM5, INTERPB, and other programs. NCEP Final Analysis (FNL) Data, which is a six-hourly, global, 1-degree grid dataset, was used in REGRID to generate first-guess pressure-level fields on model grids. Routine sounding (six-hourly) and surface (three-hourly) observational data from China Meteorological Data Sharing Service System were used in LITTLE_R to perform objective analysis to improve the first-guess fields, and in MM5 to perform four-dimensional data assimilation (FDDA) to improve the simulation result.

The dimension and resolution of the domain are 75×75 and 30 km, respectively. Thirty-three vertical layers were available and the model top was 100 hPa. The scheme selection and parameters for MM5 are listed in Table 1. The forecast time is 72 h. The model outputs data at 1 h intervals, including horizontal wind speed, vertical wind speed, temperature, precipitation, and other elements. Output data from MM5 were transformed to GrADS grid data by MM5toGrADS to display in the Grid Analysis and Display System (GrADS).

Other observation weather data was collected from published research papers [41,43,44].

Trajectory analysis

Likely sources and landing area of migrating *N. lugens* were estimated by constructing backward/forward trajectories, based on the following assumptions: (i) the planthoppers are displaced downwind [20–23,40,45]; (ii) take-off takes place mostly at dusk and partly at dawn [20–23,40,45] ; (iii) the migrants are likely to land at any time; and (iv) the planthoppers cannot fly in an atmosphere of below 16.5°C [28,40,46]. The backward trajectories from light trap locations were calculated hourly during 0700 h to 0600 h, with initial 7 heights of 500, 750, 1000, 1250, 1500,

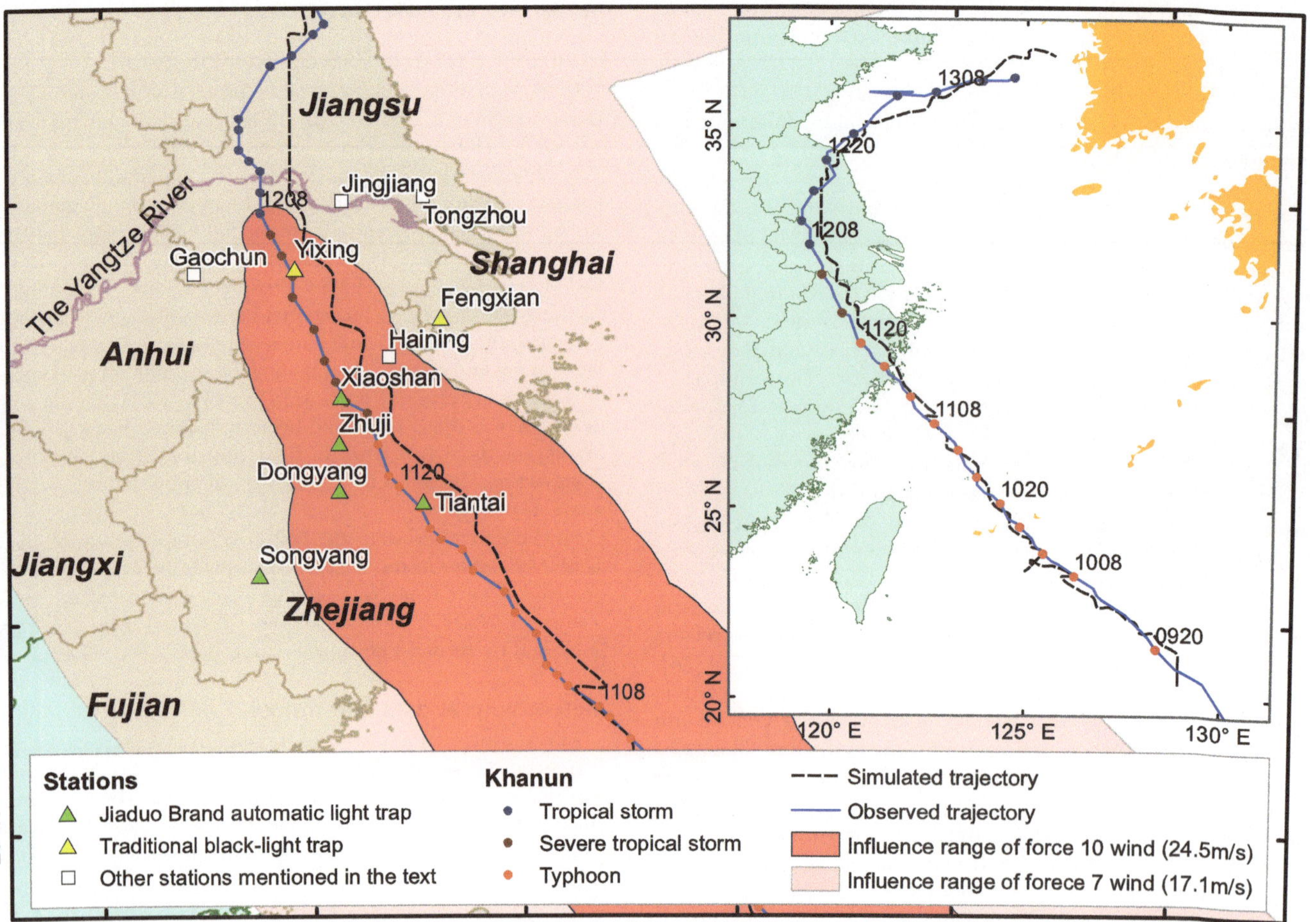

Figure 1. The trajectory and influence range of typhoon Khanun. Note: The numbers near track points represent time formatted as Day-House DDTT; for example, 0708 was at 0800 h on Sept. 7.

Table 1. Selection of scheme and parameters for MM5.

Items	Domain1
Location	30°N, 119°E
The number of grid points	75×75
Distance (km) between grid points	30
Grid start point of nest relative to its mother	1, 1
Layers	33
Map projection	Lambert
3D FDDA (Four-Dimensional Data Assimilation)	True
SFC FDDA	True
Explicit moisture schemes	Mix phase
Nesting type	One-way nest
Cumulus parameterization	Grell
Planetary boundary layer scheme	MRF PBL
Radiation scheme	RRTM
Model Start Time	0800 h on Sept. 8 and 11 (BJT)
Forecast Time	72 h

1750, and 2000 m above mean sea level. Thus, 168 trajectories (24×7) would be calculated for each location in one day. These backward trajectories were terminated when the air temperature at a point falling below 16.5C or at the take-off time, which was about 1800 h. The dawn take-off was not considered as its number is much lower than that at dusk [20–23]. The trajectories were calculated up to 36 h [13]. Hence, the flight duration is (i) below 13 h if migrants take off at dusk of the same night; (ii) 13–24 h if migrants take off at the previous dusk and lands at daytime; and (iii) above 24 h if migrants take off at the previous dusk and lands at night. A trajectory endpoint is considered to be a likely source if it is located in a place that satisfies the following conditions: (i) under fine weather at dusk (the air temperature at 2 m above ground greater than 17°C, the hourly precipitation less than 0.1 mm and the wind speed at 10 m above ground less than 4 m/s, based on the output data from MM5) [20–23]; (ii) in a rice planting area where the crops are at later growth stage [13]. The calculation setup of forward trajectory was as same as that of backward trajectory except the initial time at dusk (about 1800 h).

Backward and forward trajectories were calculated based on the method introduced by Zhu and Liao (1992) with wind field and temperature data that were transformed from the MM5 output by MM5toGrADS [47]. In this method, the displacement of particle at (i, j, k) is calculated using the following equations:

$$D_x = \frac{1}{2}(u_1 + u_2)(t_2 - t_1)$$
$$D_y = \frac{1}{2}(v_1 + v_2)(t_2 - t_1) \quad (1)$$
$$D_z = \frac{1}{2}(w_1 + w_2)(t_2 - t_1)$$

where D_x, D_y, and D_z denote the displacements in the x, y, and z dimensions, respectively, and (u_1, v_1, w_1) and (u_2, v_2, w_2) are wind velocities outputted by MM5 at times t_1 and t_2, respectively.

First, let (u_2, v_2, w_2) be equal to (u_1, v_1, w_1). The position of particles at t_2 $(i+\mu_x, j+\mu_y, k+\mu_z)$ was calculated using equation (1). The wind velocity $(u_{i+\mu x}, v_{j+\mu y}, w_{k+\mu z})$ at position $(i+\mu_x, j+\mu_y, k+\mu_z)$ at time t_2 was calculated using the Lagrange interpolation formula. (u_2, v_2, w_2) were then replaced by $(u_{i+\mu x}, v_{j+\mu y}, w_{k+\mu z})$, that is, $(u_{i+\mu x}, v_{j+\mu y}, w_{k+\mu z})$ is the zero-order approximation for (u_2, v_2, w_2). Using equation (1) and the Lagrange interpolation formula iteratively, a series of approximations for (u_2, v_2, w_2) could be calculated until the requirements listed by equation (2) were achieved.

$$[(\Delta\overline{\mu_x})^{n+1} - (\Delta\overline{\mu_x})^n]^2 + [(\Delta\overline{\mu_y})^{n+1} - (\Delta\overline{\mu_y})^n]^2 + [(\Delta\overline{\mu_z})^{n+1} - (\Delta\overline{\mu_z})^n]^2 < \varepsilon \quad (2)$$

where $\Delta\overline{\mu_x}$, $\Delta\overline{\mu_y}$ and $\Delta\overline{\mu_z}$ denote nondimensional displacements and are equal to the displacement divided by the grid distance in the x, y, and z dimensions, respectively. n and n+1 denote the n and n+1 order approximations, respectively. ε, the accuracy, was equal to 10^{-6} in this study.

The program for calculating trajectory was written in FORTRAN and run under Fedora 13 (Fedora Project, http://fedoraproject.org/).

Ethics statement

No specific permits were required for the described field studies. The brown planthopper *Nilaparvata lugens* (Stål) is a major pest of rice in Asia, and lots of manpower and resources were used to control this pest every year.

In this study, it was confirmed that (i) the location is not privately-owned or protected in any way, and (ii) the field studies did not involve endangered or protected species.

Results

The influence on the *N. lugens* migration from the typhoon Khanun had been examined according to the spatial relation of the study area and the typhoon. The possible sources of the immigrant *N. lugens* population were investigated with the MM5 model and ancillary information.

Outline of typhoon Khanun and *N. lugens* occurrence in East China

Typhoon Khanun, the fifteenth to hit China in 2005, was formed in the eastern ocean of the Philippine Islands on Sept. 7 and made landfall the southeastern Zhejiang at 1451 h on Sept. 11. The Khanun quickly weakened as it moved over land. It went back to the sea again in the northern Jiangsu at 2000 h on Sept. 12 and then disappeared in the eastern Yellow Sea on Sept. 13. Zhejiang, Shanghai, Jiangsu, Anhui, and Shandong suffered strong winds, high waves, and heavy rains as the Khanun passed over those regions (Fig. 1).

The simulated trajectory of the typhoon by MM5 was similar to the observed. The error between the simulated and the observed trajectory on land was greater than that on the sea as land topography is more complicated. At the time when routine surface observation data exited, like at 0200, 0500, 0800 h, the point of simulated trajectory was much closer to that of the observed due to the improvement made by LITTLE_R and FDDA modules. (Fig. 1). Compared with the routine sounding data, the simulated wind speeds and directions agreed generally with the observed values (Fig. 2). At the take-off time of *N. lugens*, 1800 h, there was no observational meteorological data to compare. However, the simulated winds near surface at 1800 h were seemingly credible as they were close to the observed surface wind at 1700 h (Fig. 3).

The Khanun landing coincided with the peak of the (migratory) macropterous adults of *N. lugens* in the Yangtze Delta region. From the field observations, the percentage of macropters in total adults was increasing up to ~70% in early September (Table 2). The captures of *N. lugens* by the light traps in this region also indicated this trend (Fig. 4), though it is hardly to tell if they were from local population or immigrants (see later analysis). The deep valleys of light-trap capture occurred on Sept. 10 (to 12) were simply due to the impact on the light-trap efficiency from the strong winds and torrential rains of the Khanun. However, the number of *N. lugens* adults in field increased in most counties in this region right after the Khanun passed (Table 2). For example, in Zhejiang and Jiangsu, there were five out of seven counties where the number of *N. lugens* in 'systematic observation' paddies increased during the typhoon period. Among these five counties except Jingjiang, the number increased by nearly 2 folds and higher. In the other two counties, the total numbers of *N. lugens* in 'systematic observation' paddies decreased, but the numbers of adults increased markedly. Likewise, there were six out of eight counties where the total number of *N. lugens* in farmer's paddies increased. However, the increase of *N. lugens* was small in Tiantai, Xiaoshan and Tongzhou, and was obvious in Zhuji, Yixing and Jingjiang. In Shanghai, the numbers of *N. lugens* in both 'systematic observation' paddies and farmer's paddies were decreased (Table 2). Above all, the local population of *N. lugens* seemed increased by the Khanun.

Before the approach of the Khanun

On Sept. 10, the Yangtze River Delta was affected gradually by the Khanun after its center passed the location of 24.7°N, 124.6°E on the East China Sea at 1800 h. The Yangtze Delta region was within the northwest of the typhoon periphery, where the inverted trough produced the convergent airflow and wind shear (Figs. 1 and 2A). At dusk the weather was fine and the wind speed at 10 m above ground was less than 4 m/s (Fig. 3A and 5A). Under such conditions, the macropterous adult of *N. lugens* was able to emigrate.

Most light traps in this region experienced dramatic drop on *N. lugens* capture on the night against the increase trend of the past week (Fig. 4). An exceptional catch occurred in Tiantai where the *N. lugens* capture of 26240 individuals is nearly 10 folds of that (2776) on the previous night. There was a lack of distant upwind sources from the analysis of backward trajectories which began from the region (Tiantai, Fengxian and Yixing) and mostly ended in the sea after 3 h flight duration (Fig. 6A). This might explain why most light traps caught less and indicate that the majority adults of *N. lugens* from this region did take-off and fly away. However, the huge captures in Tiantai suggest the place be under a different situation. These captures must have been the airborne population from local or nearby region forced down by the typhoon rain zone approached. On the other hand, the other light traps were further away from the approaching typhoon center and

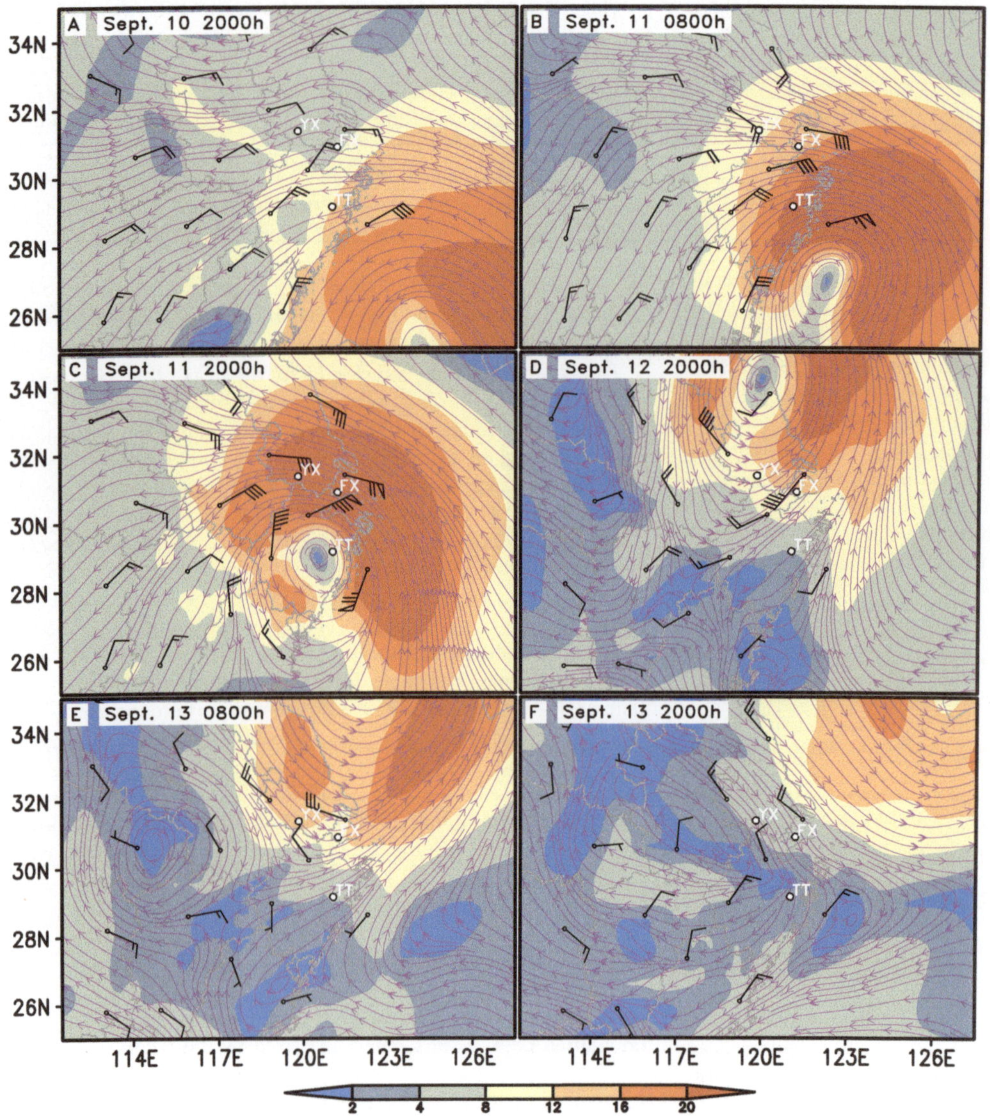

Figure 2. The simulated streamline and isotach of horizontal wind field (shaded), and the observed winds (shafts & barbs, whole barbs indicated 4 m/s) at 850 hPa. Note: TT-Tiantai, FX-Fengxian and YX-Yixing.

thus the airborne population could have flown away from the region.

In the vicinity of the Khanun

The Khanun was on the land of eastern China for nearly 30 h from 1450 h on Sept. 11 to 2000 h on Sept. 12. The Khanun made landfall over Luqiao District, Taizhou City, Zhejiang, moved to the northwest, entered Jiangsu at 0415 h in the next early morning, continued in the same direction until the mid-day and then turned towards to the northeast and went to the sea in the evening. The typhoon center had passed near Tiantai, Xiaoshan and Yixing at about 1900, 0000 and 0500 h, respectively. Since the afternoon of Sept. 11, the influence range of the force 10 wind (>24.5 m/s) extended gradually over all the seven PPS except Songyang and Fengxian (Fig. 1).

The Khanun's landfall induced torrential rain in Zhejiang, Shanghai, southern Jiangsu and southeastern Anhui from Sept. 10 to 11. In Zhejiang, about 158 observation stations obtained precipitation of above 100 mm, 61 stations recorded precipitation of above 200 mm, and 10 stations recorded precipitation of above 300 mm in 48 h between 0800 h Sept. 10 and 0800 h Sept. 12. Meanwhile, wind had strengthened in eastern Zhejiang, Shanghai, Jiangsu, and southeastern Anhui since the morning of Sept. 11. Maximum wind speeds ranged from 17.2 m/s to 32.6 m/s, even over 32.6 m/s in some places (Figs. 2B, 2C, 3B, and 5B).

The weather was not suitable for *N. lugens* to take off from this study region at the dusk of Sept. 11, when the typhoon center was near Tiantai and other 6 PPS were located in the gale zone of the typhoon. All the 7 PPS except Tiantai suffered strong winds and heavy rains. Average wind speeds at 10 m above ground ranged from 9 to 11 m/s. However, the simulated hourly precipitation ranged only from 0.2 to 6.1 mm, which was much less than the observed values. By contrast, it was calm in Tiantai when the typhoon eye zone was located nearby at dusk (Figs. 1, 3B, and 5B). Nevertheless, the eye zone was very small, with a diameter only a few kilometres, and it moved very fast. Thus, there was little chance for *N. lugens* to take off in this area, and therefore this region cannot be the source of migrants if there were on the night.

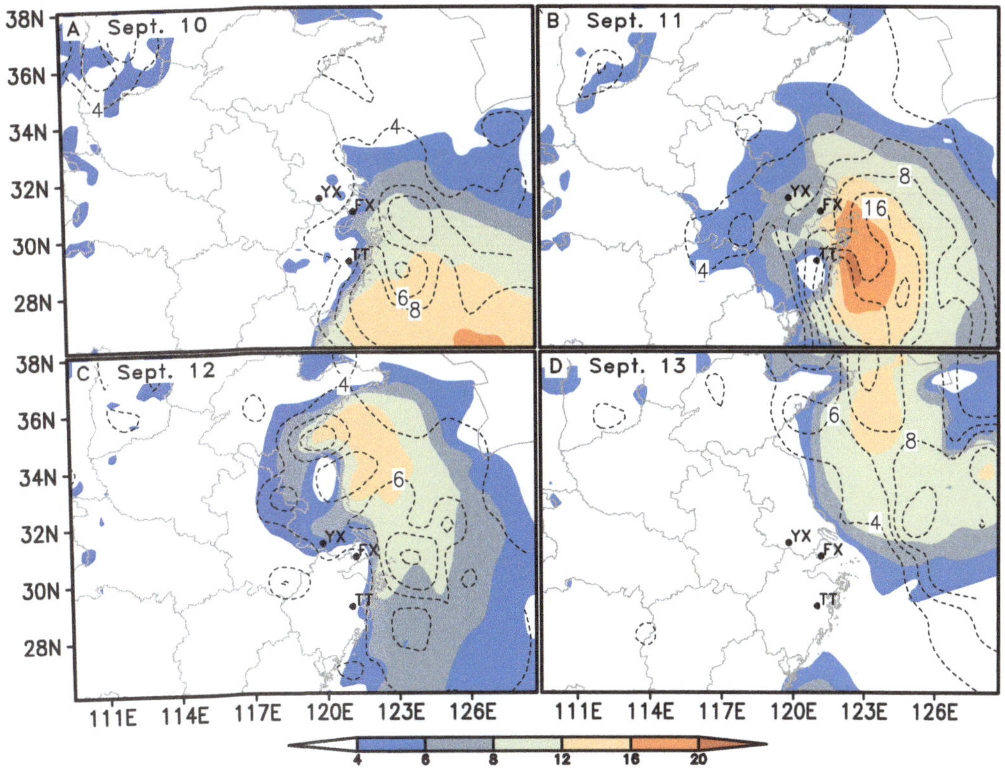

Figure 3. The simulated horizontal wind speeds at 10 m above ground at 1800 h (shaded, m/s) and the observed surface wind speeds at 1700 h (dotted line, m/s). Note: TT-Tiantai, FX-Fengxian and YX-Yixing.

All these 7 light traps caught only a few or null *N. lugens* demonstrated that the poor weather conditions prohibited this micro insect's active flight (Fig. 4).

Given that a mass of *N. lugens* airborne was forced to land on the night by the rain, the back trajectories calculated accordingly were nearly all ended in the sea from Tiantai, Fengxian and Yixing. This suggested that if there were some immigrants, they could only have come from nearby areas (Fig. 6B). However, as strong winds and heavy rains must have stopped *N. lugens* from taking off at dusk in nearby regions (Figs. 3B and 5B), therefore, there must have no migrants of *N. lugens* landed in this study region on the night of Sept. 11.

In the wake of the Khanun

On the next day (Sept. 12) after the Khanun made landfall, its center passed the study region in the early morning (Fig. 1). In the reset of the day and the following day (Sept. 13), the study area was in the rear of the typhoon (Fig. 2D, E, F), where convergence was predominated.

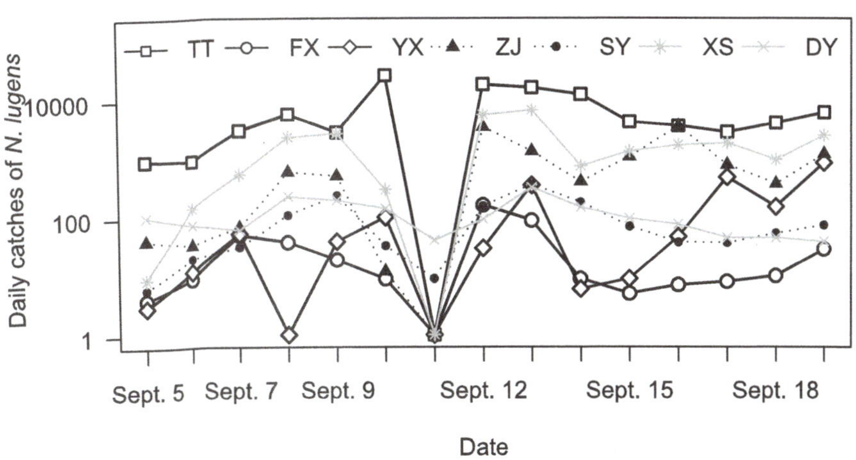

Figure 4. The daily light trap catches of *N. lugens* in Tiantai (TT), Fengxian (FX), Yixing (YX), Zhuji (ZJ), Songyang (SY), Xiaoshan (XS) and Dongyang (DY) in mid Sept., 2005.

Table 2. Population density of *N. lugens* before & after the typhoon Khanun in 2005.

Province	County	Paddy type*	The number of *N. lugens* per 100 hills					Increase or not
			Before Khanun			After Khanun		
			Aug. 30	Sept. 5	Sept. 10	Sept. 15	Sept. 20	
Zhejiang	Xiaoshan	S	235	1055	1985	16420	22975	↑
	Haining	S	314	163	-	1603	1986	↑
		AS(M)	16(13)	34(12)	-	102(19)	130(66)	↑
	Dongyang	S	174	811	2004	1168	578	↓
		AS(M)§	257(252)	32(22)	20(16)	1300(1260)	488(420)	↑
Jiangsu	Yixing	S	650	10660	13500	7440	-	↓
		AS(M)	-	510(70)	75(70)	210(190)	-	↑
	Tongzhou	S	2522	1284	1730	5000	2445	↑
	Gaochun	S	1740	2660	3900	7080	18260	↑
	Jingjiang	S	507	475	10930	12713	21888	↑
Shanghai	Fengxian	S	622	9439	11372	6431	9993	↓
Zhejiang	Tiantai	F	-	2856	3877	4644	-	↑
		AF(M)	-	499(441)	286(209)	364(291)	-	↑
	Zhuji	F	1710	1333	5902	12422	12372	↑
		AF(M)	446(223)	602(429)	812(654)	857(712)	1159(1063)	↑
	Xiaoshan	F	25	23	100	390	367	↑
		AF(M)	23(23)	17(17)	22(22)	57(57)	67(63)	↑
	Dongyang	F	670	2250	3700	2210	1130	↓
Jiangsu	Yixing	F	28.9	1232	1918	7130	-	↑
	Tongzhou	F	376	517	858	894	2527	↑
	Gaochun	F§	1211	2315	1410	997	678	↓
	Jingjiang	F§	116	-	473	1511	4651	↑
Shanghai	Fengxian	F	467	2766	5030	3606	3575	↓

*S-Systematic investigation paddies which are moderately fertile with routine cultural practices and no pesticide use to control pests during the rice growing season; F-Famers' paddies; AS-Adults in systematic investigation paddies; AF-Adults in farmer's paddies; M-Macropterous adult.

§The number of rice planthoppers, including *N. lugens* and *Sogatella furcifera* (Horváth), however, *N. lugens* account for more 80%.

On Sept. 12. At dusk, the southern part of the study region was in favourable conditions for *N. lugens* to take off. Fair weather came after the rain, and the winds weakened after the Khanun had passed. It was calm in Zhejiang and Shanghai (Figs. 3C and 5C). However, the average wind speeds at 10 m above ground in Jiangsu was still greater than 4 m/s and thus *N. lugens* take-off was probably suppressed.

The analysis of backward trajectory revealed that the south of the study region might have been the destination other than the source of *N. lugens* migration. Light-trap catches increased sharply in Songyang, Tiantai, Xiaoshan, Zhuji, Fengxian and Yixing on the night (Fig. 4). The endpoints of backward trajectories which originated from Tiantai with flight durations less than 13 h were located in its south-southwest of mostly less than 100 km area. The endpoints with flight durations over 12 h were mostly located at the borders between Zhejiang and neighbouring provinces (Jiangxi, Anhui, Jiangsu and Fujian) (Fig. 6C). However, most of Zhejiang, southern Anhui, southern Jiangsu, northern Jiangxi, and northern Fujian suffered strong winds and heavy rain at dusk on Sept. 11 (Figs. 3B and 5B). Thus these areas must be eliminated from the likely sources. Therefore, catches in Tiantai must have come from short distances where *N. lugens* took off at dusk on the same night. In Fengxian, the endpoints of backward trajectories with flight duration less than 13 h were located in the southwest at the border of Anhui and Zhejiang, where mass take-off of planthoppers was reported on the night as mid-season rice matured and was harvested [13,16]; while the other longer trajectories were ended in the sea (Fig. 6C). Likewise, the endpoints of backward trajectories from Yixing indicated that the catches of *N. lugens* might have come from as far as up to the northeastern Anhui, northern Jiangsu and southern Shandong (Fig. 6C). However these areas was still under very strong winds at 1800 h, and therefore there should not have too many *N. lugens* able to take off (Fig. 3C). Thus, the catches in Yixing must have come from the surrounding area as well with flight durations less than 12 h (the flight distance in a straight line less than 200 km) (Fig. 6C).

The analysis of forward trajectory supported the above estimation. In the southern Anhui, it was calm and no rainfall at the dusk on the day (Figs. 3C and 5C). The forward trajectories from the this area suggested that *N. lugens* could have arrived in the southern Jiangsu, northern Zhejiang and Shanghai within 12 h flight (Fig. 7). This agreed very well with the estimates of backward trajectory and indicated that the southern Anhui could have been the source area for immigrants in Yixing and Fengxian on the night.

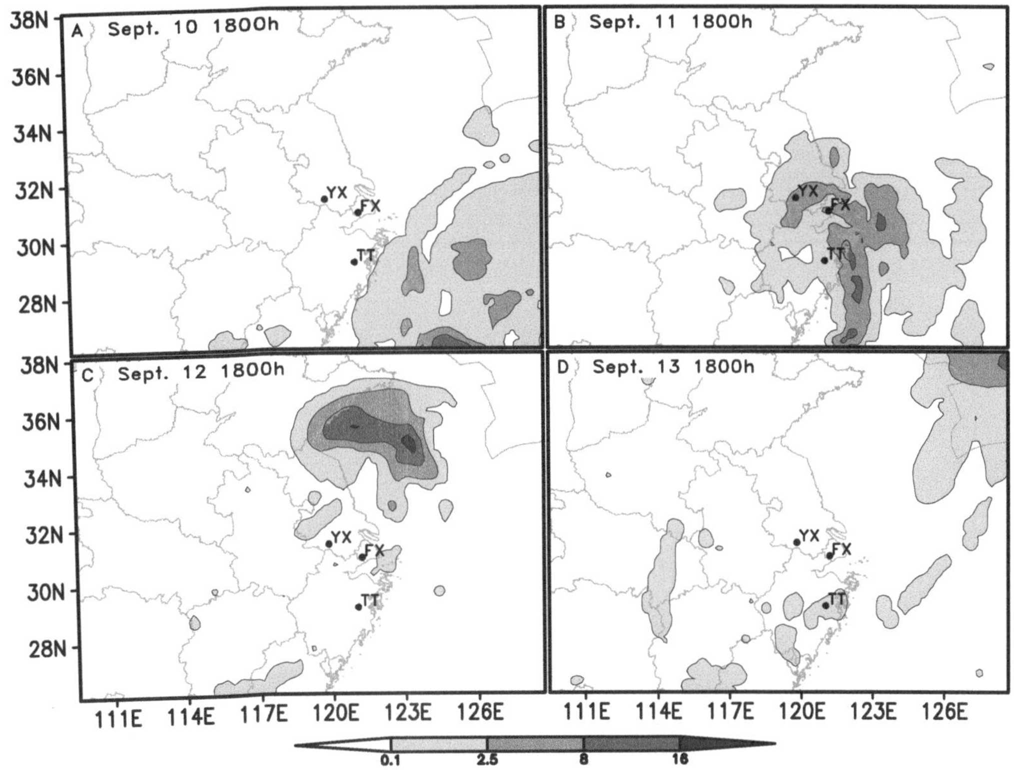

Figure 5. The simulated precipitation (mm) at 1800 h. Note: TT-Tiantai, FX-Fengxian and YX-Yixing.

On Sept. 13. The typhoon lost impact on the study region gradually on the day. It had changed to an extratropical cyclone in the Yellow Sea since 0800 h on Sept. 13, and disappeared after 2000 h (Fig. 1). The study region was still in the rear of the typhoon at daytime (Fig. 2E), but under the downdrafts of a continental high pressure which moved slowly towards the southeast due to an anticyclonic shear at nighttime (Fig. 2F).

In contrast to the previous day, the northern part of the study region might be favourable for *N. lugens* to take off at dusk. It was the clear weather with no rainfall and less than 4 m/s wind at 10 m above the ground except the southeastern part of the study region (Tiantai) where was raining lightly at the dusk (Figs. 3D and 5D)

The large numbers of light-trap capture of *N. lugens* indicated more immigrants on the night (Fig. 4). The endpoints of backward trajectories which originated from Tiantai with a flight duration less than 13 h were located in its surrounding area, whereas the endpoints with a flight duration over 12 h were located in Anhui, western and southern Zhejiang (Fig. 6D). Among these regions, the northern Anhui could be excluded as it suffered strong winds at 1800 h on Sept. 12 (Fig. 3C). Therefore, the likely source of the catches in Tiantai was Zhejiang and southern Anhui. The forward trajectories also confirmed that *N. lugens* from the southern Anhui could have got to Tiantai in the daytime on Sept. 13 after more than 12 h flight (Fig. 7). In Yixing and Fengxian, the endpoints of backward trajectories with flight duration under 13 h were located from the northwest in the southern Jiangsu, whereas the endpoints over 12 h were located in Shandong and the sea (Fig. 6D). When the Khanun was on the sea again the Shandong Peninsula suffered rough weather at dusk on Sept. 12 (Figs. 3C and 5C), and few *N. lugens* were able to take off. Therefore, catches in Yixing and Fengxian might have come from the southern Jiangsu, a nearby area (straight-line distance of straight line no more than 200 km), where *N. lugens* might have taken off at dusk on Sept. 13.

From the above study, it was evident that there was much more influx than efflux of *N. lugens* migrants in the Yangtze Delta region during the typhoon Khanun's last 4-day life of Sept. 10–13 2005. Apart from the small amount of adults might have emigrated on the night of Sept. 10 and possibly no migration activity occurred on the next night, a much large amount of *N. lugens* adults must have immigrated on the nights of Sept. 12 and 13. In addition to the residual population, the unusual immigrants during the typhoon Khanun could be the driving force for later outbreak in this region in 2005.

Discussion

In this study, the return migration of *N. lugens* under the influence of the Khanun was reconstructed using numerical simulations. MM5 has been successfully used to simulate meteorological background in insect migration studies [14,28,48,49]. Although typhoons are difficult to forecast due to their complicated structures and variable routes, MM5 can be used to simulate typhoon processes well based on reanalysis data [50–52]. In this study, the simulated trajectory of the typhoon was much better than the previous one obtained by Kou et al. (2009) due to LITTLE_R and FDDA applied with the routine sounding and surface observational data [52]. In spite of this, the results of precipitation intensity prediction by MM5 have not been very satisfactory [51,52]. It was observed this limitation in this study as well—the precipitation obtained from the numerical simulations was much less than the actual measures [43]. However this would not invalidate the results obtained from this study as wind and

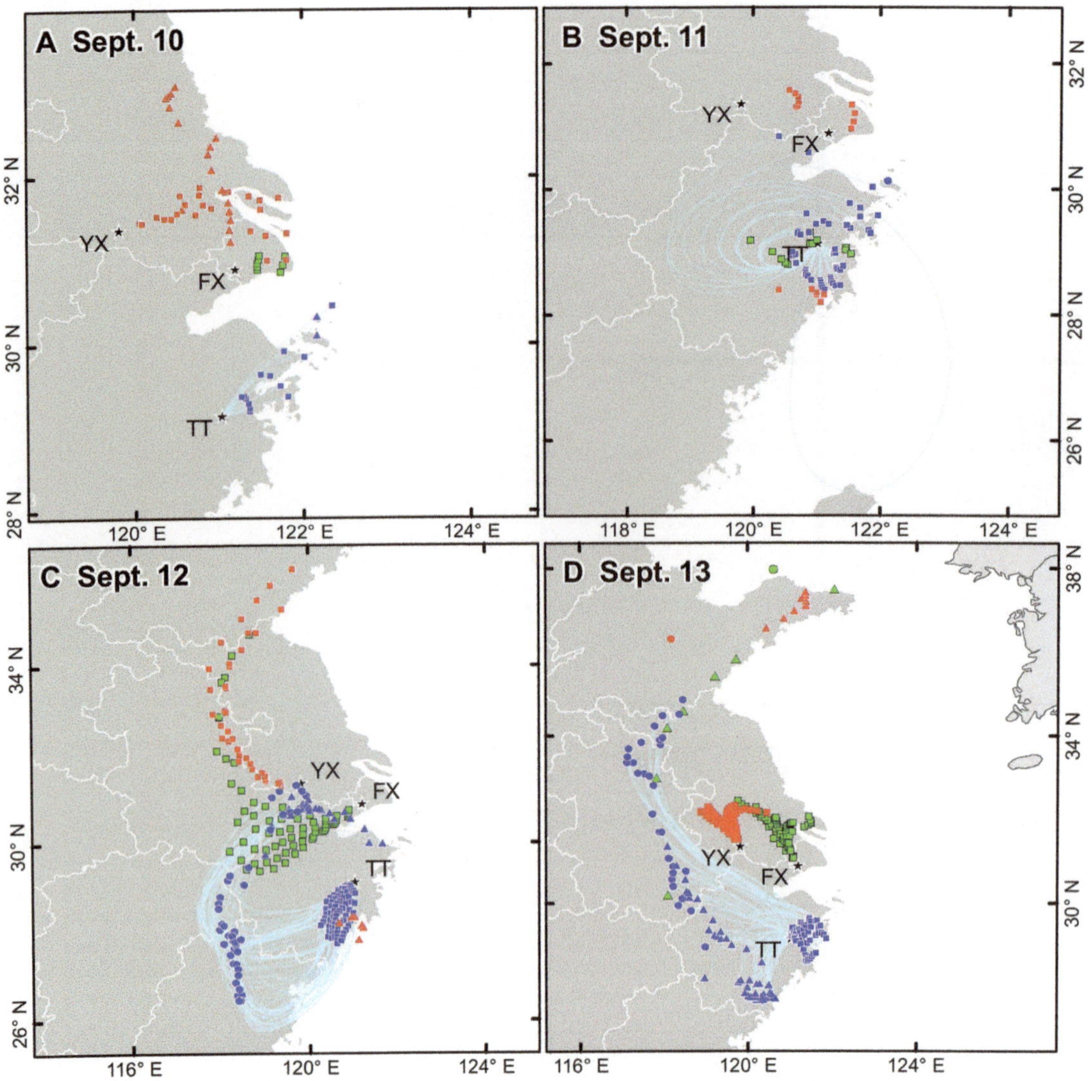

Figure 6. Endpoints of backward trajectories starting from Tiantai (TT), Fengxian (FX) and Yixing (YX) during Sept. 10–13. Note: (i) The endpoints with blue fill color started from Tiantai, those with green fill color started from Fengxian and those with red fill color started from Yixing. (ii)The endpoints in square (■): the migration duration was less than 13 h, and takes off at 1800 h on the same day; in triangle (▲): the migration duration was 13–24 h, and takes off at 1800 h on the previous day. (●): the migration duration was above 24 h, and takes off at 1800 h on the previous day.

temperature are two main factors affecting the transportation of migrating population while rain affects take-off and landing.

Insect migration means persistent and straightened out movement with temporary inhibition of station-keeping response [4,53]. Generally, *N. lugens* ascend actively straight to sky after take-off, and their phototaxis is inhibited [21,54–56]. Thus, the emigrating planthoppers will not be caught by the local light-traps. During emigration periods, the catches by local light traps are always small except if the emigration is suppressed by bad weather, such as rainfall and downdrafts [13]. This phenomenon has been observed by authors in Yongfu, Guangxi Autonomous Region in last five years (unpublished data) and was evident from this study region on the night of Sept. 10. However, the inhibition of phototaxis lasts only a short time during the ascent process, and the emigrating planthoppers were observed to be caught by local light traps at 33 m above ground [54,56]. Therefore, planthoppers could be attacted by nearby light traps if they pass the ascent stage.

This study revealed the cause of *N. lugens* later outbreak in the Yangtze Delta region in 2005. Considering the long-distance migration of *N. lugens* often is over 300 km, even 1000 km, the immigrants of *N. lugens* in the Yangtze River Delta during the typhoon Khanun were mainly located in the "nearby" regions of each station (i.e. mostly within 100 km, with a few exceeding 200 km but all less than 300 km) [13,23,57]. In other words, the migration of *N. lugens* from the Yangtze River Delta was confined in this area itself by the typhoon Khanun. Compared with their usual emigration destinations (Hunan, Jiangxi and Southern Anhui) in this time of year, *N. lugens* population in the Yangtze River Delta itself did not emigrate successfully. The analyses of trajectory also suggested that the immigrants from the southern Anhui landed in the study region (such as Zhuji, Xiaoshan, Tiantai, Yixing and Fengxian) after the Khanun passed and the resulting population increased (Table 2). This result was agreed with the previous study of Yao et al. (2007), which showed that: (i) the typhoon Khanun resulted in numerous *N. lugens* from the

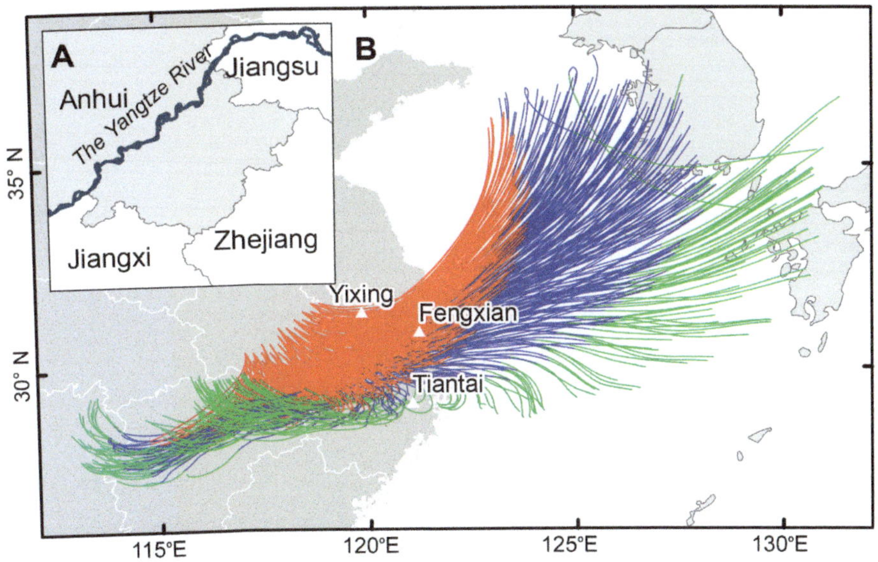

Figure 7. Forward trajectories starting from Southern Anhui at 1800 h on Sept. 12. Note: (i) Start points (+) of forward trajectories from Southern Anhui presented in Fig. (A); (ii) The red lines showed trajectories with flight duration below 12 h, viz. the range that migrants could reach if they flight within one night; (iii) The blue lines showed trajectories with flight duration of 13–24 h, viz. the range that migrants could reach in the daytime of the next day; (iv) The green lines showed trajectories with flight duration of 25–36 h, viz. the range that migrants could reach on the next night.

Jiang-Huai Valley landing in Hangjiahu area (located in North Zhejiang, and including Hangzhou, Jiaxing and Huzhou) after the typhoon had passed; and (ii) these immigrants caused the population to increased suddenly [58]. However, the source area of catches in this study was more definitive and more to the southern than in the previous study due to the backward trajectory analysis based on the more accurate meteorological background. This also indicated that typhoons on the East China Sea may not always favour the southward migration of *N. lugens* [35]. The uncommon later immigration of *N. lugens* generated much higher field population in this region.

Insect migration could be terminated by atmospheric environment (such as downdrafts, rain and cold temperature) and other factors (such as topography) [3,22,24,25,29–31]. There has hitherto been only a few examples of landing process observed by radar. In this study, convergence was seemingly the most important factor causing *N. lugens* to be concentrated and deposited. Radar observations have shown that concentrations of migrants form at convergences [2,4], with densities often increasing as much as 10 to 100 folds. Moreover, this interpretation was also suggested in the previous studies on insect migration influenced by typhoon [58,59]. In the present study, the failure of *N. lugens* emigration in the Yangtze River Delta was also due to convergence, and under its influence, the emigrants were concentrated and deposited after a short flight duration. In addition, rainfall, strong wind and downdrafts also could suppress insect emigration [13]. In the present case, wind stopped *N. lugens* to take off at dusk on Sept. 12 in Yingxing, while rainfall and downdrafts stopped *N. lugens* to take off at dusk on Sept. 13 in Tiantai.

Overall, this study showed that the return migration of *N. lugens* in the Yangtze River Delta was disrupted and unsuccessful. However, the insects which failed to emigrate could still survive in this region. Nevertheless, the denser population of *N. lugens* would have resulted in faster deterioration of the nutritional status of rice plants and reduction in the survival of the insects. The offspring of the residual and immigrated population would have grown slowly due to deteriorated habitats and declining temperature [60,61]. Given their slow development, the offspring may have even failed to reach the adult stages and emigrate before rice plants were harvested in mid and late October. Eventually, the offspring of *N. lugens* that failed to emigrate in mid September could not survive and integrate into the overwintering population. Nonetheless, the typhoon Khanun might have caused more damage to crops than its direct impact in this region. After all, this study revealed that the spatial redistribution of *N. lugens* was changed under the influence of the typhoon, leading to populations remaining high in the Yangtze River Delta, and exacerbating outbreaks of this pest, and even affecting the population itself in next year. Compared with the previous work, this study provided more details and improved the understanding on insect migration influenced by typhoon [58,59].

Acknowledgments

We thank the plant protection stations of Jiangsu Province, Shanghai Municipality and Zhejiang Province for providing insect scouting data. We are grateful to two anonymous referees for such valuable and detailed suggestions and pertinent comments on every drafts. We also thank Dr. H. K. Wang and Dr. H. Q. Feng for their constructive critics and useful suggestions that significantly improved the quality of the original paper.

Author Contributions

Conceived and designed the experiments: GH BPZ. Performed the experiments: GH FL MHL WCL WGX XHJ. Analyzed the data: GH FL MHL. Wrote the paper: GH BPZ.

References

1. Holland RA, Wikelski M, Wilcove D (2006) How and why do insects migrate? Science 313: 794–796.
2. Drake VA, Farrow RA (1988) The influence of atmospheric structure and motions on insect migration. Annu Rev Entomol 33: 183–210.
3. Westbrook JK, Isard SA (1999) Atmospheric scales of biotic dispersal. Agr Forest Entomol 97: 263–274.
4. Chapman JW, Drake VA (2010) Insect migration. In: Breed M.D. and Moore J., (eds.) Encyclopedia of Animal Behavior, volume 2, pp. 161–166 Oxford: Academic Press.
5. Chapman JW, Nesbit RL, Burgin LE, Reynolds DR, Smith AD, et al. (2010) Flight orientation behaviors promote optimal migration trajectories in high-flying insects. Science, 327, 82–685.
6. Chapman JW, Klaassen RHG, Drake VA, Fossette S, Hays GC, et al. (2011) Animal orientation strategies for movement in flows. Curr. Biol 21: 861–870.
7. Zhang ZT (1992) Insect migration and migration arena. China Plant Protection 18: 48–50.
8. Deveson ED, Drake VA, Hunter DM, Walker PW, Wang HK (2005) Evidence from traditional and new technologies for northward migrations of Australian plague locusts (*Chortoicetes terminifera*) (Walker) (Orthoptera: Acrididae) to western Queensland. Austral Ecology 30: 920–935.
9. Rosenberg J, Burt PJA (1999) Windborne displacements of desert locusts from Africa to the Caribbean and South America. Aerobiologia 15:167–175.
10. Cheng XN, Wu JC, Ma F (2003) Brown planthopper: occurrence and control. China Agricultural Press, Beijing.
11. Bottrell DG, Schoenly KG (2012) Resurrecting the ghost of green revolutions past: The brown planthopper as a recurring threat to high-yielding rice production in tropical Asia. Journal of Asia-Pacific Entomology 15: 122–140.
12. Zhai BP (2011) Rice planthoppers: A China problem under the international perspectives. Chinese Journal of Applied Entomology 48: 1184–1193.
13. Cheng SN, Chen JC, Xi X, Yang LM, Zhu ZL, et al. (1979) Studies on the migrations of brown planthopper *Nilaparvata lugens* Stål. Acta Entomologica Sinica 22: 1–21.
14. Otuka A, Matsumura M, Watanabe T, Dinh TV (2008) A migration analysis for rice planthoppers, *Sogatella furcifera* (Horváth) and *Nilaparvata lugens* (Stål) (Homoptera: Delphacidae), emigrating from northern Vietnam from April to May. Appl Entomol Zool 43: 527–534.
15. Kisimoto R, Sogawa K (1995) Migration of the Brown Planthopper *Nilaparvata lugens* and the White-backed Planthopper *Sogatella furcifera* in East Asia: the role of weather and climate. pp. 93–104 In Drake VA, Gatehouse AG. (Eds) Insect migration: Tracking resources through space and time. Cambridge University Press, New York.
16. Hu G, Cheng XN, Qi GJ, Wang FY, Lu F, et al. (2011) Rice planting systems, global warming and outbreaks of *Nilaparvata lugens* (Stål). B Entomol Res 101: 187–199.
17. Guo R, Zhao ZH (2006) Crop pests. pp. 376–378 in Ministry of Agriculture of China (Ed.) China Agriculture Yearbook. China Agricultural Press, Beijing.
18. Zhai BP, Cheng JA (2006) The conference summary of workshop on the two primary migratory pests of rice, rice planthopper and rice leaf roller, in 2006. Chinese Bulletin of Entomology 43: 585–588.
19. Chen RC, Wu JR, Zhu SD, Zhang JX (1984) Flight capacity of the brown planthopper *Nilaparvata lugens* Stål. Acta Entomologica Sinica 27: 121–127.
20. Ohkubo N, Kisimoto R (1971) Diurnal periodicity of flight behaviour of the brown planthopper, *Nilaparvata lugens* Stål in the 4th and 5th emergence periods. Jpn J App Entomol Z 15: 8–16.
21. Chen RC, Cheng XN (1980) The take-off behavior of brown planthopper (*Nilaparvata lugens* Stål) and its synchronous relations to the biological rhythm and environmental factors. Journal of Nanjing Agricultural University 2: 42–49.
22. Kisimoto R (1984) Meteorological conditions inducing long-distance immigration of the brown planthopper, *Nilaparvata lugens* (Stål). Chinese Journal of Entomology 4: 39–48.
23. Riley JR, Cheng XN, Zhang XX, Reynolds DR, Xu GM, et al. (1991) The long-distance migration of *Nilaparvata lugens* (Stål) (Delphacidae) in China: radar observations of mass return flight in the autumn. Ecol Entomol 16: 471–489.
24. Jiang GH, Tan HQ, Shen WZ, Cheng XN, Chen RC (1981) The relation between long-distance northward migration of the brown planthopper (*Nilaparvata lugens* Stål) and synoptic weather conditions. Acta Entomologica Sinica 24: 251–261.
25. Jiang GH, Tan HQ, Shen WZ, Cheng XN, Chen RC (1982). The relation between long-distance southward migration of the brown planthopper (*Nilaparvata lugens* Stål) and synoptic weather conditions. Acta Entomologica Sinica 25: 147–155.
26. Watanabe T, Seino H (1991) Correlation between the immigration area of rice planthoppers and the low-level jet stream in Japan. Appl Ent Zool 26: 457–462.
27. Feng CH, Zhai BP, Zhang XX, Tang JY (2002) Climatology of low- level jet and northward migration of rice planthoppers. Acta Ecologica Sinica 22: 559–565.
28. Otuka A, Watanabe T, Suzuki Y, Matsumura M, Furuno A, et al. (2005) Real-time prediction system for migration of rice planthoppers *Sogatella furcifera* (Horváth) and *Nilaparvata lugens* (Stål) (Homoptera: Delphacidae). Appl Entomol Zool 40: 221–229.
29. Tan HQ, Mao RZ, Cheng JY, Yao HF (1984) Relation between landing of brown planthoppers (*Nilaparvata lugens* Stål) in long distance migration and vertical air stream and rail fall. Journal of Nanjing Agricultural College 7: 18–25.
30. Cai WH, Lin TZ, Wu MY (1998) On the atmospheric circulation conditions associated with the migration and alighting of brow planthopper in Fujian Province. Acta Phytophylacica Sinica 25: 325–329.
31. Hu G, Bao YX, Wang JQ, Zhai BP (2007) Case studies on the landing mechanism s of the brown planthoppers *Nilaparvata lugens* (Stål). Acta Ecologica Sinica, 27: 5068–5075.
32. Liu CH (1984) Study on the long-distance migration of the brown planthopper in Taiwan. Chinese J Entomol 4:49–54.
33. Pender J. (1994) Migration of the brown planthopper, *Nilaparvata lugens* (Stål.) with special reference to synoptic meteorology. Grana 33:112–115.
34. Jiangsu Agricultural College, Jiangsu Meteorological Bureau (1977) Study on the migration of brown planthopper in Jiangsu Province in 1976. Jiangsu Agricultural Science 4: 50–57.
35. Wang YC, Hu GW, Xie MX (1982) A study of uppers winds analysis for the migratory path of the white-back plant hopper and the brown plant hopper in China. Acta Phytophylacica Sinica 9: 73–82.
36. Bao YX, Yan ML, Yuan CS, Li JJ (2008) Case studies of the influence of Haitang Typhoon (0505) on the catastrophic immigrations of *Nilaparvata lugens* (Stål). Scientia Meteorologica Sinica 27: 450–455.
37. Cheng JA, Zhu ZR (2006) Analysis on the key factors causing the outbreak of brown planthopper in Yangtze Area, China in 2005. China Plant Protection 32: 1–4.
38. Jiang YP, Li J, Yang QZ, Wang ZG, Gong CG, et al. (2006). The characteristic and cause of the outbreak of brow planthopper in Shanghai in 2005. China Plant Protection 32: 96–97.
39. Mu LF, Liu YC, Zhu FG, Ding HH, Teng JH, et al. (2006) Outbreak of later brown plant hopper, *Nilaparvata lugens*, in Wujiang City of 2005 and its control strategies. Chinese Bulletin of Entomology 43: 706–708.
40. Riley JR, Reynolds DR, Smith AD, Rosenberg LJ, Cheng XN, et al. (1994) Observation on the autumn migration of *Nilaparvata lugens* (Homoptera: Delphacidae) and other pests in east central China. B Entomol Res 84: 389–402.
41. China meteorological administration (2007) Yearbook of tropical cyclone 2005. Beijing: China Meteorological Press.
42. Qian DL, Wang HD, Jiang XH (2006) The outbreak and control of brown planthopper on late rice in Zhejiang Province in 2005. Zhejiang Agricultural Science 6: 678–680.
43. Zeng XX, Lv J, Shen Y (2006) Analysis on strong wind caused by landing typhoon 0515 in Zhejiang. Marine Forecasts 22: 13–20.
44. Zhao K, Sun Y, Wang SQ (2006) Diagnostic analysis on dynamic of landing typhoon 0515 (Khanun) with Jiangsu heavy rain course. Journal of Oceanography in Taiwan Strait 25: 414–418.
45. Deng WX (1981) A general survey on seasonal migrations of *Nilaparvata lugens* (Stål) and *Sogatella furcifera* (Horvath) (Homoptera Delphacidae). Acta Phytophylacica Sinica 8: 73–81.
46. Ohkubo N (1973) Experimental studies on the flight of planthoppers by the tethered flight technique. I: Characteristics of flight of the brown planthopper *Nilaparvata lugens* Stål. Acta Entomologica Sinica 27: 121–127.
47. Zhu YQ, Liao DX (1992) An investigation for computing three-dimensional trajectory. Quarterly Journal of Applied Meteorology 3: 328–333.
48. Bao YX, Zhai BP, Cheng XN (2005) Numerical simulation of the migration parameters of the Brown Planthopper, *Nilaparvata lugens* (Stål). Acta Ecologica Sinica 25: 1107–1114.
49. Otuka A, Dudhia J, Watanabe T, Furuno A (2005) A new trajectory analysis method for migratory planthoppers, *Sogatella furcifera* (Horváth) (Homoptera: Delphacidae) and *Nilaparvata lugens* (Stål), using an advanced weather forecast model. Agr Forest Entomol 7: 1–9.
50. Chen LS, Xu XD, Luo ZX, Wang JZ (2002) Introduction to tropical cyclone dynamics. China Meteorological Press, Beijing.
51. Xing JY, Song XJ, Yang XL, Zhao H (2006) Numerical simulation experiments on typhoon Khanun. Marine Forecasts 23: 1–9.
52. Kou YY, Shen TL, Xu HM. (2009) The numerical simulation of typhoon 'Khanun' and analysis of non-symmetric structure. Scientia Meteorologica Sinica 29: 606–610.
53. Kennedy JS, 1985 Migration, behavioural and ecological. In Migration: Mechanisms and Adaptive Significance, Rankin MA editors, pp. 5–26. Contributions in Marine Science, vol. 27 (Suppl.). Port Aransas, Texas: Marine Science Institute, The University of Texas at Austin.
54. Ye ZC, He SM, Lu LQ, Zhen HG, Feng BC, et al. (1981) The observation of the take-off behaviour of brown planthopper (*Nilaparvata lugens* Stål). Chinese Bulletin of Entomology 18: 97–100.
55. Zhong DL (1984) The relationship between the take-off behaviour of brown planthopper (*Nilaparvata lugens* Stål) and weather condition. Chinese Bulletin of Entomology 22: 97–99.
56. Zhao SF (1992) Behavioral mechanism of phototactic response of natural planthopper population (*Sogatella furcifera* and *Nilaparvata lugens*). Acta Agriculturae Jiangxi 4: 74–79.
57. Nanjing agricultural college, Guangdong academy of agriculture, Chengzhou prefecture institute of agriculture, Guilin prefecture bureau of agriculture (1981)

Test on the releasing and recapturing of marked planthoppers, (*Sogatella furcifera* and *Nilaparvata lugens*), Acta Ecologica Sinica 1981, 1:49–53.
58. Yao DH, Chen XF, Yao ST, Chen QH. (2007) Analysis of Effects of Typhoon Kanu (2005) on the Sudden Increase of *Nilaparvata lugens*, Chinese Journal of Agrometeorology 28(3): 347–349.
59. Wang CH, Zhai BP, Bao YX (2009) Effect of typhoon Haitang airflow field on the northward migration route of rice brown planthopper. Chinese Journal of Applied Ecology 20:2506–2512.
60. Chen RC, Qi LZ, Cheng XN, Ding ZZ, Wu ZL (1986) Studies on the population dynamics of brown planthopper *Nilaparvata lugens* (Stål) I: Effects of temperature and diet conditions on the growth of experimental population. Journal of Nanjing Agricultural University 3: 23–33.
61. Zhu SD, Lu ZQ, Hang SB, Xu H (1994) Studies on regulative effects of temperature on the population of brown planthopper, *Nilaparvata lugens* Stål. Entomological Journal of East China 3: 53–59.

4

Agro-Environmental Determinants of Avian Influenza Circulation: A Multisite Study in Thailand, Vietnam and Madagascar

Mathilde C. Paul[1,2]*, Marius Gilbert[3,4], Stéphanie Desvaux[1,5], Harena Rasamoelina Andriamanivo[6,7], Marisa Peyre[1], Nguyen Viet Khong[8], Weerapong Thanapongtharm[9], Véronique Chevalier[1]

1 Centre de coopération internationale en recherche agronomique pour le développement (CIRAD), UR AGIRs, Montpellier, France, **2** Université de Toulouse, INP-ENVT, UMR ENVT INRA 1225 IHAP, Toulouse, France, **3** Biological Control and Spatial Ecology, Université Libre de Bruxelles, Brussels, Belgium, **4** Fonds National de la Recherche Scientifique, Brussels, Belgium, **5** Direction Régionale de l'Alimentation, de l'Agriculture et de la Forêt de Languedoc- Roussillon, Montpellier, France, **6** FOFIFA-DRZV, Antananarivo, Madagascar, **7** Faculté de Médecine d'Antananarivo - Département Vétérinaire, Antananarivo, Madagascar, **8** National Institute of Veterinary Research, Hanoi, Vietnam, **9** Department of Livestock Development, Bangkok, Thailand

Abstract

Outbreaks of highly pathogenic avian influenza have occurred and have been studied in a variety of ecological systems. However, differences in the spatial resolution, geographical extent, units of analysis and risk factors examined in these studies prevent their quantitative comparison. This study aimed to develop a high-resolution, comparative study of a common set of agro-environmental determinants of avian influenza viruses (AIV) in domestic poultry in four different environments: (1) lower-Northern Thailand, where H5N1 circulated in 2004–2005, (2) the Red River Delta in Vietnam, where H5N1 is circulating widely, (3) the Vietnam highlands, where sporadic H5N1 outbreaks have occurred, and (4) the Lake Alaotra region in Madagascar, which features remarkable similarities with Asian agro-ecosystems and where low pathogenic avian influenza viruses have been found. We analyzed H5N1 outbreak data in Thailand in parallel with serological data collected on the H5 subtype in Vietnam and on low pathogenic AIV in Madagascar. Several agro-environmental covariates were examined: poultry densities, landscape dominated by rice cultivation, proximity to a water body or major road, and human population density. Relationships between covariates and AIV circulation were explored using spatial generalized linear models. We found that AIV prevalence was negatively associated with distance to the closest water body in the Red River Delta, Vietnam highlands and Madagascar. We also found a positive association between AIV and duck density in the Vietnam highlands and Thailand, and with rice landscapes in Thailand and Madagascar. Our findings confirm the important role of wetlands-rice-ducks ecosystems in the epidemiology of AI in diverse settings. Variables influencing circulation of the H5 subtype in Southeast Asia played a similar role for low pathogenic AIV in Madagascar, indicating that this area may be at risk if a highly virulent strain is introduced.

Editor: Edward Goldstein, Harvard School of Public Health, United States of America

Funding: This work was part of the FSP GRIPAVI project and was funded by the French Ministry of Foreign and European Affairs. The authors thank the French Research Agency project EcoFlu (ANR-06-SEST-12) and the French-Thai PHC program which provided additional support. The funders had no role in study design, data collection and analysis, decision to publish, or preparation of the manuscript.

Competing Interests: The authors have declared that no competing interests exist.

* Email: m.paul@envt.fr

Introduction

A new, atypical influenza virus infection caused by the H7N9 subtype [1] emerged in March 2013 in Eastern China. As of 25 October 2013, it has been associated with 137 confirmed human infections and 45 related deaths [2]. This event raises renewed questions regarding the potential of avian influenza viruses to infect humans and of human pandemicity [3]. Controlling the spread of avian influenza viruses (AIV) in poultry may contribute to reducing the risk for human infection by limiting poultry-to-human transmission and preventing the emergence of a viral form with efficient human-to-human transmission [4]. To achieve this goal, a better understanding of the local conditions that favor the circulation of avian influenza viruses in poultry is necessary.

The degree to which AIV spread and are maintained in poultry populations is highly variable and may partly be explained by variations in landscape or agro-environmental features around farms [5]. Three main drivers for the spatial distribution of Highly Pathogenic Avian Influenza (HPAI) H5N1 outbreaks have been identified in previous studies [6]. The first is density of domestic ducks, which can shed the virus with minimal clinical signs and spread it silently in poultry populations [7]. The second is anthropogenic variables, as the risk of HPAI H5N1 has been found to be greater in areas that have high human population densities and are located close to transportation networks [8,9]. This may be explained by a higher probability of outbreak detection in these areas, but also by an increased virus transmission through movements of contaminated poultry or fomites. The third driver consists of water-related variables [10–13] because AIV can persist in water for extended periods of time [14].

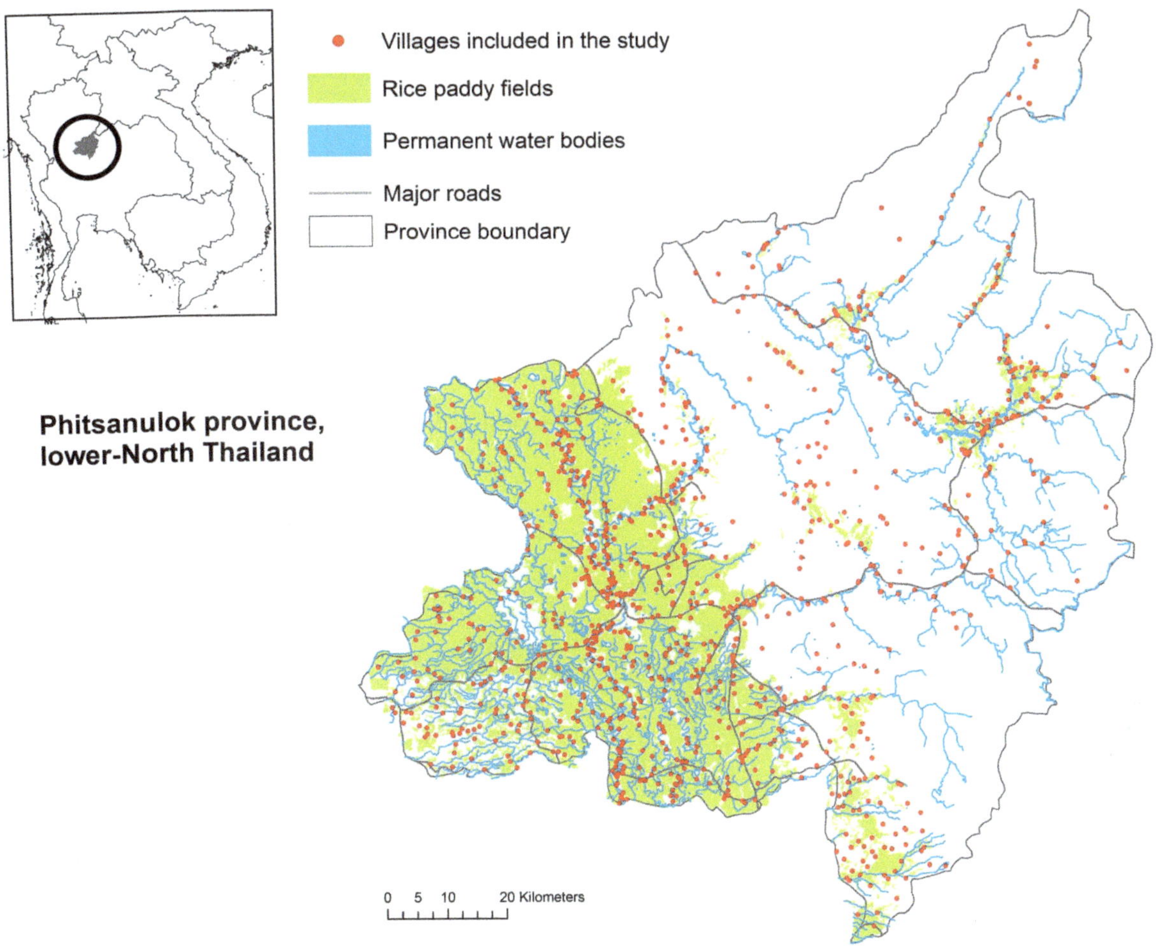

Figure 1. Study area in lower-Northern Thailand. Phitsanulok province and location of the 1032 villages included in the study.

Many studies have examined associations between agro-environmental factors and the presence of avian influenza outbreaks [5,8,9,11–13,15–25]. The comparison, based on field data, of risk factors across countries would be useful for analyzing similarities and differences in transmission patterns and subsequently improving detection and control policies. However, such a comparison is complicated by several factors. First, the studies involved used different data collection protocols and analytical methods, rendering comparisons across countries difficult. Second, while correlations between agro-environmental determinants and AIV persistence have been fairly well-studied at a broad scale, the processes operating at a fine geographic scale are complex and remain poorly documented [6]. Third, although the geographic co-distribution of H7N9 outbreaks with H5N1 outbreaks in China suggests that some areas might be a common ground for transmission of emerging low and highly pathogenic AIV [26], the current lack of studies on low pathogenic AIV renders it difficult to draw conclusions.

In our multisite study, we aimed to identify and compare the local combinations of environmental factors associated with avian influenza circulation in different settings. In parallel, we applied the same analytic framework to study fine-scale data collected from four contrasting study sites. One was in Thailand, where HPAI viruses have circulated in the past, two were in Vietnam, where HPAI viruses have and continue to circulate, and the last was in Madagascar, where HPAI viruses have not yet been detected. Despite remarkably similar features with Asian agro-ecosystems, only low pathogenic avian influenza circulation has been found in Madagascar to date [27]. Findings from this study should help to identify local combinations of risk factors for AIV circulation, and thus could be useful for tailoring prevention, surveillance and control at a very fine spatial scale.

Data and Methods

General characteristics of the study sites

The present work is part of a large-scale research project, "GRIPAVI", launched in 2007 to improve understanding of the ecological and epidemiological factors involved in the maintenance and spread of avian influenza viruses in tropical countries. In collaboration with national and international research groups, field surveys were conducted over five years in six study sites located in Vietnam, Mauritania, Mali, Ethiopia, Madagascar and Zimbabwe [28]. To develop a comparative approach, we selected two GRIPAVI sites with similar agro-ecosystems based on rice cultivation and duck farming: the Red River Delta in Vietnam and Lake Alaotra in Madagascar. We added two other study sites featuring similar ecological characteristics, one in the Vietnam highlands, and the other in lower-Northern Thailand, where we previously had conducted extensive field work to gather information on AIV [22,23].

The study thus covered a total of four sites in three countries (Thailand, Vietnam and Madagascar). These sites all featured agro-ecosystems with rice cultivation and duck farming, but varied in terms of human population density, poultry density, farming techniques (extensive or intensive) and AIV situations.

In Thailand, the study area was the province of Phitsanulok (Figure 1), which recorded the highest number of HPAI H5N1 outbreaks in chickens during the 2004–2005 epidemics. This province is located in the Yom-Nan River basin in the lower-Northern region, where the H5N1 virus re-emerged in 2008 and where there is a strong need for studies on the conditions of HPAI outbreak occurrence [29]. Landscapes of this province include plain areas characterized by intensive rice cultivation, with multiple rice crops per year, and large flocks of free-grazing ducks raised on rice fields. The eastern part of Phitsanulok province also covers sparsely populated, higher altitude areas with forests and diverse agricultural crops.

In Vietnam, two study sites were selected in two regions presenting contrasting agro-ecological features. The first study site was in the Red River Delta, which is characterized by low elevations and a combination of duck farming with intensive rice cultivation. This site was chosen because HPAI H5N1 outbreaks in Northern Vietnam mainly were concentrated in this type of agro-ecological area [30]. Two provinces were selected: Bac Giang and Thai Binh (Figure 2). The second site was Ha Giang province in the Vietnam highlands (Figure 2), a mountainous area characterized by low human and poultry densities. Rice cultivation is mainly extensive, with only once rice crop cultivated each year. The site was chosen because this part of Vietnam has been studied little yet may be exposed to an increased risk of AIV introduction due to its proximity to the Chinese border [23].

The fourth study site was the Lake Alaotra region in Madagascar (Figure 3). Madagascar is an Indian Ocean island lying 400 km off the eastern coast of southern Africa. Located 750 m above sea level, the Lake Alaotra region constitutes the largest wetland area in Madagascar, and is a rich habitat for wild birds [31]. It is the largest rice production basin in the country and also is the site of important poultry production activities. There are small-scale chicken farms and ducks and geese are allowed to graze on rice paddies. The following five municipalities located on the lakeside were included in the study: Andromba, Imerimandroso, Ampitatsimo, Anororo, and Amparafaravola.

Data on avian influenza

The data collected in the four study sites differed due to the distinctive features of each site. H5N1 data was collected in Thailand, H5 seroprevalence data at multiple time points in Vietnam, and cross-sectional low-pathogenic AIV seroprevalence data in Madagascar. Due to differences in the sampling protocols of the preliminary surveys used on the four sites, we had to re-arrange the data and redefine the epidemiological unit. We worked at the finest unit for which spatial coordinates were available. This unit was the village in Thailand and the two sites in Vietnam (Red River Delta and the highlands), and the farm in Madagascar. Outbreaks and prevalence data collected at various dates across study sites (Figure S1) were aggregated over time with the aim to obtain an overall view of the AIV situation for each study site.

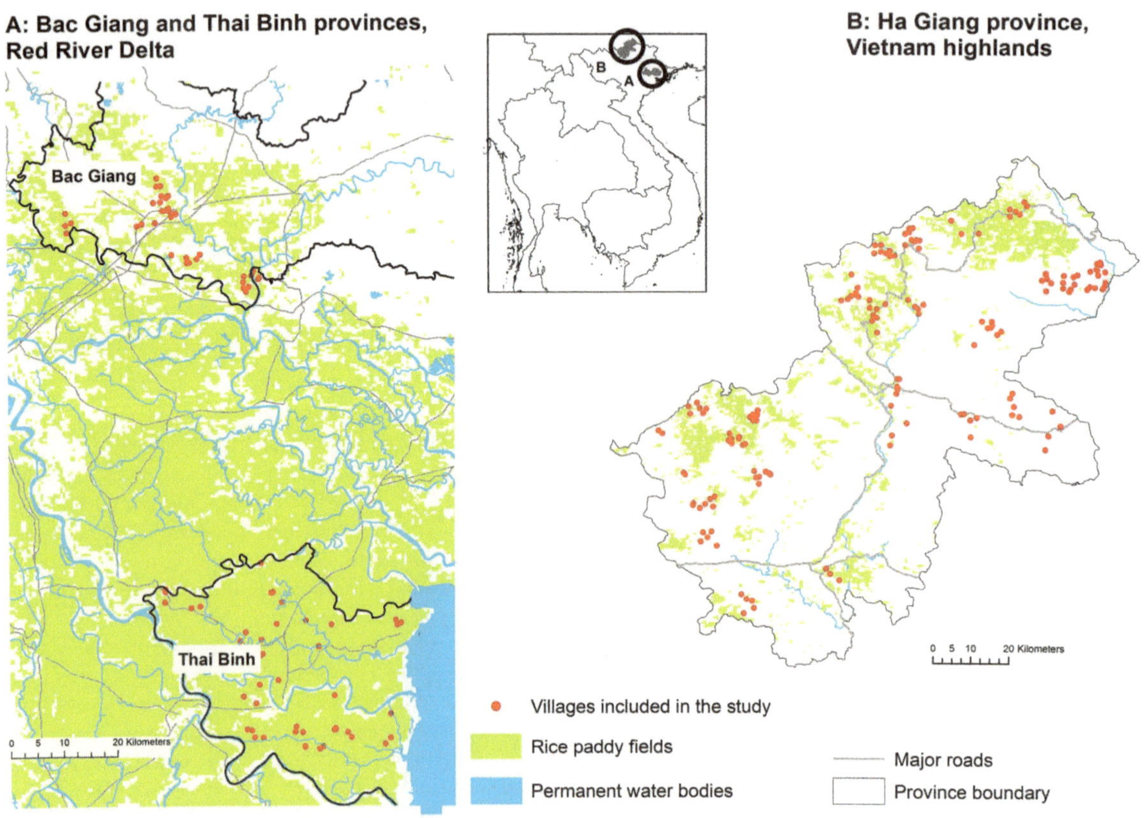

Figure 2. Study area in Vietnam. 2A: Bac Giang and Thai Binh provinces, in the Red River Delta, and location of the 83 villages included in the study. 2B: Ha Giang province, in the Vietnam highlands, and location of the 167 villages included in the study.

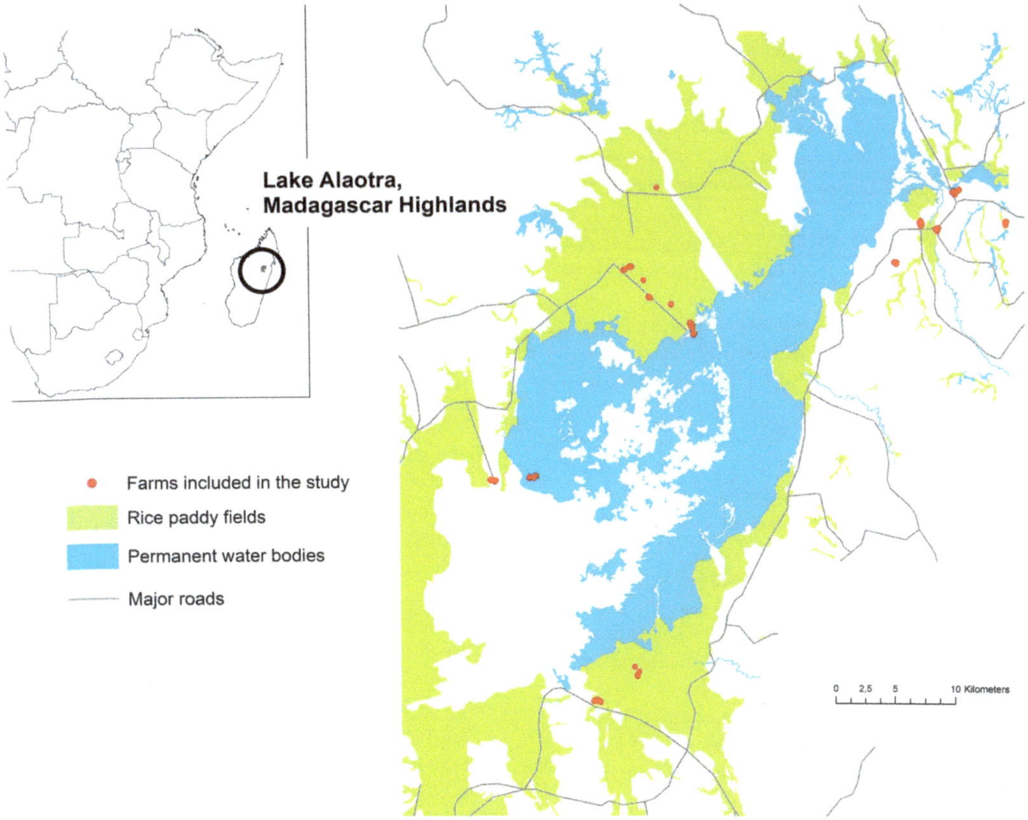

Figure 3. Study area in Madagascar highlands. Lake Alaotra and the 147 farms included in the study.

In Thailand, data on HPAI H5N1 outbreaks were collected by the Department of Livestock Development (DLD). We restricted the dataset to the second wave of the epidemic, from July 2004 to May 2005, when the surveillance system guaranteed good detection. In addition to routine surveillance and mandatory clinical reporting, two intensive active surveillance programs known as 'X-ray surveys' were implemented, with volunteers conducting door-to-door investigations to check poultry nationwide [21]. The virus was confirmed in sick or dead birds and cloacal samples from poultry and wild birds by diagnostic laboratories using reverse-transcriptase polymerase chain reaction and virus isolation [9]. All of the 1032 villages listed in the Phitsanulok administrative database were included in our comparative study.

In Vietnam and Madagascar, data on avian influenza were obtained from previous serological surveys using multistage sampling strategies that have been described elsewhere [23,27,32]. Briefly, two repeated cross-sectional surveys were conducted in the Red River Delta (Vietnam). They were based on a stratified, one-stage clustered design with a random selection of the clusters (the flocks for semi-commercial farms or the villages for backyard poultry) and a random selection of the birds within each cluster. Poultry samples were collected in December 2008, January, March and June 2009 (first survey), then monthly from March to June 2011 (second survey). In total, 3234 sera were collected from unvaccinated chickens and ducks in 83 villages and tested using a HI (hemagglutination inhibition) H5 test. In the Vietnam highlands, a serological survey was carried out with poultry samples collected monthly from April 2005 to September 2006. Within each district of Ha Giang province, four communes were chosen to represent all of the ethnic groups and to cover a wide range of environmental conditions and farming systems. Six to eight villages per commune, 6–8 farmers per village, and one or two birds in each household then were randomly selected. Our study finally included 1531 chicken sera collected in 167 villages and tested using an AIV type A ELISA competition test and a H5 pseudoviral particles assay. In Madagascar, the serological survey was based on a stratified, two-stage random sampling scheme with villages and farmers as the first- and second-degree units. A sampling fraction of 10% of villages within municipalities, and 30% of farmers within villages was set as the sampling objective. Our database included 980 sera collected in August 2008 (n = 276) and May 2009 (n = 704) from chickens, ducks and geese on 147 farms located in 14 villages. Sera were tested against avian influenza type A using an ELISA competition test (IDVET©).

Data on putative spatial risk factors

This study aimed to examine in detail a limited number of risk factors common to all four study sites for which very fine-scale data were available. The variables selected for this study were those which have been found to be significant in various countries according to a recent literature review [6]: poultry (chicken and duck) densities around villages, landscape with predominant rice cultivation, distance to the closest water body and major road, and human population density around villages.

Poultry census data were collected in Thailand at the village level during the X-ray survey organized in February 2005 by the DLD. Chicken and duck densities (birds/km^2) were calculated using a village area estimated from Delaunay triangulation between village centroids. In Vietnam, chicken and duck densities were estimated by extracting mean values from poultry density rasters [33] in a 1-km radius buffer around village centroids. In

Lake Alaotra (Madagascar), poultry density were calculated at the subdistrict level (3rd administrative level), from census data collected in May–June 2008 [27]. Gridded human population datasets were available at a 100-m resolution in all countries for the year 2010 (http://www.afripop.org/; http://www.asiapop.org/). We extracted from these raster layers the mean values of human population density in a 1-km buffer around each study location.

Layers of the road networks were obtained from the Royal Thai Survey Department in Thailand, the National Cartography House in Vietnam, and from DIVA-GIS (http://www.diva-gis.org/) in Madagascar. Euclidian distances from all study locations to the closest major road were calculated. Environmental variables were obtained for the four sites through diverse means. In Thailand, a land cover map including rice fields, rivers, lakes, canals and ponds was provided by the Royal Thai Survey Department. In Vietnam, a series of MODIS images collected in 2005 at 500-m resolution (https://lpdaac.usgs.gov/products/modis_products_table/mod09a1) were processed to map water bodies and rice paddy fields, using methods described elsewhere [34,35]. In Madagascar, one Landsat 7 enhanced thematic mapper plus (ETM+) satellite image centered on Lake Alaotra, of 30-m resolution and dated from March 2007 was analyzed using supervised classification [36]. Landcover classifications in Vietnam and Madagascar were validated by field visits [16,36]. The Euclidian distances from each study location (village or farm) to the nearest permanent body of water were calculated. The percentage of area occupied by rice paddy fields in a 1-km radius around points was calculated. Before modeling, distances as well as human and poultry densities were log-transformed to improve the fit of the linear relationship between these variables and the outcome. The rice variable was dichotomized to discriminate landscapes with predominant rice production (percentage of paddy fields in a 1-km neighborhood $\geq$ 0.80) from others.

All geoprocessing operations were realized using ArcGIS© software v.9.3 (Esri Inc.) and HawthsTools software v.3.27 (2002–2006 Spatial Ecology LLC).

Statistical analysis

The statistical analysis was based on spatial generalized linear models, which were run using the MASS package of the R software. Due to differences in data collection, the dependent variable was binomial (village infected/not infected by H5N1) in Thailand, while it was the proportion of seropositive birds in Vietnam (H5 village-level seroprevalence) and Madagascar (farm-level AIV type A seroprevalence). Putative risk factors were first screened using univariate analysis. In a second step, multivariate models were run, including all of the significant covariables from the univariate analysis (p-value<0.25, Wald test). Pair-wise correlations between variables and multicollinearity were examined through values of Spearman correlation coefficient (ρ) and variance inflation factors (VIF) [37]. A stepwise backward selection was carried out until all of the remaining variables were significant ($p \leq 0.05$). As avian influenza is a contagious disease, villages or farms located close to each other may exhibit more similar values of prevalence than those located further apart. This spatial dependency between observations was accounted for by introducing a correlation structure in the univariate and multivariate models (see details in Material S1). The extent of spatial autocorrelation was specified according to the range estimated from the spline correlogram of influenza positivity (Thailand) or prevalence (Vietnam, Madagascar) data. An exponential function was selected for the correlation matrix, as indicated by the shape of the spatial correlogram. To verify whether spatial autocorrelation was correctly accounted for, we inspected the residuals of the logistic models using a Monte Carlo method. This consisted of comparing the observed variogram with variogram 'envelopes' that were computed by simulating 999 permutations of the data values across locations [38]. Goodness-of-fit of the models was evaluated by using Hosmer-Lemeshow Chi-squared test.

Results

In lower-Northern Thailand, 163 out of the 1032 villages (15.8%) in Phitsanulok province had laboratory-confirmed HPAI H5N1 outbreaks between July 2004 and 2005. In the Red River Delta (Vietnam), we found a bird-level H5 seroprevalence of 13.0% in non-vaccinated poultry. Seropositive birds were found in 47 out of the 83 villages (56.6%) included in the study. In the Vietnam highlands, bird-level H5 seroprevalence was 3.3% and 20 villages out of 167 (11.9%) had at least one H5 seropositive bird. In Madagascar, we found a bird-level seroprevalence of 14.6% for low pathogenic AIV. Seropositive birds were found on 74 out of 147 farms (50.3%). The range of spatial autocorrelation estimated from correlograms was 35 km in Thailand, 25 km in the Vietnam highlands, 10 km in the Red River Delta of Vietnam, and 1 km in Madagascar (Figure S2). The variograms computed from the observed residuals of the models lay within the 95% limits of the simulation envelopes (Figure S3). This did not show evidence of unaccounted spatial pattern in the models.

Agro-environmental characteristics contrasted between the four study sites (Table 1). Out of the 6 variables studied, 1 to 6 (that is 1 variable in the Red River Delta, 4 in the Vietnam highlands, and 5 in lower-Northern Thailand and Madagascar, respectively) were found significantly ($p<0.25$) associated with AIV in the univariate analysis (Table 2). The simultaneous introduction of chicken and duck density in the models resulted in a multicollinearity problem for the Vietnam highlands and Madagascar (VIF>5), which may partly be due to significant correlation between these two variables (ρ>0.6). Given its previously established role in the epidemiology of avian influenza [7], only duck density was finally included in the multivariate models. After backward selection, the final model included one variable in the Red River Delta of Vietnam, two in Thailand and Madagascar, and three in the Vietnam highlands (Table 3). Higher AIV prevalence was associated with a shorter distance to a water body in the Vietnam delta and highlands, as well as in Madagascar. Duck density was found positively associated with AIV in the Vietnam highlands and lower-Northern Thailand. A positive association was found between landscape with predominant rice production and AIV infection in Thailand and Madagascar. Finally, AIV prevalence decreased with increasing distance to the closest major road in the Vietnam highlands.

The Hosmer–Lemeshow goodness-of-fit tests indicated that all final models fitted the data.

Discussion

The design of this study enabled a large sample of avian influenza data coming from various agro-environmental settings to be analyzed in parallel. All of the study sites had similar agro-ecosystems featuring the farming of domestic ducks and rice cultivation. The four study sites corresponded to different AIV situations. During the data collection period, HPAI H5N1 virus was circulating widely in lower-Northern Thailand [22] as well as in the Red River Delta of Vietnam [32], while only sporadic outbreaks occurred in the Vietnam highlands [23]. In Lake Alaotra (Madagascar highlands), no HPAI outbreaks have been reported but low pathogenic AIV are known to circulate [27]. The present study is, to our knowledge, the first to use the same

Table 1. Descriptive statistics for the variables analyzed in the four study sites.

	Lower-Northern Thailand	Red River Delta, Vietnam	Vietnam highlands	Lake Alaotra, Madagascar
	Median (IQR[a])	Median (IQR)	Median (IQR)	Median (IQR)
Chicken density (birds/km^2)	261.2 (121.4–570.7)	2836.0 (1983.0–3532.0)	146.9 (79.7–239.6)	10.9 (10.3–46.6)
Duck density (birds/km^2)	4.2 (0.1–40.5)	663.8 (497.2–978.7)	29.5 (16.2–75.9)	9.2 (4.6–11.8)
Percentage of rice fields	0.5 (0.2–0.7)	1 (0.5–1)	0.2 (0–0.4)	0.4 (0.2–0.8)
Water distance (m)	265.9 (113.7–690.1)	1691.0 (811.7–2671.0)	8948.0 (4098.0–15500.0)	740.2 (250.6–2195.0)
Road distance (m)	3300.0 (852.5–6724.0)	3728.0 (1533.0–5909.0)	7897.0 (2838.0–16600.0)	594.5 (377.3–3385.0)
Human population density (persons/km^2)	100.4 (70.3–116.0)	799.8 (573.6–975.1)	83.9 (55.6–110.4)	38.2 (36.9–53.3)

[a]interquartile range.

analytical and statistical approach to examine the distribution of risk factors of avian influenza in four study sites located in three different countries. This made it possible to compare the influence of the same set of agro-environmental determinants on the circulation of avian influenza viruses in different contexts (Table S1).

On the one hand, this multisite study shows that there is a common pattern among the study sites regarding the role of local agro-environmental characteristics in AIV circulation. Indeed, despite the heterogeneity of settings and case definitions (laboratory-confirmed HPAI H5N1 outbreaks in Thailand, H5 seropositivity in Vietnam, and low pathogenic AIV seropositivity in Madagascar), the statistical significance and the sign of the associations were constant across study sites for all tested factors in the univariate analysis (Table 2) with the exception of the Red River Delta (Vietnam). The presence of a common pattern across study sites is also reinforced by the results of the multivariate analysis, which confirmed the importance of three previously identified drivers [6], namely proximity to water bodies, predominance of rice paddy fields and duck density.

On the other hand, the present study helps to refine current understanding of avian influenza by identifying site-specific combinations of agro-environmental factors. These results can be useful for better tailoring surveillance strategies to local conditions. Indeed, we found that the general pattern of risk factors which has been so far evidenced at a broad scale corresponds at a local scale to specific combinations which may differ between study sites.

In lower-Northern Thailand, the multivariate model evidenced that duck density and rice paddy fields were important drivers of HPAI H5N1 outbreaks at the village level. This finding is consistent with a previous study carried out using a coarser epidemiological unit (sub-district) in Thailand [21]. However, it contrasts with the results of a recent study by Van Boeckel et al. [13] showing that the presence of water from rivers and flooding tended to replace rice cropping intensity as a driver of AIV circulation at a fine spatial resolution (village). The authors of this study suggest that the association between rice cultivation and HPAI H5N1 outbreaks observed in Thailand actually may be mainly explained by the frequent flooding of rice fields in intensive systems, which in turn contributes to HPAI H5N1 spread through

Table 2. Results of the spatial univariate logistic models for variables associated with H5N1 confirmed outbreaks in Thailand, H5 seroprevalence in Vietnam and low pathogenic AIV seroprevalence in Madagascar.

	Lower-Northern Thailand	Red River Delta Vietnam	Vietnam highlands	Lake Alaotra, Madagascar
Chicken density[a] (birds/km^2)	0.500[b] (0.152)[c]	0.902 (1.012)	2.675 (0.985)	0.840 (0.364)
	$p^d=0.001$	$p=0.376$	$p=0.007$	$p=0.022$
Duck density[a] (birds/km^2)	0.362 (0.070)	0.717 (0.922)	2.439 (0.615)	0.819 (0.523)
	$p<0.001$	$p=0.439$	$p<0.001$	$p=0.119$
Rice predominant	1.103 (0.339)	−0.303 (0.487)	0.466 (0.493)	0.453 (0.261)
	$p=0.001$	$p=0.535$	$p=0.346$	$p=0.085$
Water distance[a] (m)	−0.227 (0.142)	−1.010 (0.457)	−0.983 (0.427)	−0.192 (0.110)
	$p=0.110$	$p=0.030$	$p=0.023$	$p=0.082$
Road distance[a] (m)	−0.151 (0.131)	−0.026 (0.390)	−0.349 (0.283)	−0.309 (0.202)
	$p=0.251$	$p=0.948$	$p=0.219$	$p=0.128$
Human population density[a] (persons/km^2)	0.336 (0.147)	−0.016 (1.123)	0.330 (0.301)	0.096 (0.551)
	$p=0.023$	$p=0.989$	$p=0.275$	$p=0.861$

[a]log-transformed variables.
[b]coefficient estimated from the spatial logistic regression model.
[c]standard-error of the coefficient estimated.
[d]*p*-value of the Wald test.

Table 3. Results of the spatial multivariate logistic models for variables associated with H5N1 confirmed outbreaks in Thailand, H5 seroprevalence in Vietnam and low pathogenic AIV seroprevalence in Madagascar.

	Lower-Northern Thailand	Red River Delta Vietnam	Vietnam highlands	Lake Alaotra, Madagascar
Duck density[a] (birds/km^2)	0.308[b] (0.074)[c]		2.541 (0.509)	
	$p^d<0.001$		$p<0.001$	
Rice predominant	0.749 (0.350)			0.926 (0.294)
	$p=0.033$			$p=0.002$
Water distance[a] (m)		−1.010 (0.457)	−1.327 (0.412)	−0.380 (0.124)
		$p=0.030$	$p=0.002$	$p=0.003$
Road distance[a] (m)			−0.530 (0.238)	
			$p=0.028$	
Human population density[a] (persons/km^2)				
$p\ \chi^{2e}$	0.279	0.997	0.577	0.307

[a]log-transformed variables.
[b]coefficient estimated from the spatial logistic regression model.
[c]standard-error of the coefficient estimated.
[d]*p*-value of the Wald test.
[e]*p*-value of Hosmer-Lemeshow Chi-squared test.

waterborne transmission [13]. The discrepancy between our results and the Van Boeckel et al. study may derive from a difference in the explanatory variable, i.e. distance to water bodies *vs.* proportion of area covered by lakes, rivers or floods in a 1 km neighborhood around each village location. In addition, the variable we used did not capture flooded areas which coincided with HPAI H5N1 outbreaks and which were specifically targeted by Van Boeckel et al. [13,39].

In the Red River Delta of Vietnam, we found that higher H5 prevalence at the village level was associated with a shorter distance to water. This is worth noting as none of the studies carried out so far in Vietnam have found this association significant [12,16,19]. In the Red River Delta, water bodies may be contaminated by AIV through the feces of ducks which range near them or through infected dead birds thrown into the water by farmers [16]. Contaminated drinking water was proved to be an efficient route by which AIV can be transmitted to poultry [40]. We did not evidence any other predictors in this study site. This may be partly explained by the impact of massive vaccination campaigns which have been implemented from the end of 2005 in the Red River Delta [32]. Vaccination campaigns may have disturbed the overall spatial pattern of H5 circulation in poultry, rendering its analysis very challenging. Although only non-vaccinated poultry were selected for our analysis, they benefited from "herd immunity", and were thus partially protected against infection. This may be the reason why we did not find an association between poultry densities and AIV circulation. We also did not find any association between predominance of rice cultivation and H5 circulation. Rice is cultivated throughout the Red River Delta, and data showed little variability in the percentage of rice fields present around villages (Table 1). While rice cultivation alone did not appear to be a significant variable in our study, local variations in the number of rice cycles practiced per year may be influencing the spatial pattern of H5 avian influenza circulation in the study area [18]. As we intended this study to be a comparative work, we decided not to test the effect of rice cropping intensity because this would not have been meaningful for the Vietnam highlands and Madagascar, where only one rice cycle is produced each year. This question nonetheless should be investigated further. It could be integrated into much needed studies of the spatial pattern of H5 circulation in the presence of vaccination in Vietnam.

In contrast with the Red River Delta, we identified an effect of duck density on H5 circulation in the Vietnam highlands. Combined with observations from lower-Northern Thailand, this finding confirms that in the absence of poultry vaccination, domestic ducks play a key role in the circulation of H5 viruses. As in the Red River Delta, we also found in the Vietnam highlands an increased H5 seroprevalence at shorter distance to water bodies. In addition, distance to major roads was found to be a key variable, and one which was unique to the situation of the Vietnam highlands. In this site, which is characterized by sparse habitations and traditional poultry farming systems, the national road which connects the study area (Ha Giang province) to China may have favored H5 virus introduction from Yunnan. HPAI H5N1 is indeed considered endemic in China despite vaccination programs. The illegal import of live poultry from China is important, and veterinary controls at the border have limited impact on this trade. The risk of direct or indirect exposure of the Vietnamese poultry population to the HPAI H5N1 virus released by infected poultry illegally imported from China has been assessed as high [41].

It is striking that in Lake Alaotra, Madagascar, several risk factors found in Southeast Asia also are present and play a significant role in the circulation of low pathogenic avian influenza viruses. Indeed, we found that variations of low pathogenic AIV seroprevalence in Lake Alaotra were largely correlated with distance to wetlands and predominance of rice fields around poultry farms. Experimental studies have proven that low pathogenic AIV can remain infective in water for longer periods of time than HPAI H5N1 viruses [14]. It is probable that water from the lake, rivers and rice paddy fields play an important role in the transmission dynamics of low pathogenic AIV in the poultry population of Lake Alaotra through an oral-fecal route. In Lake Alaotra, AIV circulation also may be favored on farms located close to water bodies by the presence in these areas of important wild waterfowls congregations [36], which are a natural reservoir of low pathogenic AIV [42]. Another interesting characteristic of Lake Alaotra is the presence of flocks of ducks and geese, which are brought during the day to graze on rice paddy fields like in

Southeast Asia. We thus expected to observe an association between duck density and AIV seroprevalence in Madagascar similar to that found in the Southeast Asian sites in relation to H5 subtype. However, we did not find this association. This may be due to an insufficient resolution of the duck density variable, which was collected at the commune level and varied little between farms in Madagascar. Another possible explanation is that the strength of the association of duck density with low pathogenic AIV is lower than that with HPAI H5N1 and the statistical power of our study did not allow its detection. An alternative interpretation of our results is that, as suggested by a recent study on HPAI H5N1 in Thailand [13], duck density may be associated with the risk of AIV only when ducks are raised in intensive systems, or when their numbers are large for extensive systems. In Madagascar, flock size is considerably lower than in Vietnam or Thailand (10 ducks on average in Madagascar *vs* 1000 to 4000 in Thailand and Vietnam), and the low duck density (median: 9.2 ducks/km^2, interquartile range IQR: 4.6–11.8) may not reach the threshold that would be necessary to influence AIV seroprevalence in poultry populations [43]. In contrast, in the Vietnam highlands, where duck farming also is mainly extensive but with higher densities (median: 29.5 ducks/km^2, IQR: 16.2–75.9), we found a positive association between duck density and H5 seroprevalence. In lower-Northern Thailand, although the median value (4.2 ducks/km^2) was the lowest among the study sites, duck density had a wide range (IQR: 0.1–40.5) and included both extensive and large-scale intensive systems. Here also we found an effect of duck density on the probability of an HPAI H5N1 outbreak. Our observations from three contrasting agro-ecosystems thus support the hypothesis that for extensive systems, duck density influences AIV circulation only when it is above a threshold which may allow frequent contacts between ducks and between ducks and chickens. Massive vaccination campaigns carried out in the Red River Delta, Vietnam, impede from generalizing this hypothesis to this study site and thus to intensive duck farming systems.

It is noteworthy that we did not find avian influenza to be associated with human population density in any of the four study sites. This contrasts with previous studies carried out at the village level in Indonesia [8] and Thailand [13]. We considered two possible explanations. Firstly, the association that has been observed between the risk of HPAI H5N1 outbreaks and human population density may be a reflection of the increased detection sensitivity of surveillance in highly populated areas. This might be particularly applicable to studies based on avian influenza data collected through passive surveillance systems. It is reasonable to assume outbreaks are more likely to be detected and reported in more densely populated places. The fact that we did not observe an association between AIV circulation and human population density may simply confirm that such detection bias was limited in our study, which had a much more limited spatial extent. This hypothesis sounds reasonable in Vietnam and Madagascar, as AIV data were collected using cross-sectional serological surveys designed without any restriction on village size or accessibility. This hypothesis also makes sense in Thailand, where H5N1 outbreaks were detected by comprehensive active detection surveys involving several hundred thousand volunteers searching door-to-door for evidence of infection in even the smallest villages and remote areas. Secondly, human population density has also been used as a surrogate for poultry markets and areas of intensive poultry trading activities that may in turn support an increased risk of AIV transmission through flows of contaminated poultry or fomites [21]. From our study, one also could hypothesize that human population density may not be a relevant proxy for poultry trading activities in fine-scale studies focusing on a limited geographic area. When working with fine resolution data (farm or village level) on small study areas, it may be necessary to use instead variables which reflect more closely the organization of poultry trade on the ground. The distance to the closest wet markets or the presence of a poultry trader in the village [16] could constitute better risk indicators for avian influenza than human population density in this case.

This study had some limitations. Firstly, each study site presented specific constraints. In Vietnam and Madagascar, the presence of antibodies against AIV was evaluated using commercial Elisa tests with imperfect performance; this may have resulted in the overestimation of seroprevalences [24]. In Madagascar, the poultry density data may have had an insufficient resolution (commune) given the scale of analysis (farm). In addition, the study protocol in the Red River Delta (Vietnam) included 83 villages, resulting in a limited statistical power. This may have led us to erroneously conclude that some potential risk factors did not have effect on avian influenza prevalence in this site when they did. Secondly, not all data were collected according to the same procedures across the four study sites. This limited the possibility of a pooled analysis. In lower-Northern Thailand and Vietnam highlands, we believe that the data collection process allowed to obtain a fair picture of AIV circulation over a one year period. Indeed, it is acknowledged [44] that HPAI H5N1 surveillance in Thailand was comprehensive during the study period we considered (second wave of epidemics). In Vietnam highlands, despite time variations in their numbers, poultry samples were taken monthly over more than one year and provided an acceptable overall view. In the two other study sites, however, we captured only a partial picture of the AIV situation with poultry sera collected in August and May only in Madagascar, and from December to July in the Red River Delta (Vietnam). A seasonal pattern has been observed for occurrence of HPAI H5N1, with higher risk of HPAI H5N1 outbreaks from November to January and April to June [45]. Temporal variations were also previously described for seroprevalence of low pathogenic avian influenza, including H9N2 [46] and cannot be excluded in Madagascar highlands [36]. Given this discrepancy in data collection across study sites, we may thus have compared the yearly situation captured in Thailand and Vietnam highlands, with a "high-risk" period in the Red River Delta (Vietnam) and with an AIV situation corresponding to the dry season only in Madagascar. The temporal pattern of AIV circulation in poultry could be explained by different factors including varying contacts between poultry and wild birds, virus survival under climatic conditions, proportion of land covered by flooding [39], but also seasonality of poultry production and trade [36]. Despite the fact that data on the distribution in space and time of both AIV and all these factors are very difficult to obtain at high resolution, further studies should seek to include simultaneously agro-environmental, climatic and poultry trade related variables to eliminate any possible confounding. The difference in the size of the study areas also made it difficult to compare the range of spatial autocorrelation for avian influenza between sites. However, it is noteworthy that the extent of the spatial correlation we found in Thailand (35 km) and Vietnam (10 and 25 km) is consistent with results from previous studies carried out in the same countries [30,47]. The AIV data were nevertheless obtained from good-quality protocols which guaranteed the representativeness of the respective poultry populations in each study site [23,27,32]. Great care was also undertaken to obtain the highest quality ecological data in an effort to offset potential uncertainties associated with multi-country ecological comparisons.

Despite these limitations, to our knowledge this analysis is the first to compare risk factors for avian influenza across a variety of countries using fine-scale field data. We found a great deal of consistency across much of the study sites. Results confirmed the importance of wetlands-rice-ducks agro-ecosystems in the epidemiology of H5 avian influenza in Southeast Asia, and also showed that a similar agro-ecosystem contributed to the circulation of low pathogenic AIV in Madagascar. Mapping these agro-ecosystems could be useful to identify hot spots of AIV circulation across various countries. Moreover, this multisite study revealed that the relative contribution of these risk factors in the circulation of AIV differed between local environmental conditions. The analysis of agro-environmental variables collected at a very fine scale may allow the identification of villages and farms presenting a high risk for AIV circulation, and thus could help to tailor surveillance and control measures to local conditions.

Supporting Information

Figure S1 Temporal distribution of HPAI H5N1 confirmed outbreaks in lower-Northern Thailand (A), and of percentages of poultry samples taken in the Red River Delta (B) and Vietnam highlands (C).

Figure S2 Spatial correlogram of H5N1 presence or absence in lower-Northern Thailand (A), H5 seroprevalence in the Red River Delta, Vietnam (B), H5 seroprevalence in the Vietnam highlands (C) and low pathogenic AIV seroprevalence in Lake Alaotra, Madagascar highlands (D).

Figure S3 Omnidirectional variograms computed using the standardized residuals derived from the spatial generalized linear model used in lower-Northern Thailand (A), in the Red River Delta, Vietnam (B), in the Vietnam highlands (C) and in Lake Alaotra, Madagascar highlands (D). The dashed lines show the pointwise 95% limits constructed from the Monte Carlo 999 simulations; circles represent the empirical variogram.

Table S1 Summary statistics for the variables examined in the four study sites: lower-Northern Thailand (A), the Red River Delta, Vietnam (B), the Vietnam highlands (C), and Lake Alaotra, Madagascar highlands (D).

Material S1 Description of the spatial generalized linear model used to study the effect of agro-environmental determinants on avian influenza circulation.

Acknowledgments

We are grateful to the FOFIFA-DRZV, the Official Malagasy Veterinary Services and AVSF (Madagascar), the National Institute of Animal Husbandry (Vietnam), the Department of Livestock Development and the Provincial Livestock Office of Phitsanulok (Thailand). We also thank all of the farmers and staff involved in the field studies. We thank Annelise Tran and Laure Guerrini for analyzing remote sensing data in Vietnam and Madagascar. We thank Cécile Berthouly-Salazar and Carlène Trevennec for contributing to data collection in the Vietnam highlands. We also acknowledge Agnès Waret-Szkuta for fruitful discussions, and Rémi Servien for statistical advice. We thank Grace Delobel for revising our written English.

Author Contributions

Conceived and designed the experiments: MCP SD HRA MP VC. Performed the experiments: MCP SD HRA MP VC. Analyzed the data: MCP VC. Contributed reagents/materials/analysis tools: NVK WT. Wrote the paper: MCP MG VC.

References

1. Gao R, Cao B, Hu Y, Feng Z, Wang D, et al. (2013) Human Infection with a Novel Avian-Origin Influenza A (H7N9) Virus. New England Journal of Medicine 368: 1888–1897.
2. WHO (2014) Number of confirmed human cases of avian influenza A(H7N9) reported to WHO - Report 10.
3. Morens DM, Taubenberger JK, Fauci AS (2013) H7N9 avian influenza A virus and the perpetual challenge of potential human pandemicity. MBio 4.
4. Alexander DJ (2007) An overview of the epidemiology of avian influenza. Vaccine 25: 5637–5644.
5. Hogerwerf L, Wallace R, Ottaviani D, Slingenbergh J, Prosser D, et al. (2010) Persistence of Highly Pathogenic Avian Influenza H5N1 Virus Defined by Agro-Ecological Niche. EcoHealth.
6. Gilbert M, Pfeiffer DU (2012) Risk factor modelling of the spatio-temporal patterns of highly pathogenic avian influenza (HPAIV) H5N1: a review. Spatial and Spatio-temporal Epidemiology 3: 173–183.
7. Kim JK, Negovetich NJ, Forrest HL, Webster RG (2009) Ducks: the "Trojan horses" of H5N1 influenza. Influenza and Other Respiratory Viruses 3: 121–128.
8. Loth L, Gilbert M, Wu J, Czarnecki C, Hidayat M, et al. (2011) Identifying risk factors of highly pathogenic avian influenza (H5N1 subtype) in Indonesia. Preventive Veterinary Medicine 102: 50–58.
9. Tiensin T, Ahmed Syed Sayeem U, Rojanasthien S, Songserm T, Ratanakorn P, et al. (2009) Ecologic Risk Factor Investigation of Clusters of Avian Influenza A (H5N1) Virus Infection in Thailand. The Journal of Infectious Diseases 199: 1735–1743.
10. Ahmed SSU, Ersboll AK, Biswas PK, Christensen JP, Hannan ASMA, et al. (2012) Ecological Determinants of Highly Pathogenic Avian Influenza (H5N1) Outbreaks in Bangladesh. PLoS ONE 7: e33938.
11. Martin V, Pfeiffer DU, Zhou X, Xiao X, Prosser DJ, et al. (2011) Spatial Distribution and Risk Factors of Highly Pathogenic Avian Influenza (HPAI) H5N1 in China. PLoS Pathogens 7: e1001308.
12. Pfeiffer DU, Minh PQ, Martin V, Epprecht M, Otte MJ (2007) An analysis of the spatial and temporal patterns of highly pathogenic avian influenza occurrence in Vietnam using national surveillance data. Veterinary Journal 174: 302–309.
13. Van Boeckel TP, Thanapongtharm W, Robinson T, Biradar CM, Xiao X, et al. (2012) Improving Risk Models for Avian Influenza: The Role of Intensive Poultry Farming and Flooded Land during the 2004 Thailand Epidemic. PLoS ONE 7: e49528.
14. Brown JD, Goekjian G, Poulson R, Valeika S, Stallknecht DE (2009) Avian influenza virus in water: Infectivity is dependent on pH, salinity and temperature. Veterinary Microbiology 136: 20–26.
15. Cecchi G, Ilemobade A, Le Brun Y, Hogerwerf L, Slingenbergh J (2008) Agro-ecological features of the introduction and spread of the highly pathogenic avian influenza (HPAI) H5N1 in northern Nigeria. Geospatial Health 3: 7–16.
16. Desvaux S, Grosbois V, Pham TTH, Fenwick S, Tollis S, et al. (2011) Risk Factors of Highly Pathogenic Avian Influenza H5N1 Occurrence at the Village and Farm Levels in the Red River Delta Region in Vietnam. Transboundary and Emerging Diseases 58: 492–502.
17. Gilbert M, Chaitaweesub P, Parakamawongsa T, Premashthira S, Tiensin T, et al. (2006) Free-grazing ducks and Highly Pathogenic Avian Influenza, Thailand. Emerging Infectious Diseases 12: 227–234.
18. Gilbert M, Xiao X, Pfeiffer DU, Epprecht M, Boles S, et al. (2008) Mapping H5N1 highly pathogenic avian influenza risk in Southeast Asia. Proceedings of the National Academy of Sciences of the United States of America 105: 4769–4774.
19. Henning J, Pfeiffer DU, le Vu T (2009) Risk factors and characteristics of H5N1 Highly Pathogenic Avian Influenza (HPAI) post-vaccination outbreaks. Veterinary Research 40: 15.
20. Loth L, Gilbert M, Osmani MG, Kalam AM, Xiao X (2010) Risk factors and clusters of Highly Pathogenic Avian Influenza H5N1 outbreaks in Bangladesh. Preventive Veterinary Medicine 96: 104–113.
21. Paul M, Tavornpanich S, Abrial D, Gasqui P, Charras-Garrido M, et al. (2010) Anthropogenic factors and the risk of Highly Pathogenic Avian Influenza H5N1: prospects from a spatial-based model. Veterinary Research 41.

22. Paul M, Wongnarkpet S, Gasqui P, Poolkhet C, Thongratsakul S, et al. (2011) Risk factors for highly pathogenic avian influenza (HPAI) H5N1 infection in backyard chicken farms, Thailand. Acta Tropica 118: 209–216.
23. Trevennec K, Chevalier V, Grosbois V, Garcia JM, Thu HH, et al. (2011) Looking for avian influenza in remote areas. A case study in Northern Vietnam. Acta Tropica 120: 160–166.
24. Ward M, Maftei D, Apostu C, Suru A (2008) Environmental and anthropogenic risk factors for highly pathogenic avian influenza subtype H5N1 outbreaks in Romania, 2005−2006. Veterinary Research Communications.
25. Williams RAJ, Fasina FO, Peterson AT (2008) Predictable ecology and geography of avian influenza (H5N1) transmission in Nigeria and West Africa. Transactions of the Royal Society of Tropical Medicine and Hygiene 102: 471–479.
26. Wang L, Zhang W, Magalhaes RJ, Clements AC, Hu W, et al. (2013) Geographic co-distribution of influenza virus subtypes H7N9 and H5N1 in humans, China. Emerging Infectious Diseases 19: 1898–1900.
27. Rasamoelina Andriamanivo H, Lancelot R, Maminiaina OF, Rakotondrafara TF, Jourdan M, et al. (2012) Risk factors for avian influenza and Newcastle disease in smallholder farming systems, Madagascar highlands. Preventive Veterinary Medicine 104: 114–124.
28. De Visscher M-N (2013) Avian influenza and Newcastle disease: from the role of farming practices and wild bird ecology to disease surveillance and control. EMPRES-Animal Health 360 42: 26–29.
29. Chaichoune K, Wiriyarat W, Thitithanyanont A, Phonarknguen R, Sariya L, et al. (2009) Indigenous sources of 2007–2008 H5N1 avian influenza outbreaks in Thailand. 216–222.
30. Minh PQ, Morris RS, Schauer B, Stevenson M, Benschop J, et al. (2009) Spatio-temporal epidemiology of highly pathogenic avian influenza outbreaks in the two deltas of Vietnam during 2003–2007. Preventive Veterinary Medicine 89: 16–24.
31. International B (2012) Important Bird Areas factsheet: Lake Alaotra.
32. Desvaux S, Grosbois V, Pham TTH, Dao DT, Nguyen TD, et al. (2012) Evaluation of the vaccination efficacy against H5N1 in domestic poultry in the Red River Delta in Vietnam. Epidemiology and Infection 141: 776–788.
33. Van Boeckel TP, Prosser D, Franceschini G, Biradar C, Wint W, et al. (2011) Modelling the distribution of domestic ducks in Monsoon Asia. Agriculture, Ecosystems and Environment 141: 373–380.
34. Xiao X, Boles S, Frolking S, Li C, Babu JY, et al. (2006) Mapping paddy rice agriculture in South and Southeast Asia using multi-temporal MODIS images. Remote Sensing of Environment 100: 95–113.
35. Tran A, Goutard F, Chamaillé L, Baghdadi N, Lo Seen D (2010) Remote sensing and avian influenza: A review of image processing methods for extracting key variables affecting avian influenza virus survival in water from Earth Observation satellites. International Journal of Applied Earth Observation and Geoinformation 12: 1–8.
36. Guerrini L, Paul MC, Leger L, Rasamoelina Andriamanivo H, Maminiaina OF, et al. (2014) Landscape attributes driving avian influenza virus circulation in Lake Alaotra region, Madagascar. Geospatial Health 8: 445–453.
37. Dohoo I, Martin W, Stryhn H (2003) Veterinary Epidemiologic Research: AVC Inc., Charlottetown, Prince Edward Island, Canada. 865 p.
38. Diggle PJ, Ribeiro PJ (2007) Model-based Geostatistics: Springer. 228 p.
39. Thanapongtharm W, Van Boeckel TP, Biradar C, Xiao XM, Gilbert M (2013) Rivers and flooded areas identified by medium-resolution remote sensing improve risk prediction of the highly pathogenic avian influenza H5N1 in Thailand. Geospatial Health 8: 193–201.
40. Claes G, Marche S, Dewulf J, Van Den Berg T, Lambrecht B (2013) An experimental model to analyse the risk of introduction of a duck-originated H5 low-pathogenic avian influenza virus in poultry through close contact and contaminative transmission. Epidemiology and Infection: 1–12.
41. Desvaux S (2012) Epidemiology of the highly pathogenic avian influenza H5N1 in Northern Vietnam: Applications for surveillance and control. Murdoch University.
42. Olsen B, Munster VJ, Wallensten A, Waldenström J, Osterhaus ADME, et al. (2006) Global Patterns of Influenza A Virus in Wild Birds. Science 312: 384–388.
43. Gilbert M, Newman S, Takekawa J, Loth L, Biradar C, et al. (2011) Flying Over an Infected Landscape: Distribution of Highly Pathogenic Avian Influenza H5N1 Risk in South Asia and Satellite Tracking of Wild Waterfowl. EcoHealth: 1–11.
44. Goutard FL, Paul M, Tavornpanich S, Houisse I, Chanachai K, et al. (2012) Optimizing early detection of avian influenza H5N1 in backyard and free-range poultry production systems in Thailand. Preventive Veterinary Medicine 105: 223–234.
45. Tran CC, Yost RS, Yanagida JF, Saksena S, Fox J, et al. (2013) Spatio-Temporal Occurrence Modeling of Highly Pathogenic Avian Influenza Subtype H5N1: A Case Study in the Red River Delta, Vietnam. ISPRS International Journal of Geo-Information 2: 1106–1121.
46. Tombari W, Paul M, Bettaieb J, Larbi I, Nsiri J, et al. (2013) Risk Factors and Characteristics of Low Pathogenic Avian Influenza Virus Isolated from Commercial Poultry in Tunisia. PLoS ONE 8: e53524.
47. Souris M, Gonzalez J-P, Shanmugasundaram J, Corvest V, Kittayapong P (2010) Retrospective space-time analysis of H5N1 Avian Influenza emergence in Thailand. International Journal of Health Geographics 9: 3.

5

Remotely-Sensed Indicators of N-Related Biomass Allocation in *Schoenoplectus acutus*

Jessica L. O'Connell[1], Kristin B. Byrd[2], Maggi Kelly[1]*

1 Department of Environmental Sciences, Policy and Management, University of California, Berkeley, Berkeley, California, United States of America, **2** United States Geological Survey, Menlo Park, California, United States of America

Abstract

Coastal marshes depend on belowground biomass of roots and rhizomes to contribute to peat and soil organic carbon, accrete soil and alleviate flooding as sea level rises. For nutrient-limited plants, eutrophication has either reduced or stimulated belowground biomass depending on plant biomass allocation response to fertilization. Within a freshwater wetland impoundment receiving minimal sediments, we used experimental plots to explore growth models for a common freshwater macrophyte, *Schoenoplectus acutus*. We used N-addition and control plots (4 each) to test whether remotely sensed vegetation indices could predict leaf N concentration, root:shoot ratios and belowground biomass of *S. acutus*. Following 5 months of summer growth, we harvested whole plants, measured leaf N and total plant biomass of all above and belowground vegetation. Prior to harvest, we simulated measurement of plant spectral reflectance over 164 hyperspectral Hyperion satellite bands (350–2500 nm) with a portable spectroradiometer. N-addition did not alter whole plant, but reduced belowground biomass 36% and increased aboveground biomass 71%. We correlated leaf N concentration with known N-related spectral regions using all possible normalized difference (ND), simple band ratio (SR) and first order derivative ND (FDN) and SR (FDS) vegetation indices. $FDN_{1235,\ 549}$ was most strongly correlated with leaf N concentration and also was related to belowground biomass, the first demonstration of spectral indices and belowground biomass relationships. While *S. acutus* exhibited balanced growth (reduced root:shoot ratio with respect to nutrient addition), our methods also might relate N-enrichment to biomass point estimates for plants with isometric root growth. For isometric growth, foliar N indices will scale equivalently with above and belowground biomass. Leaf N vegetation indices should aid in scaling-up field estimates of biomass and assist regional monitoring.

Editor: John F. Valentine, Dauphin Island Sea Lab, United States of America

Funding: This research was supported by K. Byrd's NASA New Investigator Program in Earth Sciences Grant Number: NNH10A086I. The funders had no role in study design, data collection and analysis, decision to publish, or preparation of the manuscript.

Competing Interests: The authors have declared that no competing interests exist.

* E-mail: maggi@berkeley.edu

Introduction

Belowground biomass of wetland emergent plants is the dominant source of soil organic carbon and peat in wetlands [1,2] and contributes to long-term carbon sequestration [3]. In addition to soil organic carbon storage, wetland belowground biomass also contributes to soil stability and accretion by adding soil organic matter volume [4,5]. Many coastal wetlands may subside below sea level because of a combination of surface water drainage, salt water intrusion, marsh compaction (peat collapse), sediment starvation, and global sea level rise [6–11]. As wetlands subside, they become excessively inundated, resulting in flooding, plant death, and marsh loss. Wetland loss can be problematic because wetlands provide many functions and services to society, such as flood water mitigation, storm surge attenuation, belowground carbon storage, water filtration and purification, and habitat for wetland biota [12]. Belowground biomass may consequently maintain or build wetland elevation to prevent subsidence below sea level [13]. Therefore, tools for predicting and promoting wetland plant belowground biomass can support wetland management for resilience and provision of ecosystem benefits.

Two competing hypotheses explain relationships between below and aboveground biomass. The balanced-growth hypothesis is an ecological model which suggests, all else being equal, plants allocate growth towards the most limiting resource, e.g. towards shoots when light is limited, or towards roots when nutrients or water are limited [14,15]. This ecological concept also has been called optimal partitioning theory [16]. Alternatively, above and belowground biomass ratios may conform with the isometric allocation model, which suggests that above and belowground biomass scale equivalently across multiple environmental conditions [16]. Plant growth can correspond to a balanced-growth model [17–19] or an isometric allocation model [20,21]. In both situations, leaf N concentration can inform belowground biomass estimates. In the balanced growth case, additional nutrients, such as N and P, increase shoot growth but do not increase belowground biomass equally, therefore leaf N scales with root:shoot ratios. Alternatively, plants responding isometrically to nutrient addition will have stimulated below and aboveground biomass and constant root:shoot ratios.

While much progress has been made estimating aboveground biomass from optical remote-sensing technology [22,23], monitoring belowground biomass remains challenging. Belowground biomass measures are difficult because biomass ratios in plants sometimes vary with access to nutrients, light, competition, temperature, and phylogenetic history [14]. As a result, while above and belowground biomass usually are correlated, sometimes root:shoot ratios fluctuate greatly across vegetation types and environments [24].

Past studies indicate the potential for remote sensing to detect eutrophic conditions. In particular, remote sensing has been commonly used to estimate foliar nitrogen concentration in aboveground plant tissue [22,25–31]. For nutrient limited species, plants grown in high N environments may concentrate N in aboveground tissues at greater levels than otherwise [32,33]. When combined with appropriate growth models (balanced or isometric growth), leaf N might also provide evidence of belowground biomass when species specific field data are available to calibrate site models.

Several analytical techniques have been developed to relate hyperspectral data to canopy N concentration, such as band-depth analysis [34], spectral matching techniques [35], and partial least squares regression [22]. However vegetation indices correlated to plant characteristics through empirical methods, are a simple, straightforward and rapid method for detecting the biochemical absorption signal while minimizing background effects [36–38]. Therefore, we consider the use of vegetation indices derived from freely-available Hyperion hyperspectral imagery as a potential tool for land management. Hyperion is a hyperspectral satellite sensor on the National Aeronautics and Space Administration Earth Observing One Satellite (NASA EO-1), and has been successfully applied to map plant biophysical and biochemical characteristics [30,39,40].

Here we explore plant growth models in a novel conservation context: freshwater peat marshes within California's Sacramento-San Joaquin Delta region (Figure 1). The Delta is a wetland complex responsible for collecting a vast quantity of water for human use. Historically, the Delta was a 1400-km^2 tidal marsh network [41] and contained a large area of emergent marshes with peat 2–15 m deep, a substantial carbon pool [42]. Wetland dewatering, peat decomposition, soil compaction and crop exploitation resulted in land subsidence of up to 8 m throughout the Delta [42].

We use an experimental approach to estimate biomass allocation of *Schoenoplectus acutus*, hardstem bulrush, an emergent macrophyte common to the Delta, and to freshwater marshes throughout North America [43]. We use our experimental model to draw relationships across simulated eutrophication (N addition), field measurements of whole plant biomass production, and remotely-sensed vegetation indices of leaf nitrogen concentration. Our main questions include 1) Can percent total N concentration in leaves be used to estimate belowground biomass? and 2) Do remotely-sensed vegetation indices indicate belowground plant response to eutrophic conditions? Vegetation indices may help identify eutrophic wetlands and assist with monitoring wetland nutrient cycling. Relationships between leaf N and belowground biomass might assist estimation of spatial variation in roots and rhizomes across and within sites, supporting regional carbon monitoring efforts. Such estimates also may identify areas where marsh subsidence or accretion are likely across broad areas.

Methods

We installed 18 plant growth chambers in the northeast unit of an impounded wetland north of Mayberry Slough on Sherman island, CA, USA (Lat: 38.0533 Long: −121.7682). The Mayberry wetland is 122 ha of restored freshwater emergent marsh with 4 impounded units (Figure 1). Water levels within this wetland are managed by the California Department of Water Resources. Water flows from south to north from Mayberry Slough through the impoundments via water control structures.

Growth chambers were constructed by tacking landscaping cloth over a cylindrical frame constructed of bamboo (diameter: 30.5 cm, height: 135 cm). Growth chambers allowed water exchange with the surrounding wetland. Growth chambers were installed within the impoundment pond, 1 m from the impoundment shoreline, on 28 May 2012. Differences in pond bottom geomorphology resulted in differences (±15 cm) in water depth surrounding growth chambers. Each growth chamber was filled with 45 L of planting media (sphagnum peat moss), raising the effective water depth. Within each growth chamber, we planted 1-year-old *S. acutus* (from California Flora Nursery, Fulton, CA, USA, www.calfloranursery.com; propagules collected from the Russian River watershed, CA, USA, 140 km northwest of Mayberry Slough). We used growth chambers to allow accurate measurement of whole plant biomass. Growth chambers facilitate extraction of the entire root and rhizome system.

Plants were randomly assigned to control or N-addition treatments and installed in growth chambers. Growth chambers for these treatments were physically separated by a small incomplete levy, but open to the larger impoundment. Control treatment plants were also located upstream of the N-addition treatment in the wetland impoundment. This design ensured both treatments were similar hydrologically, except that the control treatment did not receive N additions. Within each chamber, we measured number of stems, stem height, stem diameter, and number of inflorescences monthly. The growth form of *S. acutus* is graminoid, consisting of one long cylindrical stem terminating in an inflorescence (when present) [43]. Therefore stem and inflorescence measurements adequately capture aboveground growth. We also measured water depth inside and outside of growth chambers.

Plants within the N-addition treatment were fertilized with custom blend controlled release fertilizer (Apex, 16-10-13, Simplot Professional Products Lathrop, CA, USA, apexfertilizer.com). Fertilizer was coated granules, designed for controlled nutrient release without influence from media type, moisture level, pH, or microbial activity. We added 50 g of Apex on 28 May 2012 and again on 2 July 2012. In addition, 5.65 g of uncoated granular fertilizer (Rootblast, 2-1-2, Rootblast international, Canton, Ohio, USA) were added on 28 May, 11 July, 2 August, and 25 September 2012. In total, 16.5 g inorganic N as nitrate (NO_3), 4.5 g P as phosphate (PO_4^{-3}) and 13.5 g K as potash (K_2O) were added. This represents a 4:1 ratio of N:P by weight, as observed in the most eutrophic conditions of San Francisco Bay Estuary [44]. A fertilizer blend of NPK was added because increases in these nutrients commonly co-occur in eutrophic estuaries [44,45]. Nutrient additions were added monthly because nutrient pulses (as opposed to continuous high concentrations) are common in natural conditions, where nutrients fluctuate temporally [32,46,47].

Crawfish (*Pacifastacus*, *Procambarus* or *Orconectes* sp.) [48] colonized 4 growth chambers per treatment on 2 August 2012. Entry holes created by crawfish allowed planting media to escape, flooding plants and causing mortality. We harvested all other plants on 1 October 2012 (N = 4 per treatment). All subsequent analyses were focused on these harvested plants. Plants were sorted into aboveground biomass (stems, leaves and inflorescences) and belowground biomass (roots and rhizomes). We did not separate live from dead roots because dead roots were fine roots. We could not reliably determine the transition from live to dead fine roots. All plant biomass was oven dried at 40°C until all plant tissue

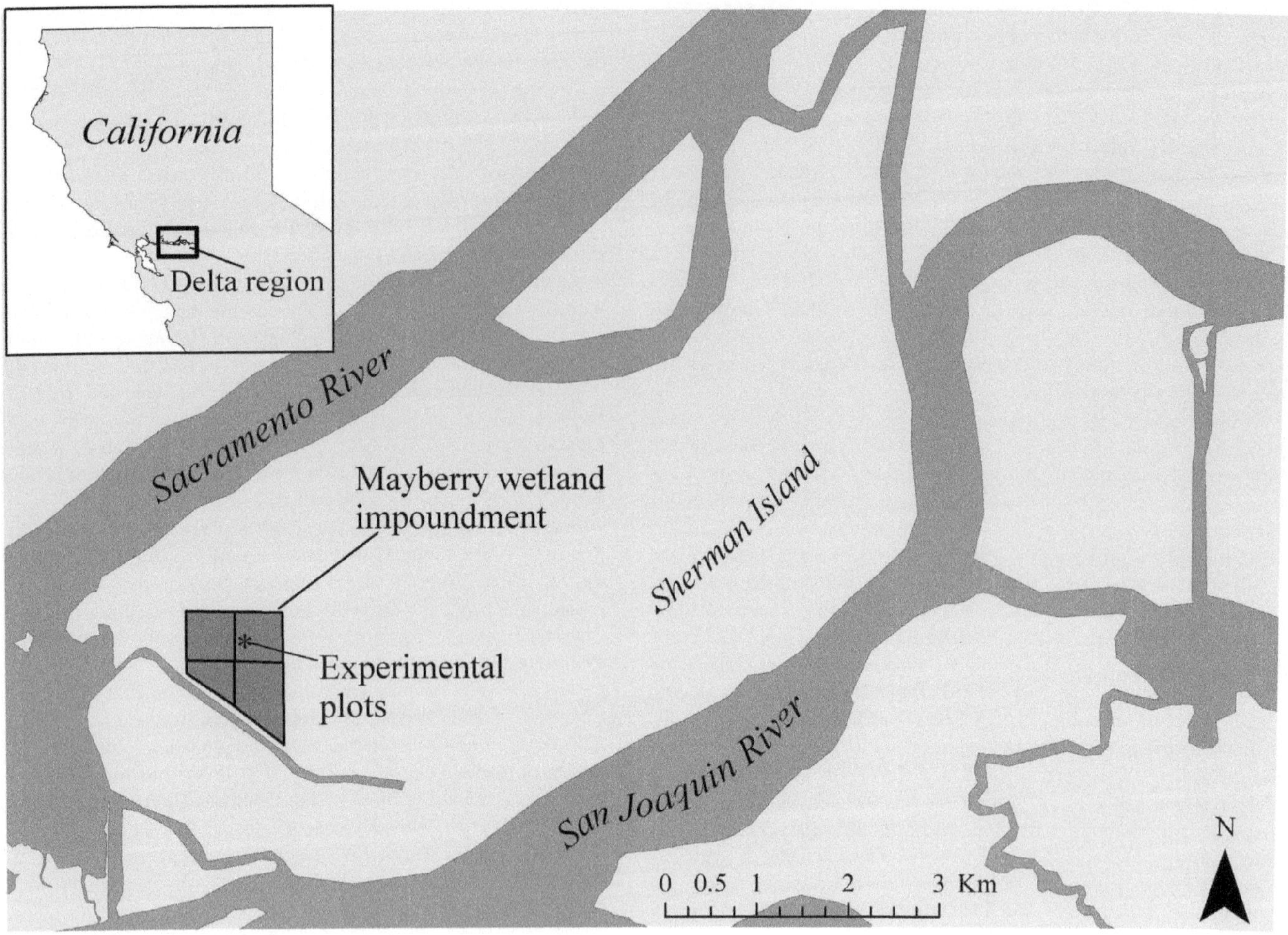

Figure 1. Study site. The Sacramento-San Joaquin River confluence and associated Delta within California's Central Valley, USA. Locations of the Mayberry wetland impoundment and experimental site are indicated.

water was removed. Subsequent analyses focused on oven-dried weights. For simplicity, we defined root:shoot ratio as the ratio of all harvested above and belowground biomass.

Estimates of Plant Biophysical and Biochemical Parameters

We used end of season above and belowground biomass as our biomass estimates. Oven-dried plant leaves were measured for total N (percent by weight) at the UC Davis analytical laboratory (Davis, California, USA; http://anlab.ucdavis.edu/). The Davis laboratory uses sample combustion coupled with thermal conductivity/IR detection (LECO FP-528 and TruSpec CN Analyzers) to measure total N. We estimated total N content (g) in leaves as: % leaf total N*aboveground biomass (g).

We used permutation methods to assess one-sided hypotheses, specifically that N addition increased whole plant and aboveground biomass, number of inflorescences, leaf N concentration (%) and leaf N (g), but reduced belowground biomass (total belowground biomass, and roots and rhizomes separately) and also reduced root:shoot ratios. We used all possible permutations (the COIN package in R, version 2.15.1, The R Foundation for Statistical Computing). Permutation tests are useful because their only assumption is that data are independent [49]. Permutation methods compute the probability that two samples come from the same population by generating a reference distribution for many permutations of the data [49]. We further evaluated whether leaf N concentration had significant relationships with root:shoot ratios using 10,000 bootstrap replicates with bias corrected, accelerated confidence intervals for linear models in program R.

Vegetation Indices of Foliar Nitrogen Concentration

Canopy reflectance spectra were obtained over each plot near midday on 25 September 2012 (6 days before plant harvest) using an ASD Inc. FieldSpec Pro FR portable spectroradiometer (Analytical Spectral Devices, Inc., Boulder, CO, USA). Spectral readings were sampled every 1.4 nm over 350–1000 nm and 2 nm over 1000–2500 nm using a 25° field of view foreoptics. Readings were taken at nadir 1 m above the vegetation canopy using a 3-m optical fiber cable. Canopy reflectance was calculated as the ratio of canopy radiance to radiance measured from a calibrated white reference, Spectralon (Labsphere, Inc.), which was collected every 10 min. At each sample location, ten reflectance measurements, each an average of 12 spectra, were collected and then averaged using ViewSpec Pro 6.1.10 (Analytical Spectral Devices, Inc, Boulder, CO, USA).

Leaf N concentration can be estimated from spectral indices such as normalized difference indices (ND, equation 1) or simple band ratios (SR, equation 2) [28]. In addition, NDs and SRs calculated from first-order derivatives of the reflectance spectrum have also been related to N concentration [25,50,51]. First-order derivatives, calculated in ViewSpec Pro 6.1.10 [52] involve the calculation of the slope of the spectrum. First derivative spectra are useful for reducing the effects of multiple scattering of radiation due to sample geometry and surface roughness [53] and for enhancing absorption features and inflection points masked by interfering background absorptions and canopy background effects [54–56]. Given the background absorptions of the surrounding water, we expected that indices derived from first derivative spectra would perform well in estimating leaf nitrogen concentration.

$$ND_{ij} = (R_j - R_i)/(R_j + R_i) \quad (1)$$

$$SR_{ij} = R_j/R_j \quad (2)$$

where R_i and R_j are the reflectance values of bands i and j.

ND or SR indices using new band combinations may be more highly correlated to leaf biochemical properties than those more commonly found in the literature [57]. As a result, we tested multiple published indices (see Table 1) and new indices to determine which were most related to leaf N concentration in our experimental plots. To test new hyperspectral indices we first simulated 164 hyperspectral Hyperion bands with the field spectroradiometer data. Hyperion has a 30-m spatial resolution and 242 bands each 10 nm-wide. The simulated Hyperion bands covered the ranges 422–1300 nm, 1443–1795 nm, and 1998–2400 nm. Water absorption regions of the spectrum with high noise were removed from analysis. These field-collected simulated Hyperion bands represent finely detailed full-spectrum data and provide a point of comparison for future field studies using other remote sensing platforms.

For the simulated dataset, we calculated all possible ND and SR indices using all band combinations [57] for a total of 13,366 Hyperion indices for each index type. We also calculated the first order derivative spectra in ViewSpec Pro 6.1.10, and calculated all possible first derivative NDs (FDN) and first derivative SRs (FDS) using all band combinations for a total of 13,366 indices for each index type.

Table 1. Previously identified relationships between leaf N concentration and hyperspectral vegetation indices.

Published N Indices
$FDS_{743,1316}$ [25]
$ND_{483,503}$ [66]
$SR_{805,710}$ [77]
$SR_{800,680}$ [51]
$SR_{700,550}$ [78]
$SR_{810,560}$ [79]
REP [80]
NDNI [36]

REP (red edge position index) = $700+40(R_{RE}-R_{700})/(R_{740}-R_{700})$, $R_{RE}=(R_{670}+R_{780})/2$.

NDNI (Normalized Difference Nitrogen Index) = $[\log(1/R_{1510})-\log(1/R_{1680})]/[\log(1/R_{1510})+\log(1/R_{1680})]$.

Using Stata/SE 12.1 (StataCorp LP 1985–2011), we ran pairwise correlations to calculate the correlation (r) between all indices (published and new) and leaf N concentration. We used leaf N concentration (%) rather than leaf N (g) because the latter is confounding with aboveground biomass (i.e. as aboveground biomass increases, total weight of N increases concomitantly, but % N per unit leaf might not increase) [28]. Coefficients of determination (R^2) were calculated as $r*r$. We also compared the bands selected in these high R^2 indices against known N absorption bands to evaluate the physical basis for index selection [36,58,59]. We used 10,000 bootstrap replicates with bias corrected, accelerated confidence intervals for linear models in program R to investigate how well the best indices were related to belowground biomass and root:shoot ratio. We present results for those indices whose correlations with leaf N were greater than 80% and whose bootstrapped 95% CI's for slope in regressions with leaf N and belowground biomass did not overlap zero. Given our small sample size, we consider this latter criterion to be conservative and more reliable than p-value for estimating statistical significance.

Ethics Statement

Permission was given for site access by the California Department of Water Resources. No special permits were required for our research sampling scheme.

Results

Plant Biophysical Parameters

Whole plant biomass (belowground+aboveground) was similar between treatments (Table 2), though aboveground biomass was 71% greater in the N-addition than in the control treatment (Table 2). Conversely, belowground biomass was 36% greater in the control than in the N-addition treatment (Table 2). This resulted in a root:shoot ratio 74% greater in the control than in the N-addition treatment (Table 2). Inflorescence production also was higher in N-addition plots than in control plots (Table 2). Further, leaf N concentration was 18% and leaf N (g) was 77% greater in the N-addition treatment than in the control (Table 2). Finally, leaf N concentration had a significant negative relationship with root:shoot ratio in *S. acutus* (Table 3). We did not detect a significant relationship between below and aboveground biomass (Table 3).

Vegetation Indices for Foliar Nitrogen Concentration

Most published indices (Table 1) were poorly correlated with leaf N concentration, with the exception of $FDS_{743,1316}$ ($F_{1,6}=5.38$, $R^2=0.47$, $P=0.06$). Among all indices generated from simulated Hyperion bands, $FDN_{1235,549}$ was most correlated with leaf N concentration (Table 4). The use of green wavelengths (~500 to 570 nm) in hyperspectral indices have been found to be sensitive to N and can improve the precision and accuracy of predictions [28]. In our experiment, green indices based on spectral reflectance were not significant, though green indices using first-order derivatives of reflectance were highly significant. The difference in the green peak region between treatments is evident when comparing the average reflectance spectrum of N-addition plots to those in control plots (Figure 2). Of the other best new indices, most contain bands in the short-wave infrared region of the spectrum (Table 4). Finally, two of the best indices explaining leaf N concentration ($FDN_{1235,549}$ and $FDS_{2184,1780}$) also exhibited a positive relationship with belowground biomass and root:shoot ratio, but were unrelated to aboveground biomass (Table 3). All other top indices for leaf N (correlation>0.8) were

Table 2. Differences between *Schoenoplectus acutus* biophysical parameters within N-addition and control treatments based on permutation tests (N = 4 per treatment).

Response variable	*Z*	*P*	Control mean ± SE	N-addition mean ± SE
Whole plant biomass (g)	1.03	0.85	133.4±61.3	112.2±23.1
AG biomass (g)	−2.02	**0.02**	**9.6±2.7**	**32.9±7.6**
BG biomass (g)	1.34	*0.09*	*123.8±58.9*	*79.3±19*
Roots: rhizomes	−0.29	0.4	1.9±0.5	1.6±0.2
Roots:shoots	2.08	**0.02**	**10.4±3.3**	**2.7±0.6**
Inflorescence number	−1.13	*0.07*	*0*	*3±2.3*
Leaf N concentration	−1.60	*0.09*	*1.4±0.1*	*1.7±0.2*
Leaf N (g)	−1.68	**0.03**	**12.8±2.7**	**56.4±16.4**

$P<0.10$; **$P<0.05$**;
BG = belowground; AG = aboveground.

also related to belowground biomass, ($P<0.05$) but 95% CI's for slope overlapped zero. Therefore, we did not present these.

Discussion

Whole plant biomass was equivalent among treatments, but above and belowground biomass did not scale equivalently as predicted under isometric growth models. Further, simulated eutrophication (N-addition) directed plant growth away from roots and rhizomes and towards leaves and inflorescences. Plants may increase belowground biomass in low nutrient conditions both to increase nutrient absorptive area and to increase root storage of carbohydrates to buffer against future scarcity [60]. In our experiment, both were observed as control plots invested more in belowground biomass, while root:rhizome ratios did not differ between treatments. Thus, plants in the N-addition treatment maximized energy capture and reproduction potential but reduced their contribution towards generating stable belowground organic matter that could contribute to wetland accretion [5,13] and peat formation [2]. Our results suggest *S. acutus* conforms to the balanced-growth model in northern California wetlands, as has been observed for other plant species elsewhere [15,17–19,60]. Plants do not always exhibit balanced growth [16,20], therefore further investigations are necessary for individual species under specific growing conditions. Where plants grow isometrically, vegetation indices of foliar N may add information to aboveground biomass measures by indicating where N-enrichment increases whole plant biomass production. In our study, simulated eutrophication was detectable by vegetation indices of leaf nitrogen concentration. Further vegetation indices were related to end of season root:shoot ratio and belowground biomass in *S. acutus*. Relationships between leaf N and vegetation indices generally are species, season and site specific [28,61]. Differences in species composition, plant structure and substrate will require re-estimation of these relationships elsewhere.

Vegetation indices detected increased foliar N concentration in *S. acutus* resulting from N addition, as noted in other wetlands [37]. Plants take up nutrients through their roots, and can use them for immediate growth or to store excess nutrients for future needs [32,33,60]. Plants may increase N content by both increasing N

Table 3. Relationship of plant biophysical parameters with potential biophysical, chemical, and remote sensing-based nitrogen predictors (N = 8); 95% CI is the bootstrap bias corrected, accelerated 95% confidence interval based on 10,000 bootstrap replicates.

Model	95% CI Intercept	95% CI Slope	*Adj-R^2*	*F*	*P*
BG biomass = AG biomass	−34.2–327.9	−12.0–8.8	0.05	0.3	0.63
Root:shoot ratio = Leaf N	**6.0–53.3**	**−32.8–−1.7**	**0.49**	**7.8**	**0.03**
BG biomass = $FDN_{1235,\ 549}$	*79.6–485.2*	*11.4–−1326.4*	*0.33*	*4.4*	*0.08*
Root:shoot = $FDN_{1235,\ 549}$	*3.6–29.9*	*0.1–76.3*	*0.27*	*3.6*	*0.1*
AG biomass = $FDN_{1235,\ 549}$	−42.5–98.5	−164.6–115.1	−0.14	0.2	0.68
Leaf N = $FDN_{1235,\ 549}$	**0.6–1.1**	**−3.1–−1.4**	**0.89**	**58.1**	**<0.01**
BG biomass = $FDS_{2184,\ 1780}$	**−39.1–103.5**	**−239.6–−8.6**	**0.44**	**6.5**	**0.04**
Root:shoot = $FDS_{2184,\ 1780}$	**−4.2–5.9**	**−18.0–−1.2**	**0.54**	**9.3**	**0.02**
AG biomass = $FDS_{2184,\ 1780}$	7.5–53.7	−20.6–45.5	−0.01	0.9	0.37
Leaf N = $FDS_{2184,\ 1780}$	**2.4–3.2**	**−3.1–−1.7**	**0.88**	**53.1**	**<0.01**

$P<0.10$; **$P<0.05$**;
FDS = first derivative simple ratio index;
FDN = first derivative normalized difference index;
BG = belowground; AG = aboveground.

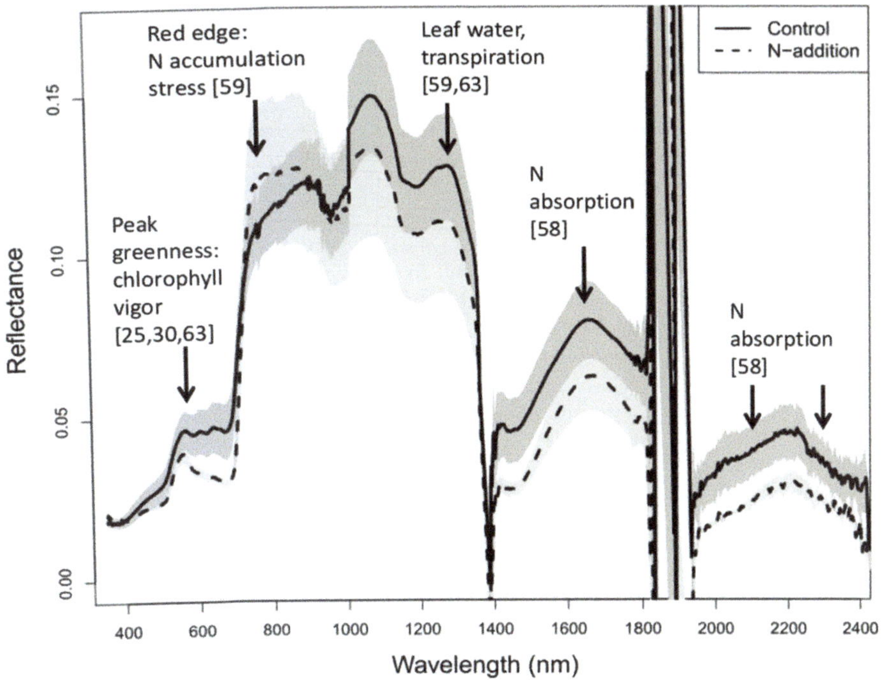

Figure 2. Reflectance spectra for treatments. Averaged reflectance spectrum for N-addition and control plants (N = 4 each). The shaded grey areas represent ± one standard error. Spectra associated with foliar N in the literature are indicated, with corresponding citations in brackets.

Table 4. Best new indices for each index type.

N	midpoint	midpoint	
Indices	**Band 1**	**Band 2**	R^2
FDN	1296	508	0.795
	1296	681	0.796
	2002	1155	0.822
	2184	1780	0.846
	1235	549	0.903
FDS	2194	1124	0.871
	2194	1235	0.877
	2194	1033	0.894
	1235	549	0.895
	2194	752	0.897
ND	1245	1205	0.685
	2274	2264	0.694
	1245	1215	0.713
	1235	1225	0.764
	2264	2244	0.793
SR	1245	1205	0.684
	2274	2264	0.692
	1245	1215	0.713
	1235	1225	0.764
	2264	2244	0.792

FDS = first derivative simple ratio index;
ND = normalized difference index;
SR = simple ratio index of;
FDN = first derivative normalized difference index.

per tissue unit or by increasing the number and length of shoots, leaves and inflorescences. Both mechanisms were observed here. Plant N concentration often is a reliable bioindicator of water quality because watersheds experience nutrient pulses, while plants integrate average conditions by absorbing nutrients from water and sediments over time [37,46,47,62].

In a data mining approach such as used in this study, there is the possibility of finding spurious correlations. Spurious correlations might be observed when multispecies responses are not accounted for, or when reflectance due to species differences in canopy architecture, differences in background effects, or scattering from leaf surface features obscure reflectance from leaf interior components, such as foliar nitrogen [61]. As a result final index selection should have a physical or biological basis [38,61]. First derivative indices were more closely related to nitrogen concentration than reflectance indices. Similar to results in Townsend et al. [30], the use of first-derivative spectra helped identify features related to foliar nutrient concentration. Bands with midpoints at 549 and 752 nm represent the slope at peak greenness and the red edge, respectively. While there are no absorption features for nitrogen in chemical bonds at shorter wavelengths, reflectance at 550 nm is related to chlorophyll concentration, and serves as a surrogate indicator of N levels and plant vigor [25,30,63]. Reflectance at the red edge is related to nitrogen accumulation and plant stress [59]. Nitrogen occurs primarily in proteins and chlorophylls of leaves, and because the two variables are moderately correlated within and across ecosystems, many researchers have associated the spectroscopic estimation of nitrogen to that of chlorophyll pigments [26]. Likewise reflectance at wavelength 1235 nm is near an absorption peak related to leaf water content [59,63]. Leaf water content is indicative of plant vigor because of leaf water relationships with plant turgor, thermoregulation and photosynthesis [26].

The first derivative spectrum is used to identify absorption features through relative rate of change of reflectance within a wavelength region, and bands related to N concentration may be adjacent to known absorption features instead of centered on these features [30]. Other top indices (Table 4) are based on bands in the shortwave infrared region near to nitrogen absorption regions of 1730, 2180, 2240 [58]. Bands past 2300 were not considered due to noise in the spectrum.

Vegetation indices of leaf N had significant relationships with root:shoot ratios and belowground biomass in *S. acutus* and might be used to identify relative differences in belowground biomass. Our results are firmly grounded in past research, which has shown that optical remote sensing can be used to estimate leaf N concentration [25–30,39,64–66], that nitrogen inputs often increase leaf N [67–70] and sometimes concomitantly reduce belowground biomass [17–19]. In isometric growth, N enrichment may stimulate above and belowground productivity equivalently [20,21,71,72]. Our study is the first to make use of remotely-sensed signals of whole-plant nitrogen response to link growth models and belowground biomass estimates.

Applications to Wetland Conservation and Monitoring

Our analytical approach involved single species experimental data. Our conclusions provide evidence that nutrient-enrichment may direct biomass allocation for *S. acutus* and can be detected by spectral reflectance signals. Applying our approach in a field setting will require overcoming additional challenges. For example, we held confounding variables equal among treatments. This allows identification of the magnitude of biomass allocation shifts from nutrient addition alone. In field settings, hydrology related parameters such as water depth, dissolved oxygen and nitrogen will co-vary [73]. Therefore field-measured vegetation indices of foliar N should represent the sum of several environmental conditions and will require further study.

An additional challenge is in addressing differential species responses to N-enrichment in diverse communities. Species invariate spectral reflectance modeling of canopy N may be unreliable [61]. To avoid this problem, researchers might map sites according to dominant cover types and explore species specific models within these. In wetlands, such an approach may be particularly useful. Wetlands have strong hydrology related environmental gradients and wetland species have specific adaptions to stress, causing species sorting. Examples include salt marshes dominated by monospecific stands of *Spartina alterniflora*, brackish marshes dominated by *Spartina patens*, or freshwater marshes dominated by several cover types that occur in distinct patches. Sites with high spatial heterogeneity in species composition may not be appropriate for our approach, but a field study is needed to substantiate this.

Estimates of belowground biomass under varying nutrient levels will assist with conservation management of freshwater marshes experiencing subsidence. In coastal marshes, decreases in belowground biomass contribute to soil instability, compaction, erosion, and marsh loss [18,19]. Coastal marshes experience frequent disturbances, such as strong storms and hurricanes, which might accelerate habitat loss where low belowground biomass results in soil instability [74]. Vegetation indices of leaf nitrogen might be used to identify areas where root growth is relatively low and therefore vulnerable to storm damage across broad regions.

Further, the Sacramento-San Joaquin River Delta region of California is experiencing large subsidence below sea level on marsh lands converted to agriculture [42]. Numerous marsh restoration projects have been proposed or are in progress across the Delta (California Delta Initiative; http://www.water.ca.gov/deltainit). Belowground biomass production within the Delta contributes to wetland accretion of 3–9 cm yr^{-1} within restored freshwater wetlands of mixed *S. acutus* and *Typha* sp. [5]. Vegetation indices should be useful for indicating spatial patterns of both low and high root growth in *S. acutus* and suggest appropriate management to meet target restoration goals. We are currently testing these indices in field settings dominated by *S. acutus* and *Typha* species in the Sacramento-San Joaquin Delta.

Estimates of belowground biomass in marshes are useful for following nutrient and carbon cycling. For example, wetlands frequently are used to treat reclaimed N-rich wastewater. Vegetation indices can help monitor canopy N concentration both spatially and temporally [28,29]. Further, belowground biomass contributes to peat formation and stable soil organic carbon complexes [2,5,75]. Estimates of belowground biomass therefore are useful for global carbon cycling models [76]. Vegetation indices such as these can be applied across broad areas and contribute to our understanding of vegetation dynamics.

Supporting Information

Dataset S1 First derivative reflectance data used in this study. Abbreviations are as follows: plot is an arbitrary number indexing samples; p_nitrogen is percent foliar N, treatment is treatment with values of N reflecting nitrogen addition and C the control; AB and BG biomass are aboveground and belowground biomass in grams; wXXX (e.g. w350–w2500) are the first derivative of reflectance at wavelength XXX, where wavelength is measured in nanometers.

Acknowledgments

We thank Susan Stitt and Carol Mladinich, Don Mulcahy, Dr. James Bartolome, Jason Joines, Caitlyn Chew, Adam McClure and Jenny Palomino for their assistance with equipment, supplies, lab space and field labor.

Any use of trade, firm, or product names is for descriptive purposes only and does not imply endorsement by the U.S. Government.

Author Contributions

Conceived and designed the experiments: JO KB MK. Performed the experiments: JO KB. Analyzed the data: JO KB. Contributed reagents/materials/analysis tools: JO KB MK. Wrote the paper: JO.

References

1. Rasse DP, Rumpel C, Dignac M-F (2005) Is soil carbon mostly root carbon? Mechanisms for a specific stabilisation. Plant Soil 269: 341–356.
2. Moore PD (1987) Ecological and hydrological aspects of peat formation. Geol Soc Spec Publ 32: 7–15. doi:10.1144/GSL.SP.1987.032.01.02.
3. Gorham E, Lehman C, Dyke A, Clymo D, Janssens J (2012) Long-term carbon sequestration in North American peatlands. Quat Sci Review 58: 77–82. doi:10.1016/j.quascirev.2012.09.018.
4. Nyman JA, Delaune RD, Roberts HH, Patrick WH (1993) Relationship between vegetation and soil formation in a rapidly submerging coastal marsh. Mar Ecol Prog Ser 96: 269–279.
5. Miller RL, Fram M, Fujii R, Wheeler G (2008) Subsidence reversal in a re-established wetland in the Sacramento-San Joaquin Delta, California, USA. San Francisco Estuary and Watershed Science 6. Available: http://escholarship.org/uc/item/5j76502x. Accessed 15 January 2013.
6. Titus JG (1988) Sea level rise and wetland loss: an overview. Greenhouse Effect, Sea Level Rise, and Coastal Wetlands. Washington, DC, USA: US Environmental Protection Agency, Vol. 186. 1–33. Available: http://papers.risingsea.net/federal_reports/sea-level-rise-and-wetlands-chap1-Titus.pdf.
7. DeLaune RD, Nyman JA, Patrick WH Jr (1994) Peat collapse, ponding and wetland loss in a rapidly submerging coastal marsh. J Coastal Res 10: 1021–1030. doi:10.2307/4298293.
8. Törnqvist TE, Wallace DJ, Storms JEA, Wallinga J, Dam RL van, et al. (2008) Mississippi Delta subsidence primarily caused by compaction of Holocene strata. Nat Geosci 1: 173–176. doi:10.1038/ngeo129.
9. Deverel SJ, Leighton DA (2010) Historic, recent, and future subside, Sacramento-San Joaquin Delta, California, USA. San Francisco Estuary and Watershed Science 8. Available: http://escholarship.org/uc/item/7xd4x0xw. Accessed 12 January 2013.
10. Nungesser MK (2011) Reading the landscape: temporal and spatial changes in a patterned peatland. Wetlands Ecol Manage 19: 475–493. doi:10.1007/s11273-011-9229-z.
11. Stralberg D, Brennan M, Callaway JC, Wood JK, Schile LM, et al. (2011) Evaluating tidal marsh sustainability in the face of sea-level rise: A hybrid modeling approach applied to San Francisco Bay. PLoS ONE 6: e27388. doi:10.1371/journal.pone.0027388.
12. Zedler JB, Kercher S (2005) Wetland resources: Status, trends, ecosystem services, and restorability. Annu Rev Environ Resour 30: 39–74. doi:10.1146/annurev.energy.30.050504.144248.
13. Nyman JA, Walters RJ, Delaune RD, Patrick JWH (2006) Marsh vertical accretion via vegetative growth. Estuar Coast Shelf Sci 69: 370–380.
14. Poorter H, Niklas KJ, Reich PB, Oleksyn J, Poot P, et al. (2012) Biomass allocation to leaves, stems and roots: meta-analyses of interspecific variation and environmental control. New Phytol 193: 30–50. doi:10.1111/j.1469-8137.2011.03952.x.
15. Shipley B, Meziane D (2002) The balanced-growth hypothesis and the allometry of leaf and root biomass allocation. Funct Ecol 16: 326–331. doi:10.2307/826585.
16. Mccarthy MC, Enquist BJ (2007) Consistency between an allometric approach and optimal partitioning theory in global patterns of plant biomass allocation. Funct Ecol 21: 713–720. doi:10.1111/j.1365-2435.2007.01276.x.
17. Darby FA, Turner RE (2008) Below- and aboveground biomass of Spartina alterniflora: Response to nutrient addition in a Louisiana salt marsh. Estuaries Coast 31: 326–334.
18. Turner RE, Howes BL, Teal JM, Milan CS, Swenson EM, et al. (2009) Salt marshes and eutrophication: An unsustainable outcome. Limnol Oceanogr 54: 1634–1642. doi:10.4319/lo.2009.54.5.1634.
19. Deegan LA, Johnson DS, Warren RS, Peterson BJ, Fleeger JW, et al. (2012) Coastal eutrophication as a driver of salt marsh loss. Nature 490: 388–392. doi:10.1038/nature11533.
20. Yang Y, Fang J, Ji C, Han W (2009) Above- and belowground biomass allocation in Tibetan grasslands. J Veg Sci 20: 177–184. doi:10.1111/j.1654-1103.2009.05566.x.
21. Zhao G, Liu J, Wang Y, Dou J, Dong X (2009) Effects of elevated CO2 concentration and nitrogen supply on biomass and active carbon of freshwater marsh after two growing seasons in Sanjiang Plain, Northeast China. J Environ Sci 21: 1393–1399. doi:10.1016/S1001-0742(08)62431-6.
22. Smith M-L, Ollinger SV, Martin ME, Aber JD, Hallett RA, et al. (2002) Direct estimation of aboveground forest productivity through hyperspectral remote sensing of canopy nitrogen. Ecol Appl 12: 1286–1302. doi:10.1890/1051-0761(2002)012[1286:DEOAFP]2.0.CO;2.
23. Lu DS (2006) The potential and challenge of remote sensing-based biomass estimation. Int J Remote Sens 27: 1297–1328.
24. Mokany K, Raison RJ, Prokushkin AS (2006) Critical analysis of root: shoot ratios in terrestrial biomes. Glob Chang Biol 12: 84–96. doi:10.1111/j.1365-2486.2005.001043.x.
25. Abdel-Rahman EM, Ahmed FB, van den Berg M (2010) Estimation of sugarcane leaf nitrogen concentration using in situ spectroscopy. Int J Appl Earth Obs Geoinf 125: S52–S57.
26. Kokaly RF, Asner GP, Ollinger SV, Martin ME, Wessman CA (2009) Characterizing canopy biochemistry from imaging spectroscopy and its application to ecosystem studies. Remote Sens Environ 113, Supplement 1: S78–S91. doi:10.1016/j.rse.2008.10.018.
27. Martin ME, Plourde LC, Ollinger SV, Smith M-L, McNeil BE (2008) A generalizable method for remote sensing of canopy nitrogen across a wide range of forest ecosystems. Remote Sens Environ 112: 3511–3519. doi:10.1016/j.rse.2008.04.008.
28. Stroppiana D, Fava F, Baschetti M, Brivio PA (2012) Estimation of nitrogen content in crops and pastures using hyperspectral vegetation indices. In: Thenkabail PS, Lyon JG, Huete A, editors. Hyperspectral remote sensing of vegetation. Boca Raton: CRC Press. 245–262. Available: http://www.crcnetbase.com/doi/pdf/10.1201/b11222-16.
29. Tian YC, Yao X, Yang J, Cao WX, Hannaway DB, et al. (2011) Assessing newly developed and published vegetation indices for estimating rice leaf nitrogen concentration with ground- and space-based hyperspectral reflectance. Field Crops Res 120: 299–310. doi:10.1016/j.fcr.2010.11.002.
30. Townsend PA, Foster JR, Chastain RA, Currie WS (2003) Application of imaging spectroscopy to mapping canopy nitrogen in the forests of the central Appalachian Mountains using Hyperion and AVIRIS. IEEE Trans Geosci Remote Sens 41: 1347–1354. doi:10.1109/tgrs.2003.813205.
31. Wessman CA, Aber JD, Peterson DL, Melillo JM (1988) Remote sensing of canopy chemistry and nitrogen cycling in temperate forest ecosystems. Nature 335: 154–156. doi:10.1038/335154a0.
32. Boyer KE, Fong P, Vance RR, Ambrose RF (2001) Salicornia virginica in a southern California salt marsh: Seasonal patterns and a nutrient-enrichment experiment. Wetlands 21: 315–326. doi:10.1672/0277-5212(2001)021[0315:SVIASC]2.0.CO;2.
33. Cohen RA, Fong P (2005) Experimental evidence supports the use of δ 15 N content of the opportunistic green macroalga Enteromorpha intestinalis (chlorophyta) to determine nitrogen sources to estuaries. J Phycology 41: 287–293. doi:10.1111/j.1529-8817.2005.04022.x.
34. Kokaly RF, Clark RN (1999) Spectroscopic determination of leaf biochemistry using band depth analysis of absorption features and stepwise multiple linear regression. Remote Sens Environ 67: 267–287. doi:10.1016/S0034-4257(98)00084-4.
35. Gao BC, Goetz AFH (1995) Retrieval of equivalent water thickness and information related to biochemical components of vegetation canopies from AVIRIS data. Remote Sens Environ 52: 155–162.
36. Serrano L, Peñuelas J, Ustin SL (2002) Remote sensing of nitrogen and lignin in Mediterranean vegetation from AVIRIS data: Decomposing biochemical from structural signals. Remote Sens Environ 81: 355–364. doi:10.1016/S0034-4257(02)00011-1.
37. Siciliano D, Wasson K, Potts DC, Olsen RC (2008) Evaluating hyperspectral imaging of wetland vegetation as a tool for detecting estuarine nutrient enrichment. Remote Sens Environ 112: 4020–4033. doi:10.1016/j.rse.2008.05.019.
38. Thenkabail PS, Mariotto I, Gumma MK, Middleton EM, Landis DR, et al. (2013) Selection of hyperspectral narrowbands (HNBs) and composition of hyperspectral twoband vegetation indices (HVIs) for biophysical characterization and discrimination of crop types using field reflectance and Hyperion/EO-1 data. Int J Appl Earth Obs Geoinf 6: 427–439. doi:10.1109/JSTARS.2013.2252601.
39. Abdel-Rahman EM, Ahmed FB, Ismail R (2013) Random forest regression and spectral band selection for estimating sugarcane leaf nitrogen concentration using EO-1 Hyperion hyperspectral data. Int J Remote Sens 34: 712–728. doi:10.1080/01431161.2012.713142.
40. Sims NC, Culvenor D, Newnham G, Coops NC, Hopmans P (2013) Towards the Operational Use of Satellite Hyperspectral Image Data for Mapping Nutrient Status and Fertilizer Requirements in Australian Plantation Forests. Selected Topics in Applied Earth Observations and Remote Sensing, IEEE Journal of 6: 320–328. doi:10.1109/jstars.2013.2251610.

41. Prokopovich NP (1985) Subsidence of peat in California and Florida. Bull Assoc Eng Geol 22: 395–420.
42. Drexler JZ, Fontaine CS de, Deverel SJ (2009) The legacy of wetland drainage on the remaining peat in the Sacramento – San Joaquin Delta, California, USA. Wetlands 29: 372–386. doi:10.1672/08-97.1.
43. USDA NRCS (2010) The PLANTS Database. National Plant Data Team. Available: http://plants.usda.gov. Accessed 30 September 2010.
44. Wankle SD, Kendall C, Francis CA, Paytan A (2006) Nitrogen sources and cycling in the San Francisco Bay estuary: a nitrate dual isotopic composition approach. Limnol Oceanogr 51: 1654–1664.
45. Environmental Protection Agency (2002) Mid-Atlantic integrated assessment (MAIA) estuaries 1997–1998. Summary Report. Narragansett, Rhode Island, USA: U.S. Environmental Protection Agency, Office of Research and Development, National Health and Environmental Effects Research Laboratory, Atlantic Ecology Division.
46. Cohen RA, Fong P (2006) Using opportunistic green macroalgae as indicators of nitrogen supply and sources to estuaries. Ecol Appl 16: 1405–1420. doi:10.1890/1051-0761(2006)016[1405:UOGMAI]2.0.CO;2.
47. Sheppard JK, Carter AB, McKenzie LJ, Pitcher CR, Coles RG (2008) Spatial patterns of sub-tidal seagrasses and their tissue nutrients in the Torres Strait, northern Australia: Implications for management. Cont Shelf Res 28: 2282–2291. doi:10.1016/j.csr.2008.03.033.
48. Carnegie Museum of Natural History (2005) Crayfish Links at Carnegie Museum of Natural History. Available: http://iz.carnegiemnh.org/index.htm. Accessed 20 November 2012.
49. Ernst MD (2004) Permutation methods: a basis for exact inference. Statist Sci 19: 676–685. doi:10.1214/088342304000000396.
50. Miphokasap P, Honda K, Vaiphasa C, Souris M, Nagai M (2012) Estimating canopy nitrogen concentration in sugarcane using field imaging spectroscopy. Remote Sensing 4: 1651–1670. doi:10.3390/rs4061651.
51. Ollinger SV (2011) Sources of variability in canopy reflectance and the convergent properties of plants. New Phytol 189: 375–394. doi:10.1111/j.1469-8137.2010.03536.x.
52. ASD Inc (2008) ViewSpec ProTM User Manual. Boulder, CO: ASD Inc., 24 p.
53. De Jong SM (1998) Imaging spectrometry for monitoring tree damage caused by volcanic activity in the Long Valley caldera, California. ITC Journal 1: 1–10.
54. Dawson TP, Curran PJ (1998) Technical note A new technique for interpolating the reflectance red edge position. Int J Remote Sens 19: 2133–2139. doi:10.1080/014311698214910.
55. Demetriades-Shah TH, Steven MD, Clark JA (1990) High resolution derivative spectra in remote sensing. Remote Sens Environ 33: 55–64. doi:10.1016/0034-4257(90)90055-Q.
56. Kochubey SM, Kazantsev TA (2012) Derivative vegetation indices as a new approach in remote sensing of vegetation. Front Earth Sci 6: 188–195. doi:10.1007/s11707-012-0325-z.
57. Thenkabail PS, Enclona EA, Ashton MS, Van Der Meer B (2004) Accuracy assessments of hyperspectral waveband performance for vegetation analysis applications. Remote Sens Environ 91: 354.
58. Curran PJ (1989) Remote sensing of foliar chemistry. Remote Sens Environ 30: 271–278. doi:10.1016/0034-4257(89)90069-2.
59. Thenkabail PS, Lyon JG, Huete A (2012) Advances in Hyperspectral Remote Sensing of Vegetation and Agricultural Croplands. In: Thenkabail PS, Lyon JG, Huete A, editors. Hyperspectral Remote Sensing of Vegetation. Boca Raton, FL: Taylor and Francis Group. 28–29.
60. Kobe RK, Iyer M, Walters MB (2010) Optimal partitioning theory revisited: Nonstructural carbohydrates dominate root mass responses to nitrogen. Ecology 91: 166–179. doi:10.1890/09-0027.1.
61. Knyazikhin Y, Schull MA, Stenberg P, Mõttus M, Rautiainen M, et al. (2013) Hyperspectral remote sensing of foliar nitrogen content. Proc Natl Acad Sci U S A 110: E185–E192. doi:10.1073/pnas.1210196109.
62. Suwandana E, Kawamura K, Sakuno Y, Evri M, Lesmana AH (2012) Hyperspectral Reflectance Response of Seagrass (Enhalus acoroides) and Brown Algae (Sargassum sp.) to Nutrient Enrichment at Laboratory Scale. J Coastal Res 28: 956–963. doi:10.2112/jcoastres-d-11-00222.1.
63. Thenkabail PS, Smith RB, De-Pauw E (2002) Evaluation of narrowband and broadband vegetation indices for determining optimal hyperspectral wavebands for agricultural crop characteristics. Photogrammetric Engineering and Remote Sensing 68: 607–621.
64. Mokhele TA, Ahmed FB (2010) Estimation of leaf nitrogen and silicon using hyperspectral remote sensing. J Appl Remote Sens 4. Available: :// WOS:000286065800001.
65. Ryu C, Suguri M, Umeda M (2009) Model for predicting the nitrogen content of rice at panicle initiation stage using data from airborne hyperspectral remote sensing. Biosyst Eng 104: 465–475. doi:10.1016/j.biosystemseng.2009.09.002.
66. Stroppiana D, Boschetti M, Brivio PA, Bacchi S (2009) Plant nitrogen concentration in paddy rice from field canopy hyperspectral radiometry. Field Crops Res 111: 119–129.
67. Sartoris JJ, Thullen JS, Barber LB, Salas DE (1999) Investigation of nitrogen transformations in a southern California constructed wastewater treatment wetland. Ecol Eng 14: 49–65. doi:10.1016/S0925-8574(99)00019-1.
68. Bedford BL, Walbridge MR, Aldous A (1999) Patterns in nutrient availability and plant diversity of temperate North American wetlands. Ecology 80: 2151–2169. doi:10.1890/0012-9658(1999)080[2151:PINAAP]2.0.CO;2.
69. Kao JT, Titus JE, Zhu W-X (2003) Differential nitrogen and phosphorus retention by five wetland plant species. Wetlands 23: 979–987. doi:10.1672/0277-5212(2003)023[0979:DNAPRB]2.0.CO;2.
70. Larkin DJ, Lishawa SC, Tuchman NC (2012) Appropriation of nitrogen by the invasive cattail Typha × glauca. Aquatic Botany 100: 62–66. doi:10.1016/j.aquabot.2012.03.001.
71. Merino JH, Huval D, Nyman AJ (2010) Implication of nutrient and salinity interaction on the productivity of Spartina patens. Wetlands Ecol Manage 18: 111–117. doi:10.1007/s11273-008-9124-4.
72. Morris JT, Shaffer GP, Nyman JA (2013) Brinson review: Perspectives on the influence of nutrients on the sustainability of coastal wetlands. Wetlands In press: 1–14. doi:10.1007/s13157-013-0480-3.
73. Mitsch WJ, Gosselink JG (2007) Wetlands. Fourth. Hoboken, NJ, USA: John Wiley & Sons. Available: http://books.google.com/books?id = 1cSKeTCi894C.
74. Cahoon DR (2006) A review of major storm impacts on coastal wetland elevations. Estuaries Coast 29: 889–898. doi:10.1007/BF02798648.
75. Drexler JZ, Fontaine CS de, Brown TA (2009) Peat accretion histories during the past 6000 years in marshes of the Sacramento–San Joaquin Delta, CA, USA. Estuaries Coast 32: 871–892. doi:10.1007/s12237-009-9202-8.
76. Belyea LR, Malmer N (2004) Carbon sequestration in peatland: patterns and mechanisms of response to climate change. Glob Chang Biol 10: 1043–1052. doi:10.1111/j.1529-8817.2003.00783.x.
77. Ramoelo A, Skidmore AK, Cho MA, Schlerf M, Mathieu R, et al. (2012) Regional estimation of savanna grass nitrogen using the red-edge band of the spaceborne RapidEye sensor. Int J Appl Earth Obs Geoinf 19: 151–162. doi:10.1016/j.jag.2012.05.009.
78. Kim MS, Daughtry CS, Chapelle EW, McMurtrey JE (1994) The use of high spectral resolution bands for estimating absorbed photosynthetically active radiation (APAR). 229–306.
79. Xue L, Cao W, Luo W, Dai T, Zhu Y (2004) Monitoring leaf nitrogen status in rice with canopy spectral reflectance. Agron J 96: 135–142.
80. Guyot G, Baret F (1988) Utilisation de la haute résolution spectral pour suivre l'état des couverts végétaux Vol. ESA SP-287. 279–286.

6

Effects of Different Tillage and Straw Return on Soil Organic Carbon in a Rice-Wheat Rotation System

Liqun Zhu[1]*, Naijuan Hu[1], Minfang Yang[2], Xinhua Zhan[2], Zhengwen Zhang[1]

1 College of Agriculture, Nanjing Agricultural University, Nanjing, China, **2** College of Resources and Environmental Science, Nanjing Agricultural University, Nanjing, China

Abstract

Soil management practices, such as tillage method or straw return, could alter soil organic carbon (C) contents. However, the effects of tillage method or straw return on soil organic C (SOC) have showed inconsistent results in different soil/climate/cropping systems. The Yangtze River Delta of China is the main production region of rice and wheat, and rice-wheat rotation is the most important cropping system in this region. However, few studies in this region have been conducted to assess the effects of different tillage methods combined with straw return on soil labile C fractions in the rice-wheat rotation system. In this study, a field experiment was used to evaluate the effects of different tillage methods, straw return and their interaction on soil total organic C (TOC) and labile organic C fractions at three soil depths (0–7, 7–14 and 14–21 cm) for a rice-wheat rotation in Yangzhong of the Yangtze River Delta of China. Soil TOC, easily oxidizable C (EOC), dissolved organic C (DOC) and microbial biomass C (MBC) contents were measured in this study. Soil TOC and labile organic C fractions contents were significantly affected by straw returns, and were higher under straw return treatments than non-straw return at three depths. At 0–7 cm depth, soil MBC was significantly higher under plowing tillage than rotary tillage, but EOC was just opposite. Rotary tillage had significantly higher soil TOC than plowing tillage at 7–14 cm depth. However, at 14–21 cm depth, TOC, DOC and MBC were significantly higher under plowing tillage than rotary tillage except for EOC. Consequently, under short-term condition, rice and wheat straw both return in rice-wheat rotation system could increase SOC content and improve soil quality in the Yangtze River Delta.

Editor: Shuijin Hu, North Carolina State University, United States of America

Funding: This work was supported by Special Fund for Agro-scientific Research in the Public Interest (201103001). The funders had no role in study design, data collection and analysis, decision to publish, or preparation of the manuscript.

Competing Interests: The authors have declared that no competing interests exist.

* E-mail: zhulq@njau.edu.cn

Introduction

Soil organic carbon (C) has profound effects on soil physical, chemical and biological properties [1]. Maintenance of soil organic C (SOC) in cropland is important, not only for improvement of agricultural productivity but also for reduction in C emission [2]. However, short- and medium-term changes of SOC are difficult to detect because of its high temporal and spatial variability [3]. On the contrary, soil labile organic C fractions (i.e., microbial biomass C (MBC), dissolved organic C (DOC), and easily oxidizable C (EOC)) that turn over quickly can respond to soil disturbance more rapidly than total organic C (TOC) [1,3,4]. Therefore, these fractions have been suggested as early sensitive indicators of the effects of land use change on soil quality (e.g. [3,5,6]).

Agricultural practices such as tillage methods are conventionally used for loosening soils to grow crops. But long-term soil disturbance by tillage is believed to be one of the major factors reducing SOC in agriculture [7]. Frequent tillage may destroy soil organic matter (SOM) [8] and speed up the movement of SOM to deep soil layers [9]. As a consequence, agricultural practices that reduce soil degradation are essential to improve soil quality and agricultural sustainability. Crop residue plays an important role in SOC sequestration, increasing crop yield, improving soil organic matter, and reducing the greenhouse gas (e.g. [10–13]). As an important agricultural practice, straw return is often implemented with tillage in the production process. Although numerous studies have indicated that tillage methods combined with straw return had a significant effect on labile SOC fractions, the results varied under different soil/climate conditions. For example, both no-tillage and shallow tillage with residue cover had significantly higher SOC than conventional tillage without residue cover in Loess Plateau of China [14], while Wang et al. [15] reported that the difference between the treatments of plowing with straw return and no-tillage with straw return on TOC in central China was not significant. Rajan et al. [2] showed that in Chitwan Valley of Nepal, no-tillage with crop residue application at upper soil depth had distinctly higher SOC sequestration than conventional tillage with crop residue. The effects of tillage on soil labile organic C vary with regional climate [16], soil condition (e.g. [17–20]), residue management practice, and crop rotation (e.g. [21,22]). Therefore, the investigation on soil labile organic C for specific soil, climate, and cropping system is necessary to improve the soil quality.

The Yangtze River Delta of China is the main production region of rice and wheat, and rice-wheat rotation is the most important cropping system in this region [23]. The total sown area of rice and wheat in the Yangtze River Delta accounted for about 20.1% of that in China in 2011, and the total yield was 22.1% of the national yield for these two crops [24]. Many field experiments in this region (e.g. [25–27]) about the effects of tillage methods

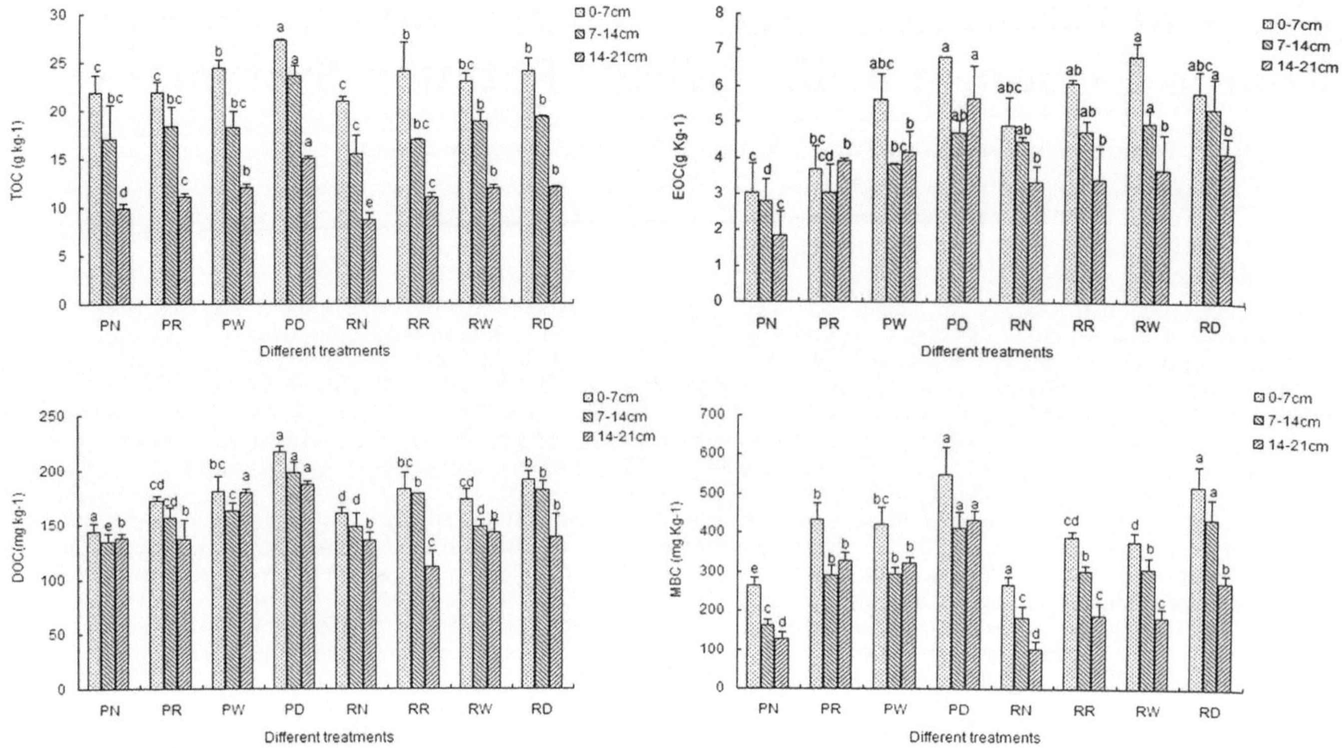

Figure 1. Effects of eight treatments on soil TOC, EOC, DOC and MBC contents at three depths.

combined with straw return on cropland ecosystem in rice-wheat rotation system have been studied during these years. However, most of them are focused on soil physical-chemical properties, soil nutrient and crop yield. To our knowledge, there is little information about the effects of different tillage and straw return on soil labile C fractions in rice-wheat rotation system. Thus, the objectives of this study were (1) to quantify the effects of tillage methods and straw return on soil TOC, MBC, DOC, EOC contents in the rice-wheat rotation system in the Yangtze River Delta, and (2) to explore an optimal management practice combination of tillage and straw return for improving the soil quality and increasing the local crop production.

Table 1. Linear correlations among soil TOC and labile organic C fractions at the 0–21 cm depth.

Index	TOC	EOC	DOC	MBC
TOC	1			
EOC	0.638**	1		
DOC	0.758**	0.684**	1	
MBC	0.741**	0.639**	0.908**	1

TOC: total organic carbon; MBC: microbial biomass carbon; DOC: dissolved organic carbon; EOC: easily oxidizable carbon.
* $P<0.05$.
** $P<0.01$.

Materials and Methods

Site Description

The experiment was conducted at Changwang Country, Youfang Town, Yangzhong City, Jiangsu Province, China (119°42′–119°58′E, 32°–2°19′N, 4–4.5 m above mean sea level) from November, 2009 to June, 2011. Access to the study site was obtained in the form of a rent contract, in which we had to confirm that our study did not involve endangered or protected species.

The experimental site had a subtropical monsoon climate with an average annual precipitation of 1000 mm, an average annual temperature of 15.1°C, and a mean annual sunshine hour of 2135 h. The soil of the experimental site was a loam and classified as an anthrosols. Rice-wheat double cropping system was the most important cropping system in the region. The main properties of soil (0–20 cm depth) sampled in November 2009 were as follows: soil organic matter 29.81 g kg^{-1}; alkali-hydrolyzale nitrogen 194.02 mg kg^{-1}; available phosphorus 13.60 mg kg^{-1}; available potassium 51.45 mg kg^{-1}; and pH 7.34.

The variety of wheat used in this study was Yangmai16 *(Triticum aestivum* L.*)* and rice was Nangeng47 (*Oryza sativa* L.).

Experimental Design and Field Managements

The experiment had a split-plot design with two tillage methods in the main plots and four straw return modes in subplots with three replications (6 m×5 m). Tillage methods included plowing tillage (P) and rotary tillage (R). Straw return modes were as follows: no straw return (N), only rice straw return (R), only wheat straw return (W), and rice and wheat straw both return (D). There were eight treatments in this study: (1) plowing tillage with no straw return (PN: rice with plowing tillage-wheat with plowing tillage); (2) plowing tillage with only rice straw return (PR: rice with

Table 2. Effects of different tillage factor and straw factor on soil TOC, EOC, DOC and MBC at 0–7 cm, 7–14 cm and 14–21 cm depth.

Soil dept(cm)	Factors	Treatments	TOC(g kg^{-1})	labile organic carbon fractions contents		
				EOC (g kg^{-1})	DOC(mg kg^{-1})	MBC(mg kg^{-1})
0–7	Tillage factor	P	23.87±2.67a	4.78±1.23a	178.36±33.74a	417.16±17.20a
		R	23.00±2.38a	5.92±1.48a	177.05±30.96a	389.22±36.56b
	Straw factor	N	21.40±1.87c	3.97±0.92b	152.25±10.76c	266.73±11.68c
		R	23.01±2.97bc	4.88±1.40ab	177.61±33.57b	411.67±27.32b
		W	23.65±1.52b	6.23±1.56a	176.92±30.87b	398.57±27.34b
		D	25.68±2.63a	6.32±1.68a	204.04±22.67a	535.79±20.31a
7–14	Tillage factor	P	19.30±1.23a	3.58±1.76b	163.05±37.30a	311.42±43.89a
		R	17.64±2.48b	4.89±1.52a	164.11±27.23a	306.50±36.09a
	Straw factor	N	16.28±3.20c	3.62±1.76c	141.49±18.49d	172.63±16.58c
		R	17.63±2.86bc	3.90±2.70bc	167.39±21.63b	315.19±20.00b
		W	18.50±1.75b	4.39±0.54ab	155.47±22.41c	313.93±14.31b
		D	21.47±2.02a	5.04±0.76a	189.98±17.71a	434.09±11.65a
14–21	Tillage factor	P	12.00±2.56a	3.89±1.42a	160.06±30.21a	280.68±58.67a
		R	10.90±2.61b	3.64±1.14a	132.00±25.31b	187.08±44.86b
	Straw factor	N	9.28±1.69d	2.58±1.18c	136.34±11.66b	114.99±18.05d
		R	10.99±1.96c	3.66±1.26b	123.67±29.32c	239.95±39.07b
		W	11.97±3.06b	3.91±1.39b	161.19±25.91a	238.92±35.06c
		D	13.56±1.75a	4.90±0.68a	162.93±26.49a	341.67±31.14a

Different letters in a line under a specific influence factor denote significant difference at the 5% level. Different capitals in a column at different soil depths and treatments present significant different at the 0.05 level. TOC: total organic carbon; MBC: microbial biomass carbon; DOC: dissolved organic carbon; EOC: easily oxidizable carbon.

plowing tillage - wheat with plowing tillage+rice straw return); (3) plowing tillage with only wheat straw return (PW: rice with plowing tillage+wheat straw return - wheat with plowing tillage); (4) plowing tillage with rice and wheat straw both return(PD: rice with plowing tillage+wheat straw return -wheat with plowing tillage+rice straw return); (5) rotary tillage with no straw return (RN: rice with rotary tillage - wheat with rotary tillage); (6) rotary tillage with only rice straw return (RR: rice with rotary tillage - wheat with rotary tillage+rice straw return); (7) rotary tillage with only wheat straw return (RW: rice with rotary tillage+wheat straw return - wheat with rotary tillage); (8) rotary tillage with rice and wheat straw both return (RD: rice with rotary tillage+wheat straw return - wheat with rotary tillage+rice straw return).

The experimental site was cultivated with a rice-wheat rotation prior to November 2009, where wheat was planted with plowing tillage from November to the following June, and rice was transplanted by plowing tillage from June to November. In this study, after wheat or rice was harvested, they were cultivated at a depth of 10–15 cm by rotary cultivation in rotary tillage plots while for the plowing tillage plots, cultivation was at a depth of 20–25 cm with a moldboard plough. Before the rice and wheat were sown, the plowing tillage plots were disked and moldboard plowed for weed control and bedding. This was followed by an application of fertilizer. For straw returned plots, the wheat and rice straw were cut into 8–10 cm after air-dried, and placed back on the surface of the soil in June or November of each year, with returned amount of 6000 kg·hm^{-2} for both wheat and rice straw.

In this study, wheat was sown on November 3, 2009 and November 24, 2010, respectively. The seed quantity was 150 kg hm^{-1} by machine. The base fertilizer applied before sowing was 135 kg·hm^{-2} pure N, 67.5 kg·hm^{-2} P_2O_5, and 67.5 kg·hm^{-2} K_2O, and topdressing was at the elongation stage with 135 kg hm^{-1} pure N. For all treatments, N was applied in the form of $CO(NH_2)_2$, and the fertilizer in wheat seasons was applied at the same rate. The wheat was harvested on June 3, 2010 and June 8, 2011, respectively. Rice was transplanted at about 3–4 seedlings per hole, 255,000 holes per hectare on June 15, 2010 and June 15, 2011, respectively. The rice was fertilized just before transplanting with a base fertilizer (120 kg hm^{-1} pure N; 60 kg hm^{-1} P_2O_5; 60 kg hm^{-1} K_2O), and at tillering stage and earing stage with a topdressing (180 kg hm^{-1} N (3:1 ratio)). The fertilizer applied in the two rice seasons were the same. The rice was harvested on November 12, 2010 and November 29, 2011, respectively. The pesticide management of both rice and wheat seasons was in accordance with the conventional, and all other management procedures were identical for the eight treatments.

Soil Sampling and Analytical Methods

Soil samples were collected by a geotome (5 cm diameter) on October 29, 2011 (just before the rice was harvested). Five random locations were chosen in each of the 24 observational plots and samples were taken from each location at three soil depths (0–7, 7–14, and 14–21 cm) separately. Soil samples from each depth were about 200 g, fully blended. The collected moist samples were ground and sieved through a 10 mesh screen. Sieved soil samples were divided into two sub-samples. One was air-dried and sieved again through 100 mesh screen for determining soil TOC and EOC. Another was immediately stored in 4°C refrigerators for determining DOC and MBC. During sieving, crop residues, root material and stones were removed.

Table 3. Affecting force analysis of different tillage and straw return and their interaction on soil TOC, EOC, DOC and MBC at 0–7 cm, 7–14 cm and 14–21 cm depth.

Soil depth(cm)	Difference source	Affecting force(%)			
		TOC	EOC	DOC	MBC
0–7	Block	10.35	8.53	6.55	0.26
	Tillage	0.34	28.06**	0.64	2.06**
	Straw return	19.69*	24.79**	38.58**	95.95**
	Straw return × tillage	10.71	4.34	4.86	0.93
	Error	58.92	34.28	49.37	0.79
7–14	Block	19.97**	5.36	11.53	0.29
	Tillage	0.02	13.03*	0.85	0.07
	Straw return	21.41**	13.16	39.21**	97.83**
	Straw return × tillage	26.30**	3.89	11.12*	0.96
	Error	32.32	64.56	37.28	0.86
14–21	Block	10.22	1.61	16.66**	0.27
	Tillage	1.05	3.21	11.77**	23.81**
	Straw return	32.35**	25.06**	35.65**	70.17**
	Straw return × tillage	14.72*	13.95*	9.60*	4.93**
	Error	41.66	56.17	26.33	0.82

The affecting force of tillage = tillage variables (square)/total variables (sum of total squares) ×100%; the affecting force of straw = straw variables (square)/total variable (sum of total squares) ×100%; the affecting force of the interaction = interaction variable (square)/total variables ((total squares of sum) ×100%. TOC: total organic carbon; MBC: microbial biomass carbon; DOC: dissolved organic carbon; EOC: easily oxidizable carbon.
* $P<0.05$.
**$P<0.01$.

Total organic C (TOC) concentration was determined by oxidation with potassium dichromate and titration with ferrous ammonium sulphate [28].

Dissolved organic C (DOC) was extracted from 10 g of moist soil with 1:2.5 ratio of soil to water at 25.8°C [29]. After shaking for 1 h and centrifuging for 10 min at 4500 r min^{-1}, the supernatant was filtered with a 0.45 mm membrane filter. The filtrate was measured by oxidation with potassium dichromate and titration with ferrous ammonium sulphate.

Microbial biomass C (MBC) was analyzed by the fumigation extraction method [30]. Each sample was weighed into two equivalent portions, one was fumigated for 24 h with ethanol-free chloroform and the other was the unfumigated control. Both fumigated and unfumigated soils were shaken for 1 h with 0.5 M K_2SO_4 (2:5 soil: extraction ratio), centrifuged and filtered.

Easily oxidizable C (EOC) was measured as described by Blair et al. [3]. Finely ground air-dried soil samples were reacted with 333 mmol L^{-1} $KMnO_4$ by shaking at 60 r min^{-1} for 1 h. The suspension was then centrifuged at 2000 r min^{-1} for 5 min. The supernatant was diluted and measured spectrophotometrically at 565 nm. All soil samples were analyzed in triplicate.

Data Analysis

The SPSS 16.0 analytical software package was used for all statistical analyses. A 2-factor analysis of variance (ANOVA) was employed for difference test among eight treatments at $P<0.05$, with separation of means by least significant difference (LSD). Correlation analysis were performed to determine correlations among soil labile organic C fractions in the 0–21 cm soil depth, and the significant probability levels of the results were given at the $P<0.05$ (*) and $P<0.01$ (**), respectively. Moreover, the affecting force analysis of tillage factor, straw factor and their interaction influence on labile organic C fractions was calculated based on the method of Leng [31]: the affecting force of tillage = tillage variables (square)/total variables (sum of total squares) × 100%; the affecting force of straw return = straw return variables (square)/total variable (sum of total squares) × 100%; the affecting force of interaction = interaction variable (square)/total variables ((total squares of sum) × 100%.

Results

Soil TOC, DOC, MBC and EOC Contents in Different Treatments

As shown in Fig. 1, the different treatments significantly affected the contents of soil TOC and labile organic C fractions, where PD generally had the highest contents of TOC, DOC, MBC and EOC at the three soil depths. Crop straw return treatments (PR, PW, PD, RR, RW, RD) had consistently higher amount of TOC and labile organic C fractions at the three soil depths than without crop straw return treatments (PN, RN). Moreover, PN had significantly lower TOC, DOC, MBC and EOC at 0–7 cm and 7–14 cm, and RN had the lowest TOC and MBC at 14–21 cm compared to other treatments (Fig. 1). Soil TOC and labile organic C fractions generally decreased with an increase in soil depth under all treatments. As expected, soil TOC and labile organic C fractions were significantly and positively correlated with each other (Table 1).

Effects of Different Tillage Methods on Soil TOC and Labile Organic C Fractions

Tillage had a significant effect on MBC at 0–7 cm soil depth, but seldom on soil TOC, DOC and EOC. Soil TOC, DOC and MBC contents were all higher under plowing tillage (P) than rotary tillage (R), while EOC was opposite at 0–7 cm soil depth (Table 2). At 7–14 cm, soil EOC under rotary tillage (R) was significantly higher than plowing tillage (P), but TOC had the contrary results, and there were no significant differences on DOC and MBC (Table 2). At 14–21 cm, soil TOC, DOC and MBC were significantly higher under plowing tillage (P) than rotary tillage (R), except EOC (Table 2).

Effects of Different Straw Return on Soil TOC and Labile Organic C Fractions

Straw return had significant effects on soil TOC and labile organic C at the three depths as shown in Table 2. In general, soil TOC and three labile organic C ranged in the following order: rice and wheat straw both return>only wheat or rice straw return>no straw return at three depths (Table 2). At 7–14 cm depth, only rice straw return in the wheat season had significantly higher DOC than only wheat straw return in the rice season (Table 2). However, at 14–21 cm depth, except for MBC, soil TOC, EOC, DOC under only rice straw return in the wheat season were lower than only wheat straw return in the rice season (Table 2). Moreover, there were significant differences in TOC and MBC among the four straw return at 14–21 cm depth (Table 2).

Affecting Force Analysis of Different Tillage, Straw Return and their Interaction on Soil TOC and Labile Organic C Fractions

Affecting force of different tillage, straw return and their interaction on soil TOC and labile organic C were different with increasing soil depth (Table 3). The affecting force of tillage increased with the increase of soil depth (Table 3). Tillage had significant affecting force on EOC and MBC at 0–7 cm depth, but seldom on TOC and DOC (Table 3). At 7–14 cm depth, the affecting force of tillage on EOC was lower than at 0–7 cm, and there was no significant affecting force on soil TOC and other labile organic C fractions (Table 3). At 14–21 cm depth, tillage had significant affecting force on DOC and MBC. However, there was no significant affecting force on TOC and EOC (Table 3).

Straw return had significant affecting force on soil TOC, DOC and MBC at the three depths, but there was no significant affecting force on EOC at 7–14 cm (Table 3). Among the four indictors, the affecting force of straw return on MBC was the greatest at the three depths, which reached 95.95%, 97.83% and 70.17%, respectively (Table 3).

The affecting force of the interaction generally increased with an increase in soil depth (Table 3). At 0–7 cm soil depth, the interaction had no significant affecting force on soil TOC and labile organic C (Table 3). Soil TOC and DOC were mainly dominated by the interaction at 7–14 cm depth, but there was no significant affecting force on MBC and EOC. At 14–21 cm depth, the interaction had significant affecting force on soil TOC and all labile organic C (Table 3).

Discussion

Suitable soil tillage practice can increase the SOC content, and improve SOC density of the plough layer [32]. The effect size of tillage methods on SOC dynamics depends on the tillage intensity [33]. Compared to conventional tillage (CT), no-tillage and reduced tillage could significantly improve the SOC content in cropland. Frequent tillage under CT easily exacerbate C-rich macroaggregates in soils broken down due to the increase of tillage intensity, then forming a large number of small aggregates with relatively low organic carbon content and free organic matter particles. Free organic matter particles have poor stability and are easy to degradation, thereby causing the loss of SOC [33,34]. In our study, at 0–7 cm soil depth, soil EOC under plowing tillage was lower than rotary tillage (Table 2). The reason could be attributed to the tillage method. Tillage increases the effect of drying–rewetting and freezing-thawing on soil, which increases macroaggregate susceptibility to disruption [21,35,36], and accelerates the labile organic C mineralization and SOM degradation, thus increasing the loss of EOC [14,37]. At 7–14 cm, rotary tillage had higher soil EOC and DOC than plowing tillage, but lower at 14–21 cm soil depth, indicating that tillage affected the vertical distribution of EOC and DOC (Table 2). The difference in soil condition after plowing tillage or rotary tillage affects the rate of straw decomposition, thereby resulting in a difference in the soil nutrient accumulation [38]. Similarly, Liu et al. [39] have found that SOM content under plowing and rotary tillage at deeper soil both were higher than that of the upper soil. The reason might be that rotary tillage and plowing tillage mixed crop straw into the deeper soil layer, making SOM well-distributed at different depths [40].

Carbon input can be increased by adopting straw return in cropland [14]. Fresh residues are C source for microbial activity and nucleation centers for aggregation when returned to cropland. The enhanced microbial activity induces the binding of residue and soil particles into macroaggregates [34,41], which could increase aggregates stability, fix the unstable C, thus improving the concentration of SOC [42] and increasing C sequestration [14]. In our research, straw return had significantly higher soil TOC and labile organic carbon fractions contents at the three soil depths than no crop straw return (Table 2). Soil TOC and labile organic C fractions in both rice and wheat straw return treatments were higher than only wheat or rice straw return (Table 2), indicating that straw return plays an very important role in increasing soil TOC and labile organic C fractions. Similar observations have been reported by other researchers [43–45]. At the three depths, soil TOC in the treatment of only wheat straw return in rice season was higher than only rice straw return in wheat season, moreover, the difference was significant ($p<0.05$) at 7–14 cm (Table 2). This was related to the relatively near-surface higher water content and favorable soil temperature during the rice growing season, resulting in relatively fast straw decomposition [46]. The decomposition of wheat straw provides enough energy and carbon source for soil microorganisms, thus increases the microorganisms' activities. Alternatively, after wheat straw return, the high temperature and humid conditions accelerates the reduction of the C/N ratio of the straw, allowing for sufficient decomposition. More nutrients are released and utilized by the crops, which therefore lightens the pressure of burning straw and improves the soil quality [47]. According to table 1, the study showed that MBC was affected by the straw return factor with an affecting force of 95.95% at 0–7 cm depth and 97.83% at 7–14 cm depth. The probable explanation maybe that crop residue might enter the labile C pool, provide substrate for the soil microorganisms, and contribute to the accumulation of labile C [48].

In our study, PD had the highest content of soil TOC at all the three soil depths (Fig. 1). The reason might be that plowing tillage made the soil and straw in the plow layer turned over quarterly, which increased the stability of the TOC content at each soil layer [49]. In addition, the rice and wheat straw were both returned under PD treatment from 2009, plowing tillage made much SOM enter into the soil and accumulate [45]. However, Tian et al. [50] found that rotary tillage with straw return had higher SOC than plowing tillage with straw return at 0–10 cm soil depth in wheat field. The diverse results might be due to the different regional climate, soil type, crop rotation and the length of study [22]. In this study, at upper soil layer, the interaction effect between tillage and straw return was not significant, but generally increased with an increase in soil depth (Table 3). Rajan et al [2] also found that single effect of residue application was not significant but its significance became apparent after its interaction with tillage system.

In our study, soil labile organic C fractions were significantly and positively correlated with TOC concentrations at 0–21 cm soil depth (Table 1). Such correlations suggested that TOC was a major determinant of soil labile organic C fractions. MBC, DOC and EOC were also significantly and positively correlated with each other in this study (Table 1). The results were consistent with Chen et al. [14], who reported similar correlations between soil TOC, labile organic C fractions (MBC, DOC, particulate organic C, EOC and hot-water extractable C), and macroaggregate C within 0–15 cm depth. Dou et al. [51] also observed the same results. MBC is the living part of SOM, which plays an important role in maintenance of soil fertility [52]. It serves as a sensitive indicator of change and future trends in organic matter level [53]. Dissolved organic C consists of organic compounds present in soil solution, acts as a substrate for microbial activity, and is the primary energy source for soil microorganisms [1]. Easily oxidizable C partly reflects enzymatic decomposition of labile SOC [54]. Therefore, it is not surprising to find the positive correlations among the labile C pools as they have a close association with each other.

Conclusions

In this study, after 2 years of a rice-wheat rotation, soil TOC and labile organic C fractions in PR, PW, PD, and RR, RW, RD were all higher than PN and RN. PD and RD had more significant effects on EOC, DOC and MBC compared to other treatments at 0–21 cm depth. Soil TOC and labile organic C fractions were highly correlated with each other. Under short-term conditions, rice and wheat straw both return in rice-wheat rotation system can increase SOC content and improve soil quality in the Yangtze River Delta, which is a suitable agricultural practice in this region under rice-wheat cropping system.

Acknowledgments

We thank Mr. Kejun Gu (Jiangsu Province Academy of Agricultural Sciences, China) for supporting the field experiment, Dr. Changqing Chen (Nanjing Agricutural University, China) for his kind help in data analysis,and Mr. Michael Rickaille (University of the West Indies) for his critical reading.

Author Contributions

Conceived and designed the experiments: LQZ. Performed the experiments: MFY NJH ZWZ. Analyzed the data: MFY. Contributed reagents/materials/analysis tools: LQZ. Wrote the paper: LQZ NJH XHZ.

References

1. Haynes RJ (2005) Labile organic matter fractions as central components of the quality of agricultural soils: an overview. Adv. Agron 85: 221–268.
2. Rajan G, Keshav RA, Zueng-Sang C, Shree CS, Khem RD (2012) Soil organic carbon sequestration as affected by tillage, crop residue, and nitrogen application in rice–wheat rotation system. Paddy Water Environ 10: 95–102.
3. Blair GJ, Lefory RDB, Lise L (1995) Soil carbon fractions based on their degree of oxidation and the development of a carbon management index for agricultural system. Aust. J. Agric. Res 46: 1459–1466.
4. Ghani A, Dexter M, Perrott WK (2003) Hot-water extractable carbon in soils: a sensitive measurement for determining impacts of fertilization, grazing and cultivation. Soil Biol. Biochem 35: 1231–1243.
5. Rudrappa L, Purakayastha TJ, Singh D, Bhadraray S (2006) Long-term manuring and fertilization effects on soil organic carbon pools in a Typic Haplustept of semi-arid sub-tropical India. Soil Till. Res 88: 180–192.
6. Yang CM, Yang LZ, Zhu OY (2005) Organic carbon and its fractions in paddy soil as affected by different nutrient and water regimes. Geoderma 124: 133–142.
7. Baker JM, Ochsner TE, Venterea RT, Griffis TJ (2007) Tillage and soil carbon sequestration–what do we really know? Agric Ecosyst Environ 118, 1–5.
8. Hernanz JL, L'opez R, Navarrete L, S'anchez-Gir'on V (2002) Long-term effects of tillage systems and rotations on soil structural stability and organic carbon stratification in semiarid central Spain. Soil Till. Res 66 (2): 129–141.
9. Shan YH, Yang LZ, Yan TM, Wang JG (2005) Downward movement of phosphorus in paddy soil installed in large-scale monolith lysimeters. Agr. Ecosyst. Environ 111 (1–4): 270–278.
10. Zhang ZJ (1998) Effects of long-term wheat-straw returning on yield of crop and soil fertility. Chinese Journal of Soil Science 29(4): 154–155.
11. Sun X, Liu Q, Wang DJ, Zhang B (2007) Effect of long-term application of straw on soil fertility. Chinese Journal of Eco-Agriculture 16(3): 587–592.
12. West TO, Post WM (2002) Soil organic carbon sequestration rates by tillage and crop rotation. Soil Science Society of American Journal 66: 1930–1946.
13. Liu SP, Nie XT, Zhang HC, Dai QG, Huo ZY, et al. (2006) Effects of tillage and straw returning on soil fertility and grain yield in a wheat-rice double cropping system. Transactions of the CSAE 22(7): 48–51.
14. Chen HQ, Hou RX, Gong YS, Li HW, Fan MS, et al. (2009) Effects of 11 years of conservation tillage on soil organic matter fractions in wheat monoculture in Loess Plateau of China. Soil Till. Res 106: 85–94.
15. Wang DD, Zhou L, Huang SQ, Li CF, Cao CG (2013) Short-term Effects of Tillage Practices and Wheat-straw Returned to the Field on Topsoil Labile Organic Carbon Fractions and Yields in Central China. Journal of Agro-Environment Science 32(4): 735–740.
16. Miller AJ, Amundson R, Burke IC, Yonker C (2004) The effect of climate and cultivation on soil organic C and N. Biogeoche mistry 67: 57–72.
17. Diekow J, Mielniczuk J, Knicker H, Bayer C, Dick DP, et al. (2005) Soil C and N stocks as affected by cropping systems and nitrogen fertilisation in a southern Brazil Acrisol managed under no-tillage for 17 years. Soil and Tillage Research 81, 87–95.
18. Galantini JA, Senesi N, Brunetti G, Rosell R (2004) In fluence of texture on organic matter distribution and quality and nitrogen and sulphur status in semiarid Pampean grassland soils of Argentina. Geoderma 123, 143–152.
19. Ouédraogo E, Mando A, Stroosnijder L (2006) Effects of tillage, organic resources and nitrogen fertiliser on soil carbon dynamics and crop nitrogen uptake in semi-arid West Africa. Soil and Tillage Res 91: 57–67.
20. Yamashita T, Feiner H, Bettina J, Helfrich M, Ludwig B (2006) Organic matter in density fractions of water-stable aggregates in silty soils: effect of land use. Soil Biology and Biochemistry 38: 3222–3234.
21. Paustian K, Collins HP, Paul EA (1997) Management controls in soil carbon. In: Paul, E.A., Paustian, K.A., Elliott, E.T., Cole, C.V. (Eds). Soil Organic Matter in Temperate Ecosystems: Long Term Experiments in North America. RC, Boca Raton, FL 15–49.
22. Puget P, Lal R (2005) Soil organic carbon and nitrogen in a Mollisol in central Ohio as affected by tillage and land use. Soil Till. Res 80, 201–213.
23. Ding LL, Cheng H, Liu ZF, Ren WW (2013) Experimental warming on the rice-wheat rotation agro-ecosystem. Plant Science Journal 31(1): 49–56.
24. Editorial Board of China Agriculture Yearbook (2012) China Agriculture Yearbook 2009, Electronic Edition. China Agriculture Press, Beijing, China.
25. Hao JH, Ding YF, Wang QS, Liu ZH, Li GH, et al. (2010) Effect of wheat crop straw application on the quality of rice population and soil properties. Journal of Nanjing Agricultural University 33(3): 13–18.
26. Zhu LQ, Zhang DW, Bian XM (2011) Effects of continuous returning straws to field and shifting different tillage methods on changes of physical-chemical properties of soil and yield components of rice. Chinese Journal of Soil Science 42(1): 81–85.
27. Liu SP, Nie XT, Zhang HC, Dai QG, Huo ZY, et al. (2006) Effects of tillage and straw returning on soil fertility and grain yield in a wheat-rice double cropping system. Transactions of the CSAE 22(7): 48–51.
28. Lu RK (1999) Soil agricultural chemistry analysis. China's agricultural science and technology press, 106–110.
29. Jiang PK, Xu QF, Xu ZH, Cao ZH (2006) Seasonal changes in soil labile organic carbon pools within a Phyllostachys praecox stand under high rate fertilization and winter mulch in subtropical China. For. Ecol. Manage 236: 30–36.
30. Vance F, Brookes P, Jenkinson D (1987) Microbial biomass measurements in forest soils: the use of the chloroform fumigation-incubation method in strongly acid soils. Soil Biochem 19: 697–702.
31. Leng SC (1992) Biological Statistic and Field Experimental Design. Beijing: China Radio& Television Press (in chinese).
32. Duan HP, Niu YZ, Bian XM (2012) Effects of tillage mode and straw return on soil organic carbon and rice yield in direct seeding rice field. Bulletin of Soil and Water Conservation 32(3): 23–27.
33. Yang JC, Han XG, Huang JH, Pan QM (2003) The dynamics of soil organic matter in cropland responding to agricultural practices. Acta Ecologica Sinica 23(4): 787–796.
34. Six J, Elliott ET, Paustian K (1999) Aggregate and soil organic matter dynamics under conventional and no-tillage systems. Soil Sci. Soc. Am. J. 63, 1350–1358.
35. Beare MH, Hendrix PF, Coleman DC (1994) Water-stable aggregates and organic matter fractions in conventional-tillage and no-tillage soils. Soil Sci. Soc. Am. J. 58, 777–786.
36. Mikha MM, Rice CW (2004) Tillage and manure effects on soil and aggregate-associated carbon and nitrogen. Soil Sci. Soc, AM. J 68: 809–816.
37. Wang J, Zhang RZ, Li AZ (2008) Effect on soil active carbon and C cool management index of different tillages. Agricultural Research in the Arid Areas 26(6): 8–12.
38. Li XJ, Zhang ZG (1999) Influence on soil floods properties of mulching straws and soil-returning straw. Territory and Natural Resources Study 1: 43–45.
39. Liu DY (2009) Physiological and ecological mechanism of stable and high yield of broadcasted rice in paddy field with high standing-stubbles under no-tillage condition. PhD: Sichuan Agricultural University.
40. Gao YJ, Zhu PL, Huang DM, Wang ZM, Li SX (2000) Long-term impact of different soil management on organic matter and total nitrogen in rice-based cropping system. Soil and Environmental Sciences 9(1): 27–30.
41. Jastrow JD (1996) Soil aggregate formation and the accrual of particulate and mineral associated organic matter. Soil Biol. Biochem. 28, 656–676.
42. Govaerts B, Sayre KD, Lichter K, Dendooven L, Deckers J (2007) Influence of permanent raised bed planting and residue management on physical and chemical soil quality in rain fed maize/wheat systems. Plant Soil 291, 39–54.
43. Stockfisch N, Forstreuter T, Ehlers W (1999) Ploughing effects on soil organic matter after twenty years of conservation tillage in Lower Saxony, Germany. Soil Till. Res 52: 91–101.

44. Chen SH, Zhu ZL, Liu DH, Shu L, Wang CQ (2008) Influence of straw mulching with no-till on soil nutrients and carbon pool management index. Plant Nutrition and Fertilizer Science 14(4): 806−809.
45. Song MW, Li AZ, Cai LQ, Zhang RS (2008) Effects of different tillage methods on soil organic carbon pool. Journal of Agro-Environment Science 27(2): 622−626.
46. Zuo YP, Jia ZK (2004) Effect of soil moisture content o n straw decomposing and its dynamic changes. Joural of Northwest Sci-Tech University of Agri. and For. (Nat. Sci. Ed.) 32(5): 61–63.
47. Dai ZG (2009) Study on nutrient release characteristics of crop residue and effect of crop residue returning on crop yield and soil fertility. PhD: Huazhong Agricultural University (in Chinese).
48. Li CF, Yue LX, Kou ZK, Zhang ZS, Wang JP, et al. (2012) Short-term effects of conservation management practices on soil labile organic carbon fractions under a rape–rice rotation in central China. Soil Till, Res 119: 31–37.
49. Zhang P, Li H, Jia ZK, Wang W, Lu WT, et al. (2011) Effects of straw returning on soil organic carbon and carbon mineralization in Semi-arid areas of southern Ningxia, China. Journal of Agro-Environment Science 30(12): 2518−2525.
50. Tian SZ, Ning YT, Wang Y, Li HJ, Zhong WL, et al. (2010) Effects of different tillage methods and straw-returning on soil organic carbon content in a winter wheat field. Chinese Journal of Applied Ecology 21(2): 373–378.
51. Dou FG, Wright AL, Hons FM (2008) Sensitivity of labile soil organic carbon to tillage in wheat-based cropping systems. Soil Sci. Soc. Am. J. 72: 1445–1453.
52. Wu TY, Schoenau JJ, Li FM, Qian PY, Malhi SS, et al. (2004) Influence of cultivation and fertilization on total organic carbon and carbon fractions in soils from the Loess Plateau of China. Soil & Till, Res 77: 59–68.
53. Gregorich EG, Ellert BH, Gregorich EG, Carter MR, Monreal CM, et al. (1994) Towards a minimum data set to assess soil organic matter quality in agricultural soils. Can. J. Soil Sci 74: 367–385.
54. Loginow W, Wisniewski W, Gonet SS, Ciescinska B (1987) Fractionation of organic carbon based on susceptibility to oxidation. Pol. J. Soil Sci 20: 47–52.

SmeltCam: Underwater Video Codend for Trawled Nets with an Application to the Distribution of the Imperiled Delta Smelt

Frederick Feyrer[1]*, Donald Portz[2], Darren Odum[3], Ken B. Newman[4], Ted Sommer[5], Dave Contreras[6], Randall Baxter[6], Steven B. Slater[6], Deanna Sereno[7], Erwin Van Nieuwenhuyse[1]

1 Bay Delta Office, Bureau of Reclamation, Sacramento, California, United States of America, **2** Technical Service Center, Bureau of Reclamation, Denver, Colorado, United States of America, **3** SureWorks LLC., Longmont, Colorado, United States of America, **4** U.S. Fish and Wildlife Service, Lodi, California, United States of America, **5** California Department of Water Resources, West Sacramento, California, United States of America, **6** California Department of Fish and Wildlife, Stockton, California, United States of America, **7** Contra Costa Water District, Concord, California, United States of America

Abstract

Studying rare and sensitive species is a challenge in conservation biology. The problem is exemplified by the case of the imperiled delta smelt *Hypomesus transpacificus*, a small delicate fish species endemic to the San Francisco Estuary, California. Persistent record-low levels of abundance and relatively high sensitivity to handling stress pose considerable challenges to studying delta smelt in the wild. To attempt to overcome these and other challenges we have developed the SmeltCam, an underwater video camera codend for trawled nets. The SmeltCam functions as an open-ended codend that automatically collects information on the number and species of fishes that pass freely through a trawled net without handling. We applied the SmeltCam to study the fine-scale distribution of juvenile delta smelt in the water column in the upper San Francisco Estuary. We learned that during flood tides delta smelt were relatively abundant throughout the water column and that during ebb tides delta smelt were significantly less abundant and occurred only in the lower half and sides of the water column. The results suggest that delta smelt manipulate their position in the water column to facilitate retention in favorable habitats. With the application of the SmeltCam we increased the survival of individual delta smelt by 72% compared to using a traditional codend, where all of the fish would have likely died due to handling stress. The SmeltCam improves upon similar previously developed silhouette photography or video recording devices and demonstrates how new technology can be developed to address important questions in conservation biology as well as lessen the negative effects associated with traditional sampling methods on imperiled species.

Editor: Christopher Fulton, The Australian National University, Australia

Funding: Funding for the research and development of the SmeltCam has been provided by the Bureau of Reclamation and the California Department of Water Resources. The California Department of Fish and Wildlife provided funding in the form of in-kind services. The State and Federal Contractors Water Agency (SFWCA) provided some additional funding to conduct the field study. SFWCA had no role in study design, data collection and analysis, decision to publish, or preparation of the manuscript.

Competing Interests: DO is employed by SureWorks LLC. SureWorks LLC intends to market the device, the SmeltCam. There are no patents, or further products in development or marketed products to declare. This does not alter the authors' adherence to all the PLOS ONE policies on sharing data and materials, as detailed online in the guide for authors.

* E-mail: ffeyrer@usbr.gov

Introduction

Conservation of endangered species faces many challenges. A key difficulty is studying and monitoring populations in which individuals are by definition low in abundance and rarely observed. Solutions to this problem often address sampling design and fitting appropriate models to data [1,2]. Potential solutions could also involve alternative methods and new technology. For example, remote photography or video methods are commonly used in ecology to address a variety of research questions [3,4,5,6] and could be applied to the study of imperiled species. In particular, underwater video systems are becoming increasingly popular for studying fishes in marine [7,8], estuarine [9,10,11] and freshwater habitats [12,13]. They are especially desirable when a key objective is to minimize or avoid the adverse effects of handling stress associated with traditional sampling methods [14,15].

The need for alternative methods to study imperiled species is exemplified by the case of the delta smelt *Hypomesus transpacificus* in the upper San Francisco Estuary. The delta smelt is a formerly abundant euryhaline pelagic fish endemic to the upper San Francisco Estuary that has experienced substantial declines in abundance (Fig. 1) [16,17]. Abundance declines have been attributed to multiple interacting factors including foodweb alterations, physical habitat loss, contaminants and water diversions [16,17,18,19]. To complicate matters further, the delta smelt is a small (maximum fork length ~ 90 mm) fish that typically dies with minimal handling stress (20); it is assumed that most individuals collected in routine monitoring surveys do not survive. Because of their small size and delicate nature, wild delta smelt have not been tagged for remote tracking or for mark-recapture studies with presently available tools or technology (but see [21] for a study on cultured delta smelt). Delta smelt live in turbid pelagic habitats [22,23] and therefore cannot be directly observed in their natural environment. Currently available hydroacoustic methods are of limited utility because three other fish species (longfin smelt *Spirinchus thaleichthys*, wakasagi *H. nipponensis*, and Mississippi silversides *Menidia audens*) co-occur with delta smelt and have a nearly identical body size and shape, which complicates species differentiation. Because of these challenges, sampling with trawled nets has been the only feasible method of studying delta smelt ecology in the wild.

Due to their extremely low abundance and delicate nature, continued study and monitoring of delta smelt poses a considerable challenge for scientists and managers. To overcome such challenges, resource agencies have invested in developing the SmeltCam, an underwater video camera codend for trawled nets (Fig. 2). Here, we describe the SmeltCam and its first application to advancing the understanding of delta smelt ecology. The purpose of our study was to better understand the fine-scale distribution of delta smelt. Our study question was: does the vertical and horizontal distribution of delta smelt vary by tide stage? The answer to this question is relevant for many reasons, including the opportunity to generate more precise population estimates of delta smelt [24]. Long-term fish monitoring in the upper San Francisco Estuary is not designed to generate actual population estimates of fishes. Rather, it generates dimensionless interannual indices of relative abundance. Sample design-based population estimates for delta smelt have been generated from the available monitoring data but are necessarily subject to its biases and limitations [24]. Three of the key issues are that sampling takes place (1) generally at center channel, (2) irrespective of tide stage, and (3) with the net towed obliquely through the water column. These issues present problems in extrapolating the trawl catches volumetrically to generate sample design-based population estimates because it is not known how delta smelt are distributed across tides vertically or horizontally in the water column. Another immediate application of our study is that the knowledge obtained on fine-scale habitat use can be used to inform behavior models examining the distribution and movements of delta smelt.

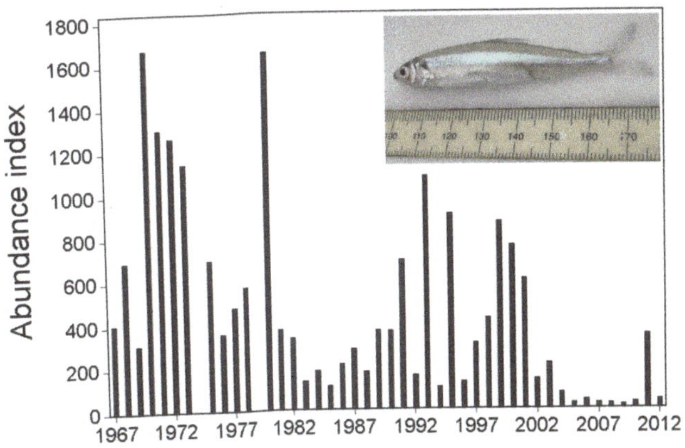

Figure 1. Time series of delta smelt abundance indices (unitless) from the California Department of Fish and Wildlife's Fall Midwater Trawl Survey. No sampling was conducted in 1974 or 1979. Inset is a photograph of a delta smelt collected during the study. Rule increments are millimeters.

Methods

SmeltCam

The SmeltCam functions as an open-ended codend that automatically collects information on the number and species of fishes that pass freely through a trawled net. Key components of the SmeltCam include a bridle system that connects to a trawled net, a water-tight electrical housing and a ballast hull (Fig. 2).

The SmeltCam body is a combination of welded sheet aluminum and machined plate aluminum. The overall dimensions of the unit are 93 cm (length)×56 cm (width)×38 cm (height). In the configuration used in our study, a 4-point bridle system with turnbuckles and shackles was used to attach the SmeltCam to the four load-bearing lines of the trawled net. The unit weighs 48.5 kg dry and valves are used to adjust and maintain water levels in portions of the hollow ballast hull in order to achieve neutral buoyancy. The addition of ballast water adds considerably to the unit's weight and necessitated a davit to lift it from and to the deck during and after deployment. The interior chamber of the SmeltCam where fish and other objects pass is 76.2 cm (length)×18.9 cm (width)×18.9 cm (height).

The starboard side of the unit contains the sealed electrical housing chamber. Within the electrical housing chamber are components and sensors that control and/or monitor positioning, physical conditions, lighting, and video functions. A global positioning system (GPS; uBlox-6 chipset, U-blox America Inc, San Jose California, USA) records position coordinates with an approximate 3 meter level of accuracy. A pressure gauge within the unit records depth with a vertical resolution of approximately 8 cm. Accelerometers measure the tilt, roll, and pitch of the device in the water. Sensors measure the interior computer and air temperature. Relative humidity is also measured to detect failure of the chamber seal and exposure of electronics to water.

The internal wall of the sealed electrical housing chamber is a 43 cm (length) × 25 cm (height) acrylic window that serves as the viewing area for the video system. The viewing area is essentially the interior chamber of the unit and thus has the same dimensions. Custom 8,000 lumen white LED is used to cast a wide swath of light to blanket the entire viewing window with even, wide angle of incidence lighting that reduces backscatter and specular reflection off of passing fish and other objects. A grayscale camera (Genie HM1400; Teledyne Dalsa, Billerica, Massachusetts, USA) mounted on the inside of the sealed electrical housing chamber captures 1600 ×1200 megapixel resolution images at a rate of sixty frames per second. The system is powered by 120 V AC 60 Hz, supplied by a 3000 W Honda portable generator, however on-board ship power is also suitable. Internal power supplies converted the 120 V AC power to 24 V DC and 12 V DC power. Power and communications to components in the sealed electrical housing chamber are provided by a simple, flexible, 3-conductor 14AWG waterproof cable, 180 m in length. Communications are sent up and down the same cable via powerline communications, in which TCP/IP packets are encoded on top of the 120 V AC power. The communications link over the 180 m distance is approximately 20 mbits/second.

Custom software has been developed to operate the system and to record data, which is operated with a standard laptop computer. A series of algorithms control object detection, tracking and identification. Object detection and tracking algorithms utilize gradient contour methods from raw image information obtained from the camera. Species identification is accomplished through algorithms in a support vector machine (SVM). The system uses several feature vectors to uniquely describe each species. The feature list includes object size, size and shape-independent list of

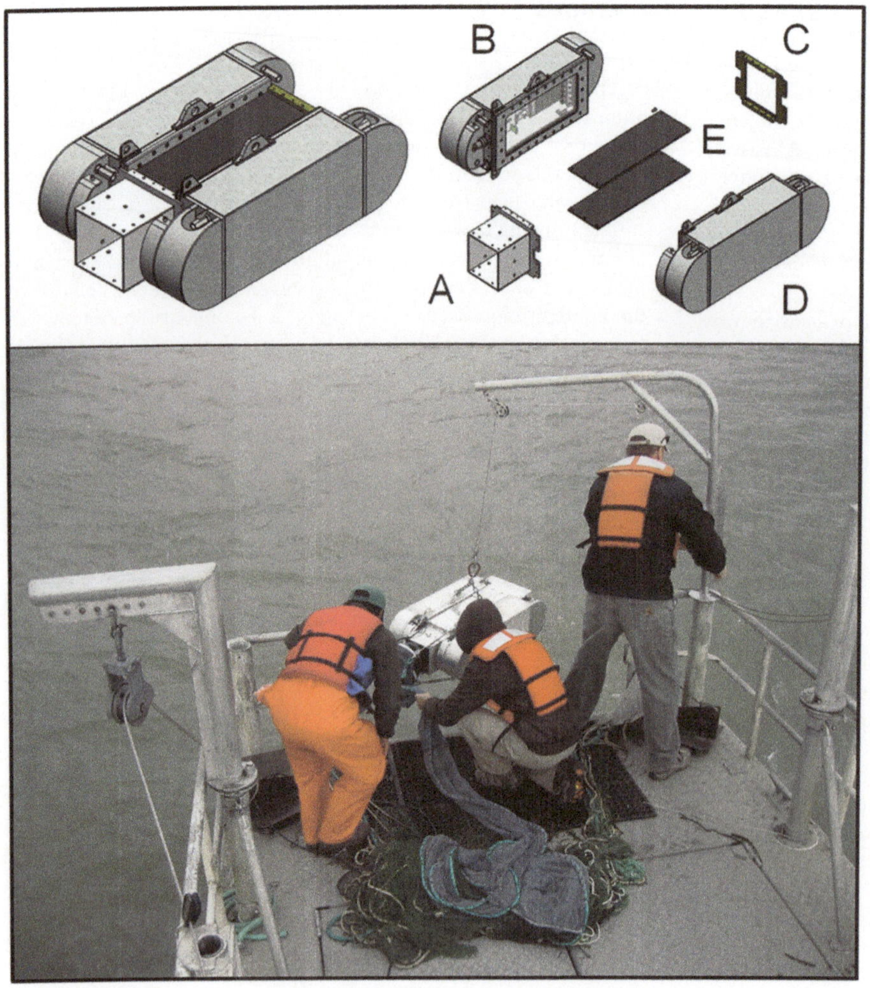

Figure 2. The SmeltCam. Upper panel is a diagram of the SmeltCam showing (A) net cowling and bow frame, (B) sealed electronics compartment, (C) stern frame, (D) ballast hull and (E) top and bottom vision tube covers. Bottom panel is a photograph of the SmeltCam being deployed from a research vessel.

shape moments (Hu moments), aspect ratio, defect from pure ellipse, RMS error (or deviation from) normalized species image, and radial local pattern. The SVM takes in all of the features and generates species identifications with an associated level of confidence for each object passing through the field of vision.

Training the SVM algorithm to identify fish species is an ongoing exercise and involves using positively identified images and metadata. Two separate cross-validation efforts were completed prior to conducting to this field study. The initial effort involved a training sample of human-identified images of delta smelt, threadfin shad *Dorosoma petenense* and American shad *Alosa sapidissima* acquired from the field in September 2011 to classify 306 images acquired in October 2011. Classification success was 88% for delta smelt, 56% for American shad and 37% for threadfin shad. We also conducted a k-fold cross validation using all fifty of the field-collected images of delta smelt obtained over the lifetime of the SmeltCam. The library of images was divided into 10 subsets (k = 10) where for each subset 10% of the images was used as a training set to identify the remaining 90% of the images. The average success rate over the 10 subsets was 91%, meaning that the algorithm could positively indentify 91% of the images that a human could positively identify. Online training is planned for the next phase of development, in which every positively identified object that passes through the device helps to improve the SVM algorithm. Algorithms are available upon request (inquiries should be sent to darren.odom@sureworksllc.com). While the algorithm works relatively well, it is continually being improved. Hence for this study we reviewed each image obtained during sampling and provided a relatively subjective human-assigned level of confidence for each species identification.

All system components and live video from the camera can be monitored in real time on board the research vessel and simultaneously written to file. Ultimately, each fish passing through the field of vision is given a species identification with an associated level of confidence, and all other sensor data is also recorded including date, time, GPS coordinates and depth. All images are also recorded and, as in the case of our study, can be reviewed for accuracy.

Delta Smelt Ecology

The delta smelt was listed as a threatened species under both the California and Federal Endangered Species Acts in 1993. The listing status was changed to endangered by California in 2009. In 2010, a Federal status review determined endangered status was warranted but precluded by other higher priority listing actions.

Delta smelt abundance has been variable but has exhibited a substantial long-term decline (Fig. 1) [16,24,25]. The delta smelt is one of four fish species in the estuary which have exhibited further step-declines in about 2002 and have remained near all time record lows for the last decade, defining an era in the ecosystem known as the pelagic organism decline [18]. Long-term trends in abundance of delta smelt and other fishes are generated from data collected by the California Department of Fish and Wildlife's (CDFW) Fall Midwater Trawl Survey (FMT), which has been conducted each year since 1967, with the exception of 1974 and 1979.

The delta smelt is an opportunistic carnivore that feeds primarily on planktonic copepods, cladocerans, mysids, and amphipods. It is primarily an annual species with very few individuals living and spawning a second year. Spawning takes place during spring in freshwater tidal habitats [17,26]. Young delta smelt move downstream with the tides until they reach favorable rearing habitats in the low salinity zone (~1–6) of the estuary [27], although some apparently remain in upstream reaches year-round [26]. During the summer and fall, juvenile delta smelt live primarily in the upper San Francisco Estuary associated with the low salinity zone [17,23]. In winter delta smelt migrate upstream to freshwater habitats where spawning occurs during spring.

Study Area and Design

Our study focused on the fine scale distribution of juvenile delta smelt in the upper estuary during fall. Field sampling for delta smelt was conducted under a permit granted to the Interagency Ecological Program by the U.S. Fish and Wildlife Service. We conducted our study November 27–30, 2012 in the Sacramento River adjacent to Sherman Island, approximately between routine FMT stations 704 and 705 (Fig. 3). We chose this region because: (1) long-term FMT monitoring data indicate delta smelt remain relatively abundant in this area due to suitable habitat conditions [23], (2) two other studies, one examining movements of delta smelt (J. Burau, U.S. Geological Survey, Sacramento, California, personal communication) and another examining sampling efficiency of several different trawled nets (R. Baxter, unpublished data), effectively sampled delta smelt in this area in the two months preceding our study, September-October 2012, and (3) routine FMT sampling in the three months preceding our study also detected delta smelt in the area.

Physical conditions during our study were typical for the region during the fall but dramatically changed shortly thereafter (Fig. 4). During the study a storm moved across California and produced a moderate amount of intermittent rain and south winds in the immediate study area. The storm persisted after our study and within two weeks produced what is colloquially termed "first flush" conditions, which refers to the initial onset of substantially elevated river flows and turbidity entering the estuary (Fig. 4; data obtained from the California Data Exchange Center http://cdec.water.ca.gov/). These conditions are associated with the upstream migration of delta smelt to areas where spawning ultimately occurs during spring [26,28]. Our study, therefore, observed fine-scale delta smelt distribution patterns during typical fall conditions prior to the "first flush" and the upstream migration of delta smelt.

We conducted our study using the same equipment (e.g., research vessel, net, and crew) as is normally used for the FMT. The only exception was that the SmeltCam was affixed to a slightly modified codend of the net rather than the codend being tied closed. Descriptions of the standard FMT protocol and sampling sites are readily available [23,29]. The net itself is 17.6 m long with a square mouth opening of 3.66 m in width and height. It has nine tapered panels of stretch mesh from 14.7 cm near the mouth to 1.3 cm in the codend. To generate the data used to calculate the interannual indices of relative abundance, the FMT collects samples via a 12-minute oblique trawl conducted at 100 sites distributed across the tidal freshwater to mesohaline regions of the estuary each month from September to December. To attach the SmeltCam, the codend of the net was modified by slightly adjusting the lengths of the last two mesh panels and attaching an additional panel of 0.64 cm knotless mesh measuring 74.9 cm in length sewn to a 7.6 cm-diameter vinyl collar attached to the SmeltCam housing. These modifications increased the total length of the net from 17.6 m to 17.8 m.

As mentioned above, long-term FMT data and recent research efforts helped guide our study design. The aforementioned study examining delta smelt movements influenced our experimental design with its observation that delta smelt were collected during flood tides but rarely during ebb tides in surface samples taken with a Kodiak Trawl in both 2010 and 2012. Expanding upon that observation and to more closely examine the position of delta smelt in the water column, we set up a factorial study design with three factors and two levels for each factor, thus $2^3 = 8$ possible treatments. The three factors and their corresponding levels were: (1) horizontal position in the water column (H: center of the channel versus side of the channel), (2) vertical position in the water column (V: upper half versus lower half) and (3) tidal phase (T: flood versus ebb). The response variable, fish counts, was defined as the number of delta smelt collected in a 10-minute trawl. Since counts are functions of density and volume sampled, and our interest was in how density varied by treatment combination, the volume of water filtered by the trawl was estimated using a mechanical flowmeter (model 2030R, General Oceanics, Inc.) deployed off the side of the research vessel during each trawl.

Available time and resources facilitated the day-time collection of fifty-six samples over a four day period, thereby allowing seven replicates per treatment combination. To determine the adequacy of these sample sizes, we estimated, given seven replicates and a standard deviation for fish counts of 2.1 (based on thirty-five FMT samples taken during the previously mentioned net efficiency study), that there was a 95% probability of rejecting a test of the null hypothesis of no factorial effects, i.e., the expected fish counts are the same for all treatment combinations, when at least one of the combinations had an expected catch that was three fish above (or below) that for other combinations. Further, pairwise differences in fish counts as small as three would be detected with 79% probability and as large as four would be detected with 95% probability.

We used a combination of GIS (geographic information system) and GPS to select and occupy sampling locations in the Sacramento River channel in order to achieve our study objective. GIS (ArcGIS 9.3.1, ESRI, Redlands, California, USA) was used to generate a total of twenty-one possible equidistant trawling lanes oriented longitudinally in the channel; the first and last seven lanes represented the sides of the channel while the middle seven lanes represented the center of the channel (Fig. 3). Sides of the channel were treated as a single unit and the specific side sampled was determined randomly. The number of lanes and their spacing were generated so that trawling in one lane would have no effect on adjacent lanes. The lanes were loaded into a GPS unit and tracked by the research vessel during sampling. Water depth averaged 10.5 m during ebb tides and 10.1 m during flood tides. Sampling depth (upper half versus lower half of the water column) was achieved by maintaining the net either above or below mid-depth during a trawl, targeting ¼ or ¾ depth (~2.6 or 7.9 m),

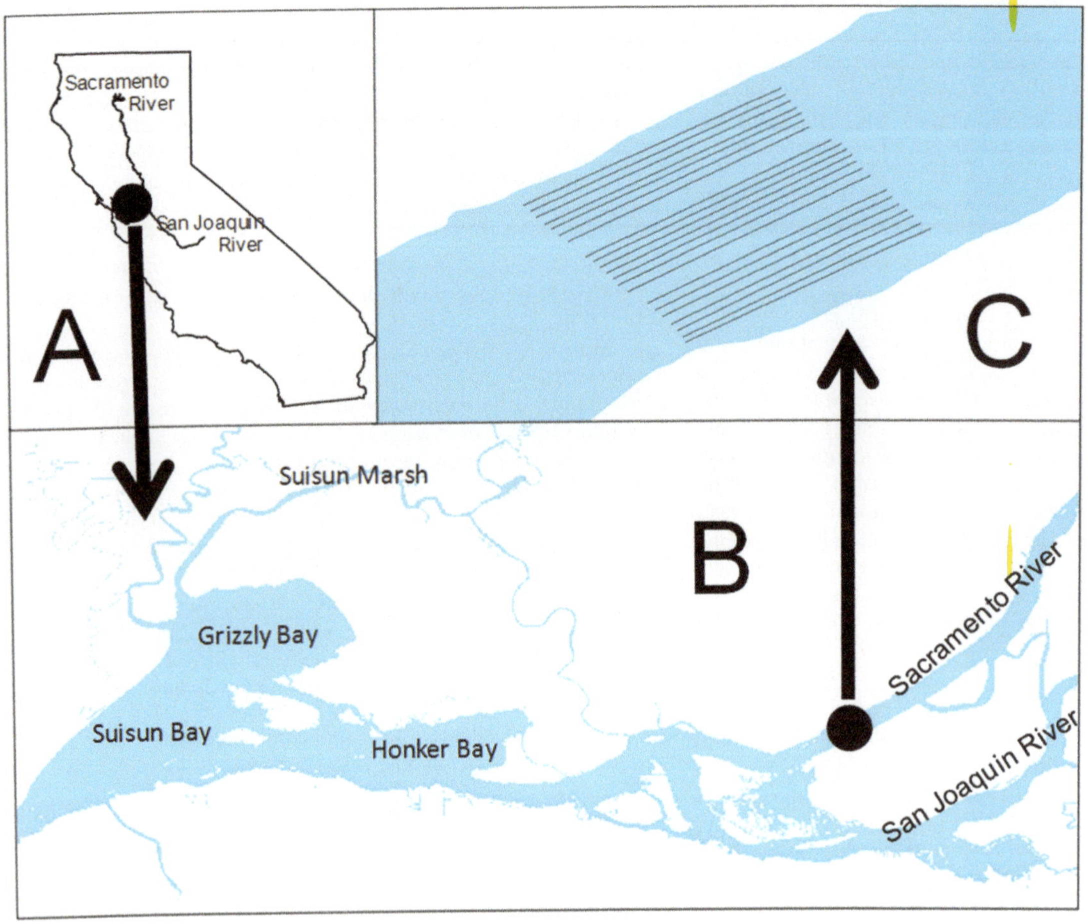

Figure 3. Map of the study area showing the (A) location of the upper San Francisco Estuary in California, (B) location of the study site in the upper estuary, and (C) orientation of trawling lane transects in the tidal Sacramento River channel.

respectively. Water depth was determined by a boat-mounted sonar unit. Net depth was adjusted with the length of cable between the net and the boat and determined by the SmeltCam's depth sensor, which was monitored in real-time during trawls.

A ten-minute sample was recorded for each trawl once the net and SmeltCam were positioned at the appropriate depth. The water filtered by the net during the time it took to go from the surface to the appropriate depth at deployment and then back again at retrieval was not considered part of the sample and was not recorded. For consistency, trawls were done so that the net was towed longitudinally in the channel against the current (i.e., upstream during ebb tides and downstream during flood tides). The order in which lanes and depths were sampled was randomly generated. Sampling necessarily had to follow the order of the tides. We examined forecasted tidal velocities generated from the CALSIM Hydrologic Model [30] to appropriately arrange sample collection around the tides. Sampling took place only during daylight hours, consistent with FMT protocol.

We measured water temperature (°C), salinity, turbidity (NTU), Chl *a* concentration (μg/L), pH and dissolved oxygen concentration (mg/L) immediately preceding and following each trawl in both the upper and lower half of the water column. Spot measurements were taken with a handheld YSI multiparameter sonde rigged with a communication cable long enough to reach the appropriate depth (YSI Inc, Yellow Springs, Ohio).

To statistically evaluate the effects of the three main factors (tide, horizontal and vertical position in the water column) on delta smelt density and the water quality variables, we fit several models commonly used for count data. In particular, we fit log linear Poisson models, models allowing for overdispersion (the quasi-Poisson and negative binomial), and models allowing for excess zeros (the zero inflated negative binomial), where overdispersion and excess zeros are with reference to the Poisson distribution [31]. Model fitting was done using the statistical computing environment R, version 2.15.1 [32], along with the R package 'pscl' [33,34]. To make between model comparisons, we calculated AIC values, $AIC = 2*k - 2*\log(\text{Likelihood})$, where $k =$ the number of parameters. AIC simultaneously quantifies goodness of fit, as defined by the likelihood of the data, and model complexity (as measured by k), and models with the smallest AIC values are considered preferable [35]. For each model, P-values for factors and factor combination were also calculated to assess the significance of particular factors.

Results

We collected 52 samples during the four days of field study; mechanical problems with the research vessel prohibited us from completing our intended number of replicates for each treatment (Table 1). In total we collected 30 samples during flood tides and 22 samples during ebb tides. Due to variations in tidal velocities

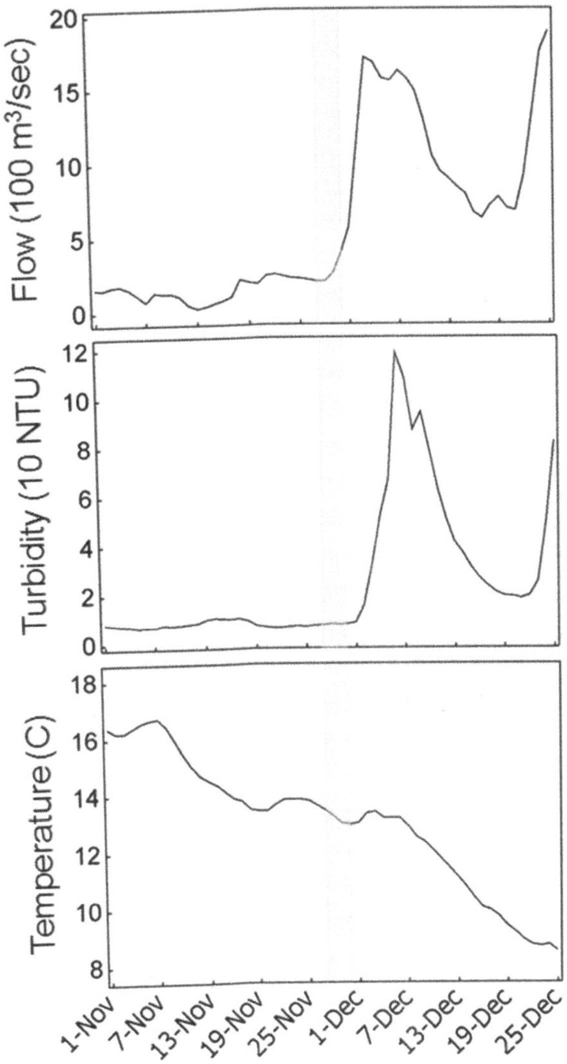

Figure 4. Seasonal time series of flow, turbidity and water temperature with the study period shaded in gray.

the volume of water sampled per 10-minute trawl varied from 4,388 m^3 to 8,057 m^3, but on average was comparable across the eight treatments (Table 1).

Water temperature averaged 13.8°C and varied by less than 1°C during the entire study (minimum = 13.5°C, maximum = 14.2°C). Dissolved oxygen concentration ranged from 8.4 mg/L to 8.9 mg/L, pH ranged from 7.4 to 7.8 and chlorophyll *a* concentration ranged from 1.2 µg/L to 3.5 µg/L. Given their low variability, these water quality variables were not included in subsequent analyses. Salinity averaged 2.6 and ranged from 0.4 to 5.3. As determined by a standard generalized linear model, salinity differed significantly ($P<0.05$) with horizontal position across the channel and vertical position in the water column. Salinity averaged about one unit higher in the center of the channel versus the side of the channel, and also on the bottom half of the water column versus the upper half of the water column (Fig. 5). Salinity did not differ across tides because we sampled the full tidal cycle. Consequently, salinity values expectedly overlapped during ebb and flood tides. Turbidity averaged 15.7 NTU and ranged from 7.6 NTU to 80 NTU. Turbidity exhibited statistically significant differences ($P<0.05$) among all combinations of factors and their interactions except for the tide:horizontal position, horizontal position:vertical position and tide:horizontal position:vertical position interactions. The most striking pattern with turbidity was that it was higher in the lower half of the water column, and substantially higher during flood tides (Fig. 5). Both salinity and turbidity were lowest during ebb tides in the center of the channel in the upper half of the water column.

We collected 352 individual fish comprised of 6 different species during our study: green sturgeon *Acipenser medirostris* (1), starry flounder *Platichthys stellatus* (1), American shad (23), striped bass *Morone saxatilis* (43), threadfin shad (87), and delta smelt (197). Count data on all of the species except for delta smelt were insufficient for further analysis. Of the 197 delta smelt collected, 142 individuals swam through the SmeltCam (Fig. 6) while 55 were entangled in the mesh of the net. We measured the fork lengths of 29 of the 55 individuals that were found in the net; they ranged from 51 mm to 75 mm (average = 64.5 mm and standard deviation = 5.2 mm). Subsequent summaries and analyses focus on individual delta smelt observed by the SmeltCam. The level of confidence in the identifications (human-assigned) ranged from

Table 1. Average±one standard deviation of the water volume (m^3) sampled across the eight study treatments.

	Flood tide	Ebb tide
Top half of the water column		
Middle channel	5,543±312 (7)	5,860±370 (6)
Side channel	5,781±780 (8)	5,587±170 (5)
Bottom half of the water column		
Middle channel	5,616+154 (7)	6,929±683 (6)
Side channel	5,656±803 (8)	6,741±1,150 (5)

Sample sizes are provided in the parentheses.

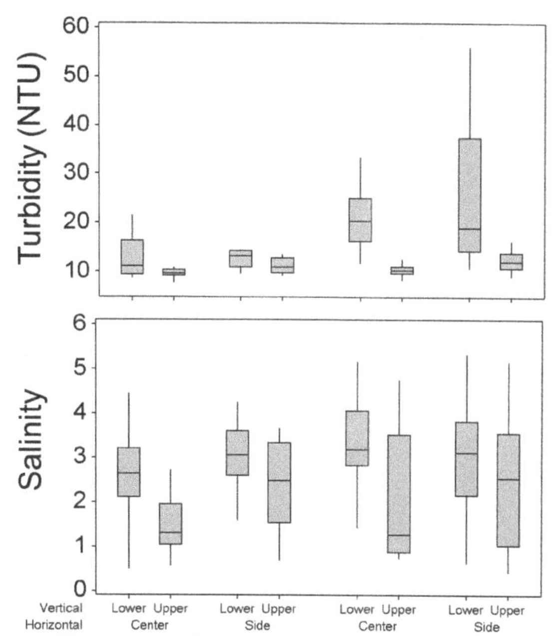

Figure 5. Box plots of turbidity and salinity by tide and position in the water column.

4%–100%, with 100% comprising the majority of the values (Fig. 7). Because there was 100% confidence in most of the identifications, the sample distribution (count of individual delta smelt per sample) did not change when examined across varying levels of confidence in species identification (Fig. 8). Nonetheless, to err on the side of caution we focus our analyses hereafter on individual delta smelt that were identified to species with 100% confidence.

Delta smelt were observed in 16 of 52 samples, thus zero counts were observed in 69% of the samples. The number of delta smelt per sample ranged from 0 to 22 (average = 2.2, standard deviation = 4.7), while density ranged from 0 to 37/10,000 m^3 (average = 3.8/10,000 m^3, standard deviation = 8.3/10,000 m^3). The average and standard deviation were more than two times higher than expected based on the aforementioned net efficiency study, which observed a mean of 0.9 delta smelt per sample and a standard deviation of 2.1.

The mean delta smelt density was 5.9/10,000 m^3 for flood tides versus 0.8/10,000 m^3 for ebb tides. For positive samples only (excluding the zero counts), the mean density was 14.9/10,000 m^3 for flood tides versus 5.0/10,000 m^3 for ebb tides. Delta smelt were observed throughout the water column on flood tides but only at the lower half and side of the channel on ebb tides (Fig. 9).

We fit the following families of models: Poisson (Po), quasi-Poisson (q-Po), negative binomial (NB), and zero-inflated negative binomial (ZINB). In each case the response variable was the observed number of delta smelt caught. We used a log link function to model the expected number of delta smelt caught. The expected number is the density times the volume sampled, and since the log of the volume sampled was handled as an offset, the factors were thus modeling expected delta smelt density. For each of the four model families, expected fish density was modeled using

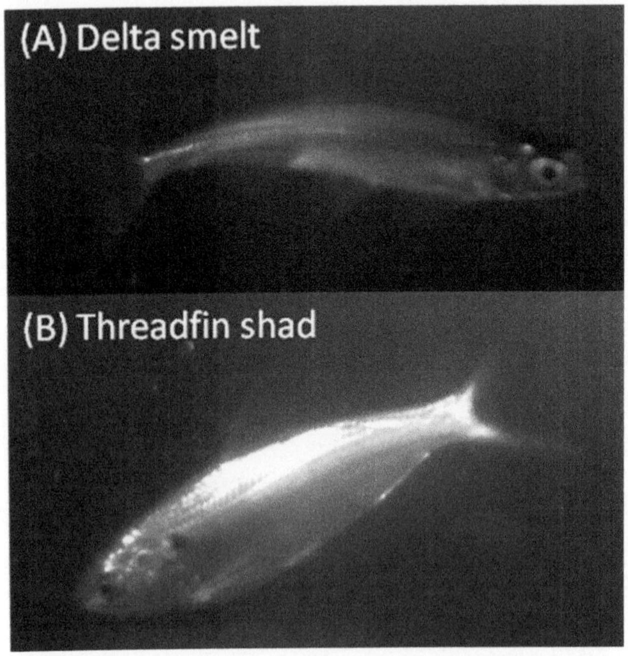

Figure 6. Examples of raw images of (A) delta smelt and (B) threadfin shad obtained by the SmeltCam during our field study. Note that in the delta smelt image all of the fins, including the adipose fin, are clearly visible and allow delta smelt to be differentiated from other species such as longfin smelt. Also note that the dorsal fin thread is visible across the caudal peduncle in the threadfin shad image.

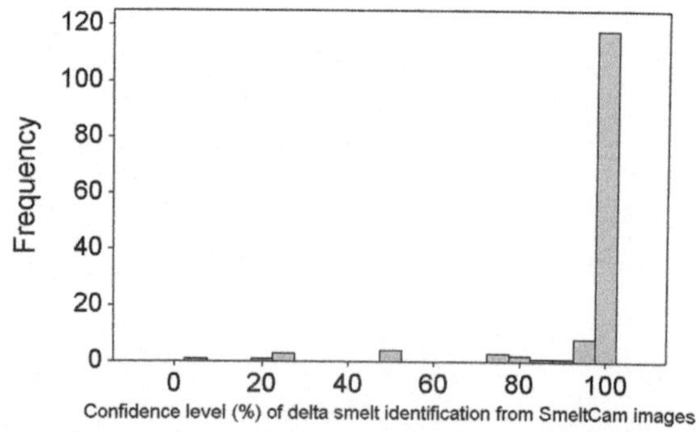

Figure 7. Frequency histogram of the confidence level (%) that delta smelt were correctly identified to species.

the following set of nested models (and the R syntax for the model formula):

1. Main effects only (H+V+T)
2. Main effects with 2-way interactions (H*V+H*T+V*T)
3. Main effects, 2-way and 3-way interactions (H*V*T)

The results are summarized in Table 2. The Poisson models found many more statistically significant factors and factor combinations than the other three families of models. However, this is due to relatively small, and likely too small, estimated variances (based on results for the other families of distributions and the fact that the Poisson models had the highest AIC values).

The quasi-Poisson model provides estimates of the inflation of the variances relative to the Poisson, e.g., overdispersion multipliers of 6.0 to 7.6. AIC values are not calculated for quasi-Poisson models because there is no likelihood. Quasi-AIC values have been developed [35] but they are of use only for comparing different quasi-Poisson models, not for comparing differences between families of distributions. Restricting attention to just the quasi-Poisson combinations, with the enlarged variance only the main effect of tide was found statistically significant for all three factorial combinations.

Amongst the negative binomial models, the main effects model has the smallest AIC value. The AIC value for the two-way interaction model is quite similar and negligible based on a rule of thumb [35], i.e., a difference of 2 units or less is not important. Given two models with negligibly different AIC values, the simpler model is preferable. Like the quasi-Poisson, the negative binomial model also increases the variance relative to the Poisson. Ver Hoef and Boveng [36] address the question of choosing between quasi-Poisson and negative binomial models and note that the key distinction is the nature of the variance function. The negative binomial variance is $\mu(1+\mu/\theta)$, the multipliers shown in Table 2 are $1/\theta$, in contrast to the quasi-Poisson $\mu\cdot\phi$, where ϕ is the overdispersion parameter. Ver Hoef and Boveng [36] suggest plotting $(y-\mu)^2$ against μ to select between the two families. Such plots were produced and the relationships were quite similar for quasi-Poisson and negative binomial and failed to indicate a preference for one family over the other. We note that the estimated coefficients for the tide effect were quite similar for quasi-Poisson (1.8) and negative binomial (2.0), so choosing between the two families does not seem critical.

Based on AIC values, the zero inflated binomial models are the best of the three families for which AIC can be calculated, with the main effects model our preference (based on the above argument on comparing AIC values). Zero inflated models are mixture

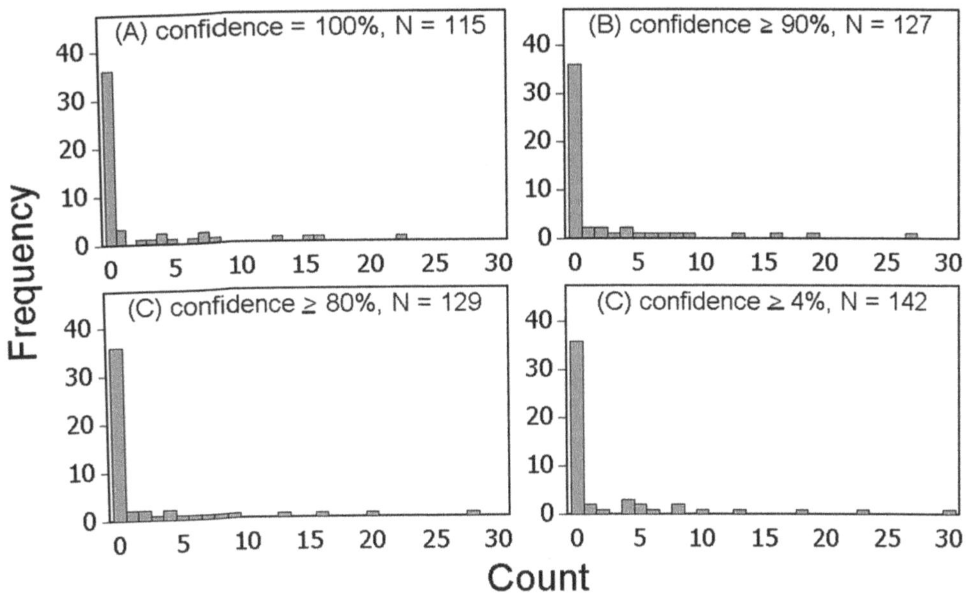

Figure 8. Frequency histograms of the count of delta smelt collected per sample for individuals that were correctly identified to species on SmeltCam images with (A) 100%, (B) ≥90%, (C) ≥80%, and (D) ≥4% confidence.

models of the following general form: π f(0)+(1- π) f(non-negative), where π is the probability that the data come from the degenerate distribution, f(0), where 0 is the only possible outcome and f(non-negative) is the probability distribution allowing 0′s and positive outcomes. The probability π was modeled according to a simple logistic model, log(π/(1- π)=Γ, while the negative binomial distribution was used for f(non-negative) with the expected counts modeled as functions of the factor levels. For all three sets of factorial combinations, π, was 0.61, i.e., there was at least a 61% probability of failing to catch any delta smelt (the probability of no delta smelt also includes the case where the negative binomial model yields a zero). This value seems reasonable given the observed 69% of zeros in the catches. Like the quasi-Poisson and negative binomial models, the sole significant factor was the main effect of the tide, with a similar coefficient of 1.7 for the flood level effect.

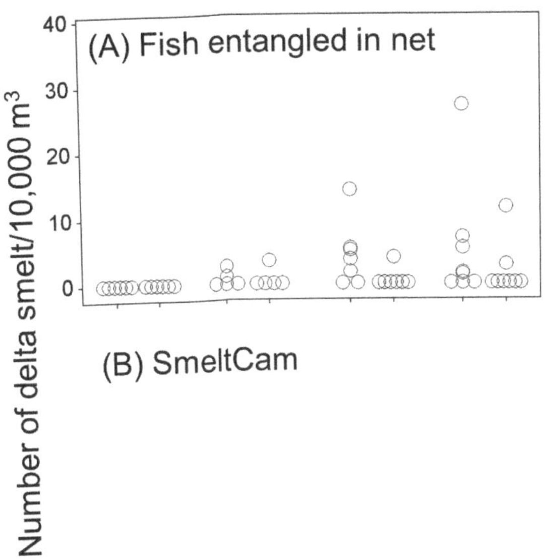

Figure 9. Individual values of delta smelt density (number/ 10,000 m³) by tide and position in the water column for fish that were (A) found entangled in the mesh of the net when it was retrieved and (B) observed by the SmeltCam.

Discussion

Our results suggest that the distribution of juvenile delta smelt in the water column varied across tides. We found that delta smelt

Table 2. Summary of model fitting results for the nested set of factors and four different families of distributions.

Factors	Po	Q-Po	NB	ZINB
H+V+T				
AIC	348	NA	164.5	162.8
Var	μ	7.6·μ	μ(1+μ·5.9)	0.15 μ (1+ μ·0.7)
Signif	H**,T**	T*	T**	T**
H:V+H:T+V:T				
AIC	313	NA	165.5	161.5
Var	μ	6.0·μ	μ(1+μ·4.8)	0.15 μ (1+ μ·0.4)
Signif	V**,T**,V:H**	T*	T**	T**
	V:T*,H:T*			
H*V*T				
AIC	314	NA	167.2	162.9
Var	μ	6.0·μ	μ(1+μ·4.7)	0.15 μ (1+ μ·0.4)
Signif	T**,H:T*	T'	–	T**

H+V+T is main effects only, H:V+H:T+V:T is main effects and 2-way interactions, and H*V*T is main effects, 2-way and 3-way interactions. Po = Poisson, Q-Po = Quasi-Poisson, NB = Negative Binomial, ZINB = Zero Inflated Negative Binomial. The Var entries refer to the variance function for the distribution families. The Signif entries denote the significant factor effects, with superscripts ′,*, and ** denoting significance at the 0.10, 0.05, and 0.01 levels, respectively. The reported AIC value for the Q-Po case is the Quasi-AIC value.

were relatively common and abundant throughout the water column during flood tides. However, during ebb tides delta smelt were less abundant, and were observed only in the lower half of the water column and sides of the channel. This pattern emerged from both the fish observed by the SmeltCam and those that were found entangled in the mesh of the net after it was retrieved. With regard to this specific pattern, there is no bias associated with not knowing the exact depth at which the fish entangled in the mesh of the net were captured because none were captured during ebb tides in the center of the channel (Fig. 9). Interestingly, variability in salinity and turbidity exhibited the same general pattern as did delta smelt, and may be the proximal reason for the distributions observed. The performance of the SmeltCam degrades when turbidity exceeds approximately 80 NTU. Turbidity reached 80 NTU in one of fifty-two samples we collected; a flood tide treatment from the lower half of the water column and side of the channel. Seven delta smelt were observed in this sample. It is possible that more delta smelt were actually present but were missed by the SmeltCam because of the elevated turbidity. If this were true, it would be consistent with the overall pattern of delta smelt distribution and would not have changed the results. Salinity and turbidity are both important components of delta smelt physical habitat [22,23]. We hypothesize that delta smelt, by simply remaining within preferred turbidity and salinity conditions across tides, could have produced much of the pattern observed. However, it appeared that delta smelt manipulate their position in the water column either through keying in on these water quality conditions or the physics underlying them.

Our results were consistent with other studies in 2010 and 2012 that found delta smelt to be abundant on flood tides but not on ebb tides (J. Burau, U.S. Geological Survey, Sacramento, California, unpublished data), although the studies were conducted under very different net flow conditions. The general consistency in results across studies in multiple years and the observation that physical habitat is a likely underlying mechanism, together provide strong evidence that delta smelt were not randomly distributed in the water column across tides. However, the extent to which the pattern observed at this location holds true at night or at other locations is uncertain. Although we did not sample at night, the delta smelt movements study did and found no difference in catch patterns compared to the day.

There is sufficient evidence to suggest that the patterns of delta smelt distribution observed in this particular location are not applicable across the entire system. Our unpublished analyses (separate independent analyses by FF, KBN and DS) of the FMT data set demonstrate a high degree of variability in delta smelt catches across tides among sampling sites. It therefore seems likely that localized physical as well as biological components of habitat influence delta smelt distribution across the system. The likelihood that delta smelt distribution in the water column across tides varies spatially in the system has important implications for the original motivation for our study. We generated information relevant to informing methods that could be developed to extrapolate survey data to generate more precise population estimates of delta smelt. However, it is clear that additional study is needed to characterize variability at other locations in order to successfully revise present methods of estimating delta smelt population size. Areas of particular interest are the broad expansive shallow water shoals located in Suisun, Grizzly and Honker bays (Fig. 3).

There is also sufficient data to suggest that the patterns of distribution we observed do not hold true for other life stages of delta smelt. A previous study of delta smelt larvae found no effect of tide on vertical distribution [37]. Similar to above, our unpublished analyses (separate independent analyses by KBN and DS) of post-larval and juvenile delta smelt long term monitoring data sets (CDFW's 20 mm and Summer Townet Surveys) demonstrate a high degree of variability in delta smelt catches across tides among sampling sites. Interestingly, previous studies demonstrated that the larvae of several native and exotic fishes (other than delta smelt) in San Francisco Estuary appeared to be behaviorally flexible in maintaining vertical position under different environmental conditions to maximize retention [38]. Together, these observations suggest that fish distribution in the water column varies according to localized habitat conditions.

Tidal movements, migrations and transport are well documented in systems worldwide and are usually associated with exploiting favorable habitats [39,40]. Invertebrates such as penaeid shrimp are well known to selectively move or migrate with tides [41] as are fish. For example, flounder larvae *P. flesus* entered the water column on flood tides to move upstream into the Elbe River Estuary, Germany [42]. Similarly, plaice larvae *Pleuronectes platessa* accomplish passive but selective horizontal transport by entering the water column during flood tides and remaining on the bottom during ebb tides [43]. Studies in a tropical tidal mangrove have also shown that fish were distributed on the bottom during ebb tides and entered the water column during flood tides to exploit intertidal habitats [44]. For delta smelt, it appears that individuals manipulate their position in the water column to facilitate either movement or retention at different life stages. As alluded to above, we believe that the patterns we observed for juvenile delta smelt facilitate retention in favorable habitats. However, upstream migration of adults and downstream migration of larvae is undoubtedly facilitated by tidal transport and net flows. As mentioned, a previous study found no effect of tide on the vertical distribution of delta smelt larvae [37], suggesting they may be passively transported downstream by net flows until reaching favorable habitats near the low salinity zone where they effectively maintain position [27], potentially by manipulating their position in the water column. As a case in point, anadromous rainbow smelt *Osmerus mordax* larvae are known to vertically migrate to maintain position in regions of high prey density [45,46]. The ability of young fishes to change their vertical distribution ontogenetically [47,48] or in response to varying net flow conditions [38] appears to be a common strategy for retention in favorable habitats.

There are several advantages and disadvantages of the SmeltCam versus similar camera systems that have been used previously to photograph objects passing through the codend of nets. Silhouette photography or video recording devices have been developed for examining plankton distributions at various scales [3,6,9,49,50]. Perhaps the biggest limitation of the present generation SmeltCam is that it has been designed for a particular size range of fish. Modifications to the system could be made for sampling large-bodied fishes with larger nets or for sampling smaller organisms such as fish larvae or other planktonic organisms. An advantage of the SmeltCam is the incorporation of new technology enabling rapid digital photography, automatic object recognition, automatic data collection, and real time observation. We are actively working on improving the system in several ways including redesigning the frame and hull to make the unit lighter and easier to handle by a single individual, enhancing software to improve and expand automatic image recognition and adding size measurements, incorporating tools to record water quality parameters in real time during sampling, and matching the unit with trawled nets with dimensions that will decrease the entanglement of fishes and improve survival.

One potential concern associated with the SmeltCam is the possibility of predation occurring inside the unit given its dimensions and light emission, both of which could potentially attract and congregate predators such as striped bass. Although we did not conduct an exhaustive investigation on the topic, we found no evidence of predation attributed to the SmeltCam during our study. Immediately upon collection we sacrificed and examined the stomach contents of 5 striped bass (that were either entangled in the net or couldn't fit through the opening of the SmeltCam) that were large enough to consume fish. None of the stomachs contained any fish remains; one stomach contained one isopod and the other four stomachs were empty.

Our study demonstrates how new technology can be developed to address key questions and uncertainties in conservation biology, and that imperiled species can be studied with relatively little harm. During our study we observed a total of 197 individual delta smelt. Of this total, 142 individuals passed through the SmeltCam alive while 55 died as a result of getting entangled in the mesh of the net. Thus, with the application of the SmeltCam in this study we increased the survival of individual delta smelt by 72% compared to using a traditional codend where all of the fish would have likely died due to handling stress. Survival is likely to increase in future studies as the SmeltCam is matched with nets with dimensions that will decrease entanglement. The SmeltCam can be affixed to virtually any type of trawled net or other fish congregating device facilitating a broad array of potential applications. The development and application of new technology such as the SmeltCam provides many new opportunities to studying imperiled species such as delta smelt and can be readily applied to other species and systems.

Acknowledgments

This study was conducted under the auspices of the Interagency Ecological Program for the San Francisco Estuary (IEP). It was made possible by the efforts of many individuals from many organizations and truly exemplified the collaborative mission of the IEP. We thank the members of the IEP Management Team, especially IEP Lead Scientist Anke Mueller-Solger and IEP Program Manager Kelly Souza, for their support. Ted Swift of the California Department of Water Resources (DWR) was instrumental in the early stages of developing the concept for the SmeltCam. Gina Benigno and Nick Van Ark of DWR provided invaluable laboratory and field support during the development of the SmeltCam. David Fullerton of the Metropolitan Water District of Southern California contributed to the development of the study design. The field work was successful because of the skills and dedication of Dave Hull, Ken Flowers and Katie Osborn from the CDFW. Karen Gerhts of DWR graciously provided the YSI sonde for the field study. Discussions with Wim Kimmerer of San Francisco State University and Pete Smith of the US Geological Survey (USGS) assisted with model fitting and interpretation. We thank Larry Brown of USGS, Peter Moyle of UC Davis, Brendan Ebner of James Cook University, and two anonymous reviewers for comments that improved the manuscript. The views and opinions expressed in this paper are those of the authors and do not necessarily reflect the official policy or position of their respective agencies.

Author Contributions

Conceived and designed the experiments: FF DP KBN TS DC RB SBS EVN. Performed the experiments: FF DP DO DC SBS. Analyzed the data: FF DP DO KBN DS. Contributed reagents/materials/analysis tools: DO DS. Wrote the paper: FF DP DO KBN TS DC RB SBS DS EVN.

References

1. Thompson W (Editor) (2004) Sampling rare or elusive species: concepts, designs, and techniques for estimating population parameters. Island Press.
2. MacKenzie DI, Nichols JD, Sutton N, Kawanishi K, Bailey LL (2005) Improving inferences in population studies of rare species that are detected imperfectly. Ecology 86: 1101–1113.
3. Lenz J, Schnack D, Petersen D, Kreikemeier J, Hermann B, et al. (1995) The Ichthyoplankton Recorder: a video recording system for in situ studies of small-scale plankton distribution patterns. ICES Journal of Marine Science: Journal du Conseil 52: 409–417.
4. Swann DE, Cutler TL (1999) Using Remote Photography in Wildlife Ecology: A Review. Wildlife Society Bulletin 27: 571–581.
5. Swann DE, Hass CC, Dalton DC Wolf SA (2004) Infrared-Triggered Cameras for Detecting Wildlife: An Evaluation and Review Wildlife Society Bulletin 32: 357–365.
6. Culverhouse PF, Williams R, Benfield M, Flood PR, Sell AF, et al. (2006) Automatic image analysis of plankton: future perspectives. Marine Ecology Progress Series 312: 297–309.
7. Watson DL, Harvey ES, Anderson MJ, Kendrick GA (2005) A comparison of temperate reef fish assemblages recorded by three underwater stereo-video techniques. Marine Biology 148: 415–425.
8. Willis TJ, Babcock RC (2000) A baited underwater video system for the determination of relative density of carnivorous reef fish. Marine and Freshwater Research 51: 755–763.
9. Olney JE, Houde ED (1993) Evaluation of in situ silhouette photography in investigations of estuarine zooplankton and ichythyoplankton. Bulletin of Marine Sciences 52: 845–872.
10. Morrison M, Carbines G (2006) Estimating the abundance and size structure of an estuarine population of the sparid *Pagrus auratus*, using a towed camera during nocturnal periods of inactivity, and comparisons with conventional sampling techniques. Fisheries Research 82: 150–161.
11. Becker A, Cowley PD, Whitfield SK (2010) Use of remote underwater video to record littoral habitat use by fish within a temporarily closed South African estuary. Journal of Experimental Marine Biology and Ecology 391: 161–168.
12. Frezza TL, Carl LM, Reid SM (2003) Evaluation of a portable underwater video camera to study fish communities in two Lake Ontario tributaries. Journal of Freshwater Ecology 18: 269–276.
13. Daum DW (2005) Monitoring fish wheel catch using event-triggered video technology. North American Journal of Fisheries Management 25: 322–328.
14. Jordan F, Jelks HL, Bortone SA, Dorazio RM (2008) Comparison of visual survey and seining methods for estimating abundance of an endangered, benthic stream fish. Environmental Biology of Fishes 81: 313–319.
15. Ellender BR, Becker A, Weyl OLF, Swartz ER (2012) Underwater video analysis as a non-destructive alternative to electrofishing for sampling imperiled headwater stream fishes. Aquatic Conservation: Marine and Freshwater Ecosystems 22: 58–65.
16. Moyle PB, Herbold B, Stevens D, Miller LW (1992) Life history and status of delta smelt in the Sacramento-San Joaquin Estuary, California. Transactions of the American Fisheries Society 121: 67–77.
17. Bennett WA (2005) Critical assessment of the delta smelt population in the San Francisco Estuary, California. San Francisco Estuary and Watershed Science. Online serial: http://repositories.cdlib.org/jmie/sfews/vol3/iss2/art1.
18. Sommer T, Armor C, Baxter R, Breuer R, Brown L, et al. (2007) The collapse of pelagic fishes in the upper San Francisco Estuary. Fisheries 32: 270–277.
19. Mac Nally, R Thompson JR, Kimmerer WJ, Feyrer F, Newman K, et al. (2010) An analysis of pelagic species decline in the upper San Francisco Estuary using multivariate autoregressive modeling. Ecological Applications 20: 1417–1430.
20. Swanson C, Mager RC, Doroshov SI, Cech JJ (1996) Use of salts, anesthetics and polymers to minimize handling and transport mortality in delta smelt. Transactions of the American Fisheries Society 125: 326–329.
21. Castillo G, Morinaka J, Lindberg J, Fujimura R, Baskerville-Bridges B, et al. (2012) Pre-screen loss and fish facility efficiency for delta smelt at the south Delta's State Water Project, California. San Francisco Estuary and Watershed Science 10(4): 1–23.
22. Feyrer F, Newman K, Nobriga M, Sommer T (2011) Modeling the effects of future outflow on the abiotic habitat of an imperiled estuarine fish. Estuaries and Coasts 34: 120–128.
23. Feyrer F, Nobriga M, Sommer T (2007) Multi-decadal trends for three declining fish species: habitat patterns and mechanisms in the San Francisco Estuary, California, U.S.A. Canadian Journal of Fisheries and Aquatic Sciences 64: 723–734.
24. Newman K (2008) Sample design-based methodology for estimating delta smelt abundance. San Francisco Estuary and Watershed Science. Online serial: http://repositories.cdlib.org/jmie/sfews/vol6/iss3/art3/.
25. Thomson J, Kimmerer W, Brown L, Newman K, Mac Nally R, et al. (2009) Bayesian change-point analysis of temporal patterns in fish abundances in the upper San Francisco Estuary. Ecological Applications 20: 1431–1448.
26. Sommer T, Nobriga M, Grimaldo L, Feyrer F, Mejia F (2011) The spawning migration of delta smelt in the upper San Francisco Estuary. San Francisco Estuary and Watershed Science (9)2 http://escholarship.org/uc/item/86m0g5sz.
27. Dege M, Brown LR (2004) Effect of outflow on spring and summer distribution and abundance of larval and juvenile fishes in the upper San Francisco Estuary. In: Feyrer F, Brown LR, Brown RL, Orsi JJ, editors. Early Life History of Fishes in the San Francisco Estuary and Watershed. American Fisheries Society, Symposium 39, Bethesda, Maryland. 49–65.
28. Grimaldo L, Sommer T, Van Ark N, Jones G, Holland E, et al. (2009) Factors affecting fish entrainment into massive water diversions in a tidal freshwater

estuary: can fish losses be managed? North American Journal of Fisheries Management 29: 1253–1270.
29. Stevens DE, Miller LW (1983) Effects of river flow on abundance of young chinook salmon, American shad, longfin smelt, and delta smelt in the Sacramento-San Joaquin River system. Transactions of the American Fisheries Society 3: 425–437.
30. Draper AJ, Munevar A, Arora S, Reyes E, Parker N, et al. (2004) CalSim: Generalized model for reservoir system analysis. Journal of Water Resource Planning and Management 130: 480–489.
31. Martin TG, Wintle BA, Rhodes JR, Kuhnert PM, Field SA, et al. (2005) Zero tolerance ecology: improving ecological inference by modelling the source of zero observations. Ecology Letters 8: 1235–1246.
32. R Development Core Team (2008) R: A language and environment for statistical computing. R Foundation for Statistical Computing, Vienna, Austria. ISBN 3-90005107-0. Available: http://www.R-project.org.
33. Zeileis A, Kleiber C, Jackman S (2008) Regression models for count data in R. Journal of Statistical Software 27(8). Available: http://www.jstatsoft.org/v27/i08/.
34. Jackman S (2012) pscl: Classes and methods for R developed in the Political Science Computational Laboratory, Stanford University. Department of Political Science, Stanford University. Stanford, California. R package version 1.04.4. Available: http://pscl.stanford.edu/.
35. Burnham KP, Anderson DR (2002) Model Selection and Multimodel Inference: A Practical Information-Theoretical Approach (2nd Edition). Springer-Verlag, New York, NY.
36. Ver Hoef JM, Boveng PL (2007) Quasi-poisson vs. negative binomial regression: how should we model overdispresed count data? Ecology 2766–2772.
37. Rockriver A (2004) Vertical distribution of larval delta smelt and striped bass near the confluence of the Sacramento and San Joaquin Rivers. In: Feyrer F, Brown LR, Brown RL, Orsi JJ, editors. Early Life History of Fishes in the San Francisco Estuary and Watershed. American Fisheries Society, Symposium 39, Bethesda, Maryland. 97–108.
38. Bennett WA, Kimmerer WJ, Burau JR (2002) Plasticity in vertical migration by native and exotic estuarine fishes in a dynamic low-salinity zone. Limnology and Oceanography 47: 1496–1507.
39. Norcross BL, Shaw RF (1984) Oceanic and estuarine transport of fish eggs and larvae: a review. Transactions of the American Fisheries Society 113: 153–165.
40. Gibson RN (2003) Go with the flow: tidal migrations in marine animals. Hydrobiologia 503: 153–161.
41. Hughes DA (1969) Responses to salinity change as a tidal transport mechanism of pink shrimp, *Penaeus duorarum*. The Biological Bulletin 136: 43–53.
42. Bos AR (1999) Tidal transport of flounder larvae (*Platichthys flesus*) in the Elbe River, Germany. Archive of Fishery and Marine Research 47: 47–60.
43. Rijnsdorp AD, Van Stralen M, Van Der Veer HW (1985) Selective tidal transport of North Sea plaice larvae *Pleuronectes platessa* in coastal nursery areas. Transactions of the American Fisheries Society 114: 461–470.
44. Krumme U (2004) Patterns in tidal migration of fish in a Brazilian mangrove channel as revealed by a split-beam echosounder. Fisheries Research 70: 1–15.
45. Dodson JJ, Dauvin JC, Ingram RG, d'Anglejan B (1989) Abundance of larval rainbow smelt (*Oserus mordax*) in relation to the maximum turbidity zone and associated macroplanktonic fauna of the middle St. Lawrence Estuary. Estuaries 12: 66–81.
46. Dauvin JC, Dodson JJ (1990) Relationship between feeding incidence and vertical and longitudinal distribution of rainbow smelt larvae (*Osmerus mordax*) in a turbid well-mixed estuary. Marine Ecology Progress Series 60: 1–12.
47. Laprise R, Dodson JJ (1989) Ontogeny and importance of tidal vertical migrations in the retention of larval smelt *Osmerus mordax* in a well-mixed estuary. Marine Ecology Progress Series 55: 101–111.
48. Laprise R, Dodson JJ (1989) Ontogentic changes in the longitudinal distribution of two species of larval fish in a turbid well-mixed estuary. Journal of Fish Biology 35(Supplement A)39–47.
49. Ortner PB, Hill LC, Edgerton HE (1981) In situ silhouette photography of Gulf Stream zooplankton. Deep Sea Research 28: 1569–1576.
50. Graham DL, Martin DL, Martin JC (2003) *In situ* quantification and analysis of large jellyfish using a novel video profiler. Marine Ecology Progress Series 254: 129–140.

Effects of Elevated CO_2 and Nitrogen Deposition on Ecosystem Carbon Fluxes on the Sanjiang Plain Wetland in Northeast China

Jianbo Wang[1,2], Tingcheng Zhu[1], Hongwei Ni[1,2]*, Haixiu Zhong[2], Xiaoling Fu[2], Jifeng Wang[2]

1 Key Laboratory for Vegetation Ecology, Ministry of Education, Institute of Grassland Science, Northeast Normal University, Changchun, Jilin Province, P. R. China,
2 Institute of Natural Resources and Ecology, Heilongjiang Academy of Sciences, Harbin, Heilongjiang Province, P.R. China

Abstract

Background: Increasing atmospheric CO_2 and nitrogen (N) deposition across the globe may affect ecosystem CO_2 exchanges and ecosystem carbon cycles. Additionally, it remains unknown how increased N deposition and N addition will alter the effects of elevated CO_2 on wetland ecosystem carbon fluxes.

Methodology/Principal Findings: Beginning in 2010, a paired, nested manipulative experimental design was used in a temperate wetland of northeastern China. The primary factor was elevated CO_2, accomplished using Open Top Chambers, and N supplied as NH_4NO_3 was the secondary factor. Gross primary productivity (GPP) was higher than ecosystem respiration (ER), leading to net carbon uptake (measured by net ecosystem CO_2 exchange, or NEE) in all four treatments over the growing season. However, their magnitude had interannual variations, which coincided with air temperature in the early growing season, with the soil temperature and with the vegetation cover. Elevated CO_2 significantly enhanced GPP and ER but overall reduced NEE because the stimulation caused by the elevated CO_2 had a greater impact on ER than on GPP. The addition of N stimulated ecosystem C fluxes in both years and ameliorated the negative impact of elevated CO_2 on NEE.

Conclusion/Significance: In this ecosystem, future elevated CO_2 may favor carbon sequestration when coupled with increasing nitrogen deposition.

Editor: Ben Bond-Lamberty, DOE Pacific Northwest National Laboratory, United States of America

Funding: This research was supported by the National Natural Science Foundation of China (30970452), the Heilongjiang Province Funds for Distinguished Young Scientists (JC03-09), Fundamental Research Funds for the Finance Department of Heilongjiang Province (2012EFA-02), and the Natural Science Foundation of Heilongjiang Province (C201032). The funders had no role in study design, data collection and analysis, decision to publish, or preparation of the manuscript.

Competing Interests: The authors have declared that no competing interests exist.

* E-mail: nihongwei2000@163.com

Introduction

Atmospheric CO_2 concentrations are predicted to double by the end of the century [1]. Increasing CO_2 concentrations usually stimulate ecosystem C uptake through photosynthesis [2,3]; however, concurrent increase in ecosystem respiration due to increasing root respiration and enhanced priming of soil organic matter may lead to ecosystem carbon losses [4,5]. The difference between photosynthetic and respiratory fluxes is called net ecosystem exchange (NEE) [6]. Thus, NEE depends on the relative changes in C uptake and release, but the extent of variance in C uptake and release in response to elevated CO_2 is uncertain.

In addition, N level are expected to vary in response to global N enrichment caused by anthropogenic activities [7]. According to one estimate, by 2050, 200 Tg N yr^{-1} will be released and then deposited into the earth's surface [8]. In general, N is a limiting factor for plant growth and net primary production. Consequently, it has been conjectured that N deposition could increase carbon dioxide uptake from the atmosphere [7,9]. However, different ecosystem respiration responses to N addition have been reported in previous studies, including increases [10–12], decreases [13,14] and no significant changes [15]. Therefore, it remains unclear whether N deposition leads to net ecosystem C sequestration [16,17].

The progressive nitrogen limitation (PNL) hypothesis [18] suggests that N additions should enhance CO_2 effects on plant productivity. Consequently, concurrent changes in atmospheric CO_2 concentration and N deposition may potentially bring about complex interactive impacts on ecosystem functioning. There has been much research on the effects of elevated atmospheric CO_2 concentrations and N deposition on the growth of individual plants, but relatively little research on the impact of ecosystem C fluxes on natural vegetation at ecosystem scale [4,19–21]. In particular, there has been no detailed study evaluating the interactive effect of elevated CO_2 and N addition on ecosystem C uxes on wetland ecosystems. Wetlands have globally significant stores of soil carbon and sink for CO_2 [22], and consequently play an important role in the global carbon cycle.

A field experiment manipulating elevated CO_2 and N deposition was conducted, beginning in May 2010, to examine the potential influences of climate change on a temperate wetland in northeast China. The objectives of this study are to address the following three questions: (1) How do elevated CO_2 and N deposition impact GPP, ER and NEE? (2) Whether elevated CO_2 and N deposition interactively affect ecosystem carbon flux on the Sanjiang Plain wetland? (3) How do environmental (e.g., soil temperature, soil water content) and biotic (e.g., vegetative cover) factors regulate the responses of the ecosystem carbon flux to elevated CO_2 and N deposition?

Materials and Methods

Ethics Statement

All work was undertaken with relevant permissions from the Honghe National Nature Reserve, China for our observational and field studies. The field studies did not involve endangered or protected species. Data will be made available upon request.

Study Site

This study was conducted at the Field Experimental Establishment of Institute of Natural Resources and Ecology—the Honghe National Nature Reserve (47°49′N, 133°40′E) on the Sanjiang Plain, a wetland located in the Heilongjiang Province of Northeastern China. A low floodplain was formed by the Heilongjiang River, the Songhuajiang River and the Wusulijiang River. Sanjiang Plain is approximately 1.04×104 km^2 in area and is N deficient, as indicated by the strong stimulation of gross primary productivity after N addition [23]. Annual N deposition is 7.6 kg hm^{-2} in this region [24]. The area has a temperate continental monsoon climate. The mean annual temperature ranges from 1.4 to 2.9°C, with the average maximum of 21–22°C occurring in July and the average minimum −18°C in January. The mean annual precipitation is 500–650 mm and 80% of rainfall occurs in summer. The frost-free period is 120–140 days. The growing season is limited to late April to early October. Meadow soil is the main soil type with 20% of soil organic matter content and 0.5% of total N at a 0–10 cm soil depth. On the Sanjiang Plain wetland, the native vegetation is characterized by the dominance of the perennial species *Calamagrostis angustifolia*, as well as the presence of *Carex appendiculata*, *Carex lasiocarpa*, *Carex pseudocuraica*, *Glyceria spiculosa* and *Lathyrus quinqunervuius*.

Open Top Chamber Design

Three pairs of Open Top Chambers (OTC), each 1.8 m high with an octagonal ground surface area of 10.78 m^2, were set up in early May 2010. They were constructed with a 5-m-wide buffer zone between them to prevent cross contamination. Light intensity in each chamber was 95% of that in open space. The precipitation intensity was identical inside and outside of the chambers. We supplied carbon dioxide to the chambers through a pipe with pinholes connected to industrial CO_2 tanks outside the chambers. We adjusted the CO_2 supply in accordance with wind speed and CO_2 concentrations by taking constant measurements with an infrared gas analyzer.

Experimental Design

The experiment used a paired, nested design with elevated CO_2 as the primary factor and nitrogen addition as the secondary factor. Three pairs of OTC were used, and within each pair one was randomly assigned to receive the elevated CO_2 treatment (700 ppm, E) and the other as the control (370 ppm, C). The CO_2 enrichment began in May 2010 and continued each year for the entire growing season (from May to September) during the daytime (6:00 a.m. to 6:00 p.m.). The CO_2 supply was shut off during the night. At each elevated CO_2 and control OTC chamber, there were two subplots, one control and the other N addition (N). To each N addition subplots, 8 g N m^{-2} yr^{-1} (N, in the form of liquid NH_4NO_3) were applied once a year in early May, while the controls were sprayed with distilled water only. Thus, there were four treatments, elevated [CO_2] and enriched N deposition (EN), elevated [CO_2] and ambient N deposition (E), ambient CO_2 and enriched N deposition (N) and ambient CO_2 and ambient N deposition as a control (C). Each treatment was replicated three times. The experiments lasted two growing seasons.

Climate Data Soil Microclimate

Climate data were obtained from a Hobo meteorological station, which was approximately 250 m distant from the experimental plots. Soil temperature (T_{soil}) at a 10 cm depth and soil water content (SWC) at a 0–10 cm soil depth were continuously recorded with an ECH_2O dielectric aquameter (EM50/R Decagon Ltd, Pullman, WA, USA) at 2-h intervals from May 2010 to September 2011 (except when the sample plots flooded in August).

Ecosystem-Level Gas Exchange Measurement

Ecosystem gas exchange was measured on 7 sampling dates during the 2010 and 2011 growing season, respectively (May to September, except when the sample plots flooded). Measurements were performed under cloudless conditions from 9:00 a.m. to 11:00 a.m. to avoid the high air temperature and humidity at noon. In May 2010, one square aluminum frame (0.5 m×0.5 m) was inserted 3 cm into the soil of each subplot, approximately 0.5 m away from the edge. The frames provided a at base between the soil surface and the CO_2 sampling chamber. When ecosystem gas exchange was measured, we turned off the CO_2 supply system at least 10 min prior so that the CO_2 concentration was consistent in chamber. Ecosystem gas exchange was measured with a transparent chamber (0.5×0.5×1.0 m) attached to an infrared gas analyzer (IRGA; LI-6400, Li-Cor, Lincoln, NE, USA), which was placed on the frame of each subplot. Four small electric fans were running continuously to promote the mix of air within the chamber during the measurement. Twelve consecutive recordings of the CO_2 concentration were taken at 10-s intervals during a 120-s period. The rates of CO_2 fluxes were calculated from the time-courses of the concentrations to determine net ecosystem exchange (NEE). Using the approach of Steduto [25], we converted the concentration change to flux-per-unit of soil area. After taking the NEE measurements, the chamber was vented, replaced on each frame and covered by an opaque cloth. The ER was then measured. Gross primary productivity (GPP) was then determined by the sum of NEE and ER (GPP = NEE+ER). Positive NEE values refer to net uptake of C by the ecosystem, and negative NEE values represent a net loss of C from the ecosystem.

Green Plant Cover

For each subplot, the percent cover estimate (percentage of ground area within a subplot covered by green vegetation) was taken. These plots were scored once each year in July.

Leaf Gas Exchange Measurement

Leaf net photosynthetic rate in the dominant plant species *C. angustifolia* was measured with an open gas-exchange system (Li-6400; Li-Cor, Lincoln, NE, USA). In each subplot, three fully expanded leaves of *C. angustifolia* were selected and measured once each year in July. Leaf gas exchange was measured in the mornings between 9:00 a.m. and 11:00 a.m. on clear days. During the measurement, leaves were illuminated at 1500 μmol m^{-2} s^{-1} using an LED light system.

Statistical Analyses

The seasonal mean values used in this study were calculated from the monthly mean values, which were first averaged using all measurements from the same month. We examined soil microclimate (T_{soil} and soil water content) and ecosystem C fluxes (NEE, ER and GPP) results for the entire 2010–2011 period and used repeated measures of ANOVAs to test for main effects and interactions, and whether these changed over time (contrasting both the measuring date and year). The between-subject effects were evaluated as elevated CO_2, N addition and their interactions, and within-subject-effects were measuring date, year and their interactions with elevated CO_2, and N addition. Regression with correction for autocorrelation and stepwise multiple linear analyses were used to examine the relationships of ecosystem C fluxes with soil temperature, soil water content and vegetation cover in the two growing seasons. Statistical analyses were conducted with SPSS software (SPSS Inc, Chicago, IL, USA), and figures were plotted with SigmaPlot 11.0 software.

Table 1. Growing season means of soil temperature (T_{soil}, °C), soil water content (SWC) and ecosystem C fluxes (NEE, ER and GPP, μmol $m^{-2}s^{-1}$) under different treatments in 2010 and 2011.

		T_{soil}	SWC	NEE	ER	GPP
2010	C	16.12±0.2	66.4±0.7	7.67±0.05	9.80±0.22	17.47±0.55
	N	16.45±0.3	66.1±1.6	8.61±0.30	11.09±0.47	19.71±0.58
	E	16.02±0.2	68.3±0.5	6.64±0.14	11.96±0.40	18.61±0.45
	EN	16.82±0.2	67.3±0.8	8.64±0.30	13.27±0.41	21.91±0.71
2011	C	14.86±0.2	72.4±1.5	4.93±0.05	5.47±0.23	10.40±0.26
	N	15.18±0.2	71.4±2.5	5.72±0.21	5.67±0.31	11.39±0.52
	E	14.52±0.3	73.8±2.6	4.10±0.14	6.39±0.24	10.49±0.38
	EN	14.92±0.4	72.6±3.5	5.57±0.21	6.78±0.28	12.35±0.49

Values represent the Means± SE. C: control plots; N: N addition; E: elevated CO_2; EN: combined elevated CO_2 and N addition.

Results

Microclimate Changes Induced by Elevated CO_2 and N Addition Treatments

In comparison with the long-term averages (1981–2010) of mean air temperature (MAT, 17.3°C) during the growing season, both 2010 (18.4°C) and 2011 (17.8°C) had higher MAT. However, during the growing season, 2010 had higher precipitation (495 mm) and 2011 had lower precipitation (364 mm) than the long-term mean precipitation (MP, 459 mm; Fig. 1). Furthermore, mean air temperature was obviously higher in 2010 than that in 2011 in the early growing season– the monthly mean temperature was higher by 0.95°C in May and 3.12°C in June (Fig. 1). Moreover, the average T_{soil} of 2010 was significantly higher than that of 2011 ($P<0.001$, Table 1, 2). N addition significantly increased T_{soil} but decreased soil water contents (SWC) ($P<0.05$). Soil water content was higher in the elevated CO_2 plots than in the control plots ($P<0.001$, Table 2).

Seasonal and Interannual Variations of Ecosystem CO_2 Fluxes

The seasonal dynamics of ecosystem C exchange followed a parabolic-like pattern with higher values in summer and lower values in spring and autumn (Fig. 2), which was similar to the air temperature change. The rates of ecosystem CO_2 flux (GPP, ER and NEE) were significantly lower in 2011 than in 2010 (Tables 1, 2; $P<0.001$). In the control plots, seasonal means for NEE, ER and GPP in 2010 were 56%, 79% and 68% greater than in 2011, respectively. The remarkable interannual variabilities in the ecosystem C fluxes might be partly attributed to differences in air temperature, especially in the early growing season.

Effects of Elevated CO_2 on Ecosystem C Fluxes

During the two years of the study, elevated CO_2 significantly decreased NEE by 15% ($P<0.01$) measured by using RMANOVA. In addition, elevated CO_2-induced decline in NEE was significantly lower in the enriched N plots than in ambient N plots. Elevated CO_2 even slightly increased NEE in the enriched N plots in 2010 (Fig. 3A).

Unlike with NEE, elevated CO_2 significantly enhanced ER and GPP by 20% and 4%, respectively (both $P<0.01$, Table 1, 2). Furthermore, there was a significant temporal shift in the ecosystem C fluxes over the entire period ($P<0.05$ for the $CO_2 \times D$, Table 2). In addition, there were significant interactions

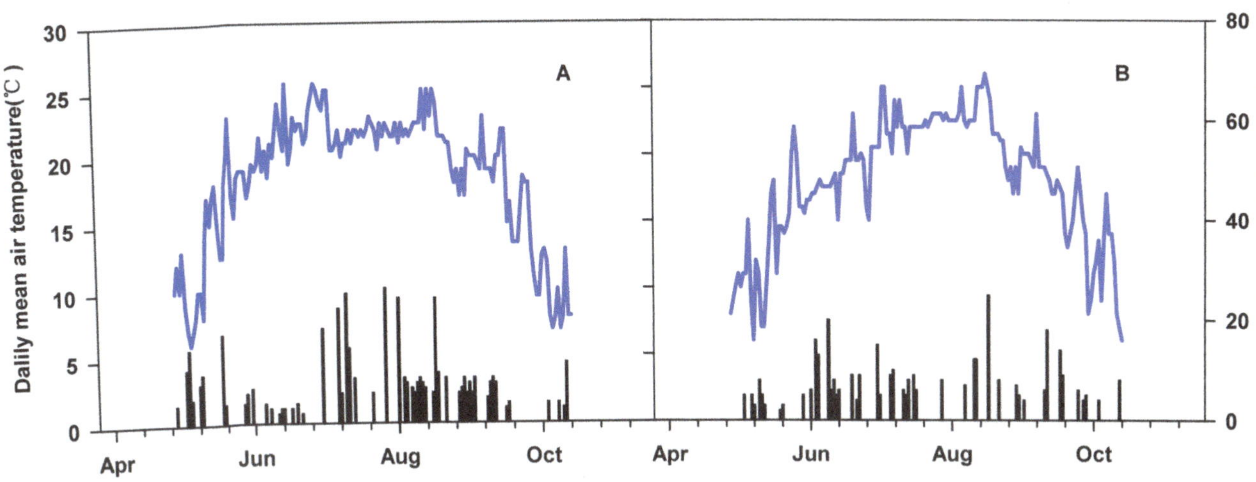

Figure 1. Daily mean air temperature (line) and daily precipitation (bars) in 2010 (A) and 2011(B) during the growing seasons. Data were obtained from a Hobo meteorological station approximately 250 m from the experimental plots.

Table 2. Results (*P* value) of repeat-measurement ANOVA on the effects of year (Y), elevated CO_2 (CO_2), N addition (N), measuring date (D), and their interactions on soil temperature (T_{soil}), soil water content (SWC), net ecosystem CO_2 exchange (NEE), ecosystem respiration (ER) and gross primary productivity (GPP).

Effect	T_{soil}	SWC	NEE	ER	GPP
D	<0.001	<0.001	<0.001	<0.001	<0.001
D×CO_2	0.567	<0.001	<0.001	<0.001	0.048
D×N	0.003	0.015	<0.001	<0.001	<0.001
D×Y×CO_2	0.038	<0.001	0.012	<0.001	<0.001
D×Y×N	0.709	0.071	<0.001	<0.001	0.016
D×CO_2×N	0.796	0.001	0.04	0.002	0.052
D×Y×CO_2×N	0.178	0.05	0.032	<0.001	0.145
Y	0.000	0.001	<0.001	<0.001	<0.001
CO_2	0.655	<0.001	0.003	<0.001	0.007
N	0.022	0.014	<0.001	0.008	<0.001
N×CO_2	0.447	0.598	0.007	0.851	0.196
Y×CO_2	0.247	0.804	0.972	0.046	0.13
Y×N	0.592	0.514	0.244	0.078	0.079
Y×CO_2×N	0.613	0.792	0.506	0.869	0.89

between elevated CO_2 and the years, which affected GPP and ER ($P<0.05$ for the CO_2×year). An increase in GPP, induced by elevated CO_2 was higher in the enriched N plots than in the ambient N plots in both years (Fig. 3 C).

Effects of N Addition on Ecosystem C Fluxes

Nitrogen addition significantly enhanced NEE, ER and GPP by 10%, 12% and 14%, respectively, (all $P<0.01$, Table 1, 2) during the two year study period. There was a significant temporal change in NEE, ER and GPP ($P<0.001$ for the N×D, Table 2) during the entire period. N-induced stimulation of seasonal average NEE and GPP was significantly higher in the elevated CO_2 plots than that in the ambient CO_2 plots in both years (Fig. 3B, D).

Interactive Effects of Elevated CO_2 and N Addition on Ecosystem C Fluxes

NEE was significantly affected by interactions between N addition and elevated CO_2, but no interactive effects on ER and GPP were detected (Table 2). In addition, there was a significant temporal shift in NEE, ER and GPP by CO_2 at ambient versus enriched N ($P=0.04$, $P=0.002$ and $P=0.052$, respectively, for the CO_2×N×D).

Vegetation Cover and Leaf -Level Gas Exchange

In agreement with the elevated CO_2 and N responses of NEE, elevated CO_2 decreased the vegetation cover by 22% in 2010 ($P<0.05$) and reduced the maximum photosynthetic rate (P_{max}) of the dominant species *C. angustifolia* by 16% in 2010 and 21% in 2011 ($P<0.05$). However, N addition enhanced vegetation cover by 26% in 2010 and 28% in 2011 ($P<0.05$) and increased P_{max} of *C. angustifolia* by 12% in 2010 and in 2011($P<0.05$). Moreover, the vegetation cover and P_{max} of *C. angustifolia* were higher in 2010 than in 2011 (Fig. 4A, B).

Controlling Factors of Ecosystem CO_2 Fluxes

Exponential relationships between ecosystem C uxes and soil temperature (T_{soil}) were found in both years (Fig. 5). T_{soil} explained 85%, 78% and 87% the variations in NEE, ER and GPP, respectively, in 2010. In 2011, soil temperature could explain 72%, 75% and 82% of the changes in NEE, ER and GPP, respectively. However, there was no significant relationship between ecosystem C fluxes and SWC (data not shown).

We found that both NEE ($R^2=0.74$ and $R^2=0.60$, $P<0.01$) and GPP ($R^2=0.36$, $P<0.01$ and $R^2=0.73$, $P<0.01$) were positively correlated with the plots' vegetation cover. By contrast, there were no significant relationships between ER and vegetation cover in both years (Fig. 6B).

Discussion

Seasonal and Interannual Variability of Ecosystem C Fluxes

The seasonal dynamics of ecosystem carbon uxes observed in this study were high in mid-summer and low in the early and late growing seasons, which reected air temperature (Fig. 1, 2). In addition, there were remarkable interannual variabilities in the ecosystem C fluxes (Fig. 2), which might be due to differences in air temperature in the early growing season. Specifically, the average air temperatures of May and of June were significantly lower in 2011 than in 2010. Consequently, a lower monthly NEE is expected in 2011, a hypothesis that would be consistent with Welp's findings [26], who found that increases in spring air temperatures caused both GPP to increase by 74% and ER by 61% during the early part of the growing season (i.e., April, May, and June) in deciduous forest. The study by Zhou [27] indicated that in a lawn ecosystem, CO_2 uptake significantly increased because of soil warming treatment. Moreover, in our study, ecosystem C fluxes were significantly exponentially related to soil temperature (Fig. 5). Less sensitive impact on ecosystem C fluxes to air temperature than soil temperature may be attributed to the wetland characteristics. In particular, the soil temperature is much lower than air temperature because the vegetation layer and the thick humus layer.

However, we found the C fluxes in 2010 are also higher than in 2011, when comparing fluxes recorded at the same soil temperature (Fig. 5), which implies that the vegetation cover partly accounted for the interannual differences in ecosystem C fluxes (Fig. 6).

In general, ecosystem C fluxes are closely related to SWC in grassland ecosystem [28,29]. However, we found no significant relationship between ecosystem C fluxes and SWC in our freshwater wetland. A lack of SWC impact on ecosystem C fluxes may be due to the wetland soil and plant roots, which are water-saturated during the growing seasons, thus, the SWC uctuations have slight effects on ecosystem C fluxes. This was consistent with some previous findings [30,31].

Elevated CO_2 Effects on Ecosystem CO_2 Fluxes

Our study provides evidence that elevated CO_2 can decrease NEE in this temperate wetland in northeast China because the changes in ER were more significant than changes in GPP for controlling the sign and magnitude of NEE (GPP and ER enhanced by 4% and 20%, respectively). This finding is consistent with a semi-arid grassland exposure to elevated CO_2 [28]. In both years of our study, both CO_2 treatments were net CO_2 sinks, but exposure to elevated CO_2 reduced CO_2 sink strength by 15% over the two years. This negative effect of elevated CO_2 on the net C uptake was in line with the trend observed in the Mojave Desert

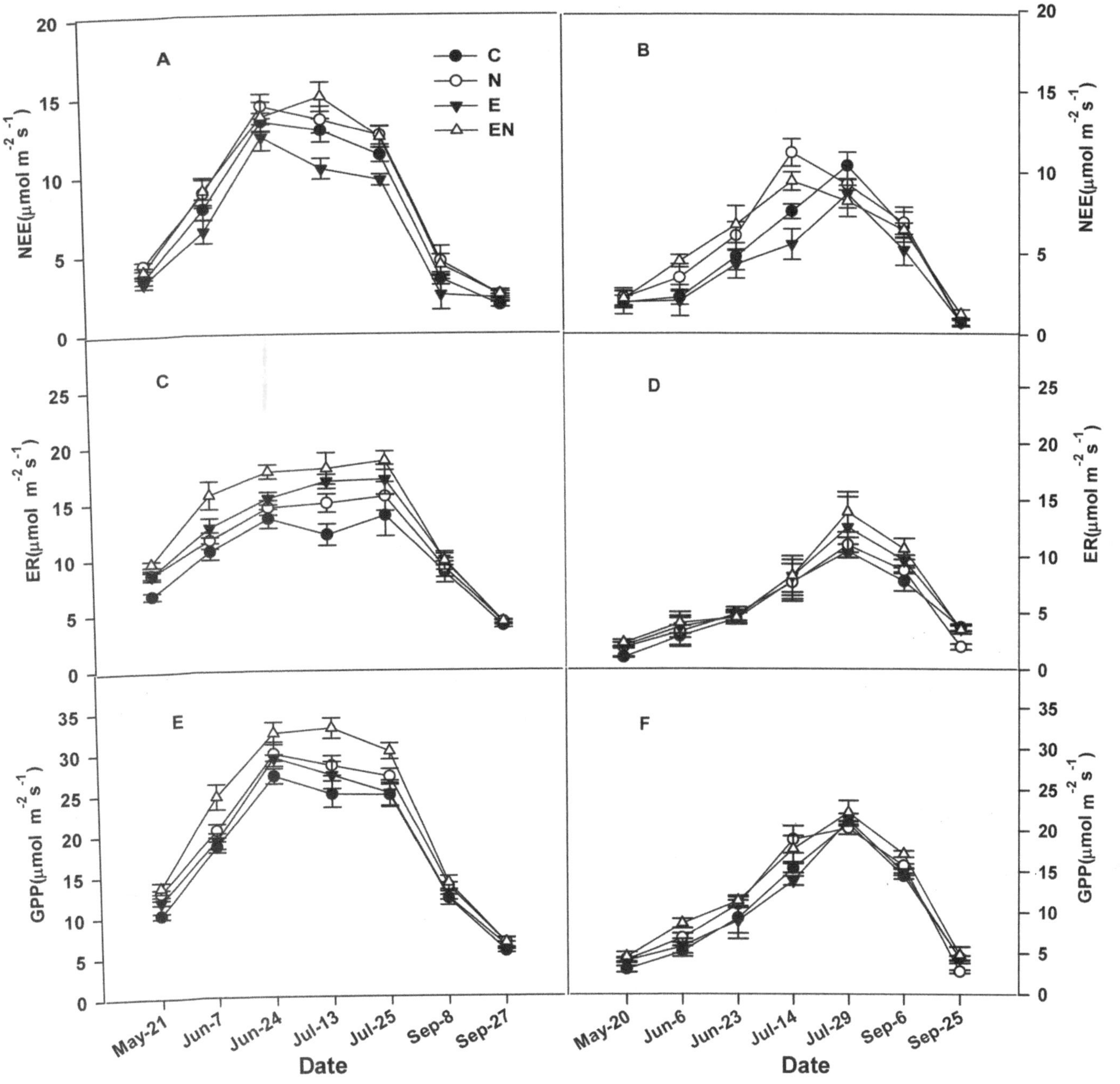

Figure 2. Seasonal dynamics of net ecosystem CO_2 exchange (NEE); ecosystem respiration (ER) and gross primary productivity (GPP) in 2010 (left) and 2011 (right) (August values are missing because of flooding on the experiment plots). C: control plots; N: N addition; E: elevated CO_2; EN: combined elevated CO_2 and N addition.

shrub ecosystem, which was exposed to elevated CO_2 for eight years, resulting in a CO_2 sink that was decreased by 30% [32]. The following two factors should, in theory, have affected the decline of NEE under elevated CO_2 in our study.

(1) Leaf photosynthetic capacity and vegetation cover. Elevated CO_2 decreased both the vegetation cover that was fumigated for two months and the light-saturated photosynthesis of the leaves of the dominant species (*C. angustifolia*) (Fig. 4), which might be a reason for the decline in NEE. In addition, elevated CO_2 decreased the chlorophyll concentration of *C. angustifolia* in cultivation experiments [33], which provided further evidence for the decline of leaf photosynthetic capacity. These results demonstrate that photosynthetic acclimation of leaf-level under elevated CO_2 may simultaneously happen at the ecosystem scale.

(2) Ecosystem C fluxes. In our study, despite the enhancd GPP response to elevated CO_2 for most of experimental period (except for 23 June and 14 July 2011), elevated CO_2 stimulated ER consistently to a greater extent compared to GPP during the experiment (Fig. 2, Table 1), resulting in reduced NEE, relative to control plots. Elevated CO_2 led to higher ER and soil respiration (data unpublished), which might be explained by increased carbon inputs, an enhanced

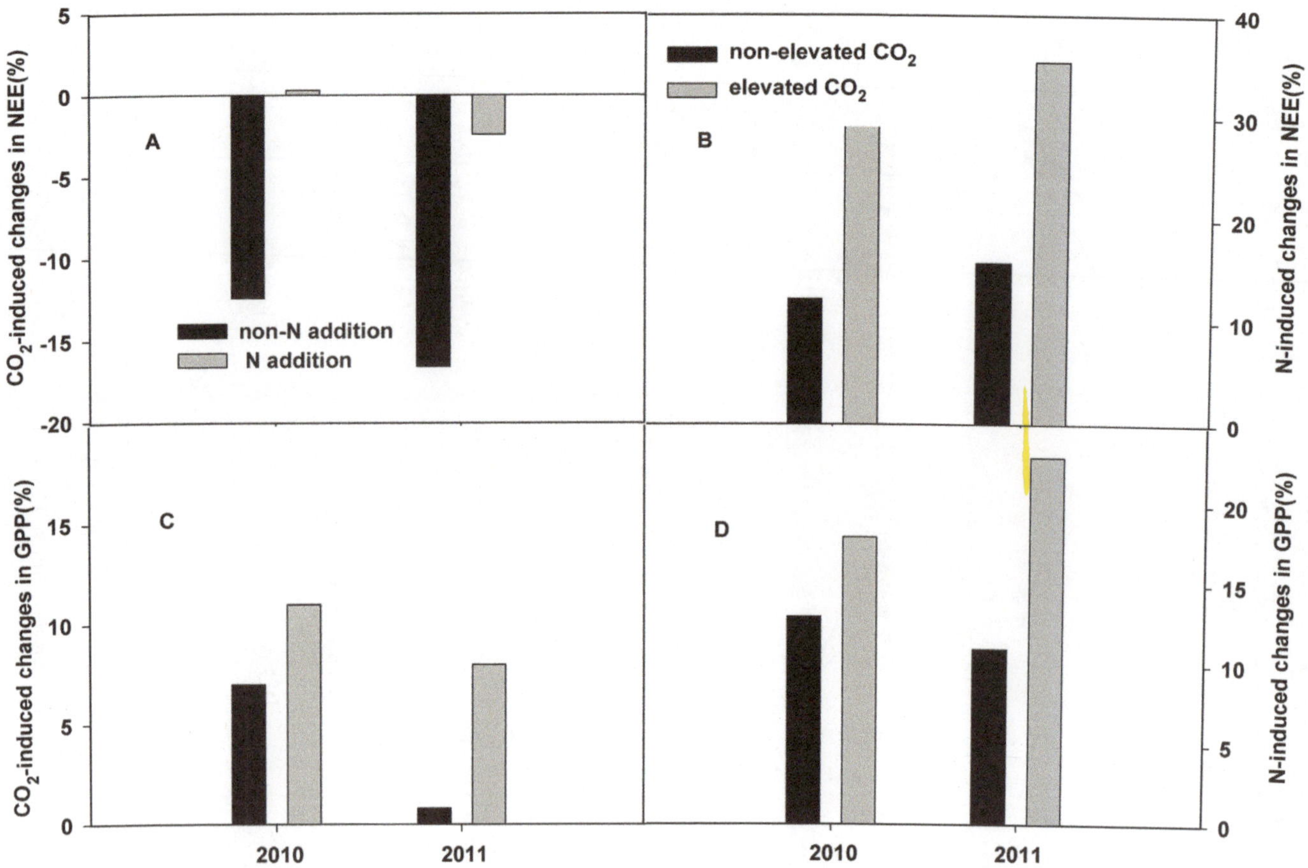

Figure 3. Elevated CO_2-induced changes in NEE (A) and GPP (C) in the non-N-addition and N-addition plots, and the N-induced changes in NEE (B) and GPP (D) in the non-elevated CO_2 and elevated CO_2 plots in 2010 and 2011. See Figure 2 for abbreviations.

turnover rate in the roots, root hairs and SOM priming [5,34–37]. Additionally, microbial biomass [38] may have also contributed to the higher respiration at elevated CO_2 level.

Our findings indicated that increased ER under elevated CO_2 casused lower NEE, compared to the control, which implies that increased root growth or activity could account for the much larger effect of elevated CO_2 on ER. Although we could not confirm the exact source of the ecosystem respiration in this experiment, these results show that the potential for C sequestration will decreased to some extent in the next several decades at this study site.

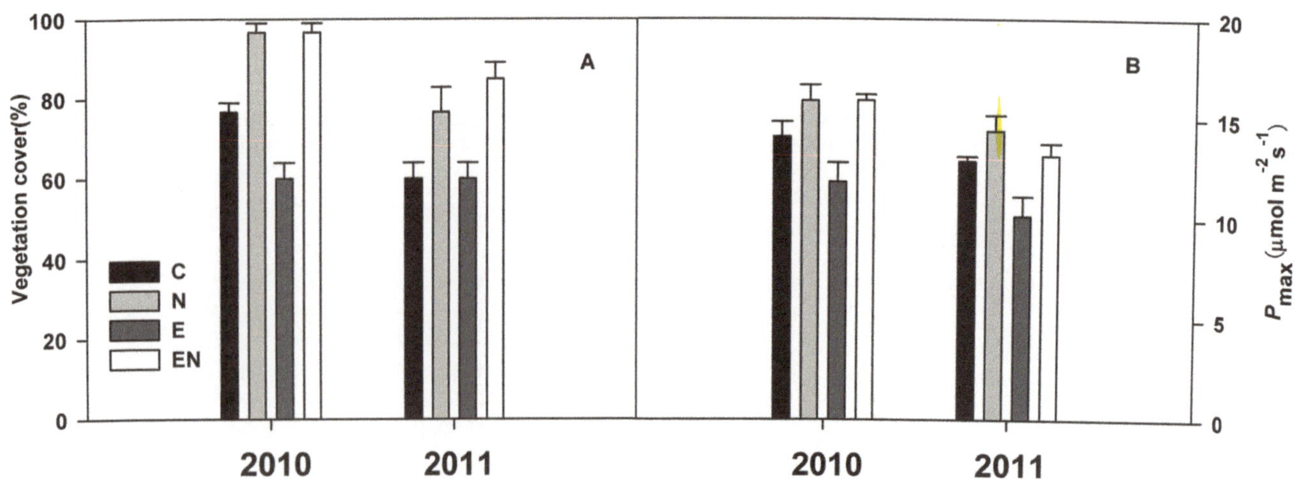

Figure 4. Responses of the vegetation cover (A) and the maximum photosynthetic rate (P_{max}) of the dominant species *Calamagrostis angustifolia* (B) to elevated CO_2 and N addition in July of 2010 and 2011. See Figure 2 for abbreviations.

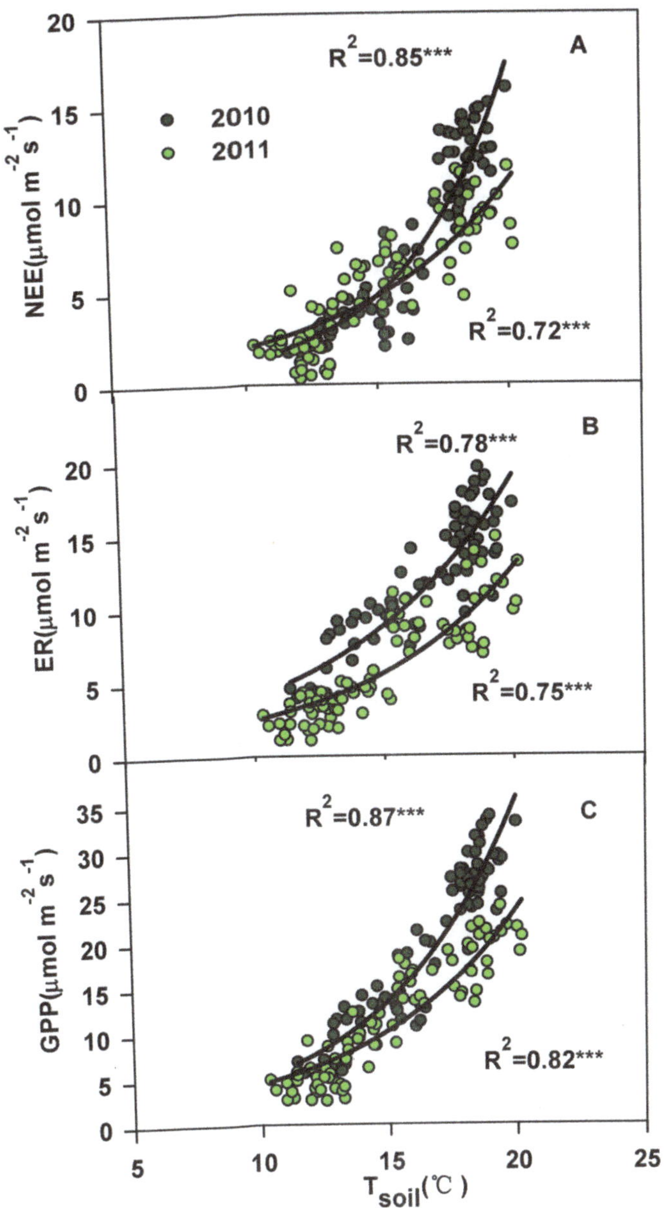

Figure 5. Temporal dependence of net ecosystem CO_2 exchange (NEE), ecosystem respiration (ER) and gross primary productivity (GPP) to soil temperature (T_{soil}, °C) across different plots in 2010 and 2011.

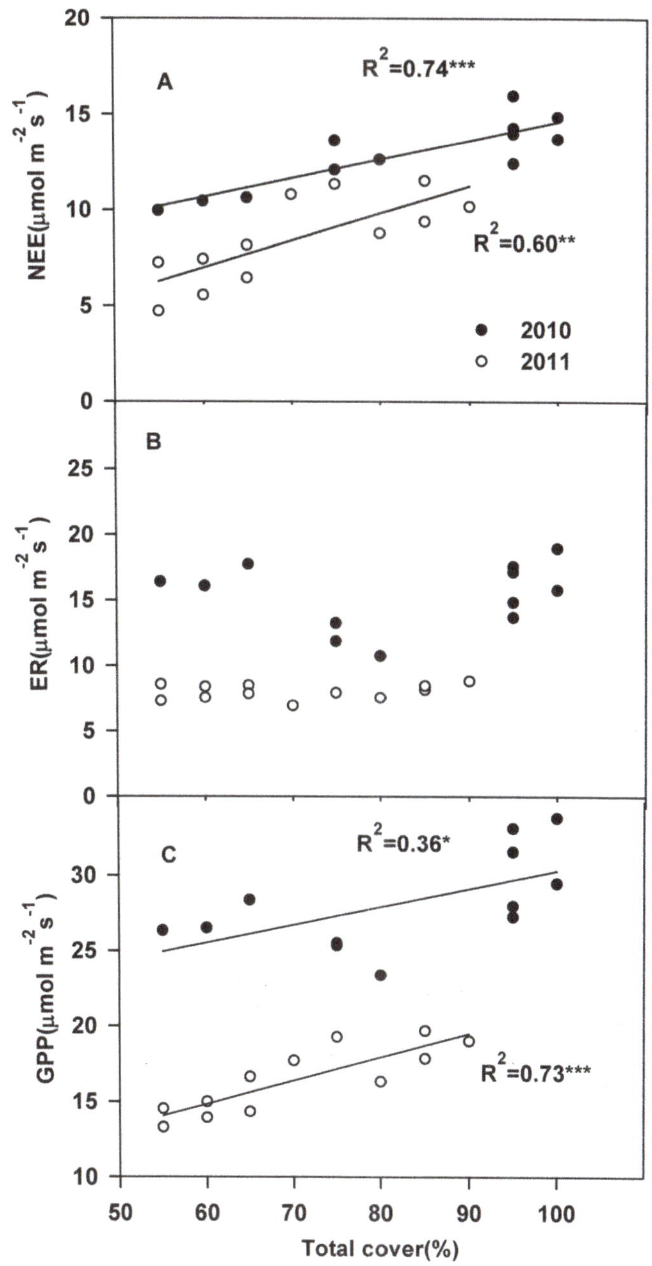

Figure 6. Correlations of net ecosystem CO_2 exchange (NEE), ecosystem respiration (ER) and gross primary productivity (GPP) with vegetation cover across different plots in July of 2010 and 2011.

N Addition Effects on Ecosystem CO_2 Fluxes

N addition significantly stimulated NEE in the two growing seasons in our ecosystem, largely due to N-induced increases in GPP, more so than those in ER (Table 1). The N-induced increases in canopy C uptake were partly caused by the stimulation of the leaf maximum photosynthetic rate of the dominant species (Fig. 4B), which was identical with another study [39], and by the higher vegetation cover on N addition plots (Fig. 4A). Stimulation in plant productivity provides more substrate for soil and plant respiration, leading to increases in ER under nitrogen fertilization. The positive effects have also been attributed to increased microbial biomass and activity [40] after fertilization. N stimulation of gross primary production on the Sanjiang Plain wetland (Fig. 2) is in accordance with meta-analysis results of positive N responses at global scales [9]. These results suggest that increased N deposition will enhance ecosystem carbon sequestration in the future.

Interactive Effect of Elevated CO_2 and N Addition on Ecosystem CO_2 Fluxes

Some research has shown that elevated CO_2 and N addition interact with each other in stimulating plant growth [18,41]. Moreover, N addition has been reported to alleviate photosynthetic acclimation [42]. Thus, elevated CO_2 and N addition might lead to potential interactive effects on ecosystem C fluxes. Oren [43] reported that soil fertility limits carbon sequestration in forest ecosystems with enriched CO_2 level. Reich [44] found that a low availability of N progressively suppresses the positive response of plant biomass to elevated CO_2 in perennial grassland. However, Aeschlimann [45] reported that there was a long-term stimulation in the net C assimilation by elevated CO_2, but high N supply led to a lower net C input than low N supply in managed grassland.

At our study sites, a significant interactive effect between elevated CO_2 and N addition on NEE ($p = 0.007$) was found during the experimental period (Table 2), and their combined effect increased NEE by 13% compared to that in the control plots (Table 1). The increased response of ecosystem CO_2 fluxes to N addition under elevated CO_2 compared to ambient CO_2 (Fig. 3), together with the interactive effects of CO_2 treatment and N treatment on NEE, indicate that elevated CO_2 triggers physiological responses to N supply at the ecosystem scale.

In addition, there were significant interactions between the sampling date and experimental treatment on ecosystem C fluxes (CO_2×D, N×D, and CO_2×N×D, Table 2), which indicated that the effect of the experimental treatment on the ecosystem C fluxes altered over time. The mechanisms responsible for the changes in the response of ecosystem C fluxes to elevated CO_2 and N addition over time may involve temporal heterogeneity in plant phenology, physiological activity, soil C and N mineralization, and root and soil organic matter decomposition and turnover [44].

We must note that there were low replication levels in our experiment. Moreover, our study was concerned with instantaneous C fluxes rather than cumulative C fluxes, which might prohibit comparison with other relevant studies. Even so, these results imply that a natural, unfertilized ecosystem experienced a decreased C sink than N-rich ecosystem at this site. Moreover, a severe N limitation in natural wetland may inhibit CO_2 response to assimilation [46] and, consequently, to the ecosystem C balance. Thus, this study demonstrated that the net ecosystem CO_2 uptake on the Sanjiang Plain wetland depends on adequate N supply.

Acknowledgments

Authors thank the constructive comments from two anonymous reviewers and the academic editor, which have greatly improved the quality of the paper. We are grateful to Daoguang Zhu for help in setting up the field facilities, Zhiyuan Yan and Rongtao Zhang for help in field work, Xiaoliang Zhou for providing meteorological data, and Li Jiang for help in data analysis.

Author Contributions

Conceived and designed the experiments: JBW HWN TCZ. Performed the experiments: JBW XLF HXZ JFW. Analyzed the data: JBW. Contributed reagents/materials/analysis tools: JBW HWN. Wrote the paper: JBW HWN TCZ.

References

1. IPCC (2007) IPCC WGI Fourth assessment report. Climatic change 2007: the physical science basis. Geneva: Intergovernmental Panel on Climate Change.
2. Saarnio S, Jarvio S, Saarinen T, Vasander H, Silvola J (2003) Minor changes in vegetation and carbon gas balance in a boreal mire under raised CO_2 or NH_4NO_3 supply. Ecosystems 6: 46–60.
3. Hungate BA, Holland EA, Jackson RB, Chapin FS, Mooneyk HA, et al. (1997) The fate of carbon in grasslands under carbon dioxide enrichment. Nature 388: 576–579.
4. Hymus GJ, Johnson DP, Dore S, Anderson H, Hinkle CR, et al. (2003) Effects of elevated atmospheric CO_2 on net ecosystem CO_2 exchange of a scrub-oak ecosystem. Global Change Biol. 9: 1802–1812.
5. Carney KM, Hungate BA, Drake BG, Megonigal JP (2007) Altered soil microbial community at elevated CO_2 leads to loss of soil carbon. Proc. Natl. Sci. USA 104: 4990–4995.
6. Randerson JT, Chapin FS, Harden J, Neff JC, Harmon ME (2002) Net ecosystem production: A comprehensive measure of net carbon accumulation by ecosystems. Ecol. Appl. 12: 937–947.
7. Gruber N, Galloway JN (2008) An earth-system perspective of the global nitrogen cycle. Nature 451: 293–296.
8. Galloway JN, Townsend AR, Erisman JW, Bekunda M, Cai ZC, et al. (2008) Transformation of the nitrogen cycle: recent trends, questions, and potential solutions. Science 320: 889–892.
9. LeBauer DS, Treseder KK (2008) Nitrogen limitation of net primary productivity in terrestrial ecosystems is globally distributed. Ecology 89: 371–379.
10. Niu SL, Wu MY, Han Y, Xia JY, Zhang Z, et al. (2010) Nitrogen effects on net ecosystem carbon exchange in a temperate steppe. Global Change Biol. 16: 144–155.
11. Xia JY, Niu SL, Wan SQ (2009) Response of ecosystem carbon exchange to warming and nitrogen addition during two hydrologically contrasting growing seasons in a temperate steppe. Global Change Biol. 15: 1544–1556.
12. Kivimäki SK, Sheppard LJ, Leith ID, Grace J (2013) Long-term enhanced nitrogen deposition increases ecosystem respiration and carbon loss from a *Sphagnum* bog in the Scottish Borders. Environ. Exp. Bot. 90: 53–61.
13. Mo JM, Zhang W, Zhu WX, Gundersen P, Fang YT, et al. (2008) Nitrogen addition reduces soil respiration in a mature tropical forest in southern China. Global Change Biol. 14: 403–412.
14. Janssens IA, Dieleman W, Luyssaert S. Subke JA, Reichstein M, et al. (2010) Reduction of forest soil respiration in response to nitrogen deposition. Nature 3: 315–322.
15. Allison SD, Czimczik CI, Treseder KK (2008) Microbial activity and soil respiration under nitrogen addition in Alaskan boreal forest. Global Change Biol. 14: 1156–1168.
16. Bradford MA, Fierer N, Jackson RB, Maddox TR, Reynolds JF (2008) Nonlinear root-derived carbon sequestration across a gradient of nitrogen and phosphorous deposition in experimental mesocosms. Global Change Biol. 14: 1113–1124.
17. Reay DS, Dentener F, Smith P, Grace J, Feely RA (2008) Global nitrogen deposition and carbon sinks. Nature Geoscience 1: 430–437.
18. Luo YQ, Su B, Currie WS, Dukes JS, Finzi A, et al. (2004) Progressive nitrogen limitation of ecosystem responses to rising atmospheric carbon dioxide. Bio Science 54: 731–739.
19. Jones MB, Donnelly A (2004) Carbon sequestration in temperate grassland ecosystems and the inuence of management, climate and elevated CO_2. New Phytol. 164: 423–439.
20. Rasse DP, Peresta G, Drake BG (2005) Seventeen years of elevated CO_2 exposure in a Chesapeake Bay Wetland: sustained but contrasting responses of plant growth and CO_2 uptake. Global Change Biol. 11: 369–377.
21. Langley JA, Megonigal P (2010) Ecosystem response to elevated CO_2 levels limited by nitrogen- induced plant species shift. Nature 466: 96–99.
22. LeMer J, Roger P (2001) Production, oxidation, emission and consumption of methane by soils: a review. Eur. J. Soil Biol. 37: 25–50.
23. Liu DY, Song CC (2008) Effects of exogenous nitrogen supply on photosynthetic characteristics and carbon sequestration potential of *Calamagrostis angustifolia* in freshwater marshes. Acta Sci. Circums. 28: 305–312.
24. Sun ZG, Liu JS, Wang JD (2007) Dynamics of nitrogen in the atmospheric wet deposition and its ecological effects in typical wetland ecosystem of Sanjiang plain. Adv.Water Sci. 18: 182–192.
25. Steduto P, Çetinkökü Ö, Albrizio R, Kanber R (2002) Automated closed-system canopy-chamber for continuous field-crop monitoring of CO_2 and H_2O fluxes. Agric. For. Meteorol. 111: 171–186.
26. Welp LR, Randerson JT, Liu HP (2007) The sensitivity of carbon uxes to spring warming and summer drought depends on plant functional type in boreal forest ecosystem. Agric. For. Meteorol. 147: 172–185.
27. Zhou XP, Wang XK, Tong L, Zhang HX, Lu F, et al. (2012) Soil warming effect on net ecosystem exchange of carbon dioxide during the transition from winter carbon source to spring carbon sink in a transition from winter carbon source to spring carbon sink in a temperate urban lawn. J. Environ. Sci. 24: 2104–2112.
28. Bachman S, White JL, Pendall E,Williams DG, Morgan JA, et al. (2009) Elevated carbon dioxide alters impacts of precipitation pulses on ecosystem photosynthesis and respiration in a semi-arid grassland. Oecologia. doi 10.1007/s00442-009-1511-x.
29. Jiang L, Guo R, Zhu TC, Niu XD, Guo JX, et al. (2012) Water- and Plant-Mediated Responses of Ecosystem Carbon Fluxes to Warming and Nitrogen Addition on the Songnen Grassland in Northeast China. PLoS ONE 7(9): e45205 doi:10.1371/journal.pone.0045205 1.
30. Song CC, Sun L, Huang Y, Wang YS, Wan ZM (2011) Carbon exchange in a freshwater marsh in the Sanjiang Plain, northeastern China. Agric. For. Meteorol. 151: 1131–1138.
31. Bonneville MC, Strachan IB, Humphreys ER, Roulet NT (2008) Net ecosystem CO_2 exchange in a temperate cattail marsh in relation to biophysical properties. Agric. For. Meteorol. 148: 69–81.

32. Jasoni RL, Smith SD, Arnone JA (2005) Net ecosystem CO_2 exchange in Mojave Desert shrublands during the eighth year of exposure to elevated CO_2. Global Change Biol. 11: 749–756.
33. Zhao GY, Liu JS, Dou JX (2008) Effects of CO_2 concentration doubling on physiological characteristics of *Calamagrostis angustifolia* in freshwater marsh. Res. Environ. Sci. 21: 134–138.
34. Casella E, Soussana JF (1997) Long-term effects of CO_2 enrichment and temperature increase on the carbon balance of a temperate grass sward. J. Exp. Bot. 48: 1309–1321.
35. Adair EC, Reich PB, Trost JJ (2011) Elevated CO_2 stimulates grassland soil respiration by increasing carbon inputs rather than by enhancing soil moisture. Global Change Biol. 17: 3546–3563.
36. Pendall E, King JY, LeCain DR, Milchunas DG, Morgan JA, et al. (2003) Elevated atmospheric CO_2 effects and soil water feedbacks on soil respiration components in a Colorado grassland. Globle. Biogeochem. Cycl. 17: 1–13.
37. Milchunas DG, Mosier AR, Morgan JA, LeCain DR, King JY, et al. (2005) Root production and tissue quality in a shortgrass steppe exposed to elevated CO_2: using a new ingrowth method. Plant Soil 268: 111–122.
38. Sowerby A, Blum H, Gray TR, Ball AS (2000) The decomposition of Lolium perenne in soils exposed to elevated CO_2: comparisons of mass loss of litter with soil respiration and soil microbial biomass. Soil Biol. Biochem.32: 1359–1366.
39. Liu DY, Song CC (2008) Response of marsh wetland plant *Calamagrostis angustifolia* to exogenous nitrogen input. Chin. J. Appl. Ecol.19: 2599–2604.
40. Lovell RD, Hatch DJ (1998) Stimulation of microbial activity following spring applications of nitrogen. Biol. Fertil. Soils 26: 28–30.
41. Hungate BA, Dukes JS, Shaw MR, LuoYQ, Field CB (2003) Nitrogen and climate change. Science 302: 1512–1513.
42. Bert G, Drake MA, Meler G (1997) More efficient plants: a consequence of rising atmospheric CO_2? Annu. Rev. Plant Physiol. Plant Mol. Biol. 48: 609–639.
43. Oren R, Ellsworth DS, Kurt HJ, Phillips N, Ewers BE, et al. (2001) Soil fertility limits carbon sequestration by forest ecosystems in a CO_2-enriched atmosphere. Nature 411: 469–472.
44. Reich PB, Hobbie SE, Lee TL, Ellsworth DS, West JB, et al. (2006) Nitrogen limitation constrains sustainability of ecosystem response to CO_2. Nature 440: 922–925.
45. Aeschlimann U, Nösbergre J, Edwards PJ (2005) Responses of net ecosystem CO_2 exchange in managed grassland to long-term CO_2 enrichment, N fertilization and plant species. Plant Cell Environ. 28: 823–833.
46. Stitt M, Krapp A (1999) The interaction between elevated carbon dioxide and nitrogen nutrition: the physiological and molecular background. Plant Cell Environ. 22: 583–621.

Restoration Enhances Wetland Biodiversity and Ecosystem Service Supply, but Results Are Context-Dependent

Paula Meli[1,2]*, José María Rey Benayas[2], Patricia Balvanera[3], Miguel Martínez Ramos[3]

1 Natura y Ecosistemas Mexicanos A.C., México DF, México, **2** Departamento de Ciencias de la Vida, Universidad de Alcalá, Alcalá de Henares, Madrid, Spain, **3** Centro de Investigaciones en Ecosistemas, Universidad Nacional Autónoma de México, Morelia, Michoacán, México

Abstract

Wetlands are valuable ecosystems because they harbor a huge biodiversity and provide key services to societies. When natural or human factors degrade wetlands, ecological restoration is often carried out to recover biodiversity and ecosystem services (ES). Although such restorations are routinely performed, we lack systematic, evidence-based assessments of their effectiveness on the recovery of biodiversity and ES. Here we performed a meta-analysis of 70 experimental studies in order to assess the effectiveness of ecological restoration and identify what factors affect it. We compared selected ecosystem performance variables between degraded and restored wetlands and between restored and natural wetlands using response ratios and random-effects categorical modeling. We assessed how context factors such as ecosystem type, main agent of degradation, restoration action, experimental design, and restoration age influenced post-restoration biodiversity and ES. Biodiversity showed excellent recovery, though the precise recovery depended strongly on the type of organisms involved. Restored wetlands showed 36% higher levels of provisioning, regulating and supporting ES than did degraded wetlands. In fact, wetlands showed levels of provisioning and cultural ES similar to those of natural wetlands; however, their levels of supporting and regulating ES were, respectively, 16% and 22% lower than in natural wetlands. Recovery of biodiversity and of ES were positively correlated, indicating a win-win restoration outcome. The extent to which restoration increased biodiversity and ES in degraded wetlands depended primarily on the main agent of degradation, restoration actions, experimental design, and ecosystem type. In contrast, the choice of specific restoration actions alone explained most differences between restored and natural wetlands. These results highlight the importance of comprehensive, multi-factorial assessment to determine the ecological status of degraded, restored and natural wetlands and thereby evaluate the effectiveness of ecological restorations. Future research on wetland restoration should also seek to identify which restoration actions work best for specific habitats.

Editor: Mehrdad Hajibabaei, University of Guelph, Canada

Funding: This research was funded primarily by a Rufford Small Grant for Nature Conservation to P.M. (40.11.09) and by grants from Pemex and the WWF-FCS Alliance to Natura y Ecosistemas Mexicanos. J.M.R.B. acknowledges support from the Spanish Ministry of Science and Education (project CGL2010-18312) and the Autonomy of Madrid (S2009AMB-1783 REMEDINAL-2). M.M.R. received sabbatical support from CONACyT, and P.B. and M.M.R. also received support from PSPA-DGAPA-Universidad Nacional Autónoma de México. The funders had no role in study design, data collection and analysis, decision to publish, or preparation of the manuscript.

Competing Interests: PEMEX is a commercial institution, it is a public-private partnership involving the Mexican government, who provided funding towards this study. There are no patents, products in development or marketed products to declare. This does not alter the authors' adherence to all the PLOS ONE policies on sharing data and materials.

* E-mail: paula@naturamexicana.org.mx

Introduction

Wetlands harbor significant biodiversity [1] and supply crucial ecosystem services (ES) [2,3], which are defined as the benefits that people obtain from ecosystems [4]. ES provided by wetlands include regulating water purification, protecting the ecosystem from soil erosion and effects of flooding, and nursing the early growth of many species essential to oceanic fisheries (**Table 1**). Although wetlands occupy less than 9% of the Earth's terrestrial surface, they contribute up to 40% of global annual renewable ES [5]. Despite their importance to human societies, wetlands are rapidly being degraded and destroyed [5], threatening the ecosystem and biodiversity on which wetland ES depend.

To compensate for their extensive degradation, wetland restoration has become common practice around the world. Several studies have reported that restoration can recover much of the biodiversity and ES lost due to degradation [6]. On the other hand, studies have called into question the effectiveness of wetland restoration, suggesting that its positive impacts depend strongly on factors such as ecosystem type and restoration actions [5]. For example, some authors have suggested that current wetland restoration methods are too slow and incomplete to allow recovery of biological structure and biogeochemical function [7]. Therefore the effectiveness of wetland restoration remains controversial, and this is in part because different studies have applied different standards to evaluate outcomes [6]. At the same time, most studies evaluating wetland restoration, including a recent meta-analysis [7], have not directly assessed ES recovery or how well restoration methods work for diverse types of organisms.

Table 1. Principal ecosystem services (ES) supplied by wetlands.

ES type[1]	*Individual ES*	*Description*
	Biogeochemical cycling	Maintenance of natural exchange or flux of material and energy between living and nonliving components of biosphere, thereby supporting climatic and biological dynamics.
Supporting	Biotic interactions	Pollination of wild species or crops; seed dispersal; preservation and maintenance of trophic chains.
	Habitat (terrestrial)	Habitat for resident and transient terrestrial populations (refugia/nursery).
	Habitat (aquatic)	Habitat for resident and transient aquatic populations (refugia/nursery).
	Plant food/raw material	The proportion of gross primary production that can be extracted as food or raw materials.
Provisioning	Animal food/raw material	The proportion of secondary production that can be extracted as food or raw materials.
	Water supply	Filtering, retention and storage of fresh water for human use (domestic, industrial, agriculture).
	Climate regulation	Regulation of the chemical composition of the atmosphere, global temperature, and other biologically mediated climatic processes at global and regional levels.
	Hydrological dynamics	Regulation of natural hydrological flows, role of land cover in regulating runoff and river discharge, and infiltration; groundwater recharge.
Regulating	Water quality	Retention and removal or breakdown of xenic nutrients and compounds; water purification.
	Regulation of extreme events	Capacity and integrity of ecosystem response to environmental fluctuation such as floods or storms, or to other extreme events.
	Regulation of soil fertility and erosion	Soil maintenance and formation, for both natural ecosystems and crops; sediment retention and prevention of erosion; shoreline stabilization; accumulation of organic matter.
	Regulation of invasive species, pests, and diseases	Regulation of invasive species populations; trophic-dynamic regulations of pest populations.
	Cultural	Contribution by ecosystems to experiences that benefit human population directly or indirectly.
Cultural	Recreation	Provision of opportunities for recreational activities.

[1]MEA (2005).

Recovering biodiversity and recovering ES can be regarded as distinct goals of wetland restoration, with a given restoration focusing on one or the other. However, assessing both types of recovery simultaneously is important for several reasons. Biodiversity and ES of restored ecosystems often do not reach pre-degradation levels or the levels of similar natural ecosystems, and recovery of biodiversity may correlate with recovery of ES [8,9]. Indeed, recovery of biodiversity may be a prerequisite for recovery of ES [7]; for instance, increasing biodiversity enhances key ES such as primary productivity [10] and soil erosion control [11]. Thus, comparable recovery of biodiversity and ES may indicate a win-win outcome for ecosystem and society alike. Additionally, assessments of wetland restoration should consider the context in which the restoration occurs, since restoration effectiveness may strongly depend on the type of ecosystem being restored, its pre-restoration condition, and the factors responsible for its degradation. By analyzing wetland restoration simultaneously in terms of biodiversity and ES, we can identify factors that affect the recovery of either or both, allowing us to develop recommendations for researchers and practitioners.

To develop an evidence-based approach for planning and assessing wetland restoration, we conducted a meta-analysis of the peer-reviewed literature to address the following four questions: (1) how much biodiversity and (2) how much of ES levels can be recovered through wetland restoration, (3) whether biodiversity and ES recovery correlate, and (4) whether the effectiveness of biodiversity and ES recovery depends on context, including ecosystem type, cause of degradation, restoration action, experimental design, and restoration age. In examining what the literature says on these questions, we hope to inform and improve efforts to restore the biodiversity and ES of degraded wetlands.

Methods

Literature search

We systematically searched the research literature to identify quantitative studies of the effects of ecological restoration on biodiversity and ES of non-marine aquatic and semi-aquatic degraded wetlands. We searched the ISI Web of Knowledge database (www.isiwebofknowledge.com), as it provides access to peer-reviewed studies. We searched studies published between 1970 and 2010 using the following string of search terms: (riparian OR river* OR lake OR mangroves OR marsh OR stream OR wetland) AND (restor* OR re-creat* OR rehabilitat* OR forest* OR reforest* OR afforest* OR plant* OR recover*) AND ((ecosystem OR environment) AND (service OR function*)). Preliminary search results were filtered to include only the following ISI-defined subject areas: "agriculture", "biodiversity and conservation", "environmental sciences and ecology", "fisheries", "forestry", "marine and freshwater biology", "plant sciences", "water resources", and "zoology". This resulted in a list of 1,931 references.

For inclusion in our meta-analysis, studies had to focus on at least one estuarine, lacustrine, palustrine, or riverine wetland, as defined by [1], as well as report the following information:

1) Quantitative assessment of passive restoration (i.e. natural regeneration) or active restoration in terms of variables related to biodiversity and/or to the supply of one or more wetland ES (**Table 1**) consistent with the framework of the Millennium Ecosystem Assessment [4], according to which biodiversity underpins all ES;
2) Comparison of restored wetland with either degraded or natural wetland;
3) Sample size of the reported data and at least a variance estimate of such data.

A total of 70 studies (**Supporting information S1**) satisfied these criteria and were included in our meta-analysis. The number of observations included in each analysis is shown in the corresponding figures.

Database building and effect size estimation

We constructed a computer database in which rows were observations and columns were properties of those observations (**Supporting information S2**; **Table S1**). For each study we extracted data on the variables used to measure the impacts of restoration (response variables). Separate databases were built for biodiversity and ES response variables. Whether we used one or the other database, or some combination of columns from both of them, depended on the specific question being addressed. Each measurement of restoration impact was recorded as a separate row in the database, even when the measurements came from the same study. Measurements were also recorded separately when the original study assumed spatially independent conditions within the same study site (e.g. measurements made near the shore vs. made on the open water of the same wetland).

We extracted data on type of wetland and ecosystem, the principal causes of degradation, specific restoration action(s) implemented, experimental design used to assess restoration outcomes, and the time elapsed since completion of the last restoration action (restoration age). All variables except restoration age were nominal and assigned to categories specifically created for our analyses (**Supporting information S3**).

Since our meta-analysis included studies differing considerably in response variables and experimental designs, we assessed the effects of restoration on biodiversity and ES relative to a control using response ratios (RRs) as the effect size metric. As an indicator of the outcome of restoration, we calculated RRs of the restored wetlands relative to reference natural wetlands [ln(Rest/Ref)] and to degraded wetlands [ln(Rest/Deg)] for each measure of the biodiversity and ES extracted from the studies. Most response variables were expected to correlate positively with biodiversity or a particular ES; for example, greater biomass was predicted to mean a higher level of supporting or provisioning ES. However, some response variables were predicted to correlate negatively with biodiversity or ES; for example, a greater concentration of a water or soil contaminant or a greater abundance of non-native species were predicted to reduce, respectively, provisioning ES and biodiversity. In these cases we inverted the sign of the RR (**Supporting information S2**).

We performed separate analyses to compare restored and degraded wetlands and to compare restored and natural wetlands [9] (**Supporting information S3**). RR calculations and statistical analyses were performed using MetaWin v2.1 [12].

Biodiversity recovery

All possible measures of biodiversity for which the included studies reported data were used to calculate RRs; these measures included (a) species, gender, taxon or family richness; and (b) indices of species abundance, diversity, similarity, and composition. Using biodiversity measures calculated for different taxonomic levels or by different formulas enabled us to screen for differences in responses to restoration at different levels of ecological complexity [9,13]. Each extracted datum was assigned to a single organism type. Data were analyzed using categorical, random-effects models because the data were most likely to satisfy the assumptions of these models [12]; the categories in the model were organism types.

To evaluate possible pseudo-replication effects, we calculated the mean RR for each of the three largest categories: macroinvertebrates, aquatic invertebrates, and vascular plants, using only one randomly selected effect size from each study. These mean RRs were similar to the means obtained when all effect sizes from each study were included, and the bias-corrected 95% bootstrap confidence interval of the reduced dataset overlapped with that of the entire dataset (**Table S2**). Therefore we retained all the data in our meta-analysis, similar to Rey Benayas *et al.* [9] and Vilà *et al.* [13].

ES recovery

Response variables were related to a wide variety of ES, so multiple RR-ES combinations were included as separate rows in the database (**Table S1**). The parallel assessment of these multiple associations allowed us to capture the simultaneous supply of several ES [14,15]. To avoid counting the same data more than once in a meta-analysis, we performed a separate meta-analysis for each ES using a random-effects model. We considered this approach suitable because we wanted to evaluate each ES separately, rather than the heterogeneity among different ES.

Correlation between biodiversity and ES recovery

We assessed the correlation between biodiversity recovery and ES recovery using the Spearman rank coefficient to quantify the correlation between the corresponding RRs. We used only RRs from studies that evaluated both biodiversity and ES, and we treated each of these studies as an independent sample. When the same study reported multiple measures of biodiversity or ES, the related RRs were averaged to generate an overall RR for biodiversity and an overall RR for ES for each study, thereby minimizing the risk of pseudo-replication. This approach led us to combine the four major ES types in order to ensure adequate sample size [9].

Context dependence of biodiversity and ES recovery

We used linear mixed-effects models to evaluate whether the effects of restoration on biodiversity and ES varied with context. Context was parameterized using four nominal fixed factors (ecosystem type, main cause of degradation, restoration action, and experimental design) and the continuous fixed factor of restoration age, defined as the decimal logarithm of the number of months between completion of the last restoration action and evaluation. We added a fifth nominal fixed factor with two levels (biodiversity or ES) because we used RRs for both biodiversity and ES recovery in the analysis. Study site was the random-effect factor and RR was the dependent variable.

We also built a second model in which we reduced the degrees of freedom by including only factor categories containing at least 30 observations. Since this reduced the average sample size in each category, we discarded this model in favor of the first. Finally, we applied a backward elimination procedure in which non-significant terms ($p<0.05$) were removed in order of decreasing p value. The selected final model contained main effects but no interactions. All model building and refinement was carried out using Data Desk v6 [16].

Results

The 70 studies analyzed here were distributed across 62 locations in 14 countries (**Supporting information S4**). Riverine wetlands were the best-represented ecosystem type (38% of studies), followed by lacustrine wetlands (27%), and finally estuarine (18%) and palustrine wetlands (17%). Nearly all studies (68) were field-based comparisons, including three passive restoration studies (4%). The remaining two studies (3%) involved one field and one greenhouse experiment.

Biodiversity recovery

Restoring degraded wetlands enhanced biodiversity by 19% (**Fig. 1a**); and biodiversity in restored wetlands did not significantly differ from that in natural wetlands (**Fig. 1b**). Restoration significantly enhanced the diversity of vertebrates (+53%), vascular plants (+45%), and terrestrial (+17%) and aquatic (+15%) invertebrates, but it had no significant effect on macroinvertebrate diversity. Restored and natural wetlands showed similar diversity of vascular plants, aquatic invertebrates, macroinvertebrates and protists. In contrast, these two types of wetlands differed significantly in the diversity of non-native vascular plants, which was 44% lower in restored wetlands, and in vertebrate diversity, which was 37% higher in restored wetlands.

ES recovery

Overall ES supply was 43% higher in restored wetlands than in degraded ones (**Fig. 2a**), but 13% lower than in natural wetlands (**Fig. 2b**). Compared to degraded wetlands, restored wetlands showed much greater supply of provisioning ES (+80%), regulating ES (+47%) and supporting ES (+40%), while the two types of wetlands showed similar supply of cultural ES. Compared to natural wetlands, restored wetlands showed similar supply of provisioning and cultural ES, but lower supply of regulating (−22%) and supporting ES (−16%).

Restoration increased most individual ES that we examined, although not to the same extent (**Fig. 2a**). Restoration increased the supply of supporting services, with increases ranging from 32% for biogeochemical cycling to 61% for biotic interactions. Increases in the supply of regulating services ranged from 31% for water quality to 176% for invasive species control. Restoration also increased both provisioning services examined in our meta-analysis: water supply (+108%) and the supply of food or raw materials of animal origin (+65%). For most individual ES that we examined, restored and natural wetlands tended to supply similar amounts (**Fig. 2b**). Exceptions, in decreasing order of difference between the two wetland types, were climate regulation, the supply of which was −30% lower in restored wetlands; provision of terrestrial habitat, −22%; regulation of fertility and soil erosion, −21%; and biogeochemical cycles, −14%.

Correlation between biodiversity and ES recovery

Biodiversity and ES response ratios positively correlated in comparisons of restored and degraded wetlands (**Fig. 3a**) and in comparisons of restored and natural wetlands (**Fig. 3b**).

Context dependence of biodiversity and ES recovery: restored vs. degraded wetlands

Comparison of restored and degraded wetlands showed that restoration effects depended on the following factors, listed in order of decreasing importance: main cause of degradation, restoration action, experimental design, and ecosystem type (**Table 2**). In contrast, restoration age did not significantly affect restoration outcomes. These results were the same for the two outcomes of biodiversity recovery and ES recovery.

Context variables explained relatively little variance (25.7%) in biodiversity and ES recovery. Nevertheless, the improvement in biodiversity and ES due to restoration varied substantially for different wetland types: salt marshes (+104%), freshwater marshes (+73%), rivers (+100%), lakes (+45%), mangroves (+33%), and streams (+9%; **Fig. S1**).

Restoration significantly ameliorated all causes of degradation that we examined, except for the presence of invasive species (**Fig. S2**). Seven of the 10 restoration actions reported by the included studies showed significant effects on biodiversity and ES supply (**Fig. S3**), with habitat creation leading to the greatest benefit (+119%), followed by soil amendment and revegetation (+91%), and passive restoration in third place (+57%). Of all restoration actions examined, exotic species removal was associated with the lowest effect size, which did not achieve statistical significance. Restoration showed significant positive effects on biodiversity and ES recovery for the three types of experimental designs in the

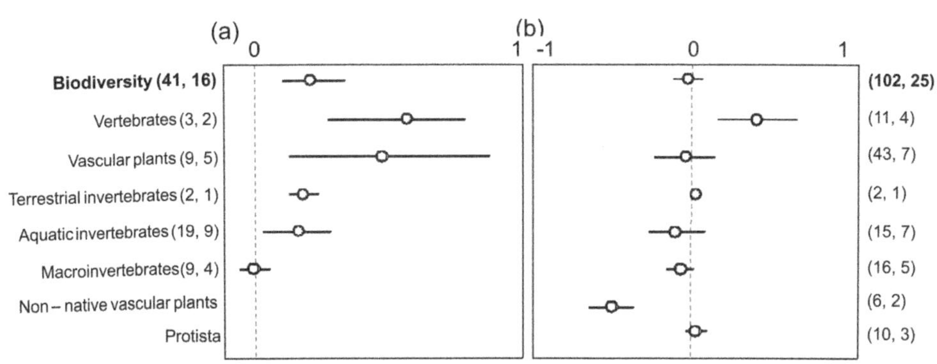

Figure 1. Mean effect size (response ratio) of ecological restoration on overall biodiversity and biodiversity of specific types of organisms in restored wetlands with respect to (a) degraded wetlands or (b) natural wetlands. Numbers in parentheses indicate the sample size (number of comparisons) followed by the numbers of studies. Bars extending from the means indicate bias-corrected 95% bootstrap confidence intervals. A mean effect size is significantly different from zero if the 95% confidence interval does not overlap with it. In comparison (a), no data were available on non-native vascular plants and protists. In comparison (b), the confidence interval for terrestrial invertebrates is not visible because it is smaller than the mean marker.

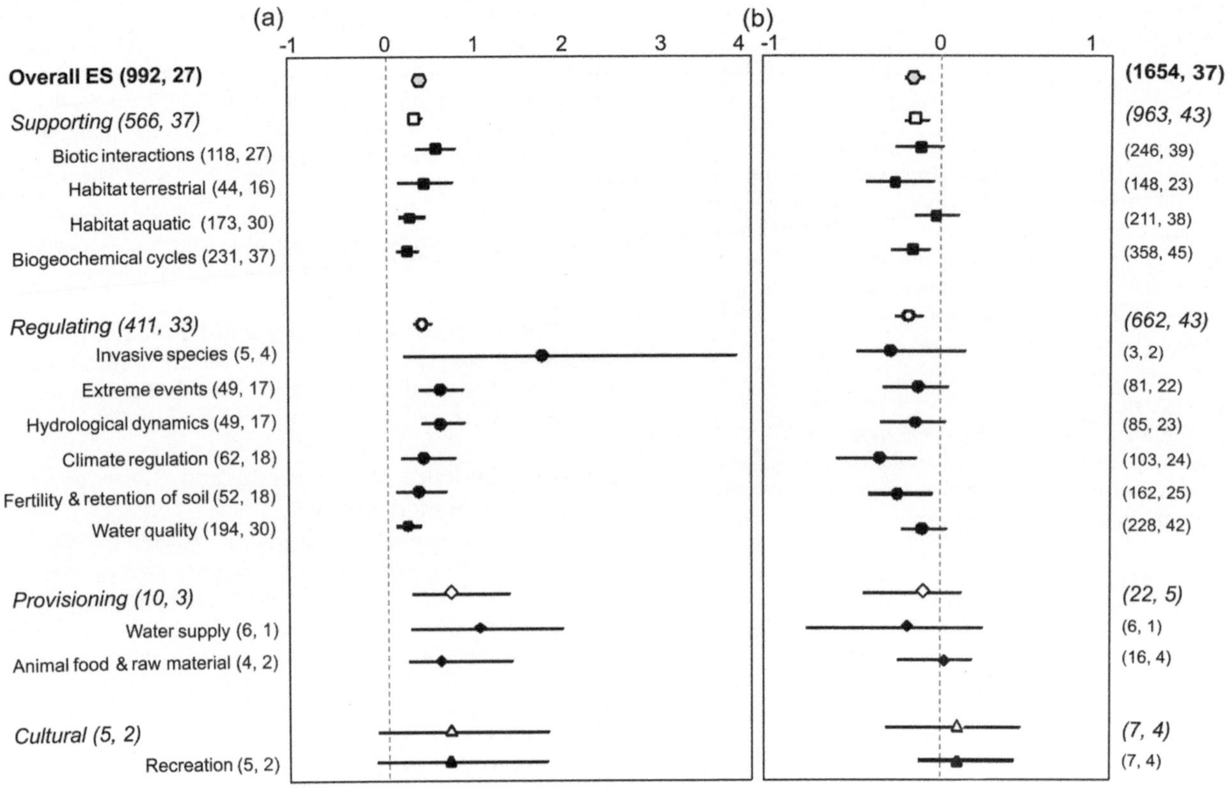

Figure 2. Mean effect size (response ratio) of ecological restoration on four major ES types defined by the MEA (2005) and on 13 individual ES (see details in Table 1) in restored wetlands with respect to (a) degraded wetlands or (b) natural wetlands. Bars extending from the means indicate bias-corrected 95% bootstrap confidence intervals. A mean effect size is significantly different from zero if the 95% confidence interval does not overlap with it. Numbers in parentheses indicate the sample size (number of comparisons) followed by the numbers of studies.

included studies: paired experiments (+61%), before-after experiments (+33%) and control-impact experiments (+22%; **Fig. S4**).

Context dependence of biodiversity and ES recovery: restored vs. natural wetlands

Comparison of restored and natural wetlands showed that restoration significantly improved recovery of biodiversity and ES supply (**Table 2**), although as before, the final model explained only a fraction of the variance (15.2%). All restoration actions led to full recovery of biodiversity and ES supply except for soil amendment and revegetation, which led to −124% lower levels of biodiversity and ES supply than in natural wetlands; passive restoration, which led to −31% lower levels; manipulation of structural heterogeneity, −15%; and hydrological dynamics, −21% (**Fig. S3**).

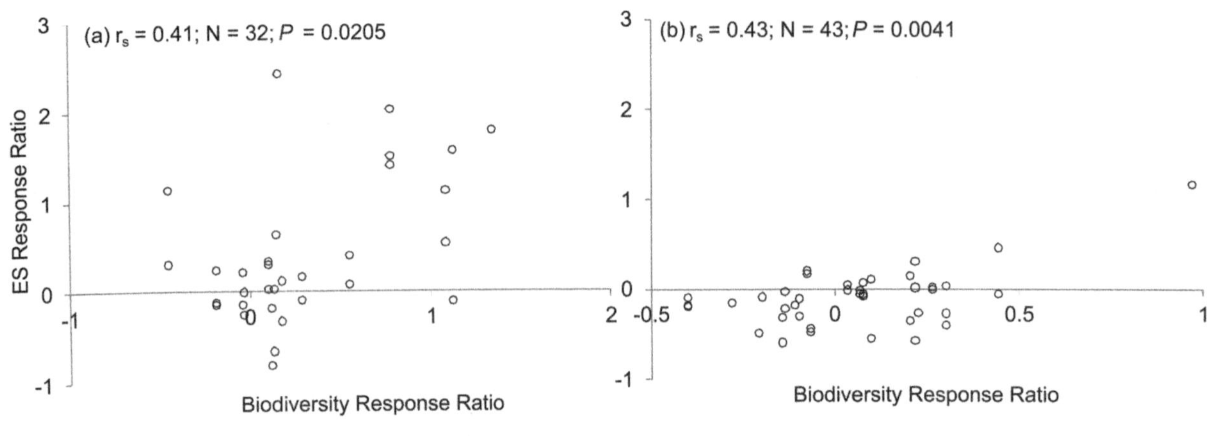

Figure 3. Spearman rank correlations between biodiversity and ES supply in restored wetlands with respect to (a) degraded wetlands or (b) natural wetlands.

Table 2. Results of mixed linear models assessing the influence of ecological context factors on the effects of restoration on biodiversity and ecosystem services of wetlands.

Factor	Wetland comparison					
	Restored vs. Degraded			*Restored vs. Natural*		
	F	*P*	Explained variance (%)	F	*P*	Explained variance (%)
ResAct	$5.32_{9,300}$	<0.0001	11.7	$2.35_{9,506}$	0.0133	3.8
DegFac	$6.03_{4,300}$	0.0001	5.9			
EcoType	$2.82_{8,300}$	0.0051	5.5			
ExpDes	$5.24_{2,300}$	0.006	2.6			
B/ES				$3.93_{3,506}$	0.0038	2.8

Abbreviations: B/ES, ratio of biodiversity to ecosystem services; DegFac, degrading factor; EcoType, ecosystem type; ExpDes, experimental design; ResAct: restoration action.

Discussion

Biodiversity recovery

Our global meta-analysis, including70 studies conducted in 14 countries, shows that wetland restoration increased biodiversity in degraded wetlands, consistent with another global meta-analysis of different ecosystem types [9]. In fact, restoration increased the biodiversity of native organisms to levels similar to those in natural wetlands. To be sure, restoration did not improve biodiversity of all organisms uniformly. Restoration increased vertebrate diversity to levels above those in natural wetlands, though this result may only be transient, since vertebrate richness can vary substantially over time [17]. Conversely, restoration led to levels of biodiversity of non-native vascular plants lower than levels in natural wetlands. Both of these outcomes may reflect the large, persistent effects of exotic plants on the habitat structure, biodiversity and functioning of wetlands [5]. In addition, wetlands dominated by exotic, invasive plants tend to support fewer native animal species and more invasive animals [5].

Greater diversity by itself is insufficient to ensure high ecosystem functioning [18]. Potentially even more important are the identities and relative proportions of species involved in the restoration process, as well as their ecological and functional properties. Unfortunately, most studies in our meta-analysis reported aggregate measures of richness or diversity but not community composition (**Supporting information S1**). Indeed a previous meta-analysis of how restoration affects major groups of organisms was restricted to calculating aggregate results for three general categories of vertebrates, macroinvertebrates, and plants [7]. Higher taxonomic and functional resolution is needed to explore the potentially quite different effects of restoration on organisms that can differ even within a class like vertebrates. Therefore, restoration studies dealing with species composition, community structure and functional ecology are urgently needed.

ES recovery

Our meta-analysis showed that restoration enhanced ES supply in degraded wetlands. The results also showed that it is more difficult to recover ES supply than to recover biodiversity; an alternative or complementary interpretation is that full recovery of ES supply takes longer than full recovery of biodiversity. Either interpretation is consistent with the meta-analysis by Rey Benayas *et al.* [9], but inconsistent with the analysis of North American wetlands by Dodds *et al.* [8].

Restoration did not enhance ES uniformly across all individual ES examined. We observed that restored wetlands provided, on average, 36% higher levels of provisioning, regulating and supporting ES than did degraded wetlands, but similar levels of cultural services. To be sure, we did not expect uniform recovery of all individual ES, given the heterogeneity of ES and wetland types included in the meta-analysis; wetlands types are known to differ in ecological dynamics, recovery rates and extents of recovery [7].

Our finding that restoration increased supply of provisioning services more than the supply of other ES may reflect the fact that, among the included studies, the desired outcomes when restoring provisioning services (e.g. abundance of target species) were generally better defined and more homogeneous than were objectives for regulating, supporting, and cultural services. Effect sizes for these last three services showed wide confidence intervals in our study, suggesting higher intra-class heterogeneity than effect sizes for provisioning services [12]. Small sample size may explain our finding that restoration did not significantly affect cultural services. Compared to natural wetlands, restored wetlands showed similar supply of provisioning and cultural services but lower supply of regulating services (mainly climate regulation, soil fertility and erosion) and supporting services (mainly biogeochemical cycles and provision of terrestrial habitat). The lower levels of climate and soil regulation, biological structure and biogeochemical cycles may reflect the intrinsically slow recovery rates reported for these surrogate variables [7]. In contrast, faster recovery rates have been reported for the water regulation variables in our study, such as hydrological dynamics and water quality, and these latter variables indeed showed full recovery.

Analysis of the ES database, which included abundance data on both non-native plant and animal species, showed that restoration increased regulation of non-native species by reducing their abundance. This result is different than our finding that restoration increased the diversity of such species, though it should be noted that the biodiversity database contained data on non-native plants but not non-native animals. The abundance of non-native species may decrease rapidly during the restoration process because these species are directly eradicated. However, a reduction in abundance, which reduces the supply of ES, does not necessarily indicate a decrease in species diversity, such as when a habitat contains several rare species in low abundance. Thus, assessment of restoration should take into account both abundance and diversity indicators.

Correlation of biodiversity and ES recovery

The relationship between biodiversity and ES supply remains poorly understood [19], yet it is crucial to work out because it has significant implications not only for restoration science but also for wider society, economics, and policy [20,21]. Our results showed that changes in biodiversity positively correlated with changes in ES supply in a variety of wetlands, ecosystem types and scales, which supports a functional role for biodiversity in the supply of ES [7,9]. This positive relationship is good news for restoration efforts, as it demonstrates the possibility of win-win scenarios for restoring biodiversity and ES. However, such win-win gains have not always proven feasible in practice, especially in restoration projects involving geographically dispersed areas [22]. Future research should explore how to optimize the synergy between biodiversity and ES supply in the design of management and conservation programs involving restoration.

The relationship between biodiversity and ES is also important because it has consequences beyond ecosystem restoration. For example, increasing plant diversity has been shown to enhance the provision of goods from plants and the regulation of erosion, invasive species and pathogens [23]; thus, recovering plant diversity may contribute to the recovery of ES beyond the immediate effects of restoration activities. Future research is needed to disentangle direct and indirect effects of restoration on biodiversity and ES, as well as clarify how the two types of effects interact.

Context dependence

Our meta-analysis identified several context factors that significantly affected biodiversity and ES recovery in restored wetlands, including ecosystem type, main cause of degradation, restoration action taken, and experimental design used to assess the restoration. This highlights the need to take context into account when evaluating the effects of wetland restoration. Particularly, examining interaction effects may generate useful insights, but the risk of multiple interactions, including two or even three factors, is too high for the relatively low statistical power of our model.

Our results also showed that biodiversity and ES recovery did not depend on restoration age. Nevertheless, they may depend on how long the restoration process took, on how many times a restoration action was repeated and on the conditions of the degraded wetland prior to restoration. Unfortunately most of the studies included in our meta-analysis did not report such data. The type and duration of interventions required in restoration depend heavily on the type and extent of ecosystem damage [24]. Future research should examine these context factors in greater detail.

Our finding that restoration effects depended on ecosystem type is consistent with an earlier meta-analysis showing that wetlands with more hydrologic flow exchange recovered faster than those that did not receive external water flow [7]. We obtained different results showing that outcomes of restoration were unrelated to flow exchange, e.g. biodiversity and ES in rivers and streams were enhanced in very different amounts. Despite these differences, the available evidence strongly indicates that the effectiveness of restoration is habitat-specific, arguing for the need for more research into how to tailor restoration projects to particular environments and how to assess their outcomes accordingly [6].

Our meta-analysis showed that only restoration action determined how close the biodiversity and ES supply of restored wetlands approached those of natural wetlands. This finding implies that unless the correct restoration action is chosen from the beginning, which is often impossible, the restored wetland may not come as close as possible to natural conditions. Applying a combination of restoration actions may therefore improve the likelihood of success.

Taken together, the results of our mixed models suggest that comparisons of degraded, restored, and reference conditions should be carried out to guide and evaluate restoration based on multiple indicators of both biodiversity and ES. These indicators should be consistent with the specific restoration goals [25], which can vary greatly depending on the context and project [26]. Our models further suggest that restoration programs should involve multiple actions to improve the likelihood of success.

Implications for wetland restoration

Comparing degraded, restored and reference conditions to guide restoration may not be feasible in many cases because the irreversibility of much of man-made ecosystem damage makes it difficult to simulate the pre-degradation condition accurately [27], and because movement of restored wetlands away from reference conditions makes it difficult to project desired outcomes [7], but it should be advisable. This highlights the need for designing restoration programs with multiple, alternative goals in mind [27,28]. These goals should take into account the social context and human values associated with decisions about wetland management and restoration. The concept of ES can be a robust guide for wetland restoration decision-making because it identifies and quantifies valuable goods and describes the processes and components that provide essential services [29]. Since several ES are difficult to measure directly, surrogate measures of ecosystem function can be used instead [30].

Accurately assessing the impact of restoration on biodiversity and ES supply requires identifying the particular ecosystem attributes in need of restoration. To capture potential differences in the restoration of individual ES, we linked the response variables to ES based on specific measures routinely included in ecological studies [31]. In addition, we evaluated the effects of response variables on multiple ES, since the variables may have indirect or unclear links to several ES that significantly affect restoration outcomes. For instance, although all plant species capture carbon, thereby increasing the supply of one ES, non-native species may have detrimental effects on other ES such as biotic interactions. A single restoration action may simultaneously affect various ES or act synergistically as a 'cascade' across trophic levels [14]. A restoration action may enhance the supply of one ES while precluding the supply of another [32], or it may generate a disservice, such as the release of greenhouse gases. Therefore, analyses of restoration data should assess both the direction and magnitude of associations between response variables and individual ES [14]. Taking into account the multiple ES associated with a restoration action facilitates the identification of tradeoffs or compromises when planning wetland restoration in which the overriding goal is optimizing multiple ES [29].

Cost plays an important role in restoration planning because it may limit the desired outcomes [33,34]. Surprisingly, the studies included in our meta-analysis did not address the issue of restoration costs. Costs are an important factor not only during restoration but also after: monitoring of wetlands following their restoration, mitigation or creation is often too brief because it is expensive to evaluate all the ecosystem functions involved.

These elements define a complex scenario for decision makers. Key to guiding decisions will be a systematic account of the relationships between wetland restoration variables and the supply of individual ES, for which the evidence base needs to be expanded. Indeed the low positive correlation between the recovery of biodiversity and ES suggests that reliable modeling of restoration outcomes will require incorporating multiple indicators that capture biodiversity, ES supply, and ecosystem processes. Such indicators should also include performance indicators that describe how much of available ES can be exploited [19], since biodiversity-related ES, for example, vary over time and space and are species-dependent. This poses a challenge for model-building, since simple models for simultaneously maximizing biodiversity and ES are unrealistic or ambitious [35], such that the two variables are not necessarily maximized in the same wetland [6]. The model that we have developed here may provide a basis for future studies that optimize biodiversity and ES supply for specific habitats and contexts.

Conclusions

Our meta-analysis strongly supports the idea that ecological restoration increases both biodiversity and ES supply in degraded wetlands, thereby benefiting the human communities that interact with and depend on them. The detailed effects of restoration depend heavily on context factors, emphasizing the need for habitat-specific planning and assessment of restorations [6]. Questions posed years ago remain largely unanswered today, such as "To what extent and over what time scale can ES be restored? [36] and "To what extent can mankind substitute for ES?" [37]. While restoration ecology is not obliged to answer these questions, exploring them may help improve the flows of ES and improve human well-being. Addressing these questions will require deepening our understanding of the links between restoration actions and changes in biophysical and ecological processes that generate ES [30]. While such research should inform and improve growing efforts to restore and mitigate loss of wetland area and loss of wetland ecosystem functions [35], they should not take importance away from efforts to conserve natural wetlands and avoid environmental degradation in the first place [8,9].

Supporting Information

Checklist S1

Supporting information S1 Studies used in the meta-analysis.

Supporting information S2 Additional information on database building.

Supporting information S3 Methodological details of the meta-analysis.

Supporting information S4 Overview of studies included in the meta-analysis.

Figure S1 Restoration effects by ecosystem type.

Figure S2 Restoration effects by main degrading factor.

Figure S3 Restoration effects by restoration action.

Figure S4 Restoration effects by experimental design.

Figure S5 PRISMA 2009 Flow Diagram.

Table S1 Linked databases used in the meta-analysis of biodiversity and ES.

Table S2 Comparison of biodiversity meta-analyses using a reduced or complete database.

Acknowledgments

We thank all authors who provided their original data to develop this work. R. Aguilar and N. Mariano provided suggestions to improve the analyses. S. Quijas provided helped to improve the figures of this ms. We are indebted to A. Chapin Rodríguez, who greatly improved the presentation of a previous version of this manuscript.

Author Contributions

Conceived and designed the experiments: JMRB PM. Performed the experiments: PM. Analyzed the data: PM JMRB PB MMR. Contributed reagents/materials/analysis tools: PM JMRB PB MMR. Wrote the paper: PM JMRB PB MMR.

References

1. Ramsar (2006) The Ramsar Convention Manual: a guide to the convention on wetlands (Ramsar, Iran, 1971), 4th ed. Gland: Ramsar Convention Secretariat. 114 p.
2. de Groot RS (1992) Functions of nature: evaluation of nature in environmental planning, management and decision making. Groningen: Wolters-Noordhoff. 315 p.
3. Costanza R, d'Arge R, de Groot R, Farber S, Grasso M, et al. (1997) The value of the world's ecosystem services and natural capital. Nat 387: 253–260.
4. Millenium Ecosystem Assesment (MA) (2005) Ecosystems and human well-being: a framework for assessment. Wetlands and water. Washington, DC: World Resources Institute. 68 p.
5. Zedler JB, Kercher S (2005) Wetland resources: status, trends, ecosystem services, and restorability. Ann Rev Environ Res 30: 39–74.
6. Zedler JB (2000) Progress in wetland restoration ecology. Trends Ecol Evol 10: 402–407.
7. Moreno-Mateos D, Power ME, Comin FA, Yockteng R (2012) Structural and functional loss in restored wetland ecosystems. PLoS Biol 10(1):e1001247. DOI:10.1371/journal.pbio.1001247.
8. Dodds WK, Wilson KC, Rehmeier RL, Knight GL, Wiggam S, et al. (2008) Comparing ecosystem goods and services provided by restored and native lands. BioSci 58: 837–845.
9. Rey Benayas JM, Newton AC, Diaz A, Bullock JM (2009) Enhancement of biodiversity and ecosystem services by ecological restoration: a meta-analysis. Sci 325: 1121–1124.
10. Cardinale BJ, Matulich KL, Hooper DU, Byrnes JE, Duffy E, et al. (2011) The functional role of producer diversity in ecosystems. Am J Bot 98: 572–592.
11. Balvanera P, Pfisterer AB, Buchmann N, Jing-Shen H, Nakashizuka T, et al. (2006) Quantifying the evidence for biodiversity effects on ecosystem functioning and services. Ecol Lett 9: 1146–1156.
12. Rosenberg MS, Adams DC, Gurevitch J (2000) Metawin: statistical software for meta-analysis. Sunderland: Sinauer Associates. 120 p.
13. Vilà M, Espinar JL, Hejda M, Hulme PE, Vojtěch J, et al. (2011) Ecological impacts of invasive alien plants: a meta-analysis of their effects on species, communities and ecosystems. Ecol Lett 14: 702–708.
14. de Bello F, Lavorel S, Díaz S, Harrington R, Cornelissen JHC, et al. (2010) Towards an assessment of multiple ecosystem processes and services via functional traits. Biod & Cons 19: 2873–2893.
15. Keeler BL, Polasky S, Brauman KA, Johnson KA, Finlay JC, et al. (2012) Linking water quality and well-being for improved assessment and valuation of ecosystem services. Proc Nat Acad Sci 109: 18619–18624.
16. KCS 2011 Data Desk v6.1. Kovach Computing Services, Wales: Anglesey.
17. Lekve K, Boulinier T, Stenseth NChr, Gjøsaeter J, Fromentin JM, et al. (2002) Spatio-temporal dynamics of species richness in coastal fish communities. Proc Roy Soc 269: 1781–1789.
18. Callaway JC (2005) The challenge of restoring functioning salt marsh ecosystems. J Coast Research 40: 24–36.
19. de Groot RS, Alkemade R, Braat L, Hein L, Willemen L (2010) Challenges in integrating the concept of ecosystem services and values in landscape planning, management and decision making. Ecol Compl 7: 260–272.

20. Worm B, Barbier EB, Beaumont N, Duffy JE, Folke C, et al. (2006) Impacts of biodiversity loss on ocean ecosystem services. Sci 314: 787–790.
21. Naidoo R, Balmford A, Costanza R, Fisher B, Green RE, et al. (2008) Global mapping of ecosystem services and conservation priorities. Proc Nat Acad Sci 105: 9495–9500.
22. Reyers B, Polasky S, Tallis H, Mooney HA, Larigaurderie A (2012) Finding common ground for biodiversity and ecosystem services. BioSci 62: 503–507.
23. Quijas S, Jackson LE, Maass M, Schmid B, Raffaelli D, et al. (2012) Plant diversity and generation of ecosystem services at the landscape scale: expert knowledge assessment. J Appl Ecol 49: 929–940.
24. Hobbs RJ, Cramer VA (2008) Restoration ecology: interventionist approaches for restoring and maintaining ecosystem function in the face of rapid environmental change. Ann Rev Environ Res 33: 39–61.
25. Hobbs RJ (2003) Ecological management and restoration: assessment, setting goals and measuring success. Ecol Manage & Res 4: S2–S3.
26. Palmer MA, Bernhardt ES, Allan JD, Lake PS, Alexander G, et al. (2005) Standards for ecologically successful river restoration. J App Ecol 42: 208–217.
27. Choi YD. (2004) Theories for ecological restoration in changing environment: toward "futuristic" restoration. Ecol Research 19: 75–81.
28. Hobbs RJ, Higgs E, Harris JA (2009) Novel ecosystems: implications for conservation and restoration. Trends Ecol Evol 24: 599–605.
29. Windhager S, Steiner F, Simmons MT, Heymann D (2010) Toward ecosystem services as a basis for design. Lands J 29: 2–10.
30. Palmer M, Filoso S (2009) Restoration of ecosystem services for environmental markets. Sci 325: 575–576.
31. Díaz S, Lavorel S, de Bello F, Quétier F, Grigulis K, et al. (2007) Incorporating plant functional diversity effects in ecosystem service assessments. Proc Nat Acad Sci U S A 104: 20684–20689.
32. Ehrenfeld JG (2000) Defining the limits of restoration: the need for realistic goals. Res Ecol 8: 2–9.
33. Aronson J, Blignaut JN, Milton SJ, Le Maitre D, Esler KJ, et al. (2010) Are socioeconomic benefits of restoration adequately quantified? A meta-analysis of recent papers (2000–2008) in Restoration Ecology and 12 other scientific journals. Res Ecol 18: 143–154.
34. de Groot RS, Blignaut J, Van der Ploeg S, Aronson J, Elmqvist T, et al. (2013) Benefits of investing in ecosystem restoration. Cons Biol DOI: 10.1111/cobi.12158.
35. Findlay SEG, Kiviat E, Nieder WC, Blair EA (2002) Functional assessment of a reference wetland set as a tool for science, management and restoration. Aq Sci 64: 107–117.
36. Daily GC (1997). Nature's services. Societal dependence on natural ecosystems. Washington, DC: Island Press. 396 p.
37. Costanza R (2000) Social goals and the valuation of ecosystem services. Ecosys 3: 4–10.

10

Outbreaks of the Brown Planthopper *Nilaparvata lugens* (Stål) in the Yangtze River Delta: Immigration or Local Reproduction?

Gao Hu[1*⁹], Fang Lu[1⁹], Bao-Ping Zhai[1], Ming-Hong Lu[2], Wan-Cai Liu[2], Feng Zhu[3], Xiang-Wen Wu[4], Gui-Hua Chen[5], Xiao-Xi Zhang[1]

1 Key Laboratory of Integrated Management of Crop Diseases and Pests (Ministry of Education), College of Plant Protection, Nanjing Agricultural University, Nanjing, China, **2** Division of Pest Forecasting, China National Agro-Tec Extension and Service Center, Beijing, China, **3** Plant Protection Station of Jiangsu Province, Nanjing, China, **4** Plant Protection Station of Shanghai City, Shanghai, China, **5** Plant Protection Station of Jinhua City, Jinhua, China

Abstract

An effective control strategy for migratory pests is difficult to implement because the cause of infestation (i.e., immigration or local reproduction) is often not established. In particular, the outbreak mechanisms of the brown planthopper, *Nilaparvata lugens* (Stål), an insect causing massive losses in rice fields in the Yangtze River Delta in China, are frequently unclear. Field surveys of *N. lugens* were performed in Jiangsu and Zhejiang Provinces in 2008 to 2010 and related historical data from 2003 onwards were collected and analyzed to clarify the cause of these infestations. Results showed that outbreaks of *N. lugens* in the Yangtze River Delta were mostly associated with an extremely high increase in population. Thus, reproduction rather than immigration from distant sources were the cause of the infestations. Although mass migration occurred late in the season (late August and early September), the source areas of *N. lugens* catches in the Yangtze River Delta were mainly located in nearby areas, including the Yangtze River Delta itself, Anhui and northern Jiangxi Provinces. These regions collectively form the lower-middle reaches of the Yangtze River, and the late migration can thus be considered as an internal bioflow within one population.

Editor: Robert B. Srygley, USDA-Agricultural Research Service, United States of America

Funding: Funding provided by the National '973' Program of China (2010CB126200), Shanghai City Sci-Tech Joint Research Project in Yangtze River Delta of Shanghai Municipal Science and Technology Commission (13395810100). The funders had no role in study design, data collection and analysis, decision to publish, or preparation of the manuscript.

Competing Interests: The authors have declared that no competing interests exist.

* E-mail: hugao@njau.edu.cn

⁹ These authors contributed equally to this work.

Introduction

Numerous migratory insects, such as *Schistocerca gregaria* (Forskal), *Chortoicetes terminifera* (Walker), *Locusta migratoria manilensis* (Meyen), and *Cnaphalocrocis medinalis* (Guenée), are pests to agricultural crops, and their population outbreaks often cause immense crop damage and serious economic problems [1–5]. The population growth of migratory insects has been attributed to migration and local reproduction. Local pest populations could grow exponentially, at least for a time, because of their high intrinsic rate of increase. In contrast, host plants could be devastated immediately after mass immigration, such as in locusts [2,3]. Therefore, determining the intrinsic cause of pest population outbreaks under the simultaneous effect of immigration and local reproduction is difficult. Here, we were specifically concerned with determining the cause of Nilaparvata lugens (Stål) infestations in the Yangtze River Delta in order to significantly improve pest forecasting and control of this pest.

N. lugens is one of the most serious rice pests in the temperate and tropical regions of East and Southeast Asia. Since 1968, *N. lugens* has greatly damaged the rice industry in China and in other countries in Asia [6,7]. In the past few years, new outbreaks of *N. lugens* have been continuously recorded. For example, the loss of rice yield in China caused by *N. lugens* was approximately 1,880,000 t in 2005 [8–10]. The Yangtze River Delta, including Shanghai Municipality, southern Jiangsu, and northern Zhejiang Provinces (Fig. 1), is one of the most important rice-producing regions in China. The occurrences of *N. lugens* outbreaks in this region, including those in 2005 and 2006, are more severe and more frequent compared with those in other regions. The problem of *N. lugens* outbreaks in the Yangtze River Delta has elicited more attention and investment compared with other areas; thus, it is the primary focus of this paper.

N. lugens has an annual return migration in East Asia. *N. lugens* cannot overwinter in temperate zones, such as mainland China, Japan, and the Korean Peninsula. Instead, an infestation is initiated by windborne spring/summer migrants from the south [6,7,11–14]. Infestation begins with a northward migration that occurs in late March every year. The distribution of *N. lugens* then further expands northward by progeny of the migrant population and can cover the entire rice-growing region in China, as well as Japan and the Korean Peninsula. From September onwards, the general direction of planthopper migration becomes predominantly southbound [15,16]. At the end of October, most *N. lugens*

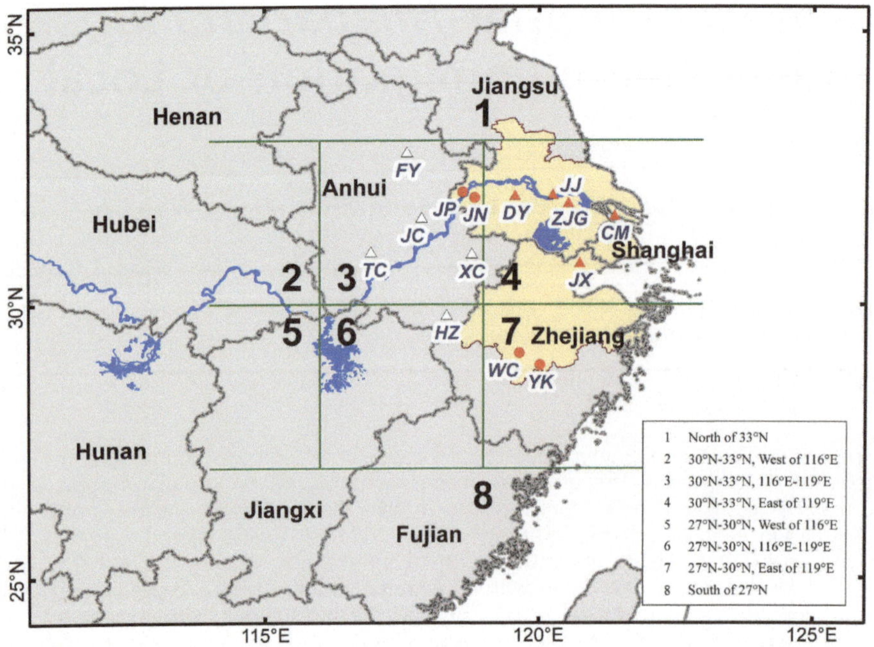

Figure 1. Location of experimental sites and the probable source area of *N. lugens* in the Yangtze River Delta. Notes: The region filled with yellow represents the Yangtze River Delta. Area 1 includes northern Jiangsu and northern Anhui; Area 2 includes southeastern Henan and eastern Hubei; Area 3 includes Southern and central Anhui; Area 4 includes the Yangtze River Delta; Area 5 includes northeastern Hunan and northwestern Jiangxi; Area 6 includes northeastern Jiangxi, northwestern Fujian, and western Zhejiang; Area 7 includes southern and central Zhejiang; Area 8 includes southern Jiangxi and southern and central Fujian. "●" represents the experimental sites in Nanjing and Jinhua City, JP-Jiangpu, JN-Jiangning, WC-Wucheng, and YK-Yongkong. "▲" represents stations in the Yangtze River Delta, CM-Chongming, DY-Danyang, JJ-Jingjiang, ZJG-Zhangjiagang, and JX-Jiaxing; "Δ" represents stations in the probable source areas of *N. lugens*, FY-Fengyang, JC-Juchao, TC-Tongcheng, XC-Xuancheng, and HZ-Huizhou.

populations are only present in the safe overwintering areas, mainly in the Indo–China Peninsula, within which a proportion of successful return migrants is present [8,11,12,14]. The immigration peaks of *N. lugens* always appear in mid- to late June and mid- to late July in the Yangtze River Delta when the early season rice ripens in the southern and northern areas of South China. *N. lugens* reproduces three generations and causes considerable damage on single-crop late-maturing rice in late September [8,12].

The lower-middle reaches of the Yangtze River are divided into two types of cultivation based on the rice planting system used. One area comprises the Yangtze River Delta, wherein single cropping rice has been dominant since 1997.The other area, including Hubei, southern and central Anhui, northern Hunan, and northern Jiangxi, consists of mixed cultivation wherein both single and double cropping techniques are used. When *N. lugens* in South China begins to emigrate because of the decline in rice availability, the large areas of single-crop rice in mixed planting areas are suitable for their continued survival and breeding. Consequently, the population rapidly grows, which leads to a mass late migration in the Yangtze River Delta where the single-crop rice is at the heading stage during late August to early September [8]. By contrast, migration before mid-August is called early migration. Hu et al. (2011) believe that outbreaks of *N. lugens* in the Yangtze River Delta are frequent because of mass late migration caused by the increased area of the single-crop rice in the mixed planting area in the lower-middle reaches of the Yangtze River [8,10,17–19].

However, *N. lugens* can reproduce two to three generations from colonizing in June and July or until the rice reaches the yellow ripe stage in early to mid-October in the Yangtze River Delta. *N. lugens* is an "r-strategy" organism; thus, the population size of this species could become very large after two or three generations. Based on this perspective, Cheng and Zhu (2006) concluded that the high growth rate of *N. lugens* is the key to outbreaks in the lower-middle reaches of the Yangtze River, wherein the immigrants provide the initial population and the high number of *N. lugens* catches by light trap in the late season comes from the local population [20]. However, this hypothesis contradicts the fact that emigrating planthoppers (produced by the build-up of local populations) are not thought to be caught by local light traps [21–24].

Field surveys of the population dynamics of *N. lugens* were performed in Jiangsu Province and Zhejiang Province from July to September (2008 to 2010) to resolve the conflicting hypotheses. The data analyzed here were from 2003 to 2010 including systematic field investigation to define the important factors of the outbreak. This study aims to provide insights into ways of forecasting and controlling the population outbreaks of *N. lugens* by studying the outbreak mechanisms of this pest in the Yangtze River Delta.

Materials and Methods

Systematic field investigation

Systematic field investigations were conducted to explore the population dynamics of *N. lugens* in the Yangtze River Delta. In each experimental site, the one or two paddies selected were moderately fertile with routine cultural practices, and, ideally, no pesticides were used to control pests during the rice-growing season. The rice varieties in most areas of the Yangtze River Delta are the local commercial varieties of japonica rice. Systematic field investigation of *N. lugens* populations was performed once every 3 d by the plant-shaking method. A plate (39 cm×29.5 cm×2 cm) was inserted at the base of the rice plants, and *N. lugens* samples were obtained by shaking the plants. The number of *N. lugens* per hill (i.e. all tillers growing from a seedling or a "clump") was counted manually [8].

Systematic field investigations were conducted in Nanjing City, Jiangsu Province from 2008 to 2010. Similar investigations were also carried out in Jinhua City, Zhejiang Province, since 2009 (Fig. 1). The experimental sites were set by the Plant Protection Station of each city; thus, the actual locations of the paddies were different for each year (Table 1). The selected paddies were leased from the local farmers. Farmers always tried to control the pest when no observers were present, and thus pesticides were applied in 2010 (Table 1).

According to the rules of investigation and forecast for *N. lugens* and *Sogatella furcifera* Horváth [25], the degree of occurrence was classified into the following five levels based on the number of *N. lugens* per 100 hills: light (<500), moderate (between 500 and 1000), high (between 1000 and 2000), outbreak (between 2001 and 3000), and severe outbreak (>3000).

Ovary dissection

Macropterous females of *N. lugens* were collected from experimental paddies and dissected every 3 d to estimate the level of ovary development. Using this method, Chen et al. (1979) [26] classified the ovarian development into five levels and related it to the migration status of the population. According to Chen et al.'s criteria, the level of ovary development of the insects gathered from the source areas or during the emigration period was at level I. By contrast, the ovaries of most female insects collected at the landing areas or during the immigration periods were mature with ovary development level III or above because the oocytes underwent continuous and immediate development upon landing of the insect [8,26,27].

Light trap

In each experimental site, a black light trap was placed less than 20 m away from a rice paddy. The trap consisted of a 20 W black light lamp with a top cover, set on top of a pole, 2 m above ground. These lamps were switched on at 1900 h (Beijing Time, same thereafter) and off at 0700 h next morning every day during the systematic field investigation. The catches in the light traps were collected every morning at 0900 h and were identified using a stereomicroscope. *N. lugens* catches were collected and counted every 2 h each night in Nanjing City in September 2010 to observe the nocturnal flight dynamics of the planthoppers.

Other light trap data and field observation data

The daily *N. lugens* light trap data and 5-day field systematic survey data during 2003 to 2010 from plant protection stations of 10 counties, including Danyang (Jiangsu), Zhangjiagang (Jiangsu), Jingjiang (Jiangsu), Chongming (Shanghai), Jiaxing (Zhejiang), Huizhou (Anhui), Tongcheng (Anhui), Xuancheng (Anhui), Fengyang (Anhui), and Juchao (Anhui), were obtained by the China National Agro-Tec Extension and Service Center (Fig. 1). The five stations in the Yangtze River Delta were studied in mid- to late August and mid- to late September since 2003 to ascertain the occurrence pattern of *N. lugens* in the Yangtze River Delta.

Traditional black light traps (20 W black light lamp) were used to catch planthoppers before 2005. From 2005 onwards, the black light traps were replaced with frequoscillation lamps (20 W black light lamp, Jiaduo Brand, Jiaduo Science, Industry and Trade Co. Ltd., China). The frequoscillation lamps were switched on at 1900 h and off at 0700 h in the next morning every day at the same time from April 1 to November 15.

Population increase rate of *N. lugens* in the Yangtze River Delta

N. lugens migrated from South China to the Yangtze River Delta before mid-August. Rice plants in the Yangtze River Delta were at the mature stage after mid-September. *N. lugens* mainly causes severe damage to rice crops in this region from mid-September. Thus, the population increase rate (*R*) in the late season was defined as the ratio of the total number of nymphs and adults when the number of early instar nymphs (instar I-III) peaked in mid- and late September to the corresponding peak in mid- and late August. The period between the early instar nymph peak in August and September was approximately equal to one generation duration of N. lugens. A high population increase rate was defined as a ratio of >40 [28,29].

Trajectory analysis

Probable source areas of migratory *N. lugens* were defined by constructing backward/forward trajectories, which were based on the following assumptions: (i) planthoppers are displaced downwind [8,12,15,21,30–32]; (ii) migration normally starts at dusk and partly at dawn [8, 12, 115, 21, 30–32]; (iii) migrants frequently concentrate at the height of approximately 1,000 m [15,32]; and (iv) the planthoppers cannot fly in an atmosphere of <16.5°C [8,15,21,30]. The NOAA ARL HYSPLIT model [8,33,34] was applied to calculate backward trajectories. The backward trajectories for light trap locations were calculated every 2 h during peak periods (1900 h to 0500 h of the next day) and terminated at the take-off time of *N. lugens* (viz. 1900 h on the day and 1900 on the day before that day), with the initial height at 500, 1,000, and 1,500 m aboveground. The duration of trajectories did not exceed 34 h [12,35,36]. An endpoint of a trajectory was considered a probable source if it was located in a rice planting area where the crop was at a later growth stage.

For the five stations (Chongming, Danyang, Jingjiang, Zhangjiagang, and Jiaxing), 90 peak days from August 21 to September

Table 1. Information on experimental paddies from 2008 to 2010.

Locations	Survey periods	Pesticide used
Jiangning, Nanjing	2008/8/3-10/6	No
Jiangning, Nanjing	2009/7/9-9/27	No
Yongkang, Jinhua	2009/7/11-9/27	No
Jiangpu, Nanjing	2010/7/31-9/23	8/3, 8/20: pesticide was applied to control rice planthopper
Wucheng, Jinhua	2010/7/11-9/25	8/13, 9/3: pesticide was applied to control rice planthopper

10 (2003 to 2010) were selected to calculate backward trajectories. The number of trajectory endpoints was counted in each 1°×1° grid to represent the source area. The whole probable source area was divided into eight zones (Fig. 1), and the probability of valid trajectory endpoints in each zone was then calculated.

Meteorological data

Data on meteorological factors (2003 to 2010), including temperature and wind, were obtained from the China Meteorological Data Service System (http://cdc.cma.gov.cn). The temperature data were expressed as the daily mean value. Wind direction data at the height of 850 hPa were obtained from the Shanghai Municipality at 0800 and 2000 h every day.

Statistical analysis

For a 5% rejection threshold, the association between population increase rate and size of light-trap catch in outbreak and non-outbreak years in five stations from 2003 to 2010 were tested using Chi-square analysis and obtained using *R* (version 3.0.0, http://www.r-project.org/). When sample sizes were less than 5, Fisher's exact test was then applied.

Ethics statement

No specific permits were required for the described field studies. The brown planthopper *N. lugens* (Stål) is a major pest of rice in Asia, and huge amounts of manpower and resources are used to control the damage it causes every year.

In this study, we confirmed the following: (i) the location is not privately owned or protected and (ii) the field studies did not involve endangered or protected species.

Results

Population dynamics of *N. lugens* in the Yangtze River Delta

The *N. lugens* population in all experimental paddies was at a low density before late July and generally peaked in the second half of September (Fig. 2). In three out of the five paddies, the *N. lugens* population exceeded 3,000 insects per 100 hills in late September. The three paddies were in Jiangning (the maximum population was 51,140 insects per 100 hills, September 20, 2008), Yongkang (15,673 insects per 100 hills, September 18, 2009), and Wucheng (6,885 insects per 100 hills, September 16, 2010). In the first two paddies, the "hopper burn" (browning of the leaves or withering of the whole plant) areas were approximately 60% and 10% (Fig. 3 and Table 2). The damage in the paddy in Wucheng was minimal. However, excluding the paddy in Wucheng, the increase rates of *N. lugens* population in the other two paddies with high density were much greater than 40, which was defined by previous studies as a high increase rate (Table 2) [29]. In the paddy in Wucheng, the population increase rate was kept low after August 10. However, pesticide control was performed twice to suppress the increase in *N. lugens* population at this site (Tables 1 and 2).

During the late season (after mid-August), three main peak periods of *N. lugens* light trap catches were recorded in the experimental sites. The first two periods were noted in late August and early September, i.e. before the population densities in the paddies reached the maximum value. However, whether these catches were immigrants from distant sources was unclear. Among these cases, the catches of *N. lugens* in Jiangpu in 2010 were the highest (3,681 specimens caught from August 21 to September 10) (Fig. 2 and Table 2). In the second peak periods (September 7 to September 9) in Jiangpu in 2010, 2,697 individuals were caught in these three nights. Up to 1,397 out of these 2,697 individuals (48.20%) were caught during 0200 h to 0500 h, and 1,114 (41.31%) were caught around midnight (2100 h to 0200 h). The individuals caught by light trap at midnight and later must have been immigrants.

In each experimental paddy, >100 macropterous females were collected and dissected during the first two peak periods of light trap catches, viz. in late August and early September. Excluding 2010, the individuals with ovaries at levels I and II accounted for >70% of the entire population, whereas those at level III and above accounted for <30% (Table 3). According to the classification criteria established by Chen et al. (1979), these values characterize an emigrating population. Thus, the macropterous adults were from relatively local populations, and the high increase rates were attributed to reproduction rather than immigration. In 2010, the individuals with ovaries at level III and above accounted for a distinctly higher proportion of the population (>30% but <60%) (Table 3). Thus, according to the classification criteria established by Chen et al. (1979) [26,27], the population consisted of local insects that continued to breed in approximately the same area or that mixed with immigrants.

These results showed that the population growth of *N. lugens* in the late season in the Yangtze River Delta was evidently caused by reproduction rather than immigration in 2008 and 2009. The outbreaks in Jiangning (2008) and Yongkong (2009) resulted from the high increase rate of the local population. In 2010, numerous immigrants maintained the large population, whereas the local population was suppressed by pesticide control.

Relationship between autumn temperature and outbreaks of *N. lugens* in the late season

The temperature in Jinhua was consistently higher than in Nanjing (Table 4). Accordingly, the population sizes of *N. lugens* were larger, and the population peaks were earlier in Jinhua (Table 2). Excluding Nanjing in 2009, the temperatures were near or above 26°C before the population densities in the paddies reached the maximum value and were suitable for the population growth of *N. lugens* (Table 4). In Nanjing in mid-September of 2009, the temperature was only 22.8°C and the *N. lugens* population may have been suppressed by these cool temperatures. The population of *N. lugens* in 2010 was controlled by pesticides, so whether a suitable temperature simulated the increase in *N. lugens* population in this season is unclear.

Occurrence of *N. lugens* in the Yangtze River Delta from 2003 to 2010

According to the field survey data in 2003 to 2010, most stations in the Yangtze River Delta suffered outbreaks in 2005, 2006, and 2007, whereas occurrences were light in other years. In the five stations (Chongming, Danyang, Jingjiang, Zhangjiagang, and Jiaxing) during these eight years, 14 out of 40 cases (35.0%) reached the outbreak level, viz. the number of *N. lugens* was >3,000 per 100 hills (Table 5).

The population increase rates of *N. lugens* in these five stations were calculated for each year (Table 5). Among the 14 outbreak cases, 10 (71.4%) occurred when population increased with a high increase rate ($R>40$), whereas only four cases (28.6%) occurred with a low increase rate (Table 6). The results of Chi-square test ($\chi 2 = 14.70$, $p<0.001$; Fisher's exact test, $p<0.001$) indicated that the outbreak of *N. lugens* in the Yangtze River Delta was significantly associated with the population increase rate.

In the five stations, the cumulative light-trap catches of *N. lugens* in early migrations were <1,000. However, the numbers in the cumulative catches during the late season varied over a wide range

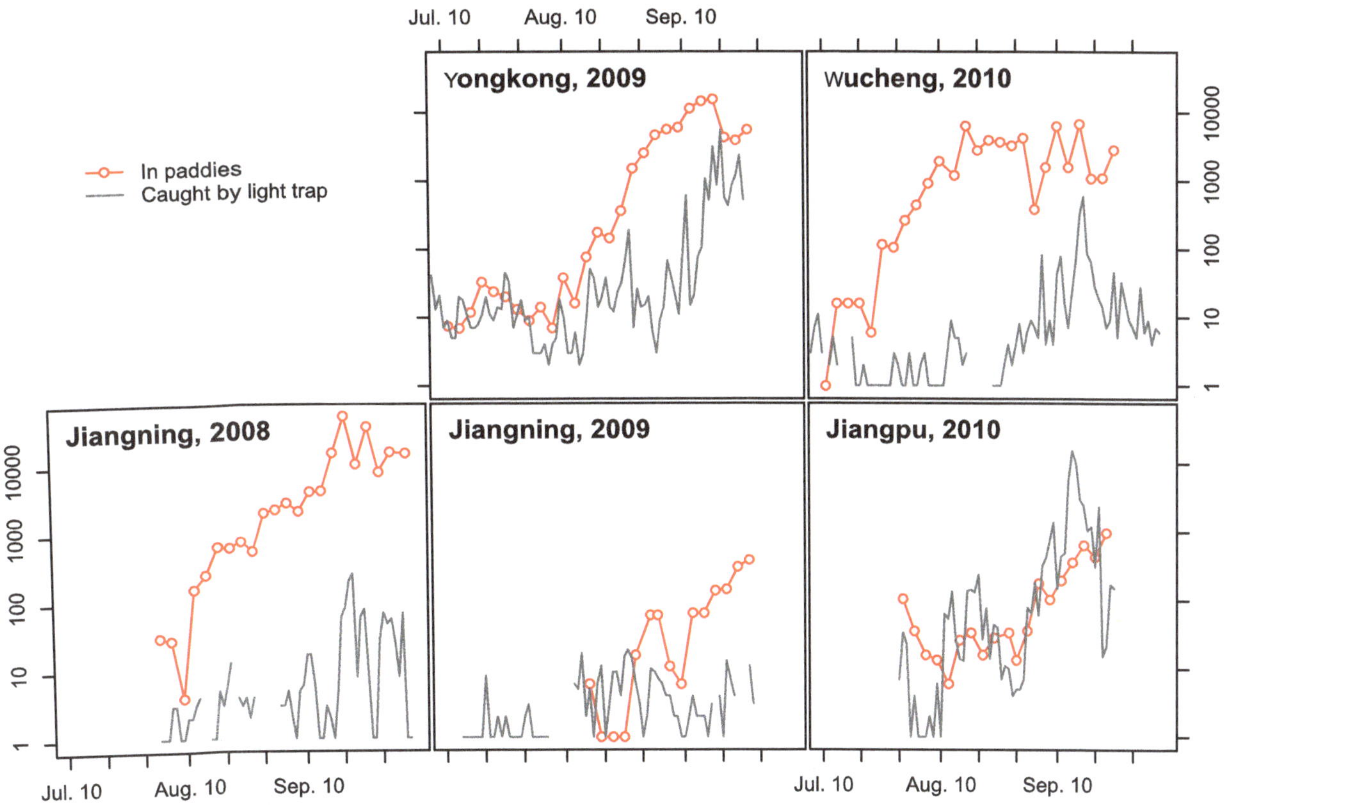

Figure 2. Number of *N. lugens* in paddies and caught by light trap at experimental sites.

(Fig. 4). In the outbreak years (2005, 2006, and 2007), mass migrations occurred during late August to early September. In 2006, the cumulative catches in the late season were the highest ever recorded at several stations, with 10,000 catches per night in peak days. The maximum catch in 2006 was 790,000, which was recorded in Zhangjiagang on August 30. In the light occurrence years, the cumulative catches in the late migration were consistently minimal. In 2003 and 2009, the catches on peak days were tens, while very few rice planthoppers were caught in the late season of 2004, 2008, and 2010 (Fig. 4). Among the 14 outbreak cases, nine cases (64.29%) occurred with numerous catches in the late season, and five cases (35.71%) occurred with a few catches (Table 6). The results of Chi-square test ($\chi 2 = 6.26$, $p = 0.012$; Fisher's exact test, $p = 0.007$) indicated that the outbreak of *N. lugens* in the Yangtze River Delta was significantly associated with light trap catches in the late season.

Among the 14 outbreak cases, five cases occurred with both high population increase rate and numerous catches in the late season, five cases occurred with only a high population increase rate, and the other four cases occurred with only numerous catches (Table 6). These five cases with only the high population increase rate occurred from 2008 to 2009 when the occurrence level of *N. lugens* was light in the whole Yangtze River Delta, and few migrants were noted during that period. Thus, the high increase rate in these five cases was confirmed to be mainly caused by reproduction. Thus, to summarize, high reproductive rates in

Figure 3. "Hopper burn" (A) and macropterous adults swarming on rice leaves (B) in Jiangning District, Nanjing City, in 2008. Note: Pictures were taken on October 2 (yellow ripe stage). Rice was harvested on October 20.

Table 2. Population increase rate of investigation fields from 2008 to 2010.

Year	Station	number of *N. lugens* per 100 hills		Proliferation multiple	Area of hopper burn (%)	Catches of *N. lugens* (Aug. 21–Sep. 10)
		Aug.	Sept.			
2008	Jiangning	645 (18)*	51140 (20)	79.29	60	58
2009	Jiangning	15 (28)	395 (27)	26.33	0	110
2009	Yongkang	124 (19)	15070 (18)	121.53	10	688
2010	Jiangpu	28 (24)	980 (23)	35.00	0	3681
2010	Wucheng	6335 (17)	6885 (16)	1.09	0	177

* The numbers in the brackets represent the date when the nymphal population was in early instars (I–III). The periods between the Aug. and Sept. dates were almost one month long and were nearly equal to the generation time of N. lugens.

local or relatively local populations seemed to be the most important factors giving rise to outbreaks.

Source areas of light trap catches in the Yangtze River Delta in the late season

Up to 1620 backward trajectories were calculated for 90 peak days in five stations, and 2,970 endpoints were obtained. After elimination of invalid endpoints (i.e. ones not ending in suitable rice cultivations), the number of available endpoints was 1,802, which accounted for 60.7% of the original 2,970.

The source areas within 2 h to 10 h durations were still in the Yangtze River Delta (83.66% of the total), followed by southern and central Anhui (10.93% of the total). The source areas within 24 h to 34 h durations were mainly in southern and central Anhui (22.71% of the total), the Yangtze River Delta (15.58%), and northern Jiangxi (14.15%). Moreover, only a few endpoints were located in northern Jiangsu, southern Zhejiang, northern Fujian, northeastern Hunan, eastern Hubei, etc. (Table 7, Fig. 5).

Whether the occurrence was light or severe, the source areas within 2 h to 10 h durations were therefore mainly in the surrounding area and partly in southern-central Anhui. The source areas in the Yangtze River Delta and southern-central Anhui accounted for 73.81% and 24.29%, respectively, in the years with light occurrence (2003 and 2009), and 86.42% and 7.19% in the years with outbreaks (2005, 2006, and 2007) (Table 7).

The source areas within 24 h to 34 h durations were different in the years with light occurrence compared to those with outbreaks. The endpoints during the light occurrence years were mostly in southwestern and western regions, including southern and central Anhui, northern Jiangxi, northeastern Hunan, northwestern Fujian, and western Zhejiang (65.33% of the total) (Table 7). The endpoints during the outbreak years were mostly distributed in southern and central Anhui, the Yangtze River Delta, northern Jiangsu, and northeastern Jiangxi (70.72% in total) (Table 7). Altogether, the source areas of *N. lugens* within 24 h to 34 h durations across all the years were mainly in Anhui and northern Jiangxi, which were located west of the Yangtze River Delta.

The interannual variation of macropterous adult number in Anhui corresponded with the interannual variation of catches in the Yangtze River Delta. For example, in the five locations (Huizhou, Tongcheng, Xuancheng, Fengyang, and Juchao) in Anhui, the numbers of macropterous adults from August 21 to September 10 were high in 2005 to 2007, when numerous catches were also obtained in the Yangtze River Delta (Fig. 6). Thus, Anhui could be the main source area of *N. lugens* in the Yangtze River Delta.

Wind directions in the late season in Yangtze River Delta

The summer (southwest) monsoon is replaced by the winter (north-east) monsoon in eastern Asia in late August and early September. In the Yangtze River Delta, the northeasterlies are gradually established at this time. However, the winds change to southwesterly/westerly under the influence of oscillation in subtropical anticyclones. For example, in Shanghai, the wind direction at 850 hPa was mainly southeast (17.6%), southwest (17.6%), and west (18.2%) in late August and was mainly northwest (20%), north (18.8%), and northeast (18.8%) in early September. During late August to early September, the probabilities of all wind directions were similar from 2003 to 2010 (Fig. 7A). However, more west or northwest winds were noted on the 21 peak days of *N. lugens* immigration from August 21 to September 10 (2003 to 2010), with

Table 3. Ovarian development of *N. lugens* in the experimental paddies from 2008 to 2010.

Year	Time	Station	Number of *N. lugens* dissected	Grade of ovarian development		
				Level I	Level II	≥ Level III
2008	8/23–9/11	Jiangning	186	168 (90.32%)	9 (4.84%)	9 (4.84%)
2009	9/9–9/18	Jiangning	111	35 (31.53%)	73 (65.77%)	3 (2.70%)
2009	8/19–9/12	Yongkang	106	68 (64.15%)	8 (7.55%)	30 (28.30%)
2010	9/9–9/16	Jiangpu	154	33 (21.43%)	47 (30.52%)	74 (48.05%)
2010	8/20–9/10	Wucheng	195	120 (61.54%)	4 (2.05%)	71 (36.41%)

Table 4. Average temperature from August 21 to September 30 every 10 d during 2008 to 2010 (°C).

Year	Nanjing				Jinhua			
	8/21–8/31	9/1–9/10	9/11–9/20	9/21–9/30	8/21–8/31	9/1–9/10	9/11–9/20	9/21–9/30
2008	25.5	24.4	25.7	23.0	-	-	-	-
2009	26.8	25.2	22.8	22.3	29.7	28.0	26.5	24.5
2010	26.7	27.1	25.5	20.1	29.3	29.2	27.9	21.4
Mean*	27.0	25.2	24.2	22.3	28.7	26.8	25.7	24.0

* Mean value for 2000–2010

33.3% (westerly) and 28.6% (northwesterly) probabilities (Fig. 7B). The results of Chi-square analysis revealed a significant difference in wind direction between migration days and other periods ($\chi2 = 24.91$, $p < 0.001$) (Table 8). Therefore, the late migration occurred more frequently under winds from the west or northwest. Considering individual seasons, during apparent late migrations in the serious outbreak years of 2005, 2006, and 2007, the frequency of wind directions was mainly west and northwest from August 21 to September 10 (Figs. 7C to 7J).

Discussion

Numerous *N. lugens* were caught by light trap in the Yangtze River Delta in the late season, but the source of these planthopper catches has not exactly been identified in previous studies. Thus, it was controversial whether "late migration" occurred, and the cause of *N. lugens* outbreak (i.e., immigration or local reproduction) has also remained unclear. Similar to some previous studies, the present study confirmed that late migration was an actual

Table 5. Number of *N. lugens* per 100 hills and population increase rate in the Yangtze River Delta from 2003–2010.

Station	Month	Number of *N. lugens*							
		2003	2004	2005	2006	2007	2008	2009	2010
Chongming	Aug.	10	28	85	38	30	12	8	86
		(20) [§]	(20)	(20)	(25)	(25)	(20)	(20)	(20)
	Sept.	319	101	357	24030	1315	291	49	622
		(20)	(20)	(20)	(20)	(20)	(15)	(15)	(15)
	R*	31.90	3.61	4.20	632.37	43.83	24.25	6.13	7.23
Danyang	Aug.	12	56	72	1797	162	170	264	54
		(20)	(15)	(25)	(20)	(20)	(31)	(31)	(31)
	Sept.	114	56	1230	6505	735	180	254	1404
		(20)	(15)	(30)	(20)	(15)	(25)	(25)	(25)
	R	9.50	1.00	17.08	3.62	4.54	1.06	0.96	26.00
Jingjiang	Aug.	42	10	318	1043	138	147	112	45
		(20)	(20)	(25)	(25)	(25)	(25)	(25)	(15)
	Sept.	30	18	21895	12348	10862	10671	5046	5298
		(20)	(25)	(20)	(25)	(25)	(25)	(20)	(20)
	R	0.71	1.80	68.85	11.84	78.71	72.59	45.05	117.73
Zhangjiagang	Aug.	46	108	210	151	13	87	54	5
		(25)	(25)	(25)	(15)	(15)	(20)	(15)	(25)
	Sept.	640	1106	965	4678	54	73	10	10
		(25)	(25)	(20)	(25)	(15)	(25)	(15)	(25)
	R	13.91	10.24	4.60	30.98	4.15	0.84	0.19	2.00
Jiaxing	Aug.	37	37	1	1891	100	13	7	186
		(15)	(20)	(20)	(20)	(20)	(20)	(15)	(25)
	Sept.	414	293	5098	17633	4377	3812	756	8242
		(20)	(20)	(20)	(20)	(15)	(25)	(15)	(25)
	R	11.19	7.92	5098.00	9.32	43.77	293.23	108.00	44.31

* R- Proliferation multiple
[§]The numbers in the brackets represent the date when the nymphal population was in early instars (I–III). The periods between the Aug. and Sept. dates were almost one month long and were nearly equal to the generation time of N. lugens.

Table 6. Frequency table comparing population increase rate and catch size for outbreak and non-outbreak situations.

	Outbreak	Non-outbreak	Total
High R and lots of catches in light trap	5	1	6
High R and few catches	5	1	6
Low R and lots of catches	4	4	8
Low R and few catches	0	20	20
Total	14	26	40

Note: R- Population increase rate

phenomenon. The four main findings of this study are the following. First, many catches in the light trap occurred at or after midnight. The flight of these individuals lasted at least 6 hours after their dusk take-off [8,12,15,21,30–32], and thus they must be immigrants. Second, the population increase rates in the field were, in many cases, near or even greater than the maximum fecundity of *N. lugens* (705 eggs per female [37]) observed in some laboratory cases, such as $R = 5{,}098$ in Jiaxing in 2005 and $R = 632$ in Chongming in 2006. Third, ovary dissection verified that the macropterous females in Nanhui, Shanghai City in 2007 were immigrants (see also [8,19]). Finally, the size of planthopper catches was related to the wind direction. Catches increased with westerly winds; thus, the major sources of the migrant insects were from the west. However, the backward trajectories in this study showed that the source areas of catches in the Yangtze River Delta in the late season were mainly located in relatively local areas, including the Yangtze River Delta itself, Anhui, and northern Jiangxi. These regions were located at a similar latitude (approximately 30°N), and the migration distances were consistently short by rice planthopper standards (mostly <300 km). When mass late migration occurred, the number of *N. lugens* in the Yangtze River Delta was consistently large, as well as that in Anhui, northern Jiangxi, and other regions at a similar latitude in eastern China. Therefore, these regions could be collectively called the lower-middle reaches of the Yangtze River. The initial population of *N. lugens* in these regions all migrated from South China in June and July. Thus, the late migration can be regarded as merely an internal flow within one population.

In the lower-middle reaches of the Yangtze River, some differences were found between the Delta itself and other areas. In the Yangtze River Delta, single-crop late-maturing rice was mostly planted, which is transplanted in mid- and late June and harvested in late October [8,18,19]. In other regions, including Hubei, southern and central Anhui, northern Hunan, and northern Jiangxi, the single-crop rice was subsequently transplanted between late May and early June and then matured in late

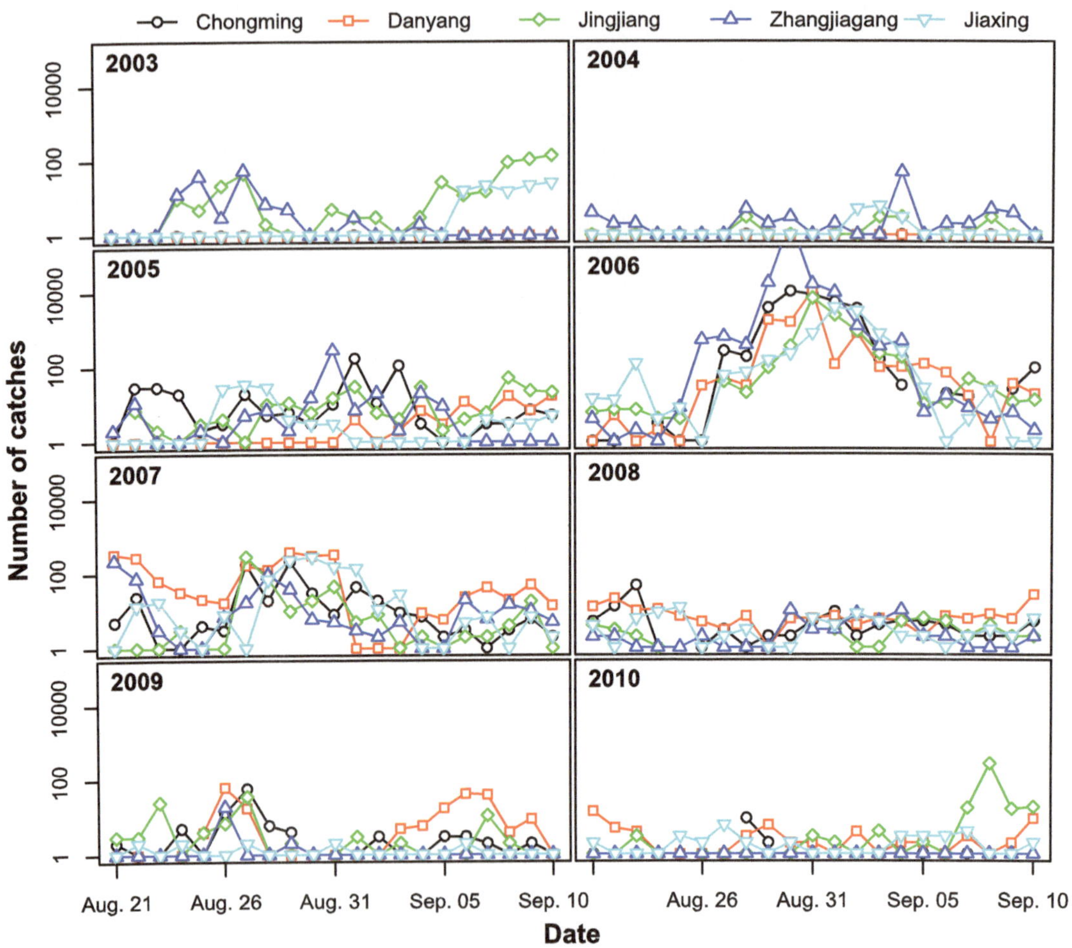

Figure 4. Daily catches of *N. lugens* from August 21 to September 10 from 2003 to2010.

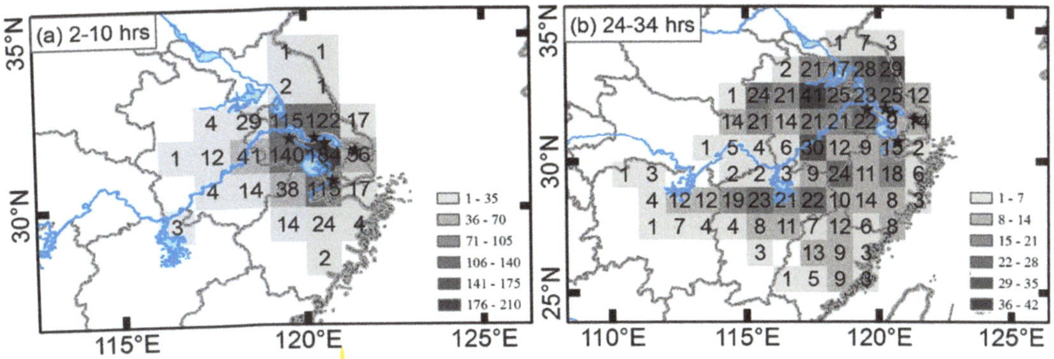

Figure 5. Probability distribution of endpoints of backward trajectories from Chongming, Danyang, Jingjiang, Jiaxing, and Zhangjiagang.

Table 7. Probability distribution of endpoints of backward trajectories from Chongming, Danyang, Jingjiang, Jiaxing, and Zhangjiagang in different available source areas.

Year	Area	2 h–10 h		24 h–34 h	
		Amount	Probability (%)	Amount	Probability (%)
All five years	1	5	0.52	108	12.84
	2	0	0	70	8.32
	3	105	10.93	191	22.71
	4	804	83.66	131	15.58
	5	0	0	102	12.13
	6	3	0.31	119	14.15
	7	44	4.58	74	8.80
	8	0	0	46	5.47
	subtotal	961	100	841	100
Light occurrence (2003 and 2009)	1	0	0	3	1.51
	2	0	0	22	11.06
	3	51	24.29	42	21.11
	4	155	73.81	21	10.55
	5	0	0	59	29.65
	6	3	1.43	29	14.57
	7	1	0.48	22	11.06
	8	0	0	1	0.50
	subtotal	210	100	199	100
Outbreak (2005, 2006, and 2007)	1	5	0.67	105	16.36
	2	0	0	48	7.48
	3	54	7.19	149	23.21
	4	649	86.42	110	17.13
	5	0	0	43	6.70
	6	0	0	90	14.02
	7	43	5.73	52	8.10
	8	0	0	45	7.01
	subtotal	751	100	642	100

Notes: Area 1 includes northern Jiangsu and northern Anhui; Area 2 includes southeastern Henan and eastern Hubei; Area 3 includes Southern and central Anhui; Area 4 includes the Yangtze River Delta; Area 5 includes northeastern Hunan and northwestern Jiangxi; Area 6 includes northeastern Jiangxi, northwestern Fujian, and western Zhejiang; Area 7 includes southern and central Zhejiang; Area 8 includes southern Jiangxi and southern and central Fujian. These probable source area were showed in Fig. 1.

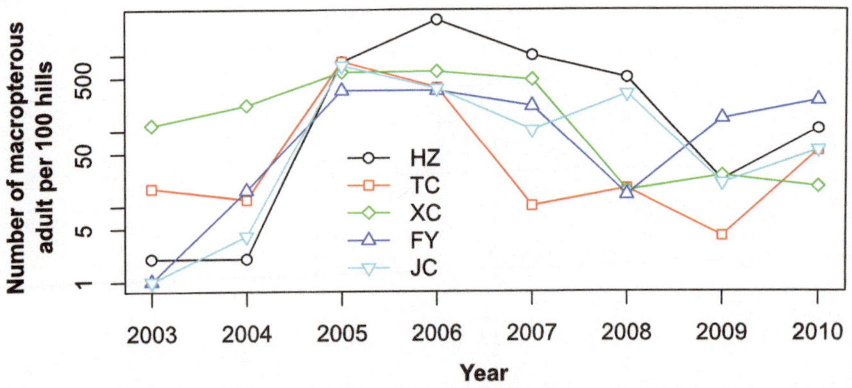

Figure 6. Peak number of *N. lugens* macropterous adults in Anhui Province from August 21 to September 10 from 2003 to 2010. Note: HZ-Huizhou, TC-Tongcheng, XC-Xuancheng, FY-Fengyang, JC-Juchao.

August, with most of the rice harvested in early and mid-September [8,18,19]. These regions are closer to South China than the Yangtze River Delta, and immigrants in June and July were greater because of shorter flight distance for the insects. Moreover, the single-crop rice plants in these regions were at the tillering to booting stage and fit for *N. lugens* whose populations were thus able to grow rapidly, whereas the rice plants in the Yangtze River Delta were still immature. Furthermore, the harvest time of rice in these regions was almost one month earlier than that in the Yangtze River Delta. Thus, large numbers *N. lugens* adults emigrated as the single-crop rice was harvested in these regions, and these insects could invade the Yangtze River Delta where the rice plants were at the heading stage and now suitable for *N. lugens*. Therefore, the late migration would exacerbate the

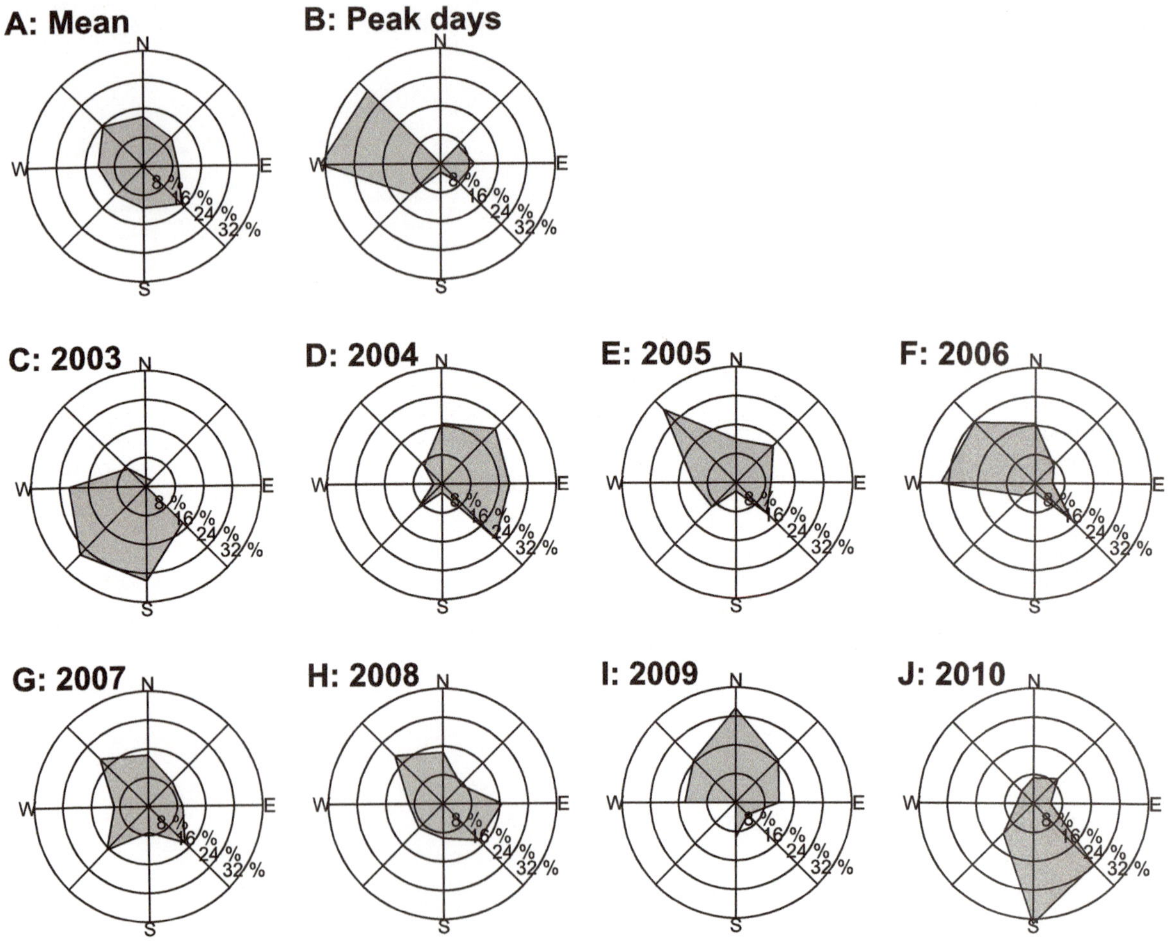

Figure 7. Frequency of wind direction at 850 hPa in Shanghai from 2003 to 2010. Note: (A) Mean value for 2003 to 2010, (B) mean value for peak days of *N. lugens* catches and (C-J) individual value for each year

Table 8. Frequency of wind direction from August 21 to September 10 from 2003 to 2010 in Shanghai.

	On peak days	On nonpeak days	Total
West and northwest wind	26	69	95
Other direction Wind	16	225	241
Total	42	294	336

outbreak of *N. lugens* in the Yangtze River Delta, and vigilance would need to be maintained. In particular, even when the *N. lugens* population in the Yangtze River Delta was suppressed by effective control strategy, mass late migration would negate the results of control efforts.

Our study showed that outbreaks of *N. lugens* mainly occurred along with abnormally high intrinsic rates of increase. Considering 17 outbreak cases (three cases in survey data and 14 cases in historical data), 12 (70.56%) occurred with high increase rates. In other cases, where the rate of increase was low, the number of *N. lugens* was large before late August. Therefore, these outbreaks of *N. lugens* could be divided into two types. One type is called the cardinal type, wherein the number of immigrants in the early season (before late August) is unusually high, thereby causing an early dense population. Thus, this type is also commonly known as the high-immigration type [8,11,38]. In the historical data used for this study, the four cases with low increase rate all occurred in 2006, and the outbreak in this year was previously analyzed as a typical case of high-immigration type [38]. In the other types of *N. lugens* outbreak, called the climate type in previous studies [8,11,38], the population was low in the early season but increased rapidly due, in general, to meteorological factors, especially temperature. In the present study, the high increase rate was mainly attributed to reproduction rather than immigration. Although mass migration was observed in the late season, the source areas were mainly located in relatively nearby areas and the late migration can be regarded as an internal flow within one population. Thus, the other cases in this study could be classified as climate type. However, a complex type could also be present, where a relatively high early immigration and favorable climatic conditions are both contributory factors [8,11,38].

Temperature could control the population of *N. lugens* by direct effects on survival, reproduction, and foraging. These effects are the most crucial factors for *N. lugens* population dynamics [37,39]. The size of the *N. lugens* population in the Yangtze River Delta was affected by temperature, as shown in the analysis of the relationship between the population dynamics of the planthopper and temperature from August to September (2008 to 2010). In the lower-middle reaches of the Yangtze River, a "cooler summer and warmer autumn" is regarded as the favored climatic condition for outbreaks of *N. lugens* [40–42], and this occurred for example in 2005 [8]. The occurrence of *N. lugens* in the Yangtze River Delta in 2005 was the most severe in the past 20 years [9,10]. Hu et al. [8] suggested that steadily warmer autumns have occurred from the 1990s in the lower-middle reaches of the Yangtze River Delta and that such conditions gradually became the norm. In this study, outbreaks of *N. lugens* in the Yangtze River Delta were mainly attributed to the high reproduction rate, thus providing further evidence on why the problems caused by *N. lugens* have become more severe.

N. lugens population can be controlled by pesticide. Limited damage was caused by *N. lugens* in the 1990s due to the use of buprofezin and imidacloprid [11], and the high level of resistance to imidacloprid resulted in chemical control failure and great yield loss in 2005 [43]. Rive cultivars with resistance to *N. lugens* can also suppress this pest significantly [6,44]. But, just as in the case of development of resistance to pesticide, this pest often adapted to resistant cultivars in a short time. The pest adapted to IR26, the first high-yielding cultivar with BPH resistance (with *Bph1* gene), within 2 or 3 years after being released to farmers [6]. High-yielding indica hybrid cultivars without the new resistant gene has promoted the population growth of *N. lugens* in southern China, especially in Hunan Province [45]. Therefore frequent outbreaks of *N. lugens* are not only related to suitable weather conditions, but also with other factors.

In summary, the outbreak mechanisms of *N. lugens* in the Yangtze River Delta were explored by surveying and evaluating historical data, and by determining the sources of catches in light traps in the late season. Although these catches supported the occurrence of migration, the source areas were found to be relatively close at hand and could be collectively considered, for control purposes, with sources in the Yangtze River Delta itself. Therefore, reproduction was the key factor contributing to the high population increase rate of N. lugens, and to the consequent outbreaks in this area of China. These results resolved the conflicting hypotheses on the source of these outbreak populations and will facilitate general control strategies of *N. lugens* in China.

Acknowledgments

We acknowledge the assistance provided by our colleagues Yue Zhao and Xingyun Fu who participated in the field investigations in 2008; Yangyang Hou, Li Wang, Yanbo Xu, He Sui, Yuan He, Yubo Zhu, and Huaijian Liao who participated in the field investigations in 2009; and Yuebo Gao, Yonghui Lei, Haibo Yang, Zhenhua Hao, Haiyan Zhang, Wei Wu, and Yuansong Fang who participated in the field investigations in 2010. We also thank the plant protection stations of Nanjing City and Jinhua City for their assistance in the maintenance of the experiment fields.

Author Contributions

Conceived and designed the experiments: GH BPZ XXZ. Performed the experiments: FL MHL WCL FZ XWW GHC. Analyzed the data: GH FL MHL. Wrote the paper: GH FL XXZ BPZ.

References

1. Pedgley DE (1993) Managing migratory insects pests - a review. Int J Pest Manage 39, 3–12.
2. Day RK, Knight JD (1995) Operational aspects of forecasting migratory pests. pp. 323–334 In Drake VA, Gatehouse AG. (Eds) *Insect migration: Tracking resources through space and time*. Cambridge University Press, New York.
3. Deveson ED, Drake VA, Hunter DM, Walker PW, Wang HK (2005) Evidence from traditional and new technologies for northward migrations of Australian plague locusts (*Chortoicetes terminifera*) (Walker) (Orthoptera: Acrididae) to western Queensland. Austral Ecology 30: 920–935.
4. Ma SC (1958) The population dynamics of the oriental migratory locust (*Locusta migratoria manilensis* Meyen) in China. Acta Entomologica Sinica 8: 1–40.
5. Chang SS, Lo ZC, Keng CG, Li GZ, Chen XL, et al. (1980) Studies on the migration of rice leaf roller *Cnaphalocrocis Medinalis* Guenée. Acta Entomologica Sinica, 23: 130–140.
6. Bottrell DG, Schoenly KG (2012) Resurrecting the ghost of green revolutions past: The brown planthopper as a recurring threat to high-yielding rice production in tropical Asia. Journal of Asia Pacific Entomology 15: 122–140.
7. Zhai BP (2011) Rice planthoppers: A China problem under the international perspectives. Chinese Journal of Applied Entomology 48: 1184–1193.
8. Hu G, Cheng XN, Qi GJ, Wang FY, Lu F, et al. (2011) Rice planting systems, global warming and outbreaks of *Nilaparvata lugens* (Stål). B Entomol Res 101: 187–199.

9. Guo R, Zhao ZH (2006) Crop pests. pp. 376–378 in Ministry of Agriculture of China (Ed.) China Agriculture Yearbook. China Agricultural Press, Beijing.
10. Zhai BP, Cheng JA (2006) The conference summary of workshop on the two primary migratory pests of rice, rice planthopper and rice leaf roller, in 2006. Chinese Bulletin of Entomology 43: 585–588.
11. Cheng XN, Wu JC, Ma F (2003) Brown planthopper: occurrence and control. China Agricultural Press, Beijing.
12. Cheng SN, Chen JC, Xi X, Yang LM, Zhu ZL, et al. (1979) Studies on the migrations of brown planthopper *Nilaparvata lugens* Stål. Acta Entomologica Sinica 22: 1–21.
13. Otuka A, Matsumura M, Watanabe T, Dinh TV (2008) A migration analysis for rice planthoppers, *Sogatella furcifera* (Horváth) and *Nilaparvata lugens* (Stål) (Homoptera: Delphacidae), emigrating from northern Vietnam from April to May. Appl Entomol Zool 43: 527–534.
14. Kisimoto R, Sogawa K (1995) Migration of the Brown Planthopper *Nilaparvata lugens* and the White-backed Planthopper *Sogatella furcifera* in East Asia: the role of weather and climate. pp. 93–104 In Drake VA, Gatehouse AG. (Eds) *Insect migration: Tracking resources through space and time.* Cambridge University Press, New York.
15. Riley JR, Cheng XN, Zhang XX, Reynolds DR, Xu GM, et al. (1991) The long-distance migration of *Nilaparvata lugens* (Stål) (Delphacidae) in China: radar observations of mass return flight in the autumn. Ecol Entomol 16: 471–489.
16. Riley JR, Reynolds DR, Smith AD, Rosenberg LJ, Cheng XN, et al. (1994) Observations on the autumn migration of *Nilaparvata lugens* (Homoptera: Delphacidae) and other pests in east central China. Bull. Entomol. Res 84, 389–402.
17. Bao YX, Li JJ, Miao QL, Shen SH, Wang JQ (2008) Simulation of atmospheric dynamical background for a great migration event of the brown planthopper (*Nilaparvata lugens* Stål): a case study. Chinese Journal of Agrometeorology 39: 347–352.
18. Lu F, Qi GJ, Chen X, Dong XG, Gui YR, et al. (2010) The late immigration of the brown planthopper *Nilaparvata lugens* (Stål) in Shanghai: Case studies from 2007. Acta Ecologica Sinica 30(12): 3215–3225.
19. Qi GJ, Xiao MK, Wu CL, Jiang C, Zhang XX, et al. (2010) Effect of the change of rice planting system on the formation of outbreak population of brown planthopper, *Nilaparvata lugens* (Stål). Acta Phytophylacica Sinica, 37:193–200.
20. Cheng JA, Zhu ZR (2006) Analysis on the key factors causing the outbreak of brown planthopper in Yangtze Area, China in 2005. China Plant Protection 32: 1–4.
21. Chen RC, Cheng XN (1980) The take-off behavior of brown planthopper (*Nilaparvata lugens* Stål) and its synchronous relations to the biological rhythm and environmental factors. Journal of Nanjing Agricultural University 2: 42–49.
22. Ye ZC, He SM, Lu LQ, Zhen HG, Feng BC, et al. (1981) The observation of the take-off behaviour of brown planthopper (*Nilaparvata lugens* Stål). Chinese Bulletin of Entomology 18: 97–100.
23. Zhong DL (1984) The relationship between the take-off behaviour of brown planthopper (*Nilaparvata lugens* Stål) and weather condition. Chinese Bulletin of Entomology 22: 97–99.
24. Zhao SF (1992) Behavioral mechanism of phototactic response of natural planthopper population (*Sogatella furcifera* and *Nilaparvata lugens*). Acta Agriculturae Jiangxi 4: 74–79.
25. (MAPRC) Ministry of Agriculture of the People's Republic of China. (1995) National standard GB/T 15794–1995: Rules for investigation and forecast of the brown planthopper (*Nilaparvata lugens* Stål) and the whitebacked planthopper (*Sogatella furcifera* Horvath). *MAPRC*, Beijing, China.
26. Chen RC, Cheng XN, Yang LM, Yin XD (1979) The ovarian development of the brown planthopper and it's relation to migration. Acta Entomologica Sinica. 22: 280–288.
27. Zheng DB, Hu G, Yang F, Du XD, Yang HB, et al. (2013) Ovarian development status and population characteristics of *Sogatella furcifera*(Horváth) and *Nilaparvata lugens* (Stål): implication for pest forecasting. J. Appl. Entomol.doi: 10.1111/jen.12067
28. Ding ZZ, Chen ML, Li PY (1985) Preliminary study on population reproduction of brown planthopper *Nilaparvata lugens* (Stål) in single-cropping late-maturing rice. Jiangsu Agricultural Sciences, 13: 1–20.
29. Chen YY, Wang MT (1996) Relationship between population reproduction of *Nilaparvata lugens* (Stål) and temperature. Jiangsu Agricultural Sciences, 24: 29–31.
30. Ohkubo N, Kisimoto R (1971) Diurnal periodicity of flight behaviour of the brown planthopper, *Nilaparvata lugens* Stål in the 4th and 5th emergence periods. Jpn. J. App. Entomol. Z. 15: 8–16.
31. Kisimoto R (1984) Meteorological conditions inducing long-distance immigration of the brown planthopper, *Nilaparvata lugens* (Stål). Chinese Journal of Entomology 4: 39–48.
32. Deng WX (1981) A general survey on seasonal migrations of *Nilaparvata lugens* (Stål) and *Sogatella furcifera* (Horváth) (Homoptera: Delphacidae) by means of airplane collections. Acta Phytophylacila Sinica. 8: 73–81.
33. Draxler RR, Hess GD (1998) An overview of the Hysplit_4 modeling system for trajectories, dispersion and deposition. Australian Meteorological Magazine. 47: 295–308.
34. Zhu M, Edward BR, David WR, Ian VM, Mark WS (2005) Low-level jet streams associated with spring aphid migration and current season spread of potato viruses in the U.S. northern Great Plains. Agricultural and Forest Meteorology. 138: 192–202.
35. Chen RC, Wu JR, Zhu SD, Zhang JX (1984) Flight capacity of the brown planthopper *Nilaparvata lugens* Stål. Acta Entomologica Sinica 27: 121–127.
36. Rosenberg LJ, Magor JI (1987) Prediction wind borne displacements of the brown planthopper *Nilaparvata lugens* from synoptic weather data. I. Long distance displacements in the northeast monsoon. Journal of Animal Ecology, 56: 39–51.
37. Li RD (1984) Population growth of the brown planthopper, *Nilaparvata lugens* Stål, as influenced by temperature. Acta Phytophylacica Sinica, 11:101–107.
38. Hu G, Xie MC, Lin ZX, Xin DY, Huang CY, et al. (2010) Are Outbreaks of *Nilaparvata lugens* (Stål) Associated with Global Warming? Environ. Entomol. 39:1705–1714.
39. Zang W, Hao SG, Wang HK, Cheng XN (1997) A simulation model of brown planthopper population dynamics in Yangtze and Huai River rice area. Journal of Nanjing Agricultural University 20, 32–38.
40. Hou TT, Huo ZG, Wu RF, Ye CL, Wang SY, et al. (2004) Impact of air temperature to the brown planthopper population in late rice crop season in Fuqing Region. Chinese Journal of Agrometeorology, 25, 28–32.
41. Pu MH, Chen JM (1979) The preliminary study of the brown planthopper occurrence degree forecasting by mathematical statistics. Plant Protection (5): 1–9.
42. Cheng JA, Zhang LG, Fan QG, Zhu ZR (1992) Simulation study on effects of temperature on population dynamics of brown planthopper. Chinese Journal of Rice Science 6, 21–26.
43. Wang YH, Gao CF, Zhu YC, Chen J, Li WH, et al. (2008) Imidacloprid susceptibility survey and selection risk assessment in field populations of *Nilaparvata lugens* (Stål) (Delphacidae). J. Econ. Entomol. 101, 515–522.
44. Brar DS, Virk PS, Jena KK, Khush GS (2009) Breeding for resistance to planthoppers in rice. pp. 401–427. In Heong KL, Hardy B (Eds.), *Planthoppers: New Threats to the Sustainability of Intensive Rice Production Systems in Asia.* Int. Rice Res. Inst, Los Baños, Philippines.
45. Sogawa K, Liu GJ, Shen JH (2003) A Review on the hyper-susceptibility of Chinese hybrid rice to insect pests. Chinese J. Rice Sci. 17:20–30.

11

A Hybrid Wetland Map for China: A Synergistic Approach Using Census and Spatially Explicit Datasets

Kun Ma[1], Liangzhi You[1,2], Junguo Liu[1,3]*, Mingxiang Zhang[1]

1 School of Nature Conservation, Beijing Forestry University, Beijing, People's Republic of China, 2 Environment and Production Technology, International Food Policy Research Institute, Washington, D.C., United States of America, 3 Ecosystems Services & Management Program, International Institute for Applied Systems Analysis, Laxenburg, Austria

Abstract

Wetlands play important ecological, economic, and cultural roles in societies around the world. However, wetland degradation has become a serious ecological issue, raising the global sustainability concern. An accurate wetland map is essential for wetland management. Here we used a fuzzy method to create a hybrid wetland map for China through the combination of five existing wetlands datasets, including four spatially explicit wetland distribution data and one wetland census. Our results show the total wetland area is 384,864 km^2, 4.08% of China's national surface area. The hybrid wetland map also shows spatial distribution of wetlands with a spatial resolution of 1 km. The reliability of the map is demonstrated by comparing it with spatially explicit datasets on lakes and reservoirs. The hybrid wetland map is by far the first wetland mapping that is consistent with the statistical data at the national and provincial levels in China. It provides a benchmark map for research on wetland protection and management. The method presented here is applicable for not only wetland mapping but also for other thematic mapping in China and beyond.

Editor: Abdisalan Mohamed Noor, Kenya Medical Research Institute - Wellcome Trust Research Programme, Kenya

Funding: This study was supported by the Special Fund for Forestry Scientific Research in the Public Interest (201204204), National Natural Science Foundation of China (91025009) and Projects of International Cooperation and Exchanges NSFC (41161140353), New Century Excellent Talents in University (NCET-09-0222), and Fundamental Research Funds for the Central Universities (HJ2010-1). The funders had no role in study design, data collection and analysis, decision to publish, or preparation of the manuscript.

Competing Interests: The authors have declared that no competing interests exist.

* E-mail: water21water@yahoo.com

Introduction

Wetlands are the "kidney of the earth": they play important ecological, economic, and cultural roles by providing such diverse ecosystem services as water supply, pollution control, nutrient recycling, groundwater recharge, soil formation, climate and flood regulation, and coastal protection [1,2,3]. Wetlands also provide recreation and tourism opportunities, and supporting vast biodiversity [4,5]. Costanza *et al.* [6] estimated the total value of global wetland ecosystems (including coastal systems and terrestrial wetlands) to be around $15.5 trillion/yr, which is approximately 47% of the total global value of ecosystem services. However, wetlands are also a vulnerable ecosystem, and have been more seriously degraded than any other ecosystem [7,8,9]. During the last century, approximately 50% of the global wetlands area has been lost [10].

In China, socioeconomic development and dietary change towards meat consumption lead to increasing demand on water resources [11,12], imposing pressures on wetland ecosystems. The main driving forces for wetland degradation include agriculture expansion, overuse of water resources and water quality deterioration, among others [13]. The Sanjiang Plain is one typical example with agriculture expansion as one major driving force as a result of the "food first" agricultural policy [14]. The Sanjiang Plain is the biggest marsh wetland in China. It is also one of the five Chinese biodiversity key areas of wetlands and fresh waters as a center of waterfowl reproduction in Northeast Asia [15]. Its marsh area decreased by 77% from 35,270 km^2 to 8,100 km^2 between 1954 and 2005 mainly as a result of agricultural expansion [14]. Due to the degradation of wetland, the species relying on wetland have been reducing significantly, and many rare waterfowls have been extinct [16]. The overuse of water resources within the river basin was largely responsible for the running dry downstream the Yellow river in the late 1990s [17]. Water quality deterioration is another reason for wetland degradation, for example in the Taihu lake in South China [18]. Wetland degradation is a serious ecological issue that has raised thesustainability concern. From 1990 to 2000, China's wetlands had a net loss of 50,360 km^2, accounting for nearly 30% of China's natural wetland area [19,20].

Accurate spatial information about wetlands is particularly important for national wetland protection and management. The rapid development and applications of Remote Sensing (RS), Geographical Information System (GIS) and Global Positioning System (GPS) have made wetland mapping feasible [21,22,23,24]. Lehner and Döll [25] created a Global Lakes and Wetlands Database (GLWD-3) by combining a variety of existing maps, data and information. Johnston and Barson [26] mapped the location and extent of wetlands and their vegetation types using Landsat TM imagery of inland wetland sites in Victoria and New South Wales, Australia. The Canadian Wildlife Service in Quebec region tested a multi-scale object-based classification method on two sites using satellite images to map wetlands in the context of the Canadian Wetland Inventory [27]. Bowen et al. [28] utilized several geospatial data sources, including high-resolution aerial imagery, digital raster graphics, and soils data to indentify and map Playa wetlands on the High Plains, Kansas, USA. In China, between 1995 and 2003, the State Forestry Administration of the People's Republic of China (SFA) conducted the first comprehensive census survey of the country's wetlands covered all of China except for Taiwan, Hong Kong, and Macao. There are other recent efforts to map China's wetlands using Landsat enhanced thematic mapper plus (ETM+) data (e.g Niu et al. 2009, Chen & Jessel 2011) [29,30]. However, considerable disagreements on wetlands distribution exist among these maps and there are large discrepancies between the maps and national and province wetland census data [29,31]. Despite the progress made in mapping wetland in different regions, a gridded wetland map consistent with census statistics is not yet available. This paper adopted a fuzzy agreement scoring approach [32] to create a new synergy map of China's wetlands using five independent wetland datasets. We also presented the validation of the final synergy map.

Methods

2.1 Data Sources

The following five wetlands datasets were used to create a new hybrid wetland map in this paper: Wetland-CAS, Wetland-BFU, Wetland-LU, the Global Lakes and Wetlands Database (GLWD-3), and the first wetland census in China. The formal four datasets are spatially explicit wetland distribution, while the census reports wetland areas by provinces (Table 1). We describe these datasets briefly below.

(1) Wetland-CAS was produced by the Institute of Remote Sensing Applications of the Chinese Academy of Sciences (CAS) and the Beijing Normal University in China and was based mainly on Landsat Enhanced Thematic Mapper Plus (ETM+) imagery. The dataset uses manual image interpretation of satellite images, aided by elevation data, soil data, land cover/land use data, and Google Earth maps. The minimum mapping unit is equivalent to 9 ha. The Wetland-CAS map was targeted at the boundary delineation of any type of wetland except those that are under agricultural use. The total area of wetland in 2000 is 359,478 km^2, contains three broadly classification, namely inland, coastal, and artificial wetlands. [29].

(2) Wetland-BFU, created by Beijing Forestry University [33], provides the location of major wetlands listed in the first wetland census survey by SFA. Obviously, the total wetlands area from this product is much larger than the statistics from the census, but Wetland-BFU provides a good reference for the location and extent of major wetlands. The map is in a shape file format, then, we converted into a raster dataset with a spatial resolution of 1 km. The total area of wetland is 427,952 km^2.

(3) The Wetland-LU was created by the China Council for International Cooperation on Environment and Development (CCICED) by extracting different wetland types from the land-use data with spatial resolution of 1 km [30]. This dataset includes two categories of wetland: rivers and lakes, and peatland and marshes. And the total area of wetland is 227,796 km^2.

(4) The GLWD-3 was developed by Lehner and Döll [25] in a partnership between the Center for Environmental Systems Research (CESR), the University of Kassel, and the World Wildlife Fund US (WWF-US). The database was generated through the use and incorporation of data derived from proprietary products of the Environmental Systems Research Institute, Inc. (ESRI), the UNEP World Conservation Monitoring Centre (UNEPWCMC), and others. GLWD-3 is a global raster map of wetlands at a spatial resolution of 30 arc seconds, equivalent to 1 km.

(5) The National Bureau of Statistics of China reports wetland census data for each province are reported by NBS [34]. The data are from the SFA's first ever wetland census. From 1995 to 2003, the SFA did an extensive national survey of wetlands with area greater than 100 ha, including lakes, marshes, coastal wetlands, reservoirs, and rivers with their average river bed (dry channels) width greater than 10 m. The field survey investigates wetland types, area and distribution. The total area of wetland is 354,855 km^2. This census attempted to provide more reliable data for wetland planning and management. SFA was responsible for the coordination and integration of wetland census. It allocated the census tasks to the Forestry Bureaus in each province, and also provided trainings for the provincial professionals and technicians to keep the census approach consistent. After receiving the census data from all the provinces, SFA also invited experts and scholars to evaluate the accuracy and reliability of the data. Hence, such a wetland census data is so far the most authoritative official data in China.

2.2 Methods

2.2.1 Synergy wetland map creation. All four wetland maps were first standardized into the same spatial resolution of 1 km. The basic idea behind the synergy approach is to give each pixel a score based on the agreement among different wetlands databases used in this analysis [35]. The approach allows the producer of the hybrid wetland map to rank the wetlands dataset, rather than giving all data the same weight, before they are combined. The more agreement among the maps, the higher the likelihood or probability that wetlands exist at the pixel [32]. Figure 1 illustrated the methodology in a series of steps.

The first step is to give priority order to the four wetland maps. This priority is based on three factors: authority, geographical specificity, and spatial resolution. The Wetland-CAS map, which was created based on remote sensing data, is a specific map focusing on China's wetlands. Hence, due to its high ranking on all three priority factors, Wetland-CAS was given the highest priority. Because the Wetland-BFU map was created based on the survey wetlands data, the most authoritative set of statistics in China, we gave it the second priority. Not a specific wetland map, the Wetland-LU is based on China's land use data with a high spatial resolution of 1 km. We therefore treated it as the third priority. The GLWD is a global map of both wetlands and lakes, which was produced based on various data sources, most published in the 1990s and 2000s. It is difficult to confirm which period this dataset exactly stands for, and so we gave it the last priority.

The second step is to create a priority rank table and give rank to each pixel. Since four different wetlands datasets were used in order of rank of confidence, there are 15 different possible combinations or confidence from 1 to 15 (Table 2). When all four wetland maps indicate a pixel as wetlands, then we give the highest rank of 1. If three maps show it is wetlands (3Y combinations), then we created the rank based on the priority orders of these three maps. In this case, for example, when GLWD (the map with the least priority) is the only map indicating no wetlands, or all other three maps with priority from 1 to 3 indicate a pixel with wetlands, we gave the highest rank for the 3Y combination (or 2, see in Table 2). But when Wetland-CAS (the map with the highest

Table 1. Comparison of Wetlands Datasets.

Item	Wetland-CAS	Wetland-BFU	Wetland-LU	GLWD-3	National Wetland Census
Year	2000	2000	2000	1990s–2000s#	1995–2003
Producer	CAS*	BFU†	CCICED‡	WWF§	NBS¶
Resolution	1 km	Shape file	2 km	30-second	For each province
Reference	Niu and Gong [2009]	BFU [2011]	CCICED [2011]	Lehner and Döll [2004]	China Statistical Yearbook [2006]
Wetlands area (km^2)	359478	427,952	227796	311000	384,855
Wetland classification	(1)River and flood plain (2)Lake (3)Marsh/swamp (4)Reservoir/pond and other man-made water bodies excluding rice field	(1)River and flood plain (2)Lake (3)Marsh, swamp, bog, fen, mire, and forest wetland (4)Coastal wetland (5)Reservoir and pond	(1)River and lake (2) Peatland and Marsh	(1)River and flood plain (2)Lake (3)Marsh, swamp, bog, fen, mire, forest/flooded forest (4)Coastal wetland (5)Reservoir	(1)River and flood plain (2)Lake (3)Marsh, swamp, bog, fen, mire, forest wetland (4)Coastal wetland (5)Reservoir and pond

Producer list:
*CAS: Chinese Academy of Science.
†BFU: Beijing Forestry University.
‡CCICED: China Council for International Cooperation on Environment and Development.
§WWF: World Wildlife Foundation/.
¶NBS: National Bureau of Statistics of China.
#1990s–2000s: GLWD-3 was produced based on various data sources, most published in the 1990s and 2000s. It is difficult to confirm which period this dataset exactly stands for.

priority) is the only map indicating wetlands, we give the lowest rank for 3Y combinations (or 5, see in Table 2). A similar approach was applied for 2Y and 1Y combination. Rank 1 implies that a pixel has the highest confident of wetland existence, while rank 15 means the lowest confidence. A synergy map was then created after implementing the ranking (Fig. S1).

The third step is to allocate statistical wetlands area to pixels. To create a raster wetland map consistent with statistics, the statistics of the wetlands area for each province must be allocated into

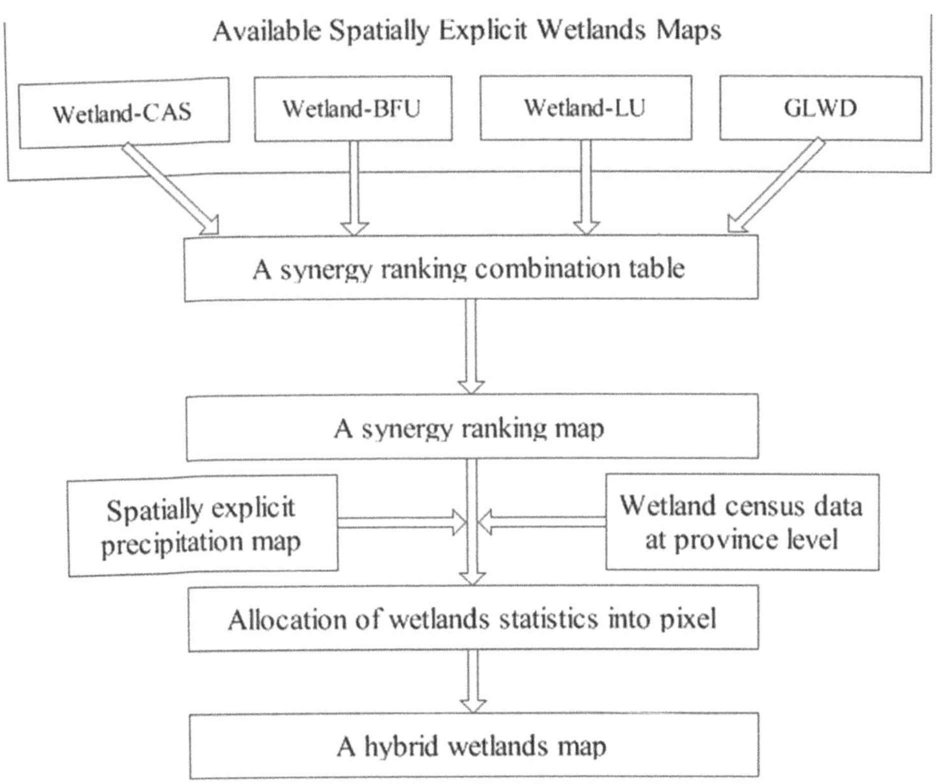

Figure 1. Overview of the methodology for creation of a wetland map.

Table 2. Synergy map ranking combinations when combining four wetlands datasets.

Rank	Wetland-CAS	Wetland-BFU	Wetland-LU	GLWD
1	Y	Y	Y	Y
2	Y	Y	Y	N
3	Y	Y	N	Y
4	Y	N	Y	Y
5	N	Y	Y	Y
6	Y	Y	N	N
7	Y	N	Y	N
8	N	Y	Y	N
9	Y	N	N	Y
10	N	Y	N	Y
11	N	N	Y	Y
12	Y	N	N	N
13	N	Y	N	N
14	N	N	Y	N
15	N	N	N	Y

Here Y means the presence of wetland and N means no wetland.

different pixels of the synergy map. There are two situations in the allocation process. In the first situation, the total area of pixels with a rank from 1 to 15 is higher than the total wetlands area. In this case, wetlands area was first allocated to the pixels with higher rank then to the pixels with lower rank. We then added the area of pixels in a province with a rank of 1. If this area were smaller than the total wetlands area in this province, we assume all these pixels with rank 1 are wetlands. Then we add the area of pixels in this province with rank of 1 and 2. This process will repeat until the total area of pixels with a rank of 1 to certain (e.g. *i*) is larger than the statistical wetlands area. At this stage, we need to decide how to allocate wetlands in pixels with the rank of *i*. Since all of these pixels have the same rank, we used the 1960–2000 average annual precipitation data as an auxiliary criterion to reorder these pixels, and assumed the pixels with higher precipitation have a higher probability of wetlands. For example, suppose the statistical area of wetland for a province is 100 km^2, the sum of rank from 1 to 5 is 90 km^2, and the rank from 1 to 6 is 120 km^2. In this case, we need to reorder the pixels of rank 6 with precipitation information and allocate the remaining 10 km^2 of wetlands according to the criterion of precipitation.

The second situation is that the total area of pixels with a rank from 1 to 15 in a province is lower than the province's statistical wetlands area. In this case, we assume all these pixels with ranks 1 to 15 are wetlands. We again use precipitation as an auxiliary criterion to rank the pixels and allocate the remaining wetlands area (the difference between the statistical wetlands area and the area of all pixels with ranks from 1 to 15).

2.2.2 Confidence index calculation. After the hybrid wetland map being created, we calculated its confidence index. Based on the synergy map ranking in Table 2, we assign the maximum confidence, C_{max}, for the highest rank (1) and the minimum confidence, C_{min}, for the lowest rank (15). Knowing that there still exists some inaccuracy even for the highest rank (where all four sources agree), we assume that C_{max} is 95% and C_{min} is 5%. The confidence index between the maximum and minimum is simple linear interpolation:

$$C_i = C_{\max} - (n-1)/(N-1) * (C_{\max} - C_{\min}) \quad (1)$$

Where C_i is the confidence index of pixel *i*, n represents the rank for that pixel and N is the lowest rank used for that province. N varies from province to province.

To get a general picture, we calculated overall confidence index for each province. This is a wetlands area-weighted confidence index:

$$C_i = \sum C_{ij} A_{ij} / \sum A_{ij} \quad (2)$$

Where C_j is the confidence for province *j*. C_{ij} is the confidence for pixel *i* in province *j*. A_{ij} is the hybrid wetlands area of pixel *i* in province *j*.

2.2.3 Agreement degree calculation. In order to further estimate the accuracy of the hybrid wetland, an agreement degree (AD) is used here to show the percentage of two independent datasets, the China's Lake Dataset and the Global Reservoir and Dam Database (GRanD), captured by the different wetland maps. The AD is calculated by

$$AD = Overlay\ area / Total\ area \quad (3)$$

Here, the *Overlay area* is the overlay area between each wetland map and the China's Lake Database or the GRanD within a province. The *Total area* is the total lake (or reservoir) area presented by the China's Lake Database (or the GRanD) in the province.

Results

3.1 Hybrid Wetland Map

A hybrid wetland map of China was created by applying the synergy wetland map creation methodology in each province (Fig. 2). This map shows that the total wetlands area is 384,864 km^2, 4.08% of the total national land surface area (excluding Taiwan, Hong Kong, and Macao). The calculation compares very well with the statistical wetlands area of 384,855.26 km^2. And the final results are very close to the wetlands area statistics in each province with a relative error range from −0.20% to 0.06% (Table S1). Tibet has the highest wetlands share, and accounts for 13.6% of the total national wetlands area. The top four provinces with the highest shares of wetland, namely Tibet, Heilongjiang, Inner Mongolia, and Qinghai, cover 46.6% of China's total wetlands area.

The CCICED divided China into 8 ecological regions based on the scheme of ecological regionalization in China. Most wetlands in China are in the country's Northwest and Northeast and in the Qinghai-Tibet Plateau. These three regions possess 62.1% of the wetlands in China (Fig. S2).

3.2 Confidence Index

The hybrid wetland map is created from multiple wetland maps and wetland census data. They all have inherent uncertainty which would affect the accuracy of the final hybrid wetlands distribution. Figure 3 shows the confidence index of the hybrid wetland map. About 19% of wetlands (in red) show the high confidence level with confidence index over 80%. The low confidence level (i.e. less than 20%, in blue) is widespread,

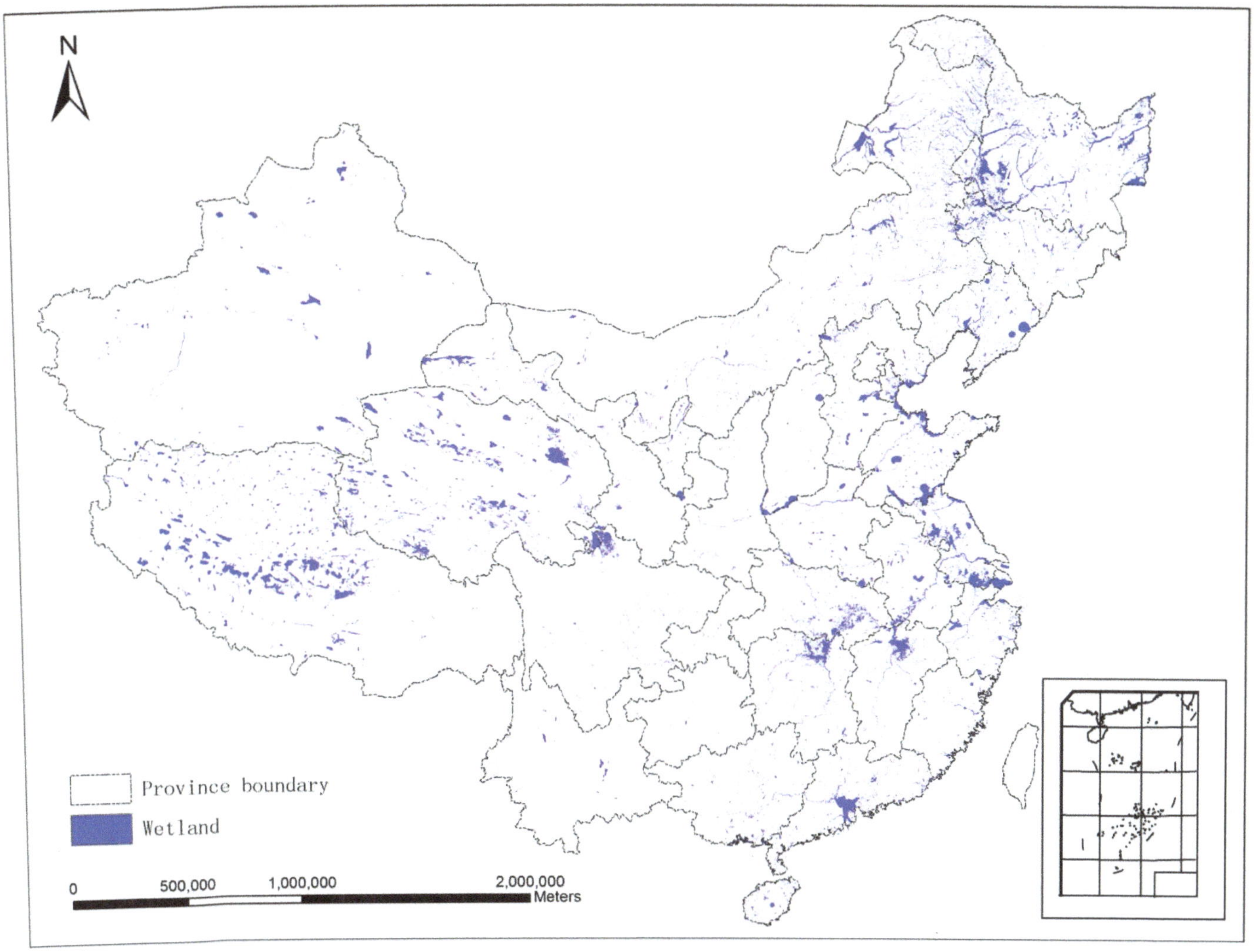

Figure 2. Hybrid wetland map of China. Wetland in Taiwan, Hong Kong and Macao are not included due to the lack of data.

covering almost 40% of wetlands area; it is in particular dominant in North China. The national average confidence index is 40.9%. There are 11 provinces with higher confidence index than the national average. Anhui is the highest province with the confidence index 68.9% (Fig. 3). While the confidence index of Guizhou province is barely over 12%, the lowest among all the 31 provinces. The provinces with low confidence index need more careful investigation in future wetlands surveys.

3.3 Comparison with Independent Datasets

In order to further estimate the accuracy of the hybrid wetland map, we compared the wetlands area with two independent data sets: (1) China's Lake Database, a shape file map created by the Chinese Academy of Sciences in 2002 that shows the location and area of lakes in China; (2) The Global Reservoir and Dam Database (GRanD), which compiles reservoirs worldwide with a storage capacity of more than 0.1 km^3.

At the national level, for reservoirs, the AD value of the hybrid wetland map is 71.2%, which is higher than all other wetland maps. For lakes, the AD value of the hybrid wetland map is 90.5%, higher than other wetland maps except for the GLWD (Fig. 4). This is because the GLWD provides a representation of the maximum global wetlands extent [27]. At the province level, a similar pattern appears (Table S2, S3).This shows that our hybrid wetland map captures the location of most of the lakes and reservoirs in China. It is a reliable product for China's wetland maps.

Discussion

We synergistically combined four existing wetland maps and census data to generate a hybrid wetland map in China. This product has been published online at the Geo-wiki website (ftp://ftp.iiasa.ac.at/pub/for/Steffen/wetland_geo_wiki/), and it is available for the public for free downloading. The comparison of the hybrid wetland map with independent lake and reservoir databases shows that our hybrid map is a reliable product. This geo-referenced wetlands data provides a benchmark of wetlands in China. This hybrid wetland map is valuable for wetlands research as well as wetlands conservation and management.

When comparing the total wetland area with these four wetland maps, the synergy map is consistent with the national and provincial wetlands statistics on wetlands area, and demonstrates spatial distribution of wetlands. Considerable differences exist when comparing our map with other available spatially explicit

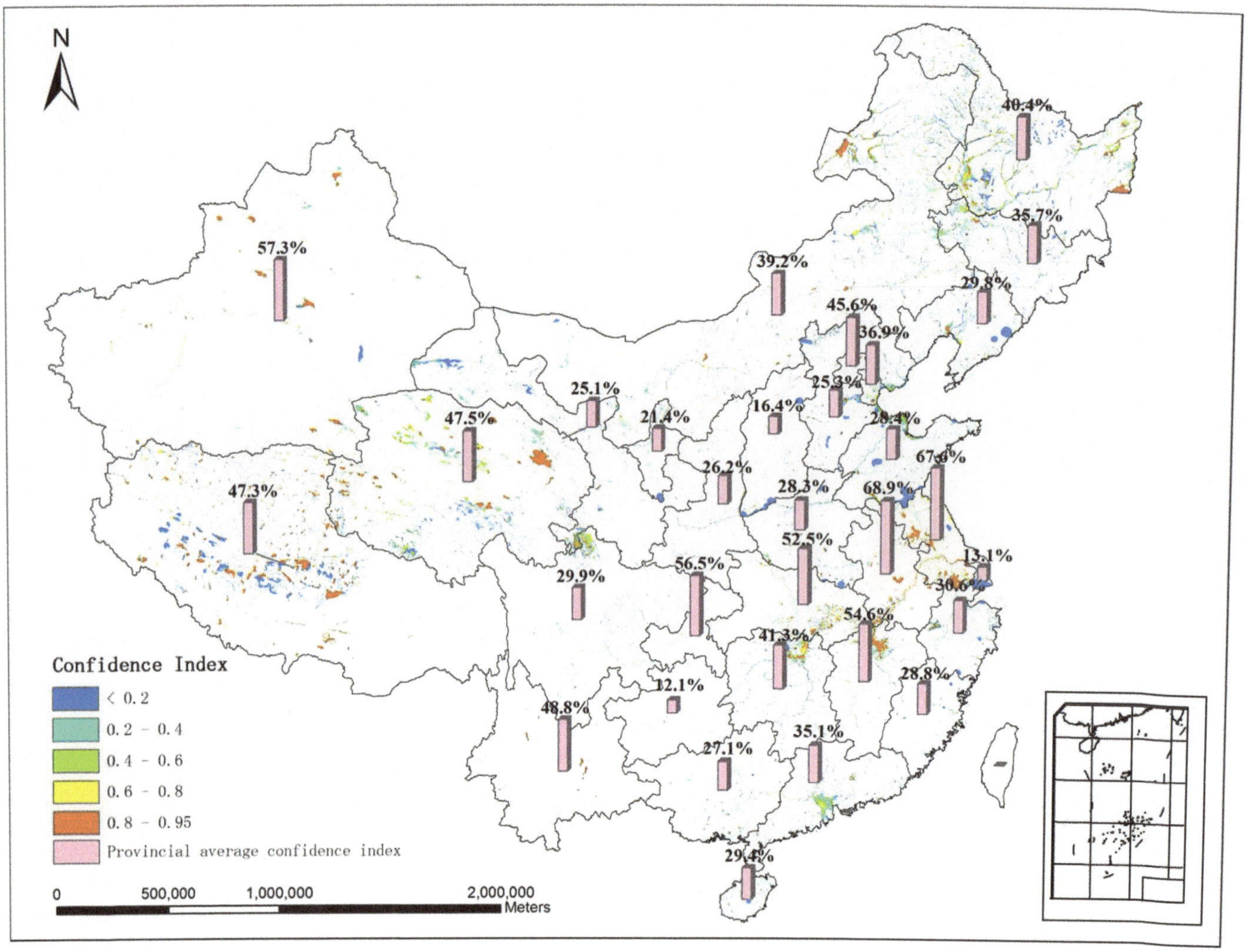

Figure 3. Confidence index map. The red color shows the high confidence index cells and the blue color shows the low confidence index cells. The cube refers to the average confidence index for each province.

wetland maps. One possible reason is the different degree of detail in mapping unit. For example, the total wetland area from our study is about 7% higher than the result from Wetland-CAS [29], which excluded rivers with width narrower than 90 m and wetlands with area smaller than 9 ha due to the spatial resolution constraint of remotely sensed data. But our study is based on the allocation of wetland area from the National Wetland Census by investigating wetlands with area greater than 100 ha.

Our approach used in this paper can reduce the uncertainties in mapping the national wetland. The previously available four wetland maps were given different priorities based on their reliability, geographical focus, and spatial resolution. The allocation of wetland area is based on the rank of each pixel, which was estimated by considering the priorities and agreement of the four wetland maps. This helps to allocate wetland to the pixels with higher confidence levels.

The result of confidence index mainly depends on the quality of input data. If consistent data with higher quality were available, the confidence index could be significantly improved. Due to the considerable disagreements and large discrepancy of these four datasets, the national average confidence index is only 40.9% with 20 provinces with lower values than the national average. For those provinces, our results indicate that careful wetland investigation is urgently needed to improve the accuracy of the wetland area there.

The method is elastic so that it can be easily applied when new wetlands data becomes available (e.g. the second wetlands inventory data that is coming soon). Generally, two spatially explicit datasets and one statistical data are the minimal requirements of datasets for applying this method to other regions. But we strongly suggest that at least three spatial explicit data and one accurate statistical data should be used in order to keep the hybrid map reliable.

A number of limitations in our methodology and results still remain. First, the hybrid wetland map was created from multiple wetland maps of different periods and resolution. We standardized the four previously available wetland maps into the same spatial resolution of 1 km. Some errors may occur during this process. Second, all wetland types in Ramsar Convention can be found in China, but we did not identify different wetland types in this paper; instead, we treated wetland as one type of land cover/use. Except for the wetland dataset from CCICED [30], all the other four datasets include five major wetland categories, e.g. river and flood plain, lake, marshes, coastal wetland, and reservoir/pond

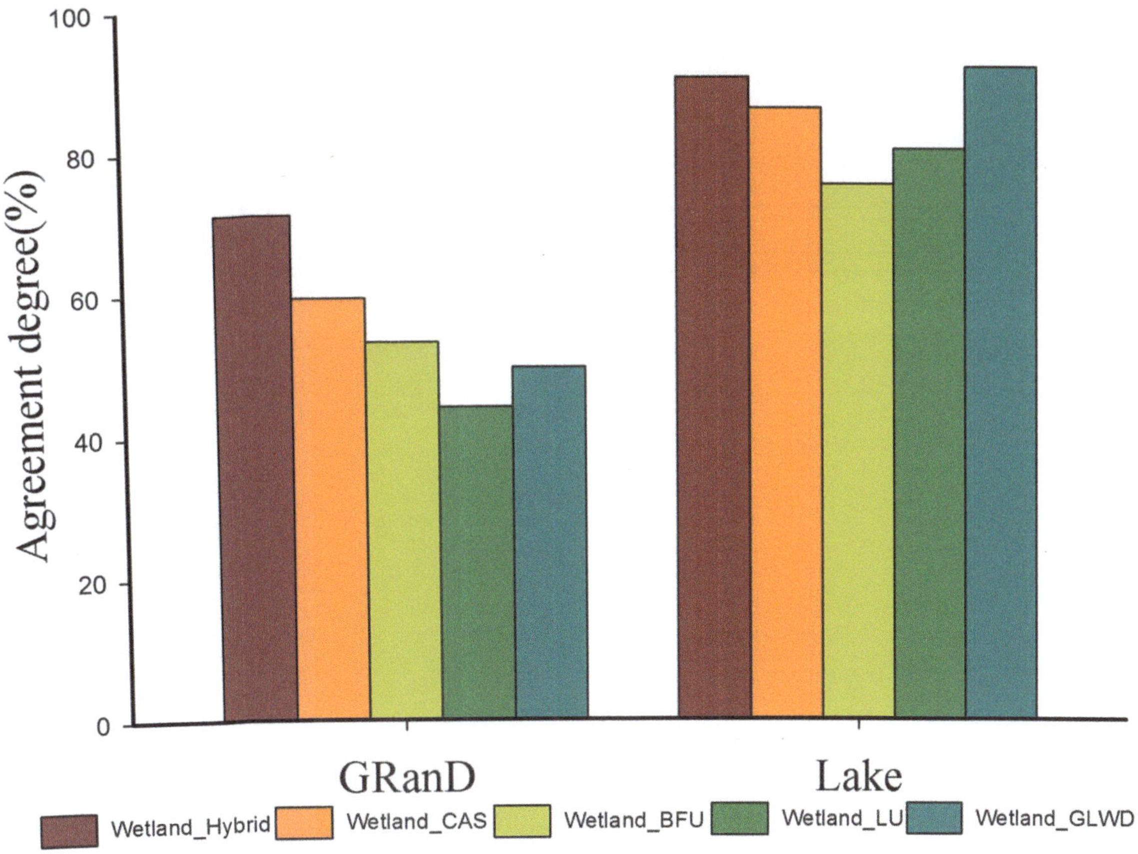

Figure 4. The comparison of wetlands area between the hybrid map and the existing four wetlands datasets with independent two datasets at national level. GRanD (Global Reservoir and Dam Database) and Lake (China's Lake Database).

(see Table 1), although the sub-categories among these five main categories are to some extent different. Nevertheless, mapping different wetland types is also valuable, and we will work on this in the next step. Third, paddy crop field is an important type of wetland, or artificial wetland. Due to the lack of data in the previously available wetland maps, we did not include the paddy rice field in the current wetland product. In addition, the mapping of paddy rice field is more precise in farmland resource inventory programs than wetland census in China [29]. Last but not least, the first wetland census just surveyed the wetland with area larger than 100 hm^2. Therefore, our current wetland area is a conservative estimate of wetland in China.

The product presented in this study is by far the first wetland mapping that is consistent with the statistical data at the national and provincial levels in China. The hybrid wetland map is valuable for wetland planning and management. The confidence level of wetland mapping can also provide insights for further wetland survey and research, e.g. more focuses should be on the regions with low values of confident indices. Applying the synergistic approach to wetland statistics in different periods can identify where and to what extent wetland is changed. Such a study is expected to be conducted when the second wetland census is available in China. The approach used in this article is applicable not only for wetland mapping but also for other thematic mapping in China and even in the rest of the world.

Supporting Information

Figure S1 Priority rank map for China's wetland. A rank with a small number (e.g. 1) indicates the confidence of wetlands existence is high, while a rank with a large number (e.g. 15) indicates the confidence of wetlands existence is low.

Figure S2 Share of wetlands area (in bracket) in different ecological regions in China.

Table S1 Comparison of wetlands area between the provincial statistics and aggregated provincial wetlands area from the hybrid wetland map.

Table S2 Agreement degree of five geo-referenced wetland maps (including the hybrid wetland map Hybrid from this study) with China's Lake Database for each province.

Table S3 Agreement degree of five geo-referenced wetland maps (including the hybrid wetland map Hybrid from this study) with Global Reservoir and Dam Database for each province.

Author Contributions

Conceived and designed the experiments: LY JL. Performed the experiments: KM LY JL. Analyzed the data: KM LY JL. Contributed reagents/materials/analysis tools: KM LY JL MZ. Wrote the paper: KM LY JL.

References

1. Woodward RT, Wui YS (2001) The economic value of wetland services: a meta-analysis. Ecological Economics 37: 257–270.
2. Leibowitz SG (2003) Isolated wetlands and their functions: an ecological perspective. Wetlands 23: 517–531.
3. Millennium EA (2005) Ecosystems and Human Well-being: Wetlands and Water Synthesis.
4. Jiang M, Lu XG, Xu LS, Chu LJ, Tong SZ (2007) Flood mitigation benefit of wetland soil–A case study in Momoge National Nature Reserve in China. Ecological Economics 61: 217–223.
5. Adhikari AR, Acharya K, Shanahan SA, Zhou XP (2011) Removal of nutrients and metals by constructed and naturally created wetlands in the Las Vegas Valley, Nevada. Environmental Monitoring and Assessment 180: 97–113.
6. Costanza R, d'Arge R, Groot Rd, Farber S, Grasso M, et al. (1997) The value of the world's ecosystem services and natural capital. nature 387: 253–260.
7. Turner RK, van den Bergh JCJM, Söderqvist T, Barendregt A, van der Straaten J, et al. (2000) Ecological-economic analysis of wetlands: scientific integration for management and policy. Ecological Economics 35: 7–23.
8. Vörösmarty CJ, McIntyre PB, Gessner MO, Dudgeon D, Prusevich A, et al. (2010) Global threats to human water security and river biodiversity. nature 467: 555–561.
9. Carpenter SR, Stanley EH, Zanden MJV (2011) State of the World's Freshwater Ecosystems: Physical, Chemical, and Biological Changes. Annual Review of Environment and Resources 36: null.
10. Zedler JB, Kercher S (2005) WETLAND RESOURCES: Status, Trends, Ecosystem Services, and Restorability. Annual Review of Environment and Resources 30: 39–74.
11. Liu JG, Savenije HHG (2008) Food consumption patterns and their effect on water requirement in China. Hydrology and Earth System Science. 12: 887–898.
12. Liu JG, Yang H, Savenije HHG (2008) China's move to higher-meat diet hits water security. Nature. 454: 397.
13. Yang Z, Chen B (2011) Systematic studies on wetlands in China. Ecological Modelling 222: 221–223.
14. Wang Z, Song K, Ma W, Ren C, Zhang B, et al. (2011) Loss and Fragmentation of Marshes in the Sanjiang Plain, Northeast China, 1954–2005. Wetlands: 1–10.
15. Chen LZ (1998) Biodiversity: Current situation and policy of conservation in China. Beijing: Science Press. 197–204.
16. Liu HY, Lu XG, Liu ZQ (2000) Landscape planning and ecology construction of wetland comprehensive protected area system in the Sanjiang Plain. Journal of Environmental Sciences. 12: 361–366.
17. Zhang JF, Sun QX (2005) Causes of wetland degradation and ecological restoration in the Yellow River Delta Region. Forestry Studies in China 7: 15–18.
18. Qu CS, Chen W, Bi J, Huang L, Li FY (2011) Ecological risk assessment of pesticide residues in Taihu Lake wetland, China. Ecological Modelling 222: 287–292.
19. Cyranoski D (2009) Putting China's wetlands on the map. nature 458: 134.
20. Gong P, Niu ZG, Cheng X, Zhao KY, Zhou DM, et al. (2010) China's wetland change (1990–2000) determined by remote sensing. SCIENCE CHINA Earth Sciences 53: 1036–1042.
21. Chopra R, Verma V, Sharma P (2001) Mapping, monitoring and conservation of Harike wetland ecosystem, Punjab, India, through remote sensing. International Journal of Remote Sensing 22: 89–98.
22. Prigent C, Matthews E, Aires F, Rossow WB (2001) Remote sensing of global wetland dynamics with multiple satellite data sets. Geophys Res Lett 28: 4631–4634.
23. Ozesmi SL, Bauer ME (2002) Satellite remote sensing of wetlands. Wetlands Ecology and Management 10: 381–402.
24. Rebelo LM, Finlayson C, Nagabhatla N (2009) Remote sensing and GIS for wetland inventory, mapping and change analysis. Journal of Environmental Management 90: 2144–2153.
25. Lehner B, Döll P (2004) Development and validation of a global database of lakes, reservoirs and wetlands. Journal of Hydrology 296: 1–22.
26. Johnston RM, Barson MM (1993) Remote sensing of Australian wetlands: An evaluation of Landsat TM data for inventory and classification. Marine and Freshwater Research 44: 235–252.
27. Grenier M, Demers AM, Labrecque S, Benoit M, Fournier RA, et al. (2007) An object-based method to map wetland using RADARSAT-1 and Landsat ETM images: test case on two sites in Quebec, Canada. Canadian Journal of Remote Sensing 33: S28–S45.
28. Bowen MW, Johnson WC, Egbert SL, Klopfenstein ST (2010) A GIS-based approach to identify and map playa wetlands on the High Plains, Kansas, USA. Wetlands 30: 675–684.
29. Niu ZG, Gong P, Cheng X, Guo JH, Wang L, et al. (2009) Geographical characteristics of China's wetlands derived from remotely sensed data. Science in China Series D: Earth Sciences 52: 723–738.
30. Chen YY, Jessel B (2011) Ecosystem services and management strategy in China. Beijing: China Environmental Science Press.
31. Yang B, Wang Y, Zhu D (1997) The Tidal Flat Resource of China. Journal of Natural Resources 12: 307–316.
32. Fritz S, You L, Bun A, See L, McCallum I, et al. (2011) Cropland for sub-Saharan Africa: A synergistic approach using five land cover data sets. Geophysical Research Letters 38: L04404.
33. BFU (2011) Mapping China's Wetland based on the first National Wetland Census. Beijing Forestry University.
34. National Bureau of Statistics of China (2006) China Statistical Yearbook (2006). Beijing: China Statistics Press.
35. Jung M, Henkel K, Herold M, Churkina G (2006) Exploiting synergies of global land cover products for carbon cycle modeling. Remote Sensing of Environment 101: 534–553.

12

Heavy Metal Pollution Characteristics of Surface Sediments in Different Aquatic Ecosystems in Eastern China

Wenzhong Tang[1]*, Baoqing Shan[1]*, Wenqiang Zhang[1], Hong Zhang[1], Lishuo Wang[1,2], Yuekui Ding[1]

1 State Key Laboratory on Environmental Aquatic Chemistry, Research Center for Eco-Environmental Sciences, Chinese Academy of Sciences, Beijing, China, **2** China University of Mining & Technology (Beijing), Beijing, China

Abstract

Aquatic ecosystems in eastern China are suffering threats from heavy metal pollution because of rapid economic development and urbanization. Heavy metals in surface sediments were determined in five different aquatic ecosystems (river, reservoir, estuary, lake, and wetland ecosystems). The average Cd, Cr, Cu, Ni, Pb, and Zn concentrations were 0.716, 118, 37.3, 32.7, 56.6, and 204 mg/kg, respectively, and the higher concentrations were mainly found in sediment samples from river ecosystems. Cd was the most anthropogenically enriched pollutant, followed by Zn and Pb, indicated by enrichment factors >1.5. According to consensus-based sediment quality guidelines, potential ecological risk indices, and risk assessment codes, all five types of aquatic ecosystems were found to be polluted with heavy metals, and the most polluted ecosystems were mainly rivers. Cd was the most serious pollutant in all five aquatic ecosystems, and it was mainly found in the exchangeable fraction (about 30% of the total Cd concentration, on average). The results indicate that heavy metal contamination, especially of Cd, in aquatic ecosystems in eastern China should be taken into account in the development of management strategies for protecting the aquatic environment.

Editor: Manuel Reigosa, University of Vigo, Spain

Funding: This work was supported by the National Natural Science Foundation of China (No. 21107126), the National Water Pollution Control Program (No. 2012ZX07203-002), and the One-Three-Fiver Program of Research Center for Eco-Environmental Sciences (No. YSW2013B02). The funders had no role in study design, data collection and analysis, decision to publish, or preparation of the manuscript.

Competing Interests: The authors have declared that no competing interests exist.

* Email: wztang@rcees.ac.cn (WZT); bqshan@rcees.ac.cn (BQS)

Introduction

Heavy metal contamination has been a worldwide environmental problem for many years [1–4]. Heavy metals are widespread, persistent, and potentially toxic in aquatic ecosystems [5–7]. After being transported into aquatic ecosystems, heavy metals can be absorbed by suspended solids and accumulated in sediments. Sediment acts as a sink for heavy metals but can re-release the heavy metals, acting as a source of pollutants and degrading the quality of the aquatic ecosystems [8,9]. Therefore, the accumulation of heavy metals in sediment is a cause of growing interest and concern. Environmental problems caused by heavy metal pollution in aquatic ecosystems have recently been extensively studied so that procedures for effectively managing these ecosystems can be developed [10–12].

Heavy metal contamination in aquatic ecosystems usually accompanies rapid economic development, in both developed and developing countries [13]. Little attention was paid to environmental problems caused by heavy metal contamination in sediments in China by researchers or the government until 2000 [4,14]. Economic development, especially industrial development, has recently been continuously and rapidly intensifying in China, particularly in eastern China. Heavy metals discharged into the air and effluents by industrial activities have caused severe pollution in aquatic ecosystems in eastern China [4,6,15]. The State Council of China approved the "12th National Five Year Plan for Comprehensive Prevention and Control of Heavy Metal Pollution" in early 2011, with the aim of controlling heavy metal contamination in the environment [16]. Eastern China is the most developed and most densely populated region in China. Rapid economic development and urbanization have caused significant amounts of pollution, including heavy metal pollution, in different aquatic ecosystems (rivers, lakes, and wetlands) in eastern China [6,13,15,16]. It is important to comprehensively understand the characteristics of heavy metal contamination in different types of aquatic ecosystems in this area to provide a reference point for the development of better procedures for controlling and managing heavy metals in the affected ecosystems.

In the study presented here, we evaluated the heavy metal pollution status of five different types of aquatic ecosystems (river, reservoir, estuary, lake, and wetland ecosystems) in eastern China. The specific goals were: 1) to determine heavy metal (Cd, Cr, Cu, Ni, Pb, and Zn) concentrations in surface sediments in twelve study areas; 2) to use enrichment factors (EFs) to determine the importance of lithogenic and anthropogenic sources of the heavy metals; and 3) to assess the risks associated with the presence of the heavy metals in the aquatic ecosystems using consensus-based sediment quality guidelines (SQGs), potential ecological risk (PER) indices, and risk assessment codes (RACs).

Materials and Methods

Ethics Statement

No specific permits were required for the field studies described here. The study areas are not privately owned or protected in any way, and the field studies did not involve endangered or protected species.

Description of the Study Areas

Twelve study areas (Fig. 1) in eastern China that have been affected by rapid industrial development and urbanization were selected to represent five typical aquatic ecosystems (reservoirs, rivers, estuaries, wetlands, and lakes). The study areas were in Heilongjiang Province, Heibei Province, Beijing Municipality, Tianjin Municipality, Zhejiang Province, Anhui Province, Jiangxi Province, and Guangdong Province. The study areas were in the region between 113° 15′ 31.08″ E and 126° 30′ 52.43″ E and between 23° 50′ 26.85″ N and 45° 46′ 18.01″ N.

The river ecosystems were the Songhuajiang River (SR), two typical rivers in the Haihe River basin (HR1 and HR2), and the Zhujiang River (ZR). The reservoir ecosystem was the Miyun Reservoir (MRe). Estuarine ecosystems were represented by the estuaries of rivers HR1 (estuary HE1) and HR2 (estuary HE2). The lake ecosystems were Chaohu Lake (CL) and Poyanghu Lake (PL), and the wetland ecosystems were the Baiyangdian Wetland (BW) and two areas in the Shijiuyang Wetland (SW1 and SW2).

Sample Collection and Analysis

Surface sediment samples were collected from the twelve study areas (Fig. 1) using a Peterson grab sampler between August and October 2013. Five sampling stations were established in each study area, and three parallel samples were taken from each station (Fig. S1 in File S1). The samples (n = 180) were air-dried in the laboratory then dried in an oven at 40°C. The dried samples were ground and passed through a 100 mesh sieve before they were analyzed.

For the total heavy metal analysis, each sediment sample (0.100 g) was digested in a 5:1 mixture of hydrofluoric acid and perchloric acid [17] in a microwave digestion system (MARS XPress; CEM, Matthews, NC, USA) using the digestion conditions shown in Table S1 in File S1. The geochemical fractionation of the heavy metals in a sample was determined using the BCR three-step sequential extraction procedure [18], to determine the exchangeable (sediment solution, carbonates, and exchangeable metals), reducible (Fe/Mn oxides), oxidizable (organic matter and sulfides), and residual (remaining non-silicate bound) fractions of the heavy metals. All of the above solutions were stored at 4°C until instrumental analysis was performed. The Cd, Cr, Cu, Fe, Ni, Pb, and Zn concentrations in the extracts were determined using an inductively coupled plasma-mass spectrometry (ICP-MS) instrument (7500a; Agilent Technologies, Santa Clara, CA, USA), which gave detection limits of 0.015–0.120 μg/L, and an inductively coupled plasma optical emission spectrometry (ICP-OES) instrument (Optima 2000DV; Perkin Elmer, Waltham, MA, USA), which gave detection limits of 0.001–0.030 mg/L. The quality control procedures used in the laboratory were analyzing a sediment reference material (GBW07302a; National Institute of Metrology, Beijing, China) and analyzing samples in triplicate. The recoveries of the heavy metals varied but were all within the range 90–105%. The analytical method gave a good level of precision, the relative standard deviation being less than 5%. Each result that is presented is the average of the measurements made at the five sampling stations in each study area.

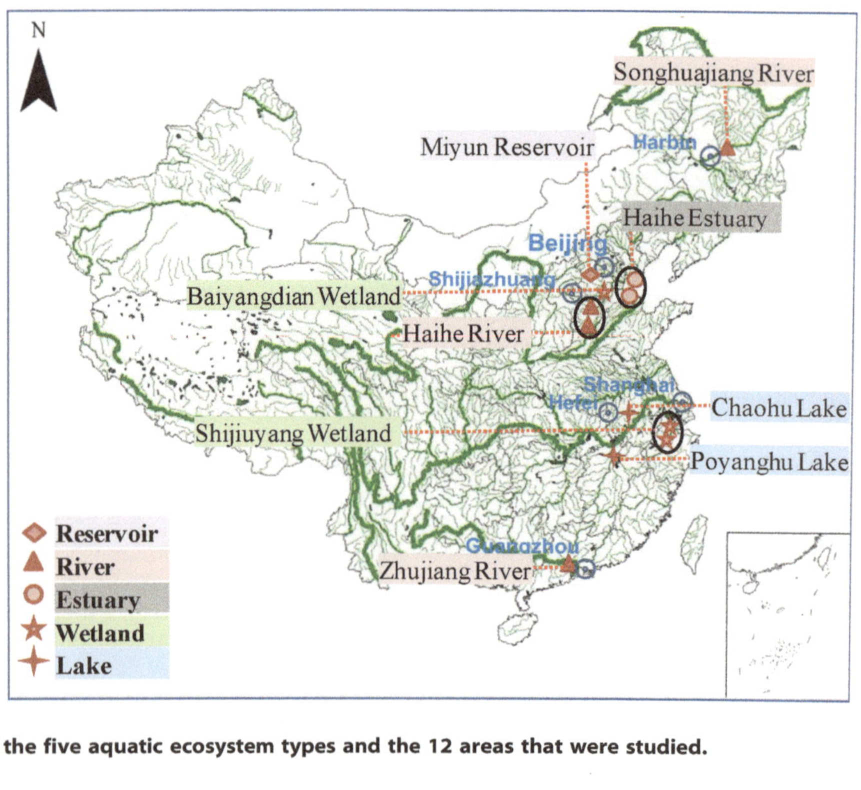

Figure 1. Locations of the five aquatic ecosystem types and the 12 areas that were studied.

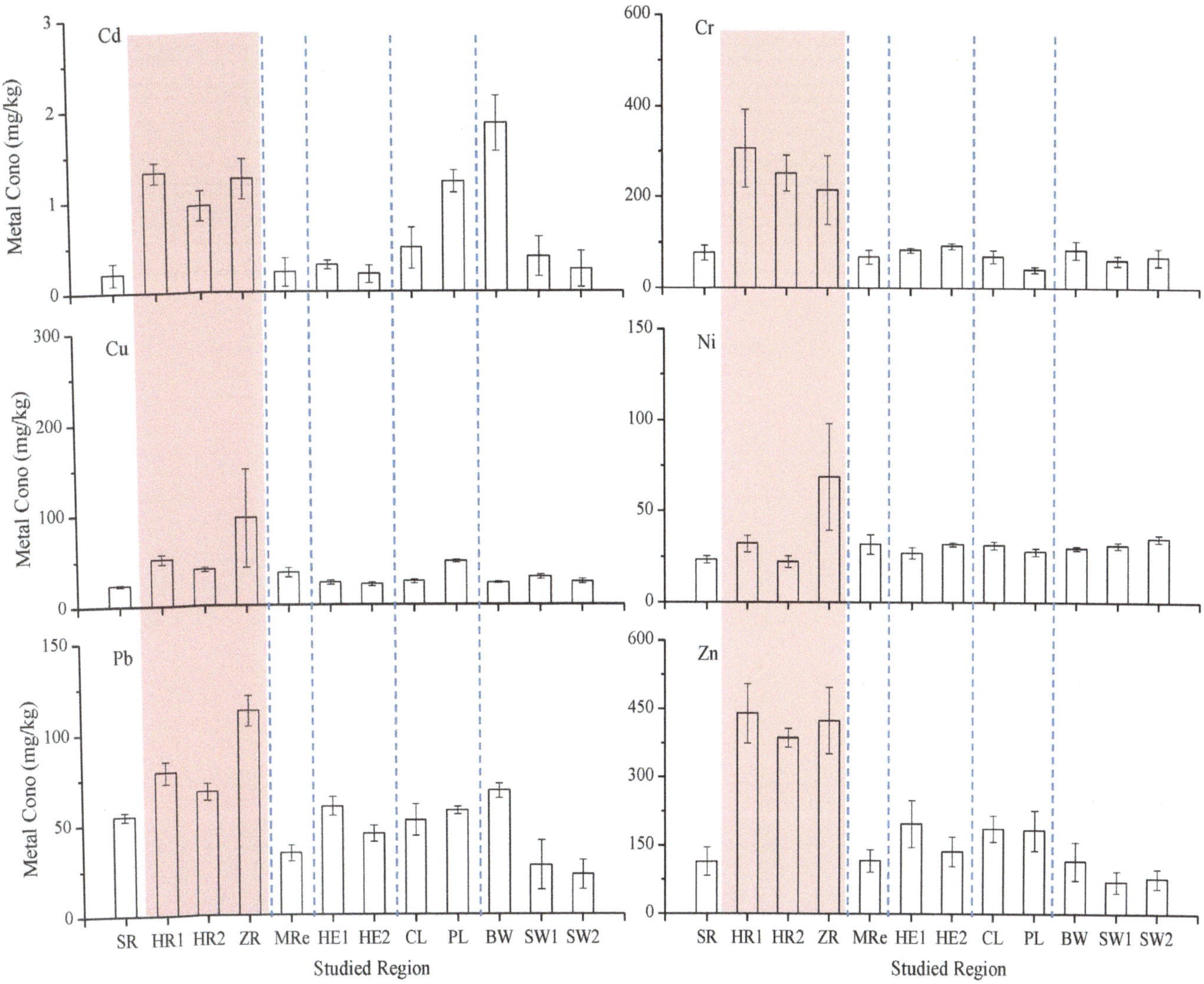

Figure 2. Heavy metal concentrations in the surface sediment samples from each of the study sites.

Enrichment Factors

The EFs for the heavy metals in the surface sediment samples were calculated using eq. 1, which was published by Zhang et al. [19], to provide information on the sources of the metal contaminants and temporal variations in the contributions of the different sources.

$$EF = [C_{n(sample)}/C_{Fe(sample)}]/[B_{n(baseline)}/B_{Fe(baseline)}] \quad (1)$$

where C_n is the metal concentration in a sediment, B_n is the background concentration of the metal, C_{Fe} is the Fe concentration in the sediment, and B_{Fe} is the background Fe concentration. The background metal concentrations in soils from each of the twelve study areas were used as the baseline concentrations [20].

Risk Assessment

The ecological risks posed by heavy metals in surface sediments were assessed using three methods, consensus-based SQGs [7], PER indices [15,21], and RACs [22]. Details of the assessment methods are provided in Table S2 in File S1.

Results and Discussion

Heavy Metal Concentrations in Surface Sediments

The heavy metal concentrations found in the surface sediments from the five types of aquatic ecosystems are shown in Fig. 2. The average Cd, Cr, Cu, Ni, Pb, and Zn concentrations in the twelve study areas were 0.716, 118, 37.3, 32.7, 56.6, and 204 mg/kg, respectively. The highest average Cd concentrations (>1.00 mg/kg) were found in surface sediments from the HR1, ZR, PL, and BW sites. The highest average Cr concentrations were found in surface sediments from the HR1, HR2, and ZR sites (307, 252, and 215 mg/kg, respectively). The highest average Cu, Ni, and Pb concentrations (96.1, 68.8, and 112 mg/kg, respectively) were found in surface sediments from the ZR site. The highest average Zn concentrations were found in surface sediments from the HR1, HR2, and ZR sites (440, 387, and 424 mg/kg, respectively). The highest total heavy metal concentrations (i.e., the sum of the Cd, Cr, Cu, Ni, Pb, and Zn concentrations) in surface sediments were mainly found in samples from river ecosystems (Fig. 1 and Fig. 2), especially samples from the HR1 and ZR sites, which contained total heavy metal concentrations of 911 and 918 mg/kg, respectively. The high heavy metal concentrations in the surface

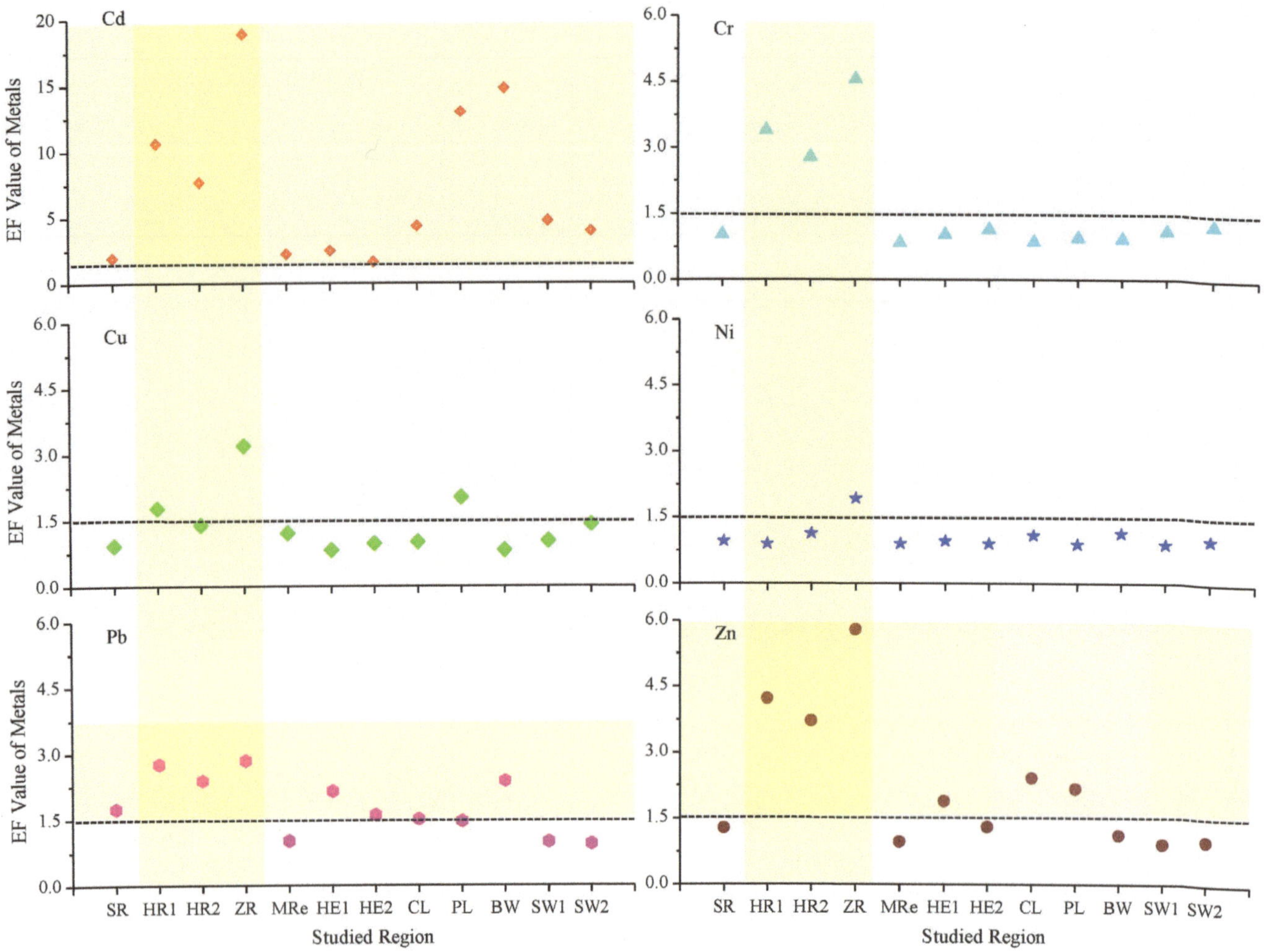

Figure 3. Heavy metal enrichment factors (EFs) in the sediment samples from the study sites.

sediments from the rivers implied that the rivers in the study area have suffered substantial inputs of heavy metals from anthropogenic sources [23].

The relationships between the concentrations of different heavy metals elements can provide information on the sources of the heavy metal [24]. Therefore, the Spearman correlation coefficients for the relationships between the heavy metal concentrations were determined (Table S3 in File S1). Positive correlations ($p \leq 0.01$) were found between the Cd and Pb concentrations and between the Pb and Zn concentrations. Significant positive correlations ($p \leq 0.05$) were also found between the Cd and Cu concentrations, the Cd and Zn concentrations, the Cr and Pb concentrations, and the Cr and Zn concentrations.

Heavy Metal Sources to the Surface Sediments

The EF is a normalization technique that is widely used to identify the proportion of a metal derived from natural environmental sources and the proportion of the metal derived from anthropogenic activities [16,25]. The EF for each metal was calculated, and is shown in Fig. 3, to determine the anthropogenic influences on the heavy metal concentrations in surface sediments in the five types of aquatic ecosystems that were studied in the study areas. The mean EF was highest for Cd (7.16), indicating that anthropogenic sources had had a stronger impact on Cd concentrations than the concentrations of the other heavy metals in the sediments in the study areas. The remaining EFs decreased in the order Zn (2.23), Pb (1.81), Cr (1.65), Cu (1.37), and Ni (1.05). The EFs were highest for the surface sediments from the river sites, and the average EFs were 3.93, 3.17, and 6.18 for the samples from the HR1, HR2, and ZR sites, respectively. The spatial distribution of the EFs for each metal was similar to the spatial distribution of the concentrations of that metal.

An EF of approximately 1 suggests that the metal may originate entirely from natural sources, such as crustal material and natural weathering processes. An EF slightly higher than 1 may not necessarily be caused by the addition of the metal from anthropogenic activities, but may be caused by natural differences in the metal concentrations in the sediment and the reference soil used for the EF calculation. Therefore, an EF of between 0.5 and 1.5 suggests that the metal may be entirely derived from crustal materials and natural weathering processes. However, an EF higher than 1.5 suggests that a significant proportion of the metal originates from anthropogenic processes [16,25,26].

The Cd, Pb, and Zn concentrations in the sediments from the aquatic ecosystems in the study areas positively correlated with each other (Table S3 in File S1) and these were found to be the most anthropogenically enriched of the heavy metals that were analyzed at most of the study sites (EF>1.5) (Fig. 3). The potential anthropogenic sources of these metals include industrial processes,

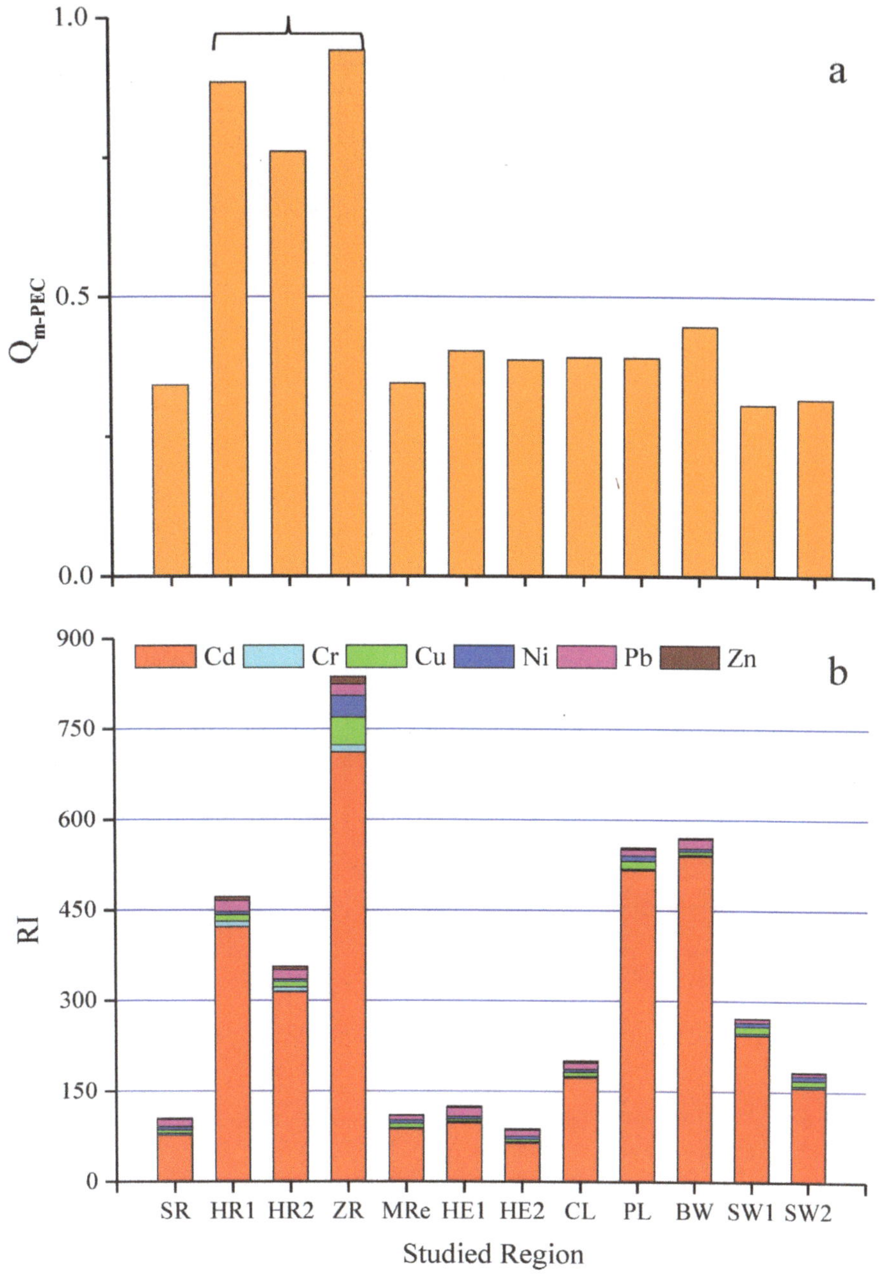

Figure 4. Heavy metal risk assessment indices for the sediment samples from each study site. (a) $Q_{m\text{-}PEC}$ and (b) comprehensive PER index (*RI*).

agriculture, and mining [16,27,28]. The mean EF for Cr in the sediments that were analyzed was slightly higher than 1.5, and this was mainly caused by high EFs at the HR1, HR2, and ZR sites (3.39, 278, and 4.55, respectively). This indicates that Cr was enriched in the sediments at the HR1, HR2, and ZR sites because of anthropogenic emissions. The mean EFs for Cu and Ni were less than 1.5, indicating that these two metals were mainly derived from natural sources, such as the underlying geological material. The EFs that were obtained may be useful indicators of the role of anthropogenic processes in the distributions of the metals in the study area. The mean EFs for Cd in the sediment samples from the five types of aquatic ecosystems were all greater than 1.5, and the Cd EFs were particularly high at the ZR (18.9), BW (14.8), and PL (13.0) sites (Fig. 3). These results indicate that Cd pollution has become a major environmental problem in aquatic ecosystems in eastern China, and this conclusion is consistent with the conclusions of other studies [4,6,16,27].

Risks Posed by Heavy Metals in Surface Sediments

The ecological risks posed to the five types of aquatic ecosystems that were studied by the heavy metals that were measured in surface sediments were assessed using consensus-based SQGs, PER indices, and RACs. The mean probable effect concentration quotients ($Q_{m\text{-}PEC}$s) for the sediments from sites HR1, HR2, and ZR were higher than 0.5 (0.885, 0.761, and 0.942, respectively), which indicates that these sediments were toxic and were seriously

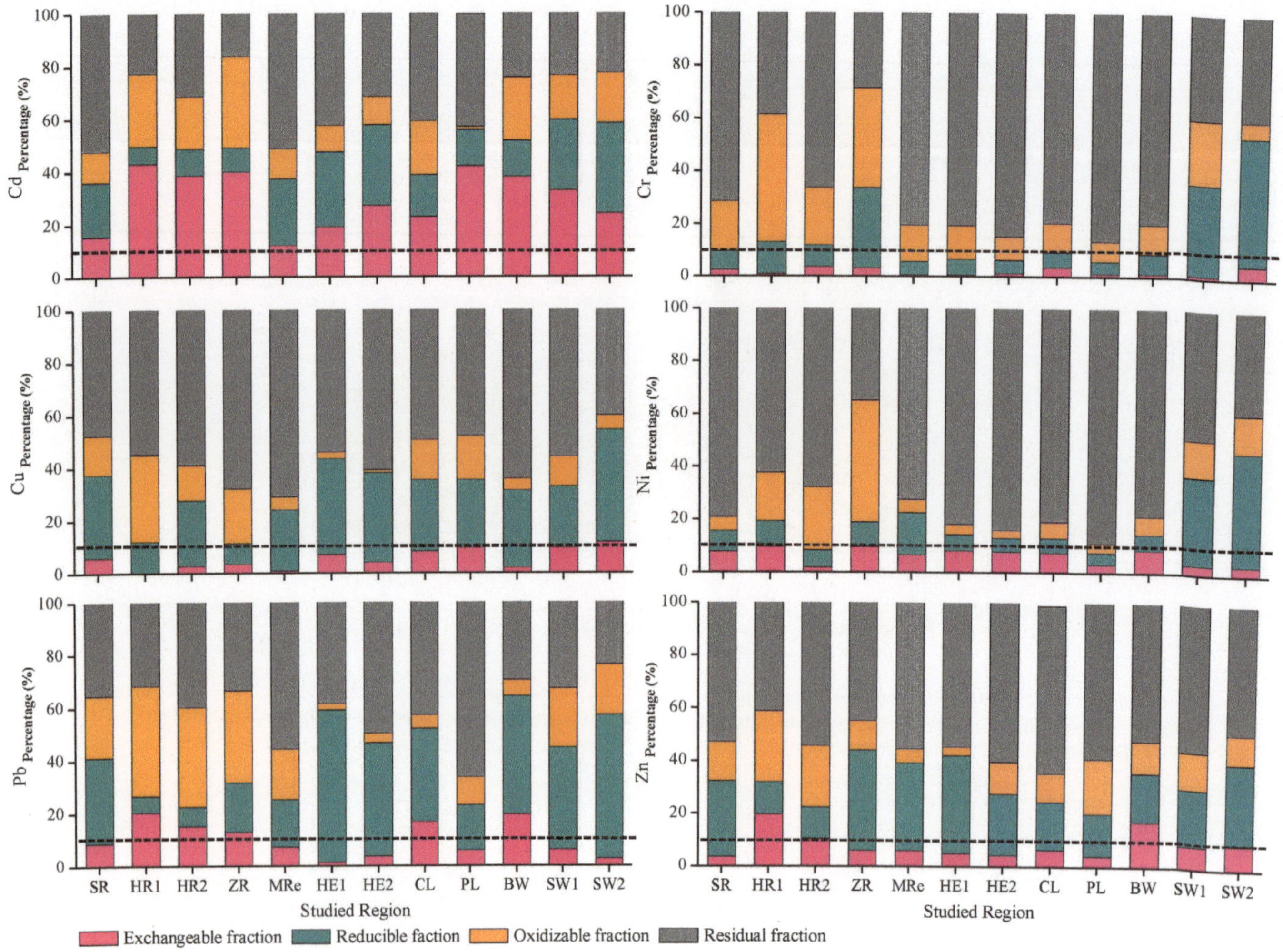

Figure 5. Proportions of the heavy metal fractions in sediments from the five aquatic ecosystem types.

contaminated with heavy metals (Fig. 4a). The PER index for an individual element (E^i_r) describes the sensitivity of a biological community to a given substance and can be used to illustrate the risk posed to that community by contaminants [4,28]. For the aquatic ecosystems in eastern China that we studied, the E^i_r values for the heavy metals decreased in the order Cd>Cu>Pb>Ni> Cr>Zn. All of the metals that we studied apart from Cd had low PER indices. The E^i_r values for Cd ranged from 64.2 (at HE2) to 711 (at ZR), and the average was 283, indicating that sediments in aquatic ecosystems in eastern China are highly contaminated with Cd, which is consistent with the conclusions we drew from the Cd EFs.

The comprehensive PER index (*RI*) distribution can be useful for identifying sites that need more attention than others because of risks associated with pollutants [16]. In this study, the mean RI values for the surface sediments were highest at the ZR, BW, and PL sites (836, 570, and 554, respectively), followed by the HR1 and HR2 sites (472 and 356, respectively), and the overall average RI was 322. According to the PER index criteria (Table S2 in File S1), the comprehensive PER for the surface sediments at the ZR site showed that this site suffers from very high-grade risks. The BW, PL, HR1, and HR2 sites all had high-grade risks, and the CL, SW1, and SW2 sites had moderate-grade risks. The sediments in the river ecosystems had higher PER indices than did the sediments from the other ecosystems, mainly resulting from Cd pollution, and this conclusion was similar to the conclusions we drew from the distributions of the heavy metal concentrations and EFs. Heavy metal pollution was found to be most severe in the river ecosystems, and this is probably because rivers, being open systems, are more vulnerable to contamination caused by anthropogenic processes. Pollutants are often discharged directly into rivers, and they can be quickly deposited in river sediments. The pollutants that are not deposited in the sediments will be transported to other ecosystems in the flowing water [6].

The Cd, Cr, Cu, Ni, Pb, and Zn speciation in the sediment samples was investigated (Fig. 5) to further evaluate the heavy metal contamination in the sediment samples. Cr, Cu, Ni, and Zn were predominantly found in the residual fraction in the studied sediments, and this fraction contained an average of more than 50% of the total metal content. In contrast, about 30% of the total Cd content and about 10% of the total Pb content were found in the exchangeable fraction of each sample. A high proportion of a metal in the exchangeable fraction indicates anthropogenic pollution [29,30]. According to the RAC classification (Table S2 in File S1), the risks posed by Cd in the surface sediments were high at the HR1, HR2, ZR, PL, BW, and SW1 sites and medium at the other sites. The risks posed by Pb in the surface sediments were found to be medium at the HR1, HR2, ZR, CL, and BW sites, and the risks posed by Zn were also found to be medium at the HR1, HR2, and BW sites. These results corroborate the EF and PER index results. Contaminants in the sediment may move into the foodchain, particularly if the contaminants are in bioavailable forms. Cd has a high transfer rate in the environment, so it can accumulate in relatively large amounts in plants without any apparent effects on the plants. Cd can, therefore, be found in most human foodstuffs, and this can cause health problems in humans [16,31,32]. Therefore, heavy metal contaminants, especially Cd, in aquatic ecosystems (particularly river ecosystems) in eastern China should be carefully monitored and controlled in the future [6,33].

Conclusions

We studied heavy metal pollutants in surface sediments from different aquatic ecosystems in eastern China. The heavy metal concentrations were usually higher in surface sediments in river ecosystems than in other ecosystems. The EFs showed that Cd, Zn, and Pb were enriched in the surface sediments more because of anthropogenic sources than were Cr, Cu, and Ni, which were predominantly derived from natural sources in most of the study areas. Using consensus-based SQGs, PER indices, and RACs, we found that all five types of aquatic ecosystems, especially river ecosystems, in the study areas have been polluted with heavy metals. The most serious pollutant in the study areas is Cd, which was mainly found in the exchangeable fraction of the sediment samples. This information could be useful for the development of effective management strategies for controlling heavy metal pollution in aquatic ecosystems in eastern China.

Supporting Information

File S1 Table S1. Microwave digestion conditions used for the sediment samples. **Table S2.** The evaluation criteria for different risk assessment methods. **Table S3.** Correlation coefficients for the heavy metals in the surface sediments from different types of aquatic ecosystems. **Figure S1.** Map showing the layouts used at the sampling stations in each type of aquatic ecosystem.

Author Contributions

Conceived and designed the experiments: WZT BQS WQZ. Performed the experiments: WZT LSW YKD. Analyzed the data: WZT BQS HZ. Contributed reagents/materials/analysis tools: WZT WQZ LSW. Wrote the paper: WZT BQS.

References

1. Iksandar IK, Keeney DR (1974) Concentration of heavy metals in sediment cores from selected Wisconsin lakes. Environ Sci Technol 8: 165–170.
2. Allengil SM, Gubala CP, Landers DH, Lasorsa BK, Crecelius EA, et al. (1997) Heavy metal accumulation in sediment and freshwater fish in US Arctic lakes. Environ Toxicol Chem 16: 733–741.
3. Kim IS, Kang KH, Johnson-Green P, Lee EJ (2003) Investigation of heavy metal accumulation in Polygonum thunbergii for phytoextraction. Environ Pollut 126: 235–243.
4. Ma Z, Chen K, Yuan Z, Bi J, Huang L (2013) Ecological risk assessment of heavy metals in surface sediments of six major Chinese freshwater lakes. J Environ Qual 42: 341–350.
5. Taweel A, Shuhaimi-Othman M, Ahmad AK (2013) Assessment of heavy metals in tilapia fish (*Oreochromis niloticus*) from the Langat River and Engineering Lake in Bangi, Malaysia, and evaluation of the health risk from tilapia consumption. Ecotox Environ Safe 93: 45–51.
6. Zhang LW, Shao HB (2013) Heavy metal pollution in sediments from aquatic ecosystems in China. Clean-Soil, Air, Water 41: 878–882.
7. MacDonald DD, Ingersoll CG, Berger T (2000) Development and evaluation of consensus-based sediment quality guidelines for freshwater ecosystems. Arch Environ Con Tox 39: 20–31.
8. Hakanson L (1980) An ecological risk index for aquatic pollution control: A sedimentological approach. Water Res 14: 975–1001.
9. Suresh G, Sutharsan P, Ramasamy V, Venkatachalapathy R (2012) Assessment of spatial distribution and potential ecological risk of the heavy metals in relation to granulometric contents of Veeranam lake sediments, India. Ecotox Environ Safe 84: 117–124.
10. Karak T, Bhattacharyya P, Paul RK, Das DK (2013) Metal accumulation, biochemical response and yield of Indian mustard grown in soil amended with rural roadside pond sediment. Ecotox Environ Safe 92: 161–173.
11. Buggy CJ, Tobin JM (2008) Seasonal and spatial distribution of metals in surface sediment of an urban estuary. Environ Pollut 155: 308–319.
12. Jiang M, Zeng G, Zhang C, Ma X, Chen M, et al. (2013) Assessment of Heavy Metal Contamination in the Surrounding Soils and Surface Sediments in Xiawangang River, Qingshuitang District. PLoS One 8: e71176.
13. Gao XL, Chen CTA (2012) Heavy metal pollution status in surface sediments of the coastal Bohai Bay. Water Res 46: 1901–1911.
14. He MC, Wang ZJ, Tang HX (1998) The chemical, toxicological and ecological studies in assessing the heavy metal pollution in Le An River, China. Water Res 32: 510–518.
15. Guo WH, Liu XB, Liu ZG, Li GF (2010) Pollution and potential ecological risk evaluation of heavy metals in the sediments around Dongjiang Harbor, Tianjin. Procedia Environ Sci 2: 729–736.
16. Tang WZ, Zhao Y, Wang C, Shan BQ, Cui JG (2013) Heavy metal contamination of overlying waters and bed sediments of Haihe Basin in China. Ecotox Environ Safe 98: 317–323.
17. Tessier A, Campbell PGC, Bisson M (1979) Sequential extraction procedure for the speciation of particulate trace metals. Anal Chem 51: 844–851.
18. Nemati K, Bakar NKA, Abas MR, Sobhanzadeh E (2011) Speciation of heavy metals by modified BCR sequential extraction procedure in different depths of sediments from Sungai Buloh, Selangor, Malaysia. J Hazard Mater 192: 402–410.
19. Zhang H, Shan BQ (2008) Historical records of heavy metal accumulation in sediments and the relationship with agricultural intensification in the Yangtze–Huaihe region, China. Sci Total Environ 399: 113–120.
20. China National Environmental Monitoring Center (CNEMC) (1990). The background concentrations of soil elements in China. Beijing: China Environmental Science Press.
21. Liu YJ, Shi GX, Mao L, Cheng G, Jiang SJ, et al. (2012) Direct and indirect influences of 8 yr of nitrogen and phosphorus fertilization on Glomeromycota in an alpine meadow ecosystem. New Phytol 194: 523–535.
22. Singh KP, Mohan D, Singh VK, Malik A (2005) Studies on distribution and fractionation of heavy metals in Gomti river sediments-a tributary of the Ganges, India. J Hydrol 312: 14–27.
23. Sondi I, Lojen S, Juračić M, Prohić E (2008) Mechanisms of land–sea interactions–the distribution of metals and sedimentary organic matter in sediments of a river-dominated Mediterranean karstic estuary. Estuar Coast Shelf S 80: 12–20.
24. Dragovic S, Mihailivic N, Gajic B (2008) Heavy metals in soils: distribution, relationship with soil characteristics and radionuclides and multivariate assessment of contamination sources. Chemosphere 72: 491–495.
25. Menge DNL, Hedin LO, Pacala SW (2012) Nitrogen and phosphorus limitation over long-term ecosystem development in terrestrial ecosystems. PLoS One 7: e42045.
26. Zhang J, Liu CL (2002) Riverine composition and estuarine geochemistry of particulate metals in China - weathering features, anthropogenic impact and chemical fluxes. Estuar Coast Shelf S 54: 1051–1070.
27. Tang WZ, Shan BQ, Zhang H, Mao ZP (2010) Heavy metal sources and associated risk in response to agricultural intensification in the estuarine sediments of Chaohu Lake Valley, East China. J Hazard Mater 176: 945–951.
28. Fu J, Zhao CP, Luo YP, Liu CS, Kyzas GZ, et al. (2014) Heavy metals in surface sediments of the Jialu River, China: Their relations to environmental factors. J Hazard Mater 270: 102–109.
29. Yang Z, Wang Y, Shen Z, Niu J, Tang Z (2009) Distribution and speciation of heavy metals in sediments from the mainstream, tributaries, and lakes of the Yangtze River catchment of Wuhan, China. J Hazard Mater 166: 1186–1194.
30. Weng H, Ma X, Fu F, Zhang J, Liu Z, et al. (2014) Transformation of heavy metal speciation during sludge drying: Mechanistic insights. J Hazard Mater 265: 96–103.
31. Lambert R, Grant C, Sauve S (2007) Cadmium and zinc in soil solution extracts following the application of phosphate fertilizers. Sci Total Environ 378: 293–305.
32. Satarug S, Garrett SH, Sens MA, Sens DA (2010) Cadmium, environmental exposure, and health outcomes. Environ Heal Persp 118: 182–190.
33. Cheng SP (2003) Heavy metal pollution in China: Origin, pattern and control. Environ Sci Pollut R 10:192–198.

13

Seasonal Dynamics of Trace Elements in Tidal Salt Marsh Soils as Affected by the Flow-Sediment Regulation Regime

Junhong Bai[1,2]*, Rong Xiao[1], Qingqing Zhao[1], Qiongqiong Lu[1], Junjing Wang[1], K. Ramesh Reddy[2]

1 State Key Laboratory of Water Environment Stimulation, School of Environment, Beijing Normal University, Beijing, P. R. China, **2** Wetland Biogeochemistry Laboratory, Soil and Water Science Department, University of Florida, Gainesville, Florida, United States of America

Abstract

Soil profiles were collected in three salt marshes with different plant species (i.e. *Phragmites australis*, *Tamarix chinensis* and *Suaeda salsa*) in the Yellow River Delta (YRD) of China during three seasons (summer and fall of 2007 and the following spring of 2008) after the flow-sediment regulation regime. Total elemental contents of As, Cd, Cu, Pb and Zn were determined using inductively coupled plasma atomic absorption spectrometry to investigate temporal variations in trace elements in soil profiles of the three salt marshes, assess the enrichment levels and ecological risks of these trace elements in three sampling seasons and identify their influencing factors. Trace elements did not change significantly along soil profiles at each site in each sampling season. The highest value for each sampling site was observed in summer and the lowest one in fall. Soils in both *P. australis* and *S. salsa* wetlands tended to have higher trace element levels than those in *T. chinensis* wetland. Compared to other elements, both Cd and As had higher enrichment factors exceeding moderate enrichment levels. However, the toxic unit (TU) values of these trace elements did not exceed probable effect levels. Correlation analysis showed that these trace elements were closely linked to soil properties such as moisture, sulfur, salinity, soil organic matter, soil texture and pH values. Principal component analysis showed that the sampling season affected by the flow-sediment regulation regime was the dominant factor influencing the distribution patterns of these trace elements in soils, and plant community type was another important factor. The findings of this study could contribute to wetland conservation and management in coastal regions affected by the hydrological engineering.

Editor: Bo Li, Fudan University, China

Funding: This work was financially supported by the National Basic Research Program (2013CB430406), the National Natural Science Foundation of China (51379012 and 51179006), the Program for New Century Excellent Talents in University (NECT-10-0235) and the Fok Ying Tung Foundation (132009). The funders had no role in study design, data collection and analysis, decision to publish, or preparation of the manuscript.

Competing Interests: The authors have declared that no competing interests exist.

* Email: junhongbai@163.com

Introduction

Sediment contamination in rivers, lakes, reservoirs, and wetlands has been widely reported in the developing countries as a result of the intense land use in agricultural and urban environment. Contamination of sediments with trace elements is a major concern [1,2]. Moreover, these trace elements can be transferred and carried downstream into wetland ecosystems, and accumulate in wetland soils [3,4]. Wetland soils serve as "source" and "sink" of these chemical pollutants [5]. For example, salt marshes and estuarine sediments can retain these metals as metal sulfides [5,6]. However, seasonal hydrological changes and water level fluctuation can affect Eh and pH of wetlands, thus resulting in mobilizing trace elements [5,7].

Several studies on large river delta have focused on source identification of trace elements and organic pollutants in surface sediments [6,8,9,10]. Some studies have shown that heavy metals can be accumulated and retained in wetland soils for a long time period [11], and they would not vary significantly over seasons in salt marshes under natural conditions [12]. The global regulation of rivers and steams by building reservoirs and dams has brought great effects on downstream ecosystems [13]. Recently, Bai et al. [4] have reported that the flow-sediment regulation contributed to trace elements (i.e. Arsenic, Cadmium, and others) accumulation in surface wetland soils of the Yellow River Delta as bed sands can carry and settle these trace elements to the downstream due to strong hydraulic flushing [12,14]. However,less information is available on the dynamics and fate of these trace elements in estuarine wetlands after the flow- sediment regulation.

The Yellow River Delta (YRD) is one of the most active regions of land-ocean interaction, and a national nature reserve was established to better protect this newly-formed wetland ecosystem and maintain biological diversity [15]. However, with the rapid development of agriculture, fisheries, and the extensive exploitation of the Shengli oilfield (It is the second largest oilfield in China and is mainly located in both sides of the Yellow River Estuary, which was originally built in 1962 and the working area has covered approximately 4.4×10^4 km^2), the Yellow River Delta was greatly impacted by the intense human activities, leading to serious wetland degradation [16]. Moreover, the flow-sediment regulation regime has shown a significant influence on trace element accumulation [4] and wetland plant distribution [16] in the Yellow River Delta.

The primary objectives of this study were: (1) to determine the dynamic changes of selected trace elements including As, Cd, Cu, Pb, and Zn in wetland soils covered by different dominant plant species (i.e., *Phragmites australis*, *Suaeda salsa* and *Tamarix chinensis*) of the Yellow River Delta; (2) to assess enrichment levels of these trace elements and determine their ecological risks using enrichment and toxic indicators; and (3) to identify the relationships among trace elements and other selected soil properties.

Materials and Methods

Ethics statement

Our study area is located in the Yellow River Delta wetland nature reserve, which is owned by the Chinese government. We obtained a specific permit from the administration bureau of the Yellow River Delta National Nature Reserve to conduct this study in the reserve. Moreover, our sampling sites were not located in any strictly protected areas and the field studies did not involve endangered or protected species.

Study area

The study area is located in the Yellow River Delta of China (37°37′48″ to 37°48′36″ N and 119°01′ to 119°21′ E) (Figure 1). It has a temperate, continental monsoon climate with distinct seasons. The annual mean air temperature is 12°C, with 196 frost-free days, and the annual mean precipitation is 552 mm [17]. Soils are characterized as loamy fine sand or coarse sand in all soil layers with clay content of 5–7% and can be classified as aquents, aquolls and fluvaquents [18]. The dominant vegetation in tidal salt marshes include herbaceous plants such as *Phragmites australis* and *Suaeda salsa* and woody plant namely *Tamarix chinensis* [17]. The vegetation succession in this delta follows the order *Suaeda salsa* → *Tamarix chinensis* → *Phragmites australis*, and plant distributions are dominantly controlled by soil salinity. The study area was consistently influenced by the flow-sediment regulation of the upstream Xiaolangdi Reservoir from June and July since 2002 [4]. Xiaolangdi Reservoir began storing water in 1999, and considerable siltation occurred in the reservoir after commissioning, with a total sediment trapping of 32.47×10^8 t from 1997 to 2007 [19]. The flow-sediment regulation of the upstream Xiaolangdi Reservoir has been enforced from June to July since 2002 [17]. For this study, three sampling sites with different dominant plant species (Table 1) were selected. The three sites are approximately 100 m away from each other and they were consistently inundated by tidal seawater (twice every day).

Sample collection and analysis

Soil samples were collected using a soil auger (4.8-cm diam.) from three sampling sites in summer (August, 2007), fall (November, 2007), and spring (April, 2008). Well-developed root systems were observed to depths of 50 cm at *P. australis* wetland and *T. chinensis* wetland, whereas the root zone was confined to depths of 15–20 cm at *S. salsa* wetland. Therefore, in each sampling season, the top 50 cm soils (sectioned into 0–10, 10–20, 20–30, and 30–50 cm) were collected with three or five replicates in *P. australis* wetland and *T. chinensis* wetland, and the top 20 cm soils (sectioned into 0–10 and 10–20 cm) were sampled with four replicates in *S. salsa* wetland. A total 120 soil samples were collected, including 36 samples in *P. australis* wetland, 60 samples in *T. chinensis* wetland, and 24 samples in *S. salsa* wetland. All soil samples were placed in polyethylene bags and brought to the laboratory, where they were air dried at room temperature for three weeks. Some air-dried soils of selected two profiles at each site in each sampling season were used for soil particle size analysis. All the other air-dried soil samples were sieved through a 2-mm nylon sieve to remove coarse debris and then ground with a pestle and mortar until all particles passed a 0.149-mm nylon sieve for the determination of soil chemical properties. Soil bulk density cores were also correspondingly collected using a 184 cm^3 cylinder from each soil layer of each soil profile, oven dried at 105°C for 24 h, and weighed for the determination of bulk density (BD) and soil moisture.

Plant samples for the aboveground parts (including stems and leaves) were also collected at each soil sampling site in summer. They were placed in paper bags after clean washing and transported to the laboratory. All plant samples were oven dried at 65°C for 48 h, and ground into fine powder for the determination of trace metals. Soil and plant samples were digested with an $HClO_4$-HNO_3-HF mixture in Teflon tubes to analyze total contents of Al, As, Cd, Cu, Pb, and Zn. The solutions of the digested samples were analyzed by inductively coupled plasma atomic absorption spectrometry (ICP-AAS). Quality assurance and quality control were assessed using duplicates, method blanks and standard reference soil materials (GBW07401) and standard reference plant materials (GBW07602) from the Chinese Academy of Measurement Sciences with each batch of samples (1 blank and 1 standard for each 10 samples). The recovery rates of samples spiked with standards ranged from 99 to 106% for soils, and from 82–119% for plants. Soil organic matter (SOM) was measured using dichromate oxidation [20]. Soil pH and salinity were determined in the supernatants of 1:5 soil and water mixtures using a pH meter or salinity meter. Soil particle size was analyzed using a Laser Particle Size Analyzer (Microtrac Inc., USA).

Enrichment factor (EF)

Enrichment factor (EF) was selected to evaluate possible anthropogenic inputs of observed sample metals[21]. EF was defined as $EF = (M/Al)_{sample}/(M/Al)_{background}$, where M_{sample} and $M_{background}$ are the determined contents of targeted elements (e.g., As, Cd, Cu, Pb, and Zn) in soil samples and their background contents, respectively; Al_{sample} and $Al_{background}$ are the determined content s of Al in soil samples and its background content, respectively; Al is used the reference element because it is a conservative lithogenic element presumed not to be enriched due to local contamination [22]. Contamination categories were classified based on EF values: EF<2, deficiency to minimal enrichment; EF = 2–5, moderate enrichment; EF = 5–20, significant enrichment; EF = 20–40, very high enrichment; EF>40, extremely high enrichment [23]. Background values for $M_{background}$ and $Al_{background}$ in this study were referenced to background contents in the loess materials of the Yellow River [24]. Background contents of 62700 $\mu g\ g^{-1}$ for Al, 10.7 $\mu g\ g^{-1}$ for As, 0.095 $\mu g\ g^{-1}$ for Cd, 21.1 $\mu g\ g^{-1}$ for Cu, 21.6 $\mu g\ g^{-1}$ for Pb and 64.5 $\mu g\ g^{-1}$ for Zn were used.

Toxic units (TUs)

Toxic units (TUs) are used to normalize the toxicities of various trace elements to allow for the comparison of the relative effects. TUs were defined as the ratios of the observed contents to the probable effect level (PEL) values [25]. PEL values represent the thresholds of chemical contents above which adverse effects are likely to occur. The PEL values for marine and estuarine ecosystem was used in this study (Table 2)[26].

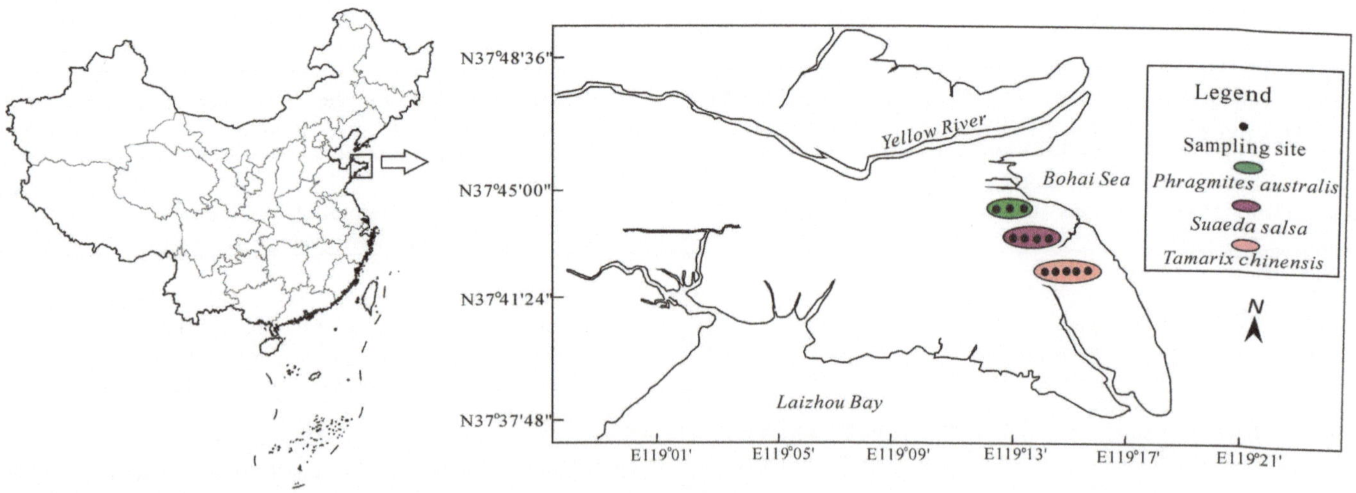

Figure 1. Location map of three sampling sites in the Yellow River Delta.

Biological Concentration Factors (BCFs)

Biological concentration factor is the ratio of metal content in the aboveground parts of plants to metal content in surrounding soils, which can more accurately describe the plant's uptake potential than plant metal content [27].

Statistical analysis

Pearson correlation analysis was performed to determine the relationships between trace elements and selected soil properties and the relationships between Al, trace elements, and sand and clay contents in selected soil profiles. Principal component analysis (PCA) was used to discriminate soil samples with similar or different contamination patterns and identify their influencing factors. Analysis of variance (ANOVA) analysis was implemented to analyze the differences in trace elements between sampling sites and seasons. Differences were considered to be significant at $p < 0.05$. Statistical analysis was performed using the SPSS 16.0 and Canoco 4.5 software packages for Windows.

Results and Discussion

Soil characteristics in three salt marshes

Selected physical-chemical properties in the top 20 cm soils of three salt marshes of the YRD are summarized in Table 2. No significant differences in soil moisture among three seasons in each sampling sites due to consistent tidal flooding condition. However, Soils in *P. australis* wetland contained higher moisture contents in spring and summer compared to *T. chinensis* wetland due to less sand contents in *P. australis* wetland ($p<0.05$; Table 2). Bulk density at each sampling site of the three was lower in spring than in summer and fall ($p<0.05$), but there were no significant differences among three sites in each season. With the exception of higher SOM level in *P. australis* wetland than those in *T. chinensis* and *S. salsa* wetlands in fall ($p<0.05$), no significant differences were observed among three sites in summer and spring. Compared to other seasons, higher SOM content appeared in *P. australis* wetland in fall, while it was higher in *T. chinensis* wetland in spring ($p<0.05$), which was likely caused by plant litter inputs and decomposition [5]. All soil samples of three sites exhibited higher salinities in spring than in summer or fall ($p<0.05$), which

Table 1. Ecological characteristics of three sampling sites in different sampling seasons.

Site	Location	Replicates	Sampling time	Plant height (m)	Plant density (m^{-2})
P. australis wetland	37°43′35.1″ to 37°43′59″N 119°12′42″ to 119°13′21″E	3	Spring	0.1–0.3	35–50
			Summer	1.8–2.2	230–280
			Fall	1.8–2.2	190–230
T. chinensis wetland	37°42′47″ to 37°42′52″N 119°13′49″ to 119°14′15″E	5	Spring	1–1.6	15–30
			Summer	1–2	15–30
			Fall	1–2	15–30
S. salsa wetland	37°43′16″ to 37°43′34″N 119°13′36″ to 119°13′53″E	4	Spring	0.05–0.15	100–220
			Summer	0.20–0.40	280–580
			Fall	0.30–0.45	136–520

Note: Spring = April; Summer = August; Fall = November.

Table 2. Summary of physical-chemical properties in the top 20 cm soils at three sampling sites during three sampling seasons.

	Moisture (%)	BD(g cm^{-3})	SOM(g kg^{-1})	Salinity(‰)	pH	Sand (%)	Clay (%)
P. australis sites							
Spring	28.8±2.9^{a1}	1.4±0.1^{a1}	6.9±0.9^{a1}	1.5±0.3^{a1}	8.4±0.1^{a1}	52.1±8.1^{a1}	5.2±0.8^{a1}
Summer	30.7±4.4^{a1}	1.8±0.1^{b1}	6.7±1.0^{a1}	0.6±0.2^{b1}	8.4±0.1^{a12}	23.8±4.1^{b1}	13.9±2.5^{b1}
Fall	28.0±2.0^{a1}	1.8±0.0^{b1}	8.2±0.6^{b1}	0.7±0.1^{b1}	8.6±0.1^{a1}	55.7±4.5^{a1}	5.0±0.2^{a1}
T. chinensis sites							
Spring	23.6±1.9^{a2}	1.5±0.1^{a1}	7.0±3.0^{a1}	3.4±1.0^{a2}	8.3±0.2^{a1}	81.1±22.2^{a12}	0.7±0.9^{a2}
Summer	23.0±1.6^{a2}	1.9±0.1^{b1}	5.2±1.8^{b1}	2.8±0.7^{ab2}	8.1±0.2^{a1}	61.34±2.8^{a2}	4.9±1.6^{b2}
Fall	25.0±2.2^{a12}	1.8±0.0^{b1}	5.9±2.2^{b2}	2.0±0.5^{b2}	8.6±0.1^{b1}	60.7±6.5^{a1}	0.3±0.2^{b1}
S. salsa sites							
Spring	26.2±1.6^{ab12}	1.5±0.1^{a1}	6.8±1.4^{a1}	1.7±0.4^{a1}	8.4±0.1^{a1}	76.2±9.1^{a2}	1.5±1.2^{a2}
Summer	29.7±4.0^{a1}	1.8±0.1^{b1}	6.5±0.5^{a1}	0.6±0.4^{b1}	8.7±0.3^{ab2}	23.12±1.8^{b1}	14.2±0.1^{b1}
Fall	23.1±2.2^{b2}	1.9±0.1^{b1}	5.5±1.3^{b2}	2.1±1.1^{a2}	8.7±0.1^{b1}	81.1±3.2^{a2}	0.6±0.3^{a1}

Spring = April; Summer = August; Fall = November. BD = Bulk density; SOM = Soil organic matter.
^{a,b} different letters represent significant differences between three seasons at each site.
^{1,2} different numbers represent significant differences between three sites in each season.

was associated with the freshwater inputs from the flow-sediment regulation [4]. Moreover, *T. chinensis* wetland showed higher soil salinity ($p<0.05$) than *P. australis* wetland or *S. salsa* wetland in three seasons as *T. chinensis* wetland was further away from the river channel (Figure 1). However, no significant differences in pH values were observed among three wetlands in spring and fall except that *T. chinensis* wetland had lower pH values ($p<0.05$) than *S. salsa* wetland in summer. Both *T. chinensis* and *S. salsa* wetlands showed higher pH values ($p<0.05$) in fall compared to spring and summer, which might be associated with weaker soil respiration in fall as the dissolved CO_2 in overlaying water could decrease pH values [28]. All sampling sites showed higher clay contents in summer than in spring and fall and sand contents were also lower in both *P. australis* and *S. salsa* wetlands in summer ($p<0.05$; Table 2). This indicated that the flow and sediment regulation in summer brought more clay and silt contents to the tidal salt marshes [4]. However, no significant differences in sand or clay contents were observed between spring and fall, which was associated with tidal seawater erosion [4]. Moreover, soils in *P. australis* and *S. salsa* wetlands contained more clay contents and less sand contents in summer compared to *T. chinensis* wetland ($p<0.05$). This might be caused by the fact that *T. chinensis* wetland is located at the farther distances from the Yellow River Chanel (Figure 1). In both spring and fall, *P. australis* wetland had higher clay contents than *T. chinensis* and *S. salsa* wetlands.

Spatial and temporal variations in trace elements in marsh soils

Profile distributions of these trace elements in three seasons are shown in Figure 2. Generally, trace element contents exhibited a decreasing tendency with depth in each season except for As, Cd and Zn in *P. australis* wetland and Cd, Cu and Zn in *S. salsa* wetland in summer. Prusty et al. [29] also reported that trace elements such as As, Cu, Pb, and Zn decreased with increasing depths in wetland ecosystems. This was greatly associated with plant cycling which lead to trace elements upwards movement through plant litters and return to surface soils [29,30]. However, clear seasonal changes in the contents of these trace elements were observed along soil profiles (Figure 2). Among three seasons, most soil profiles of three sampling sites exhibited the highest levels of As, Cd and Zn in summer, followed by spring, whereas the lowest levels in fall. As for Cu and Pb, lower levels were also observed in fall compared to summer and spring. The average contents of As, Cd and Zn in the top 20 cm soils in *P. australis* and *T. chinensis* wetlands were significantly higher in summer than in fall ($p<0.05$; Table 3). Similarly, much higher average contents of these trace elements in the top 20 cm soils were also observed in *S. salsa* wetland in summer compared to fall ($p<0.05$; Table 3). This is inconsistent with the results by Roychoudhury [12], who presented that trace metal contents in surface soils were relatively lower in summer than in winter due to higher wetland plant uptake and standing stocks of metals in summer. Additionally, Ehlken and Kirchner [31] reported that higher water contents, lower salinities and bulk densities increased the potential mobility of metals in soil solutions. However, all trace elements generally reached their higher levels in summer in this study, although all soils contained lower salinities and higher water contents in three sampling sites after the flow-sediment regulation, and then fell to the lowest levels in fall. Therefore, the higher trace element contents observed in summer might be attributed to exogenous inputs of trace metals (e.g., water and sediment inputs from the flow-sediment regulation) [4,32]. The increased plant uptake of trace elements in the late fall could explain the lower trace element contents in these marsh soils in this season, as plants possess efficient root-to-shoot

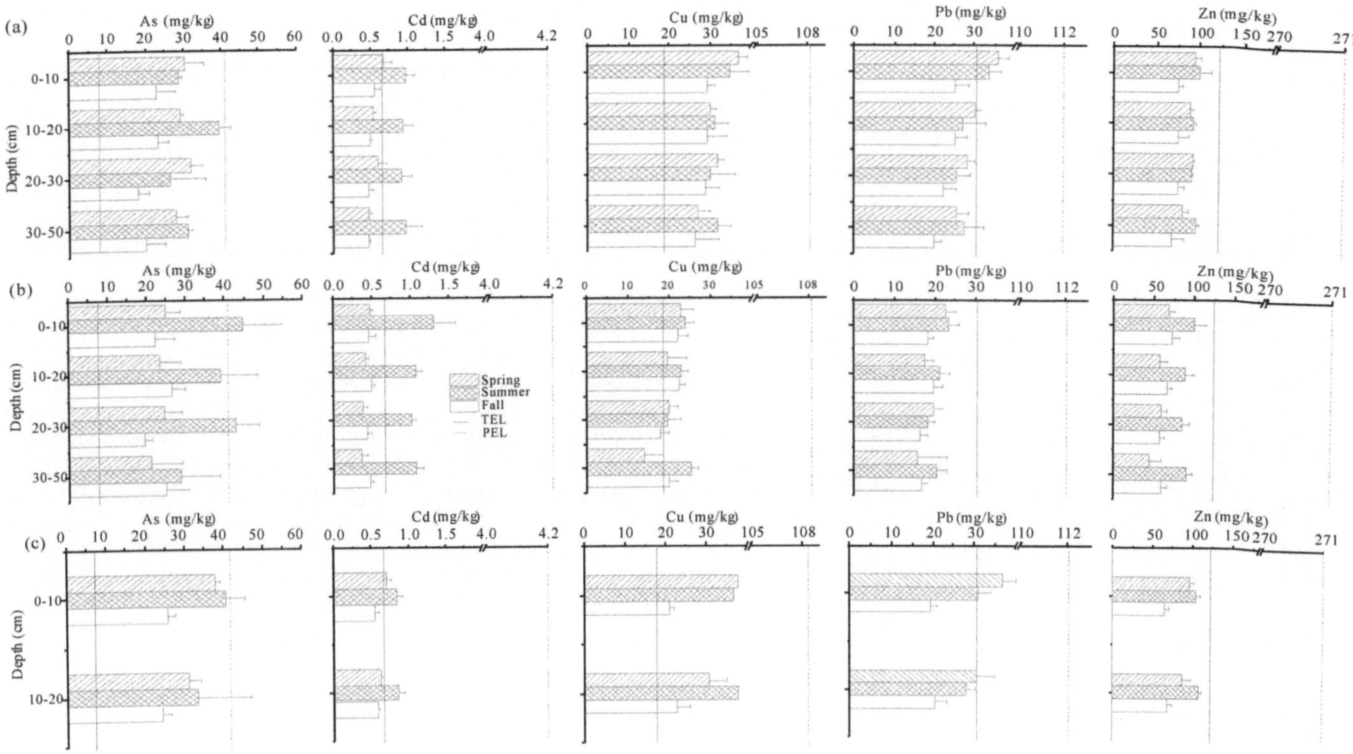

Figure 2. Profile distributions of As and heavy metals in marsh soils with *P. australis* (a), *T. chinensis* (b) and *S. salsa* (c) in three sampling seasons.

translocation systems that are activated at the end of the growing season [33].

Both Cu and Pb contents in soil profiles were generally lower in *T. chinensis* wetland than those in *P. australis* and *S. salsa* wetlands in spring and summer, whereas higher levels of As and Cd in soil profiles were observed in *T. chinensis* wetland in summer (Figure 2). Table 3 also showed that the average contents of trace elements in the top 20 cm soils were significantly higher in *P. australis* and *S. salsa* wetlands in spring compared to *T. chinensis* wetland ($p<0.05$). Meanwhile, higher levels of Cu and Pb in *P. australis* wetland and higher Cd contents in *S. salsa* wetland were observed in fall compared to other sites ($p<0.05$). In summer, *P. australis* and *S. salsa* wetlands exhibited greatly higher Cu and Pb contents than *T. chinensis* wetland ($p<0.05$, Table 3). The differences in Cu and Pb between sites or sampling seasons were most likely related to SOM, sand and clay contents (Tables 4 and 5). Additionally, lower Cu and Pb levels in the top 20 cm soils in *T. chinensis* wetland in summer compared to *S. salsa* wetland might be caused by higher plant uptake as the aboveground parts of *T. chinensis* had higher BCFs values for both Cu and Pb compared to *S. salsa* ($p<0.05$; Figure 3). However, it is difficult to explain much higher BCFs values of *T. chinensis* for Cd with higher soil Cd levels ($p<0.05$; Table 2) than *P. australis*. Therefore, the variations in trace element levels among the three sampling sites might be influenced by the multiple environmental factors including plant communities, soil properties, and water levels etc. [34,35,36]. As shown in Figure 4, all trace elements generally showed similar band-type distribution patterns, indicating higher levels of these trace elements in *P. australis* and *S. salsa* wetlands higher compared to *T. chinensis* wetland except for some soil samples with higher levels of As and Cd. These trace elements increased linearly with increasing Al content at each sampling site, which implies that these trace elements and Al might have common origin.

Assessment of trace element pollution by sediment quality guidelines (SQGs)

The extent of trace element pollution was assessed by comparing element contents in marsh soils to the sediment quality criteria developed by the Chinese Marine Sediment Quality Criteria (GB 18668-2002) [37] and the sediment quality guidelines (SQGs)[23] (Table 3). The Chinese Marine Sediment Quality Criteria indicated that approximately 85–95% of all collected soil samples in three sampling sites during three seasons exceeded the Class I criteria (suitable for fisheries and natural waters) but fell within the scope of Class II criteria (used for industrial and tourism sites) for As and Cd, and all samples were below the Class I criteria for Cu, Pb and Zn (Table 3). This supplies that most soil samples in this region have been moderately polluted by As and Cd. As shown in Figure 2, the SQGs also indicated that almost all soil samples along soil profiles at three sites fell within the range between the threshold effect levels (TELs) and probable effect levels (PELs) for As and Cu in each of the three seasons, even some samples of three sites exceed PELs in summer. Moreover, all soil profile samples at each site exceeded TELs in summer, whereas they were below TELs in both spring and fall. However, approximately 85% of soil profile samples in this region didn't exceed TELs for Pb in three seasons, and almost all soil profile samples at each site exhibited much lower Zn contents than TELs in each season. This implies higher potential risks of As, Cu and Cd in soil profiles of this region. Therefore, it is still required to control heavy metal pollution (i.e., As, Cu and Cd) in this region, as higher trace elements may make it more difficult to restore the degraded living habitat [36].

Table 3. Mean contents of As and heavy metals in the top 20 cm soils of three sampling sites during the whole sampling period (mg kg^{-1} dry weight).

	As	Cd	Cu	Pb	Zn
P. australis sites					
Summer	35.8 ± 2.4^{a1}	0.9 ± 0.1^{a1}	30.7 ± 5.5^{a1}	28.56 ± 5.3^{ab1}	95.1 ± 11.0^{a1}
Fall	23.0 ± 4.8^{b1}	0.5 ± 0.1^{b12}	29.2 ± 1.5^{a1}	25.27 ± 1.3^{b1}	74.7 ± 5.7^{b1}
Spring	29.8 ± 3.5^{c1}	0.6 ± 0.1^{b1}	33.5 ± 4.3^{a1}	32.9 ± 2.4^{ac1}	91.2 ± 10.4^{a1}
T. chinensis sites					
Summer	41.9 ± 3.2^{a2}	1.2 ± 0.2^{a2}	23.0 ± 2.3^{a2}	22.4 ± 2.8^{a2}	93.9 ± 13.1^{a2}
Fall	24.7 ± 3.4^{b1}	0.5 ± 0.0^{b2}	22.1 ± 1.4^{a2}	19.4 ± 2.8^{a2}	72.0 ± 10.7^{b1}
Spring	24.1 ± 4.5^{b2}	0.4 ± 0.0^{b2}	21.4 ± 1.6^{a2}	20.1 ± 1.8^{a2}	62.5 ± 4.5^{c2}
S. salsa sites					
Summer	40.5 ± 6.4^{a2}	0.9 ± 0.1^{a1}	37.7 ± 1.7^{a1}	29.9 ± 1.9^{a1}	106.3 ± 3.8^{a2}
Fall	25.2 ± 1.8^{b1}	0.6 ± 0.0^{b1}	22.0 ± 2.1^{b2}	19.9 ± 2.4^{b2}	$66. 9\pm6.2^{b1}$
Spring	34.5 ± 1.4^{c1}	0.7 ± 0.0^{bc3}	34.5 ± 3.7^{a1}	33.7 ± 4.5^{a1}	92.1 ± 9.5^{c1}
Sediment Quality Criteria of China*					
Class I	20	0.5	35	60	150
Class II	65	1.5	100	130	350
Class III	93	5	200	250	600
SQGs†					
TEL	7.2	0.68	18.7	30.2	124.0
PEL	41.6	4.2	108.2	112.2	271.0

a,b different letters represent significant differences between three seasons at each site.
1,2 different numbers represent significant differences between three sites in each season.
* National Standard of P.R. China [37].
† SQGs: Sediment Quality Guidelines for marine ecosystem; TEL: threshold effect level; PEL: probable effect level [23].

Enrichment factors of trace elements in marsh soils

Although the contents of trace elements were lower in the YRD than in most large rivers and estuaries [4], trace element pollution in the YRD is increasingly serious due to continuous disturbances caused by human activities and sediment movements [36,38]. Trace element contents in the upstream loess were used as indicators to determine the enrichment levels for trace elements in the soils. Figure 5 shows the proportions of enrichment levels in marsh soils at three sampling sites in each season. Cd levels in all soil samples were at the significant enrichment level in three seasons. As levels in almost 95% of soil samples were at a moderate enrichment level. Comparatively, more than 90% of soil samples showed minimal enrichment levels of Cu, Pb and Zn in three sampling sites in each season. Both *P. australis* and *T. chinensis* wetlands showed higher enrichment levels of Cd and lower enrichment of Cu, Pb, and Zn in summer than in both spring and fall, which was associated with higher BCFs of Cu, Pb and Zn and lower BCFs of As and Cd at both sites (Figure 3). Moreover, almost all EF values of As and Cd and more than 50% of EF values of Cu, Pb and Zn for all soils in three seasons exceeded 1.5, indicating that a significant portion of metal is delivered from non-crustal materials, or non-natural weathering processes, so anthropogenic sources might be an important contributor [39]. Bai et al. [4] also presented that the flow-sediment regulation could elevate the levels of trace metals.

Ecotoxicity assessments of trace elements

The potential acute toxicities of contaminants in soil samples can be estimated using the toxic unit (TU) index, which is the ratio of the determined content to the probable effect level (PEL) [26]. The TUs, the sum of toxic units ($\sum$TUs) and relative contributions of trace elements at each soil layer in the YRD are illustrated in Figure 6. The mean TUs value of each element at three sites decreased in the order of As > Zn > Cu > Pb > Cd in each of

Table 4. Relationships between soil texture and Al and heavy metals in typical profiles from three sampling sites in three sampling seasons.

	Al	As	Cd	Cu	Pb	Zn
Sand	-0.743^{**}	-0.306^{*}	-0.259^{*}	-0.805^{**}	-0.753^{**}	-0.121
Clay	0.758^{**}	0.305^{*}	0.263^{*}	0.850^{**}	0.756^{**}	0.276^{*}

* represents significant correlation at the level of $p < 0.05$;
** represents significant correlation at the level of $p < 0.01$.
n=60.

Table 5. Correlation matrix among trace elements and selected soil properties for three sampling sites.

	Moisture	Bd	SOM	Salinity	pH	S	Al	As	Cd	Cu	Pb	Zn
Moisture	1.000											
Bd	0.210*	1.000										
SOM	−0.006	−0.189*	1.000									
Salinity	0.143	−0.179	−0.151	1.000								
pH	0.238**	0.201*	−0.038	−0.376**	1.000							
S	0.034	−0.262**	0.478**	0.099	0.014	1.000						
Al	−0.028	0.212*	0.144	−0.265**	−0.141	0.140	1.000					
As	−0.097	0.020	0.037	0.009	−0.295**	0.137	0.507**	1.000				
Cd	−0.082	0.277**	0.131	−0.087	−0.346**	0.062	0.745**	0.618**	1.000			
Cu	−0.139	−0.193*	0.496**	−0.297**	0.110	0.455**	0.617**	0.343**	0.377**	1.000		
Pb	−0.147	−0.348**	0.440**	−0.203*	0.021	0.455**	0.529**	0.371**	0.347**	0.883**	1.000	
Zn	−0.155	0.021	0.278**	−0.237**	−0.101	0.262**	0.783**	0.571**	0.723**	0.772**	0.726**	1.000

* represents significant correlation at the level of $p < 0.05$;
** represents significant correlation at the level of $p < 0.01$.
n = 144.

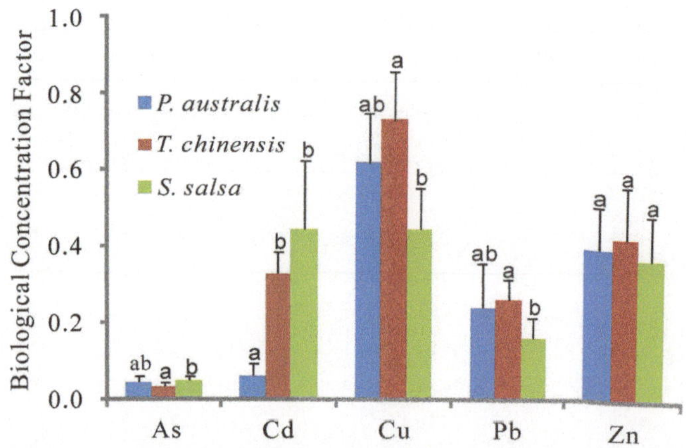

Figure 3. Biological concentration factors of the aboveground parts of plants in each site. [abc] Different letters on the error bars represent significant differences ($p<0.05$).

three seasons. As and Cd in all soil samples had much higher TUs in summer compared to spring and fall. Generally, the $\sum$TUs decreased with depth along soil profiles. The $\sum$TUs for surface soils (0–10 cm) in *S. salsa* wetland in spring or in *T. chinensis* wetland in summer were much higher than those for all the other soil samples of the three sites, but neither exceeded the moderate toxicity level ($\sum$TUs = 4;) [25]. The order of the $\sum$TUs in the top 20 cm soils of three sampling sites in spring was *S. salsa* wetland > *P. australis* wetland >*T. chinensis* wetland. Additionally, the $\sum$TUs for all soil layers at the three sites reached higher levels in summer than in fall and spring. The $\sum$TUs for each layer were higher in *P. australis* and *S. salsa* wetlands than *T. chinensis* wetland in spring, whereas they exhibited similar $\sum$TUs levels at three sites in summer and fall.

The values of the $\sum$TUs increased from spring to summer and then decreased to their lowest levels in fall. This corresponded to the temporal variations in trace element contaminations during the three sampling seasons which was closely related to the input flow containing trace elements from upstream during the flow-sediment regulation period from June to July. Tixier et al. [40] presented that the storm water exhibited the seasonally varying toxicity of trace elements with high toxicity in spring and recovery in fall, which was caused by the Terraview–Willow field storm water management facility. The $\sum$TUs for all soil layers of the three sampling sites were less than 4, suggesting low ecotoxicity in the current YRD soils. This is consistent with the result by Gao and Chen [41], who presented that the surface sediments showed no or low ecotoxicity of trace elements in the coastal Bohai Bay. However, As manifested the highest contribution ratios (ranging from 24.4 to 53.7%) to the $\sum$TUs at three sites in each of the three seasons compared to other trace elements (Figure 6). Therefore, the soils were 'As toxic' (with TUs for As greater than 0.5), indicating that As might have greatly negative effects and ecological risks on some or all of the environmental components [42].

Relationships between trace elements and soil physico-chemical properties

The correlation analysis between trace elements in the marsh soils of each sampling site was performed to assess possible co-contamination from similar sources. As shown in Table 4, significantly positive correlations among all trace elements (i.e., As, Cd, Pb, Cu, and Zn) were observed, suggesting that these trace

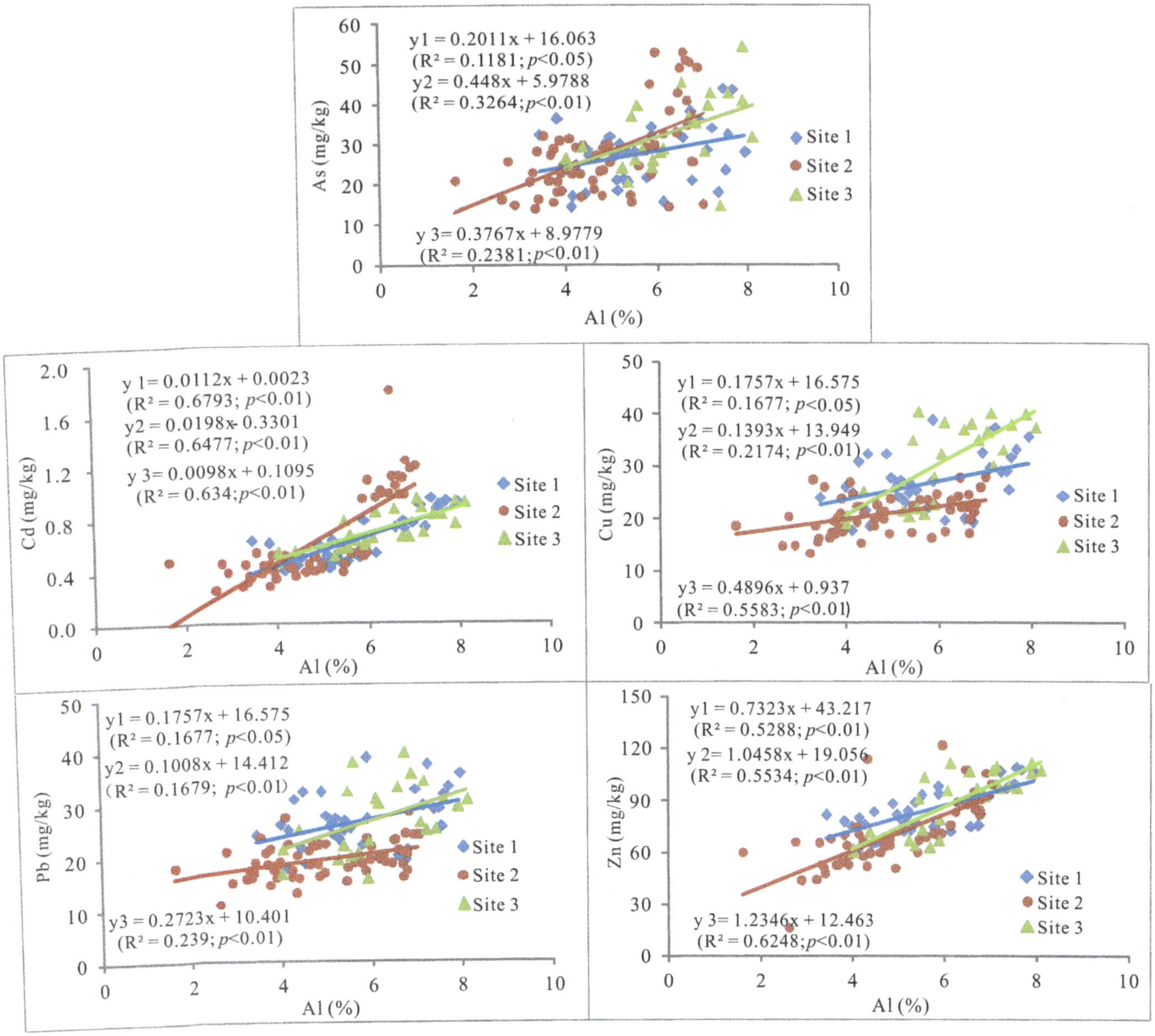

Figure 4. Relationships between the contents of aluminum (%) and trace elements (mg/kg) in all soil samples in the Yellow River Delta.

elements in marsh soils might originate from common source. Figure 4 also exhibited that the linear relationships between Al and these trace elements at each sampling site, which implies that these trace elements and Al were greatly associated with the upstream loess materials with higher Al contents of the Yellow River (62700 μg g^{-1}) [24]. Moreover, Al and these trace elements were significantly correlated with sand and clay contents (Table 4), indicating that higher clay contents in August (Table 2) would contribute to these trace element accumulation. Therefore, the upstream flow-sediment regulation in June could take main responsibility for the accumulation of these trace elements in the delta in the studied period due to no other pollution source inputs in the study area. This is also supported by the result of Bai et al. [4], who presented that the upstream flow-sediment regulation contributed to elevating trace element contents in surface soils in this delta.

Table 5 showed the negatively linear correlations between salinity and Al, Cu, Pb and Zn in the study area ($p<0.01$). This is in agreement with the results presented by Du Laing et al. [43], indicating increasing salinity might enhance the mobility of these metals in soils, as the chloro complex formation and the substitution of Ca^{2+} and Mg^{2+} in the exchange positions could promote the mobility of heavy metals [44]. Additionally, Manousaki et al. [45] found that salinity could favor metal accumulation in shoots of salt cedars, thus decreasing soil metal contents. Bulk densities exhibited negative correlations with Cu and Pb in all soils, implying that lower bulk densities in saturated substances with higher SOM favored the accumulation of both metals. Numerous studies reported that SOM and soil pH played important roles in determining the fates of heavy metals in soils [11,46]. SOM can act as a major sink for heavy metals due to its strong complexing capacity for metallic contaminants [47]. In this

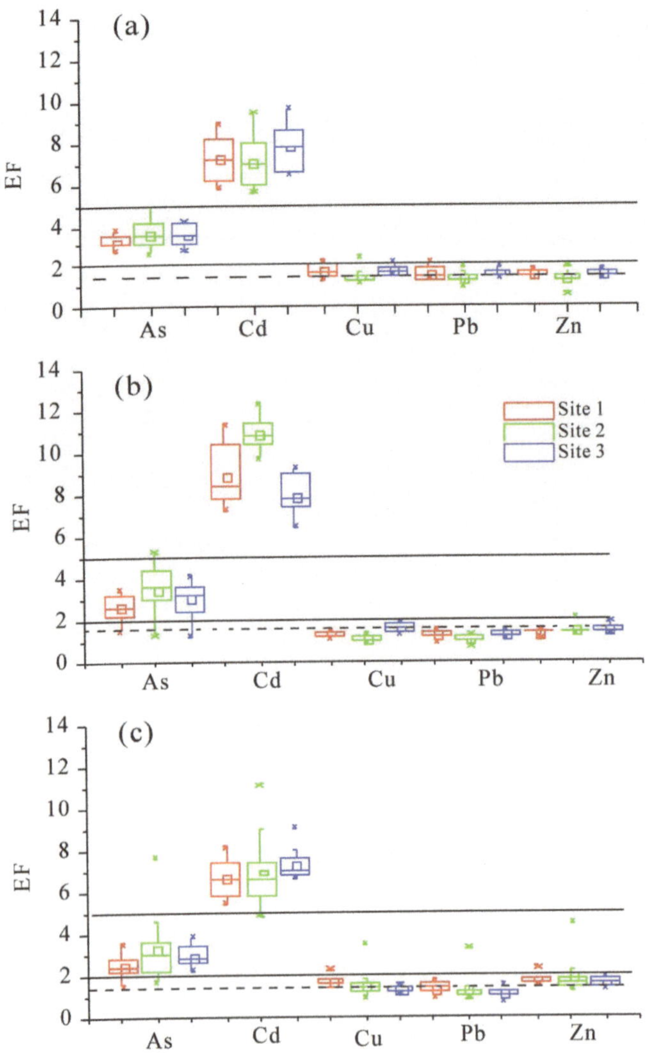

Figure 5. Box-plots of enrichment factors of trace elements in all soil samples of each site of the three in three seasons (a) Spring; (b) Summer; (c) Fall. Straight lines represent the EF = 2 or 5; dash line represents EF = 1.5.

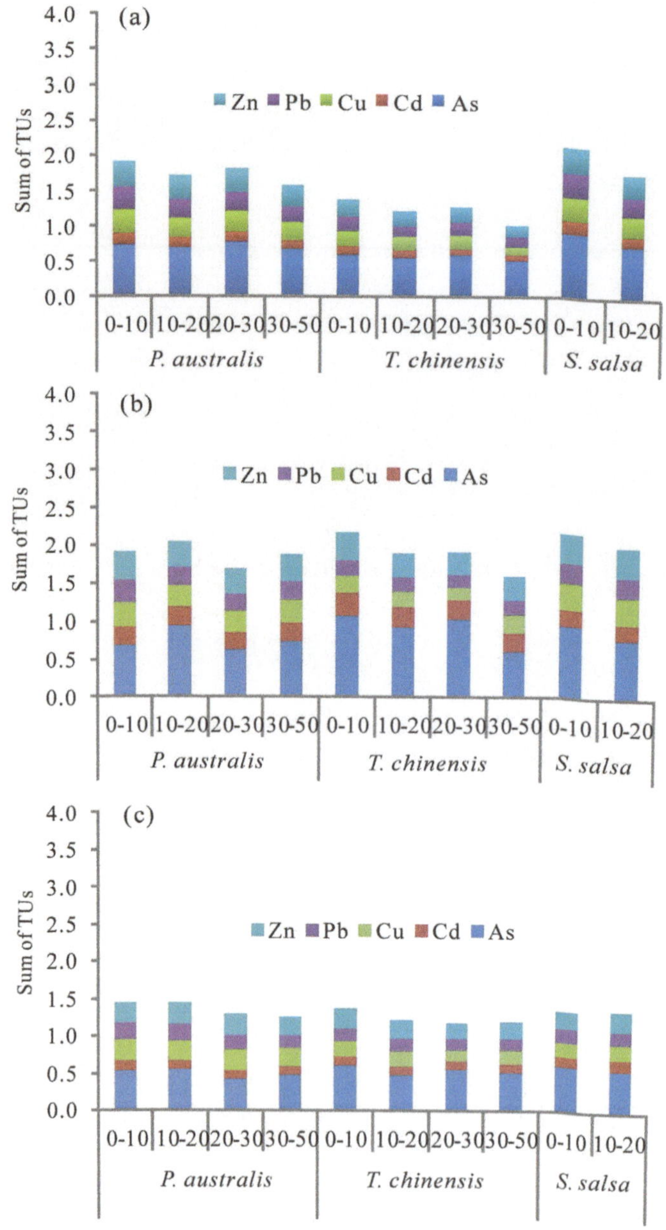

Figure 6. The sum of the toxic units (∑TUs) and toxic contribution of each metal at each soil layer of three sampling sites in three seasons (a) Spring; (b) Summer; (c) Fall.

study, SOM showed a positive linear relationship with Cu, Pb and Zn ($p<0.01$; Table 5). Most researchers presented that subalkaline environments could decrease the mobility of metals and increase the ability of soil to stabilize metals [11,36,48]. However, in this study, soil pH values were negatively correlated with As and Cd ($p<0.01$; Table 5). This was most likely due to plant uptake or leaching as alkalization will increase metal leaching at higher pH ranges (8–10) [46,49]. Sulfur exhibited significantly positive correlations with Cu, Pb and Zn ($p<0.01$) due to the deposition of metal sulphates (i.e., CuS, PbS and ZnS) under anaerobic conditions [4]. Moreover, Kumpiene et al. [46] reported that the presence of one contaminant (e.g., Cu or Pb) can decrease the stabilization efficiency of another contaminant (e.g., Zn) due to competition for sorption sites. Therefore, heavy metals levels in the marsh soils could be impacted by such factors as moisture, salinity, SOM, pH values, sulfur and the interactions between trace elements.

To assess the relationships between spatial distribution patterns of trace elements and several environmental factors (including 5 soil parameters), PCA was implemented to characterize the variations in trace element contents among sampling sites. The factor loading scores in the biplot (Figure 7) exhibited a clear separation amongst *P. australis*, *T. chinensis* and *S. salsa* wetlands and amongst spring, summer, and autumn along the first principal component (PC) accounting for 75.1% and the second PC accounting for 16.0% of the variance. As shown in Figure 7, three main groups can be clearly identified. The red plate contained the soils collected in spring, the yellow plate contained the soils collected in summer, and the blue plate included the soils collected in fall. It appears that the sampling seasons play an important role in shaping the spatial distribution patterns of As and heavy metals in marsh soils. However, the soils in *T. chinensis* wetland in spring did not fall into the red plate but entered or attached to the blue plate, suggesting that the distribution patterns

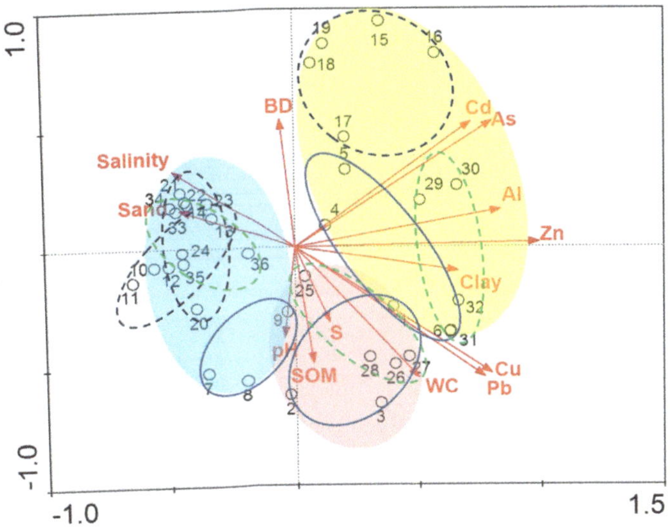

Figure 7. Ordination plots of the PCA results for surface soil samples. The direction of an arrow indicates the steepest increase in the variable, and the length indicates the strength relative to other variables. S, sulfur; BD, bulk density; SOM, soil organic matter; WC, moisture. No. 1, 2, 3 represent three replicates of *P. australis* wetland in spring; No. 4, 5, 6 represent three replicates of *P. australis* wetland in summer; No. 7, 8, 9 represent three replicates of *P. australis* wetland in fall; No. 10, 11, 12, 13, 14 represent five replicates of *T. chinensis* wetland in spring; No. 15, 16, 17, 18, 19 represent five replicates of *T. chinensis* wetland in summer; No. 20, 21, 22, 23, 24 represent five replicates of *T. chinensis* wetland in fall; No. 25, 26, 27, 28 represent four replicates of *S. salsa* wetland in spring; No. 29, 30, 31, 32 represent four replicates of *S. salsa* wetland in summer; No. 33, 34, 35, 36 represent four replicates of *S. salsa* wetland in fall.

of trace elements in those soils are more similar to those in *T. chinensis* and *S. salsa* wetlands in fall than those in *P. australis* and *S. salsa* wetlands in spring.

PCA exhibited that the strongest determinants of the distribution patterns of trace elements in spring were soil moisture, S, SOM and pH value, while the strongest influencing factors in fall were soil salinity and soil texture. Obviously, salinity, water content and soil texture controlled the distribution patterns of trace elements in *T. chinensis* and *S. salsa* wetlands, while SOM, S and soil pH values were the dominant factor in *P. australis* wetland. All trace elements with the longest arrows in the summer group testified the highest accumulations of trace elements in summer, especially in *P. australis* and *S. salsa* wetlands with herbaceous plants. The results obtained from the PCA were consistent with the results derived from the correlation analysis.

Although spatial distribution patterns of trace elements were dominantly controlled by sampling seasons due to the flow-sediment regulation, the effects of plant species on their distributions in soils could not be ignored due to different plant accumulation capabilities for different species (Figure 3). As shown in Figure 7, in the same season plate, sampling sites with the same plant community were clustered together. This was because plant roots can significantly affect plants' abilities to transfer metals [50,51], as they employed a variety of ways to alleviate the stresses from root zone pollution through increasing the rate of root exudation as a result of losses of membrane integrity or of breakdowns in normal cell metabolism [52,53]. The fixation and stabilization of trace elements by plants originating from physiological characteristics (e.g., root exudation in response to underground stress) were different between herbaceous plants (*P. australis* and *S. salsa*)) and woody plant (*T. chinensis*). Additionally, eclogical characteristics of plants (e.g., plant density) might also affect the responses of plants to trace elements, especially under the interactive impacts of water and salt in the tidal salt marshes. This implies that marsh soils with different plant communities might maintain different capabilities for immobilizing trace elements at the same pollution level (Figure 3). Further studies are needed to investigate the impacts of land cover changes on these trace elements.

Conclusions

Obvious spatial and seasonal changes in trace elements in wetland soils in the intertidal salt marshes were observed. The higher trace element contents (especially As and Cd) in summer compared to spring and fall suggested the contribution of the flow-sediment regulation (from June to July) to pollutants from upper reaches deposition and accumulation in the delta and higher self-purifying capabilities of these salt marshes. This implies that these tidal salt marshes can alleviate the effects of the upstream flow-sediment regulation. Higher trace element contents in these soils were closely correlated to soil properties such as soil moisture, salinity, sulfur, SOM, soil texture and pH value. It is noted that sampling seasons had dominant impacts on the spatial distribution patterns and ecotoxicity of trace elements in soils in this study. Therefore, seasonal changes in trace element contamination should be taken into account in tidal wetlands especially in tidal wetlands with great hydrological fluctuations. We also observed that marsh soils with herbaceous plants (*P. australis* and *S. salsa*) had higher trace element accumulation than woody plant (*T. chinensis*). Although all soil samples showed lower ecotoxicity, higher ecotoxicity of As in each of the three seasons should attract our concerns in this region. Moreover,it is important to continuously monitor these trace elements in coastal wetland soils to identify the ecological and environmental effects of hydrological engineering. The findings of this study could contribute to wetland conservation and management in coastal regions affected by the hydrological engineering. Further studies of the final fate and transport of these trace elements and plant uptake mechanism under water and salt stress in coastal wetlands will greatly contribute to trace element pollution control in coastal region.

Acknowledgments

The authors are grateful to the Editor and two anonymous reviewers for their valuable comments and suggestions. We also express our thanks to other members of our research group for their field work.

Author Contributions

Conceived and designed the experiments: JB RX. Performed the experiments: QZ QL JW. Analyzed the data: JB KR. Contributed reagents/materials/analysis tools: JB KR. Wrote the paper: JB RX KR.

References

1. Carpenter B, Haltmeier R, Wilde C (1997) Dredged sediments disposal. Journal of Water Environment Technology 9: 47–50.
2. Yeloff DE, Labadz JC, Hunt CO (2005) The suitability of a south Pennine (UK) reservoir as an archive of recent environmental change. Journal of Paleolimnology 34: 339–348.
3. Bai JH, Cui BS, Xu XF, Gao HF, Ding QY (2009) Heavy metal contamination in riverine soils upstream and downstream of a hydroelectric dam on the Lancang River, China. Environmental Engineering Science 26: 941–946.
4. Bai JH, Xiao R, Zhang K, Gao H (2012) Arsenic and heavy metal pollution in wetland soils from tidal freshwater and salt marshes before and after the

flow-sediment regulation regime in the Yellow River Delta, China. Journal of Hydrology 450–451: 244–253.
5. Reddy KR, DeLaune RD (2008) Biogeochemistry of Wetlands: Science and Applications, 1st Ed., CRC Press, Boca Raton, FL.
6. Bai JH, Xiao R, Cui BS, Zhang KJ, Wang QG, et al. (2011) Assessment of heavy metal pollution in wetland soils from the young and old reclaimed regions in the Pearl River Estuary, South China. Environmental Pollution 159: 817–824.
7. Gambrell RP (1994) Trace and toxic metals in wetlands—A review. Journal of Environmental Quality 23:883–891.
8. Grabowski LA, Houpis JLJ, Woods WI, Johnson KA (2001) Seasonal bioavailability of sediment-associated heavy metals along the Mississippi river floodplain. Chemosphere 45: 643–651.
9. Zhang WG, Feng H, Chang JN, Qu JG, Xie HX, et al. (2009) Heavy metalcontamination in surface sediments of Yangtze River intertidal zone: an assessment from different indexes. Environmental Pollution 157: 1533–1543.
10. Wang LL, Yang ZF, Niu JF, Wang JY (2009) Characterization, ecological risk assessment and source diagnostics of polycyclic aromatic hydrocarbons in water column of the Yellow River Delta, one of the most plenty biodiversity zones in the world. Journal of Hazard Materials 169: 460–465.
11. Bai JH, Yang ZF, Cui BS, Gao HF, Ding Q Y(2010) Some heavy metals distribution in wetland soils under different land use types along a typical plateau lake, China. Soil &Tillage Research 106: 344–348.
12. Roychoudhury AN (2007) Spatial and seasonal variations in depth profile of trace metals in saltmarsh sediments from Sapelo Island, Georgia, USA. Estuarry, Coast and Shelf Science 72: 675–689.
13. Brandt SA (2000) Classification of geomorphological effects downstream of dams. Catena 40: 375–401.
14. Hollert H, Keiter S, König N, Rudolf M, Ulrich M, et al. (2003) A new sediment contact assay to assess particulate-bound pollutants using Zebrafish (Daniorerio) embryos. Journal of Soils and Sediments 3: 197–207.
15. Zhao YM, Song C S (1995) Scientific Survey of the Yellow River Delta Natural Reserve. China Forestry Press, Beijing, pp. 26–30.
16. Li SN, Wang GX, Deng W, Hu YM, Hu WW (2009) Influence of hydrology process on wetland landscape pattern: A case study in the Yellow River Delta. Ecological Engineering 35: 1719–1726.
17. Cui BS, Tang N, Zhao XS, Bai JH (2009) A management-oriented valuation method to determine ecological water requirement for wetlands in the Yellow River Delta of China. Journal for Natural Conservation 17:129–141.
18. Soil Survey Staff (2010) Keys to Soil Taxonomy, 11th ed. USDA-Natural Resources Conservation Service, Washington, DC.
19. Peng J, Chen S, Dong P (2010) Temporal variation of sediment load in the Yellow River basin, China, and its impacts on the lower reaches and the river delta. Catena 83, 135–147.
20. Page AL, Miller RH, Keeney DR (1982) Methods of Soil Analysis Part 2: Chemical and Microbiological Properties, 2nd edn., American Society of Agronomy, and Soil Science Society of America, Madison. WI., pp. 539–579.
21. Selvaraj K, Ram Mohan V, Szefer P (2004) Evaluation of metal contamination in coastal sediments of the Bay of Bengal, India: geochemical and statistical approaches. Marine Pollution Bulletin 49: 174–185.
22. Chabukdhara M, Nema AK (2012) Assessment of heavy metal contamination in Hindon River sediments: A chemometric and geochemical approach. Chemosphere 87: 945–953.
23. Han Y, Du P, Cao J, Eric SP (2006) Multivariate analysis of heavy metal contamination in urban dusts of Xi'an City, China. Science of the Total Environment 355, 176–186.
24. China National Environmental Monitoring Center (CNEMC)(1990) Chinese Elemental Background Values for Soils. Chinese Environmental Science Press, Beijing.
25. Pedersen F, Sjbrnestad E, Andersen HV, Kjlholt J, Poll C (1998) Characterization of sediments from Copenhagen harbour by use of biotests. Water Science and Technology 37: 233-240.
26. Long ER, MacDonald DD (1998) Recommended uses of empirically derived, sediment quality guidelines for marine and estuarine ecosystems. Human and Ecological Risk Assessment 4: 1019–1039.
27. Madejón P, Murillo J, Maranon T, Cabrera F, Soriano M (2003) Trace element and nutrient accumulation in sunflower plants two years after the aznalcollar mine spill. Science of the Total Environment 307: 239–257.
28. Mitsch WJ, Gosselink JG (2007) Wetlands, 4th ed. John Wiley Inc, New York.
29. Prusty B, Kumar A, Azeez PA (2007) Vertical distribution of alkali and alkaline earth metals in the soil profile of a wetland-terrestrial ecosystem complex in India. Austrianl Journal of Soil Research 45: 533–542.
30. Gregorauskiene V, Kadunas V (2006) Vertical distribution patterns of trace and major elements within soil profile in Lithuania. Geological Quarterly 50: 229–237.
31. Ehlken S, Kirchner G (2002) Environmental processes affecting plant root uptake of radioactive trace elements and variability of transfer factor data: a review. Journal of Environmental Radioactivity 58: 97–112.
32. Lassabatere L, Winiarski T (2004) Retention of three heavy metals (Zn, Pb, and Cd) in a calcareous soil controlled by the modification of flow with geotextiles. Environmental Science & Technology 38: 4215–4221.
33. Bragato C, Schiavon M, Polese R, Ertani A, Pittarello M, et al. (2009) Seasonal variations of Cu, Zn, Ni and Cr concentration in Phragmites australis (Cav.) Trin ex steudel in a constructed wetland of North Italy. Desalination 246: 35–44.
34. Beth R, Ehrenfeld JG, Haggblom MM (2003) A comparison of sediment microbial communities associated with *Phragmites australis* and *Spartina alterniflora* in two brackish wetlands of New Jersey. Estuaries 26: 465–474.
35. Zehetner F, Rosenfellner U, Mentler A, Gerzabek MH (2009) Distribution of road salt residues, heavy metals and polycyclic aromatic hydrocarbons across a highway-forest interface. Water, Air and Soil Pollution 198: 125–132.
36. Nie M, Xian NX, Fu XH, Chen XF, Li B (2010) The interactive effects of petroleum-hydrocarbon spillage and plant rhizosphere on concentrations and distribution of heavy metals in sediments in the Yellow River Delta, China. Journal of Hazard Materials 174: 156–161.
37. National Standard of PR China (2002) Marine Sediment Quality (GB18668-2002).Standards Press of China, Beijing (in Chinese).
38. Zhang J, Huang WW, Martin JM (1988) Trace metals distribution in huanghe (Yellow River) estuarine sediments. Estuary Coast and Shelf Science 26: 499–516.
39. Feng H, Han X, Zhang W, Yu L (2004) A preliminary study of heavy metal contamination in Yangtze River intertidal zone due to urbanization. Marine Pollution Bulltin 49: 910–915.
40. Tixier G, Rochfort Q, Grapentine L, Marsalek J, Lafont M (2012) Spatial and seasonal toxicity in a stormwater management facility: Evidence obtained by adapting an integrated sediment quality assessment approach. Water Research 46(20): 6671–6682.
41. Gao XL, Chen CT (2012) Heavy metal pollution status in surface sediments of the coastal Bohai Bay. Water Research 46: 1901–1911.
42. Lacutusu R (2000) Appraising levels of soil contamination and pollution with heavy metals. Research Report No. 4, European Soil Bureau.
43. Du Laing G, De Vos R, Vandecasteele B, Lesage E, Tack FMG, et al. (2008) Effect of salinity on heavy metal mobility and availability in intertidal sediments of the Scheldt estuary. Estuary, Coast, Shelf Science 77: 589–602.
44. Doner HE (1978) Chloride as a factor in mobilities of Ni (II), Cu (II), and Cd (II) in soil. Soil Science Society of American Journal 42: 882–885.
45. Manousaki E, Kokkali F, Kalogerakis N (2009) Influence of salinity on lead and cadmium accumulation by the salt cedar (Tamarix smyrnensis Bunge). Journal of Chemical Technology & Biotechnology 84: 877–883.
46. Kumpiene J, Lagerkvist A, Maurice C (2008) Stabilization of As, Cr, Cu, Pb and Zn in soil using amendments-a review. Waste Management 28: 215–225.
47. Gonzalez ZI, Krachler M, Cheburkin AK, Shotyk W (2006) Spatial distribution of natural enrichments of arsenic, selenium, and uranium in a minerotrophic peatland, Gola di Lago, Canton Ticino, Switzerland. Environmental Science & Technology 40: 6568–6574.
48. Ullrich SM, Ramsey MH, Helios-Rybicka E (1999) Total and exchangeable concentrations of heavy metals in soils near Bytom, an area of Pb/Zn mining and smelting in Upper Silesia, Poland. Appllied Geochemistry 14: 187–196.
49. Hartley W, Edwards R, Lepp NW (2004) Arsenic and heavy metal mobility in iron oxide-amended contaminated soils as evaluated by short- and long-term leaching tests. Environmental Pollution 131: 495–504.
50. Audet P, Charest C (2008) Allocation plasticity and plant-metal partitioning: meta-analytical perspectives in phytoremediation. Environmental Pollution 156: 290–296.
51. Mench M, Morel JL, Guckert A, Guillet B (1988) Metal-binding with root exudates of low-molecular weight. Journal of Soil Science 39: 521–527.
52. Neumann G, Römheld V (2001) The Release of Root Exudates as Affected by the Plant's P hysiological Status. Marcel Dekker Inc, New York.
53. Chiang PN, Wang MK, Chiu CY, Chou SY (2006) Effects of cadmium amendments on low-molecular-weight organic acid exudates in rhizosphere soils of tobacco and sunflower. Environmental Toxicology 21: 479–488.

14

Response of Two Dominant Boreal Freshwater Wetland Plants to Manipulated Warming and Altered Precipitation

Yuanchun Zou[1], Guoping Wang[1], Michael Grace[2], Xiaonan Lou[1], Xiaofei Yu[1], Xianguo Lu[1]*

1 Key Lab of Wetland Ecology and Environment, Northeast Institute of Geography and Agroecology, Chinese Academy of Sciences, Changchun, China, **2** Water Studies Centre and School of Chemistry, Monash University, Clayton, Australia

Abstract

This study characterized the morphological and photosynthetic responses of two wetland plant species when they were subject to 2–6°C fluctuations in growth temperature and ±50% of precipitation, in order to predict the evolution of natural wetlands in Sanjiang Plain of North-eastern China. We investigated the morphological and photosynthetic responses of two dominant and competitive boreal freshwater wetland plants in Northeastern China to manipulation of warming (ambient, +2.0°C, +4.0°C, +6.0°C) and altered precipitation (−50%, ambient, +50%) simultaneously by incubating the plants from seedling to senescence within climate-controlled environmental chambers. Post-harvest, secondary growth of *C. angustifolia* was observed to explore intergenerational effects. The results indicated that *C. angustifolia* demonstrated a greater acclimated capacity than *G. spiculosa* to respond to climate change due to higher resistance to temperature and precipitation manipulations. The accumulated effect on aboveground biomass of post-harvest secondary growth of *C. angustifolia* was significant. These results explain the expansion of *C. angustifolia* during last 40 years and indicate the further expansion in natural boreal wetlands under a warmer and wetter future. Stability of the natural surface water table is critical for the conservation and restoration of *G. spiculosa* populations reacting to encroachment stress from *C. angustifolia* expansion.

Editor: Eric Gordon Lamb, University of Saskatchewan, Canada

Funding: The research was supported by the National Basic Research Program of China (2010CB951301-2), the National Key Technology R & D Program (2012BAC19B05), and the National Natural Science Foundation of China (41271107, 41101195) in the study design, data collection and analysis and preparation of the manuscript.

Competing Interests: The authors have declared that no competing interests exist.

* Email: luxg@iga.ac.cn

Introduction

Wetlands are among the most important terrestrial carbon pools and play an important role in global carbon cycling [1,2]. Boreal wetlands comprise about half of the total global wetland area, store about one third of the world's organic carbon in the form of living or partially decayed vegetation and soil organic carbon and contribute 34% of global atmospheric methane flux [3,4]. Due to their large and dynamic carbon pool, boreal wetlands may not be only significantly impacted by climate change, but may also provide negative (e.g. increasing organic carbon sink) or positive feedback (e.g. increasing greenhouse gas emissions) mechanisms to anthropogenic climate change [5,6] and mitigate climate warming overall.

Boreal wetlands are considered particularly vulnerable to climate change because they depend on specific climatic conditions with low temperature and high water availability [7,8]. To quantify anticipated climate effects on boreal wetland ecosystem processes, field experiments simulating climate change have been undertaken in North America and northern Europe [2,5,9], and these studies suggested that both the performance of a specific plant and the compositions of plant communities will change in different directions and at different magnitudes as a response to warming and/or changes in the depth of the surface water table that usually connected with temperature and precipitation change. Despite these efforts, species-specific eco-physiological responses of wetland plants to changes in temperature and precipitation, and especially the interactive effects of these factors, remain unclear. Simulations provide a mechanism for exploring the importance of these temperature-precipitation interactions on wetland plants under future climate conditions [10]. Multifactor experimental manipulations of temperature and precipitation in the laboratory can complement field studies by enabling controlled adjustment of covarying climate variables based on the historical data to generate relatively realistic climate change experimental scenarios [11]. The responses then could be scaled up to the level of the ecosystem when the initial structure of the wetland and the environmental driving variables are consequently well coupled.

As typical dominant boreal freshwater wetland plants in Northeast China, Siberia and Far East Russia, *Calamagrostis angustifolia* and *Glyceria spiculosa* both belong to the family Poaceae, and are often distributed as large, dense clumps in palustrine, riverine and lacustrine wetlands [12,13]. Generally, *C. angustifolia* covers saturated microhabitats while *G. spiculosa* covers the flooded microhabitats with greater surface water depth. Both plants play important ecological functions in the wetlands as primary producers for food webs, shelter for some invertebrates and major organic carbon sources for carbon sequestration. In recent years, with climate warming and surface water deficiency in some wetlands of the Sanjiang Plain, the microhabitats once covered by *G. spiculosa* have been almost replaced by *C. angustifolia* due to the drawdown of surface water depth, caused either by climate change or anthropogenic activities such as drainage.

In this paper, we simultaneously considered two major factors (temperature and precipitation) involved in climate change effects on boreal wetland ecosystem functioning. The two plant species, *C. angustifolia* and *G. spiculosa*, were incubated from seedling to senescence in environmental chambers under manipulated warming and altered precipitation conditions. The objectives of this study were: (1) to characterize the morphological and photosynthetic responses of the two plants to 2–6°C warming and ±50% of precipitation changes; and (2) to analyze the resistance and acclimation of the two plants and predict their potential distribution in the future under the projected changed climate conditions in the Sanjiang Plain wetlands of Northeast China.

Materials and Methods

Study site and target species

The study site is located in Xingkai Lake (Khanka) lacustrine wetlands of Sanjiang Plain, the largest expanse of freshwater wetlands in China. Xingkai Lake is the largest lake in Northeast Asia and a transboundary lake shared by China and Russia. According to calculations based on the 54 years of local meteorological records, the annual mean precipitation is 561 mm and the temperature ranges from −39°C in January to 36°C in July with the annual mean of 3.5°C for 1957–2010. According to a linear fitting of local meteorological records in Xingkai Lake lacustrine wetlands (Annual temperature = −52.55 + 0.028*year, $r>0.99$, $p<0.0001$; unpublished), there is a significant warming trend of 0.28°C per decade in Xingkai Lake over the past half century. This is even greater than the average level of 0.22°C per decade for the whole country [14].

The target plant species for the three-month incubation experiments were collected *in situ* from *Calamagrostis angustifolia* (45°20′59" N, 132°18′52" E) and *Glyceria spiculosa* (45°21′49" N, 132°19′26" E) communities within Xingkai Lake lacustrine wetlands. Representative and homogeneous individuals of *C. angustifolia* and *G. spiculosa* were excavated with intact roots and soils using a planting shovel in May of 2011 and 2012, respectively, when the first new shoots were apparent in springtime. The soil types were silty clay meadow marsh soil and humus marsh soil, respectively. The seedlings and soils with the depth of approximate 20 cm were cut into small blocks with rhizome based, transported to the laboratory in Northeast Institute of Geography and Agroecology within two days, and transplanted into incubation pots with 10-cm diameter and 15-cm depth with 4–5 seedlings in each pot. To minimize loss of vitality during transplantation, these potted seedlings were first incubated in a greenhouse for one month to germinate new ramets before placement into environmental chambers. The temperature in the greenhouse was set as the average air temperature of June in Xingkai Lake lacustrine wetlands. All the field studies got the permission from administrative authority of Xingkai Lake National Nature Reserve and did not involve any endangered or protected species.

Experimental design

The design between the warming and the precipitation was a two-way randomized block experiment. Four environmental chambers were deployed to control the temperature. Within each chamber, three plastic container (35 cm × 25 cm × 10 cm) were used to control water level fluctuations caused by precipitation manipulation. Each container contained three replicated pots. Before incubation, any pot growing approximate ten vigorous seedlings with similar performance (about 20 cm and 15 cm heights for *C. angustifolia* and *G. spiculosa*, respectively) was selected and randomly separated into 4 groups corresponding to the 4 temperature regimes. The ambient temperature was set as the local monthly average air temperatures in July, August and September, respectively fitted. The ambient temperature was adjusted five times per day based on the local monthly (July, August and September) average air temperatures over the time periods 1:00–4:00, 4:00–10:00, 10:00–16:00, 16:00–19:00 and 19:00–1:00. Table 1 provides the temperature and illuminance values for the plant incubations in the environmental chambers. The illuminance values were based on the field observation. These pots were exchange their positions in the chambers daily to keep the temperature and light intensity similar for a given pot. The warming scenarios were then set according to the ambient temperature regime with +2°C, +4°C and +6°C added for each interval. The lights were plant growth fluorescent lamps that are recognized as suitable for plants' growth [15].

The ambient water depths were set to mimic the natural conditions in Xingkai Lake wetlands, with no net water for *C. angustifolia* and 10 cm of water above the soil surface for *G. spiculosa*. The ambient precipitation regime was set according to the local monthly average precipitation. These seedlings were gently watered over one hour using sprinkler nozzles to simulate natural precipitation every 3 days with one tenth of the amount of monthly precipitation. The altered precipitations were set as -50% and +50% based on the ambient precipitation rates.

After the three month incubation in the environmental chambers and subsequent harvesting of all above-ground biomass, one *C. angustifolia* pot was randomly selected from the 3 replicates of each precipitation treatment container within each environmental chamber. Twelve pots were consequently selected

Table 1. Temperature and illuminance control for plants' incubation in the environmental chambers.

Time period		0:00–7:00	7:00–11:00	11:00–14:00	14:00–16:00	16:00–24:00
Illuminance (lux)		0	10000	20000	15000	5000
Temperature	July	16	20	28	24	17
(°C)	August	17	19	26	23	18
	September	11	14	19	17	12

for a secondary growth test to analyze the potential effect of climate change on the next generation of this plant species. These pots were incubated for one month in the same environmental chamber with the average temperature and precipitation in June of 17.4°C and 74.4 mm, respectively in Xingkai Lake wetlands.

Response measurements

The morphological indicators (height, stem diameter and leaf area) were observed once per month using a meter stick, vernier caliper and portable area meter (LI-3000, Li-Cor Inc., USA) from seedling to senescence, with each indicator measured by 4 replicates for *C. angustifolia* and 3 replicates for *G. spiculosa*. The photosynthetic indicators, including leaf net photosynthetic rate (Pn), leaf stomatal conductance (Gs) and leaf transpiration rate (Ts) were measured by a portable photosynthesis system (LI-6400, Li-Cor, Inc., USA) during the peak growth phase (around August), with each indicator measured by 4 replicates for *C. angustifolia* and 3 replicates for *G. spiculosa*. The chlorophyll content was determined by homogenizing and extracting the leaf tissue in 80% acetone until the green colour turn white completely, then measuring the optical absorbance at 663 and 645 nm for chlorophyll a and b, respectively using a spectrophotometer (UV-2500, Shimadzu, Japan), and calculating concentrations using the specific absorption coefficients [16,17]. The reported leaf chlorophyll content was the sum of chlorophyll a and b, with 3 replicates both for *C. angustifolia* and *G. spiculosa*.

Above-ground biomass was measured after plant senescence by drying to constant weight at 80°C. The total nitrogen content of harvested leaves was determined using a continuous flow analyzer (SAN^{++}, SKALAR, the Netherlands) after digesting 0.3000 g of homogeneous, milled material with H_2SO_4/H_2O_2, with 3 replicates both for *C. angustifolia* and *G. spiculosa*. The below-ground biomass was not measured as it is a mixture of newly developed rhizomes/roots within that growing season and the rhizomes/roots from previous growing seasons. Older material usually accounts for most of the total below-ground biomass [18].

The response measurements for the secondary growth test of *C. angustifolia* were conducted as described above. The measured indicators were height, stem diameter above the soil, leaf area of each leaf on the plant based on measuring a suitable number of leaves, Pn, Gs, Ts and above-ground biomass, with each indicator measured by 3 replicates.

Statistical analyses

Statistical analyses were performed using SPSS Statistics 21.0 (SPSS Inc., USA). The main and the interaction effects of temperature and precipitation on the two plants' morphological indicators were compared by repeated measures analysis of variance (ANOVA). For repeated measures ANOVA, Mauchly's Test of Sphericity was first examined and the Greenhouse-Geisser adjustment was adopted when the null hypothesis was rejected. The main and the interaction effects of temperature and precipitation on the two plants' photosynthetic indicators, leaf chlorophyll contents during the phase of peak growth, the above-ground biomasses and leaf nitrogen contents after senescence were compared by two-way ANOVA. One-way ANOVA was performed to compare the differences of above-ground biomasses of *C. angustifolia* and *G. spiculosa*. Differences in morphology, photosynthesis and above-ground biomass between different secondary grown *C. angustifolia* after the temperature and precipitation manipulation for one month were also compared using one-way ANOVA. The least significant difference (LSD) was used respectively to perform *post hoc* multiple comparisons when

Table 2. Repeated-measures analysis of variance (ANOVA) of morphological indicators for *Calamagrostis angustifolia* and *Glyceria spiculosa*.

Variable		Month		Month × Temperature		Month × Precipitation		Month × Temperature × Precipitation	
		F	*p*	*F*	*p*	*F*	*p*	*F*	*p*
C. angustifolia	Height	30.864	**<0.001**	0.910	0.482	2.611	0.052	2.025	**0.044**
	Basal stem diameter	0.064	0.860	0.538	0.702	0.343	0.764	0.513	0.835
	Leaf area	151.032	**<0.001**	2.449	**0.047**	0.271	0.857	0.922	0.517
G. spiculosa	Height	111.065	**<0.001**	1.038	0.412	2.203	0.116	2.520	**0.033**
	Basal stem diameter	3.165	0.051	0.412	0.867	1.426	0.240	0.839	0.612
	Leaf area	61.079	**<0.001**	1.739	0.155	3.271	**0.031**	1.363	0.239

Significant effects ($p<0.05$) are shown in bold. When the highest order of interaction effect is significant, the significant main effect is not in bold.

equal variances were assumed; otherwise, Tamhane's T2 method was used.

Results

Morphological indicators

Repeated-measures ANOVA of morphological indicators showed that the interaction effects of temperature and precipitation on the height of *C. angustifolia* and *G. spiculosa* were both significant ($F = 2.025$, $p = 0.044$; $F = 2.520$, $p = 0.033$, respectively). For both species, main effects of temperature or precipitation, and any interaction effects, on basal stem diameter were all insignificant. The main effects of temperature on leaf area of *C. angustifolia* and the effect of precipitation on that of *G. spiculosa* were both significant ($F = 2.449$, $p = 0.047$; $F = 3.271$, $p = 0.031$, respectively) (Table 2).

C. angustifolia's height increased from July to September. The estimated marginal means (as the average of the results for each month) of the height suggested that the greatest heights were observed in the ambient precipitation for each warming manipulation (Fig. 1A). Under ambient precipitation conditions (Fig. 1B), the greatest height occurred at ambient temperature (except for the early growth of July), while the lowest height was found in the +2°C manipulation (except in September). When the precipitation decreased by 50% from ambient (Fig. 1C), the greatest height occurred in the +4°C manipulation (except for the early growth of July), while the lowest height was in the +6°C manipulation. When the precipitation increased by 50% from ambient, the greatest height occurred in the +4°C manipulation, while the other three treatments produced plants of comparable height (Fig. 1D).

G. spiculosa's height also increased throughout the incubation (Fig. 2B–D). The estimated marginal means of the height suggested that ambient precipitation and 2–4°C of warming could support greater plant height (Fig. 2A). In the ambient precipitation conditions, there were no significant differences in height between the different temperature treatments (Fig. 2B). For both the −50% and +50% precipitation treatments, the greatest height occurred in the +4°C manipulation (Fig. 2C, D).

The leaf area of the two species increased first and then decreased throughout the incubation. For *C. angustifolia*, the leaf area with the +6°C manipulation was smaller than other manipulations in July. There was no significant difference in leaf area between the ambient and the +2°C manipulation in July. In

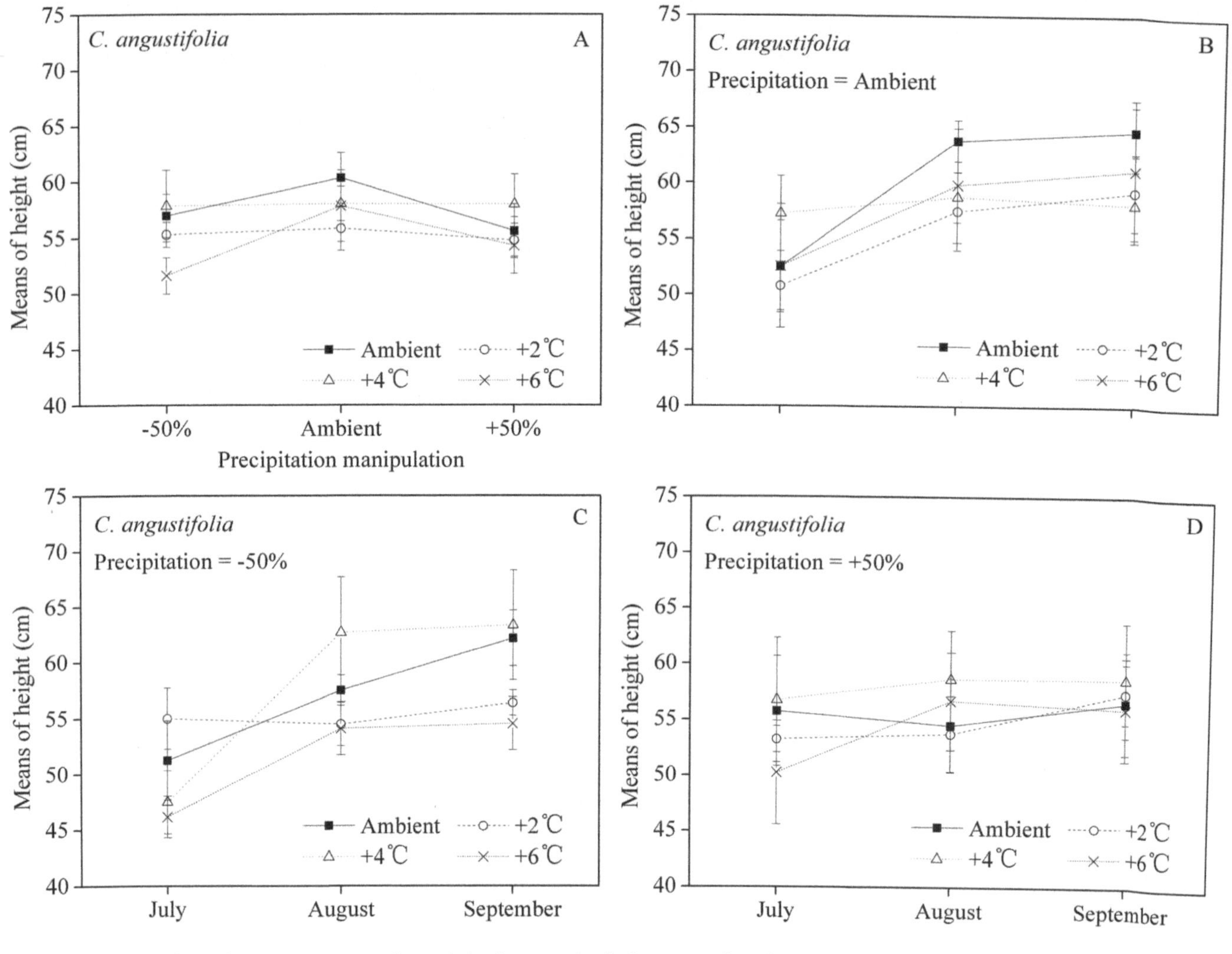

Figure 1. Interaction of temperature and precipitation manipulations on plant height (A) of *Calamagrostis angustifolia*, and the monthly estimated marginal means of height of *C. angustifolia* for different temperature treatments when precipitation was manipulated at ambient (B), −50% (C) and +50% (D) levels, respectively. The error bars represent means ± 1 standard error.

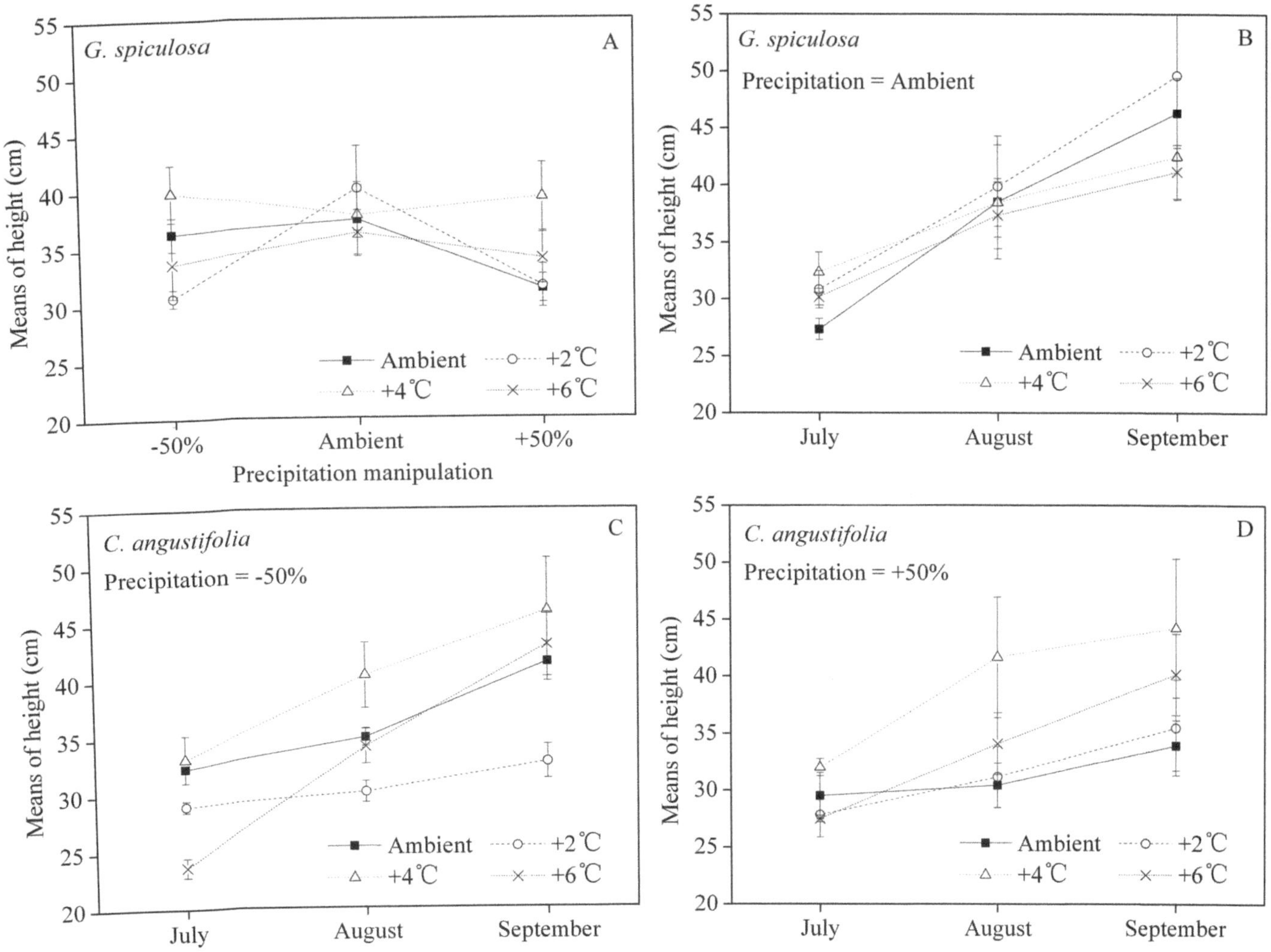

Figure 2. Interaction of temperature and precipitation manipulations on plant height (A) of *Glyceria spiculosa*, and the monthly estimated marginal means of height of *G. spiculosa* for different temperature treatments when the precipitation was manipulated at ambient (B), −50% (C) and +50% (D) levels, respectively. The error bars represent means ± 1 standard error.

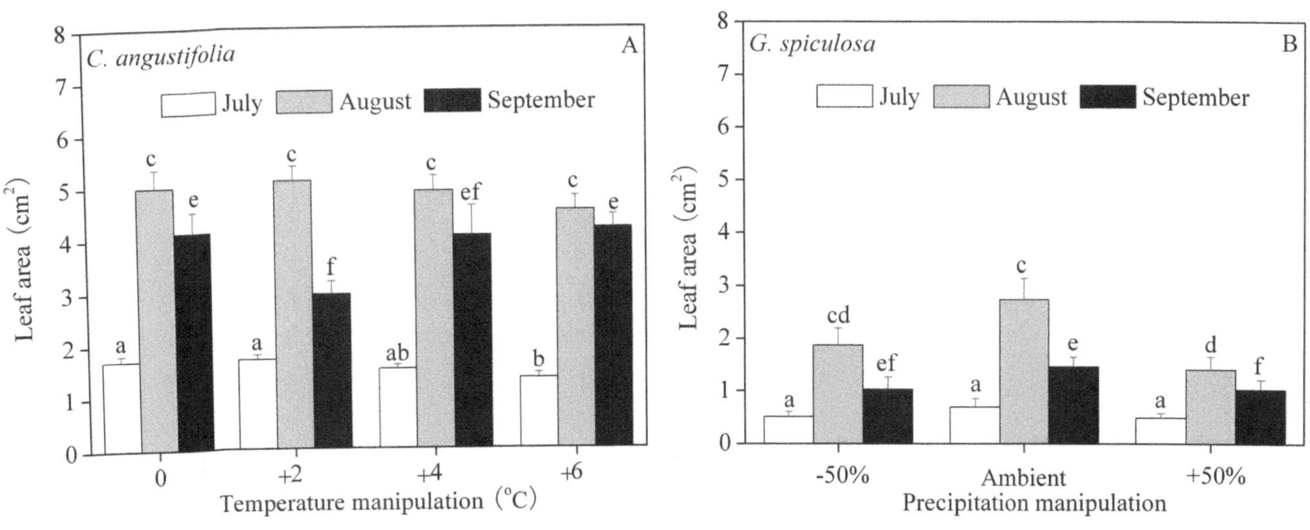

Figure 3. Leaf area of *C. angustifolia* (A) and *G. spiculosa* (B) monthly. For each month, different letters over the bars indicate significant differences ($p<0.05$) between temperature manipulations for *C. angustifolia* and precipitation manipulations for *G. spiculosa*. The letters 'a' and 'b' are for the July results, 'c' and 'd' for the August results, and 'e' and 'f' for the September results. The error bars represent means ± 1 standard error.

Table 3. Two-way ANOVA of photosynthetic indicators and leaf chlorophyll content for *C. angustifolia* and *G. spiculosa* during the phase of peak growth, and above-ground biomass and leaf total nitrogen content after harvest.

Variable		Temperature		Precipitation		Temperature × Precipitation	
		F	*p*	*F*	*p*	*F*	*p*
C. angustifolia	Pn	9.878	**<0.001**	7.516	**0.002**	4.974	**0.001**
	Gs	3.691	**0.021**	8.639	**0.001**	6.825	**<0.001**
	Ts	4.410	**0.010**	3.013	0.062	9.208	**<0.001**
	Chlorophyll	0.966	0.419	0.765	0.473	2.545	**0.037**
	Aboveground biomass	4.512	**0.012**	2.349	0.117	0.842	0.550
	Leaf total nitrogen	7.125	**0.001**	3.642	0.052	0.524	0.784
G. spiculosa	Pn	1.663	0.201	11.361	**<0.001**	2.244	0.073
	Gs	17.989	**<0.001**	1.966	0.162	3.400	**0.014**
	Ts	5.091	**0.007**	2.883	0.075	3.170	**0.020**
	Chlorophyll	4.920	**0.008**	7.153	**0.004**	2.322	**0.046**
	Aboveground biomass	4.156	**0.017**	13.637	**<0.001**	1.002	0.447
	Leaf total nitrogen	2.737	0.066	1.134	0.338	2.394	0.059

Significant effects ($p<0.05$) are shown in bold. When the highest order of interaction effect is significant, the significant main effect is not in bold.

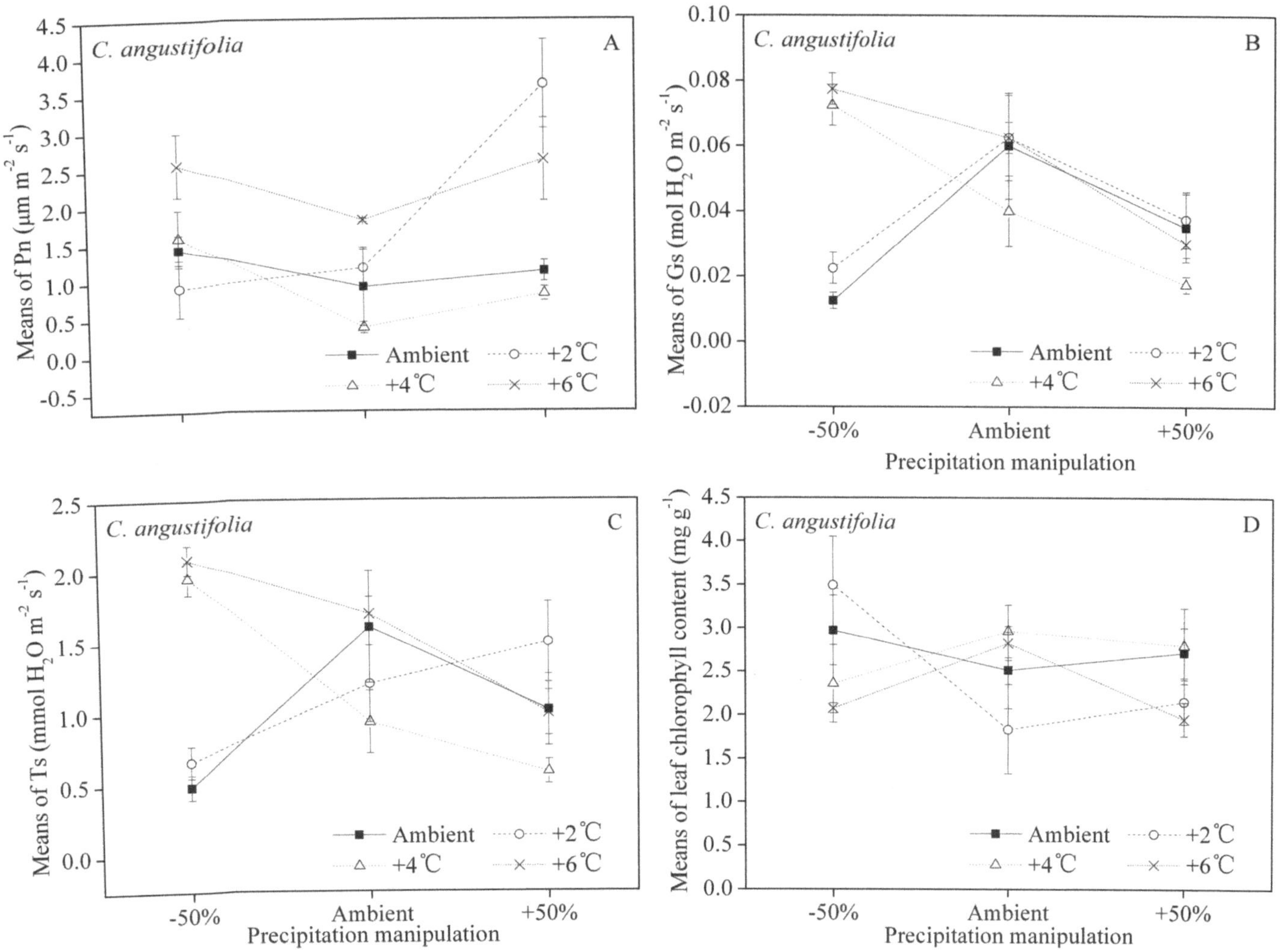

Figure 4. Interactions of temperature and precipitation manipulations on *C. angustifolia*'s leaf net photosynthetic rate (Pn, A), leaf stomatal conductance (Gs, B), leaf transpiration rate (Ts, C) and leaf chlorophyll content (D), respectively. The error bars represent means ± 1 standard error.

September, the leaf area with the +4°C manipulation was not significantly different from either the ambient or the +6°C manipulation. There was no significant difference in leaf area with different temperature manipulations in August (Fig. 3A). For *G. spiculosa*, there was no significant difference between ambient and the −50% precipitation treatments in any of the three months, while the leaf area with the 50% precipitation treatment was smaller than other treatments in August and September (Fig. 3B).

Photosynthetic indicators and leaf chlorophyll content

Two-way ANOVA of photosynthetic indicators and leaf chlorophyll content showed that the interaction effects of temperature and precipitation on these two species were both significant, except for the net photosynthetic rate (Pn) for *G. spiculosa*, which was only significantly affected by precipitation (Table 3).

For *C. angustifolia*, smaller Pn occurred under ambient precipitation except for the +2°C manipulation, and greater Pn was measured in the +6°C manipulation with the exception of the +50% precipitation treatment (Fig. 4A). The leaf stomatal conductance, Gs, decreased with the enhanced precipitation for the +4°C and +6°C manipulations (Fig. 4B). The greatest leaf transpiration rates, Ts, were found in the +4°C and +6°C manipulations under −50% precipitation (Fig. 4C). Ts rates decreased in both these temperature treatments with increasing precipitation. There was no systematic pattern in Ts at ambient temperature but the +2°C manipulation showed Ts increased with increasing precipitation. There were no consistent trends in leaf chlorophyll content with temperature or precipitation treatments (Fig. 4C) but all concentrations fell within the range 1.83–3.49 mg/g.

For *G. spiculosa*, Pn decreased with precipitation (Fig. 5A). With the exception of the +2°C treatment under the −50% precipitation regimen, the highest Gs results were found under ambient temperature (Fig. 5B). There was no consistent response in Gs to increased precipitation over the temperature treatments. There was no consistent trend in Ts across either temperature or precipitation treatments (Fig. 5C). All temperature treatments showed a decrease in leaf chlorophyll content between −50% and ambient precipitation (Fig. 5D). Increasing precipitation to 50% above ambient resulted in no consistent trend in chlorophyll with increasing temperature regime.

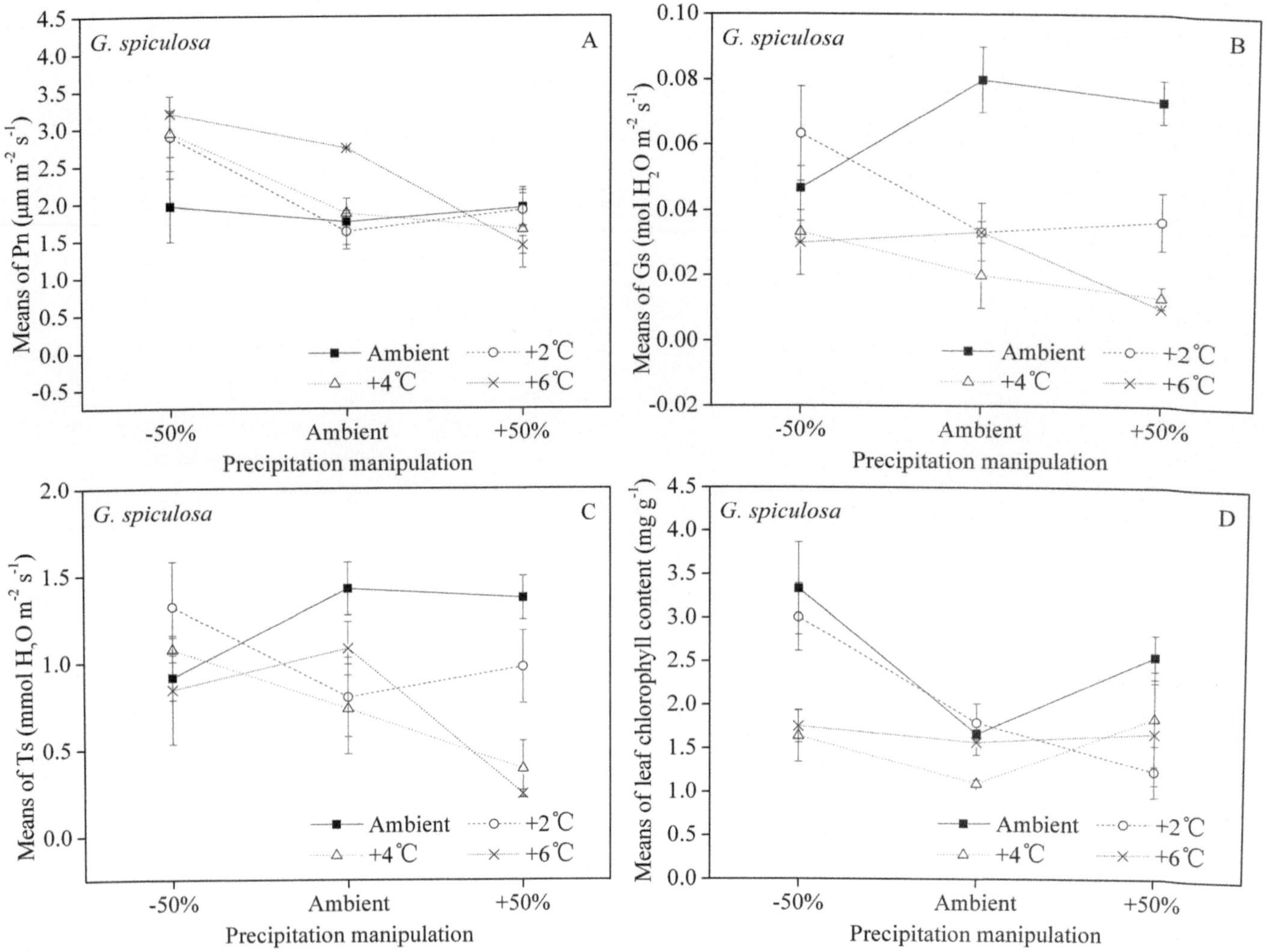

Figure 5. Main effect of precipitation manipulation on *G. spiculosa*'s leaf net photosynthetic rate (Pn, A), interactions of temperature and precipitation manipulations on leaf stomatal conductance (Gs, B), leaf transpiration rate (Ts, C) and leaf chlorophyll content (D), respectively. Different letters shared by the bars indicate significant differences ($p<0.05$) between precipitation treatments. The error bars represent means ± 1 standard error.

Above-ground biomass and leaf nitrogen content

Above-ground biomass of *C. angustifolia* was only significantly affected by temperature ($F = 4.512$, $p = 0.012$), while the effect of precipitation and the interaction effect were insignificant (Table 3). Biomass was at a maximum in the +4°C manipulation, increasing by 58% above that at ambient temperature (Fig. 6A). Leaf total nitrogen of *C. angustifolia* was also only significantly affected by temperature ($F = 7.125$, $p = 0.001$; Table 3). All three temperature increments resulted in statistically significant increases above ambient temperature, with a maximum value at +6°C, representing a 39% increase above ambient temperature (Fig. 6B).

Above-ground biomass of *G. spiculosa* was significantly affected both by temperature and precipitation ($F = 4.156$, $p = 0.017$; $F = 13.637$, $p<0.0001$, respectively), while the interaction effect was insignificant (Table 3). As noted with *C. angustifolia*, biomass was highest in the +4°C manipulation, with an increase of 37% above ambient temperature (Fig. 6C). Both increasing and decreasing precipitation had negative effects on above-ground biomass, with reductions of 29% and 32%, respectively compared to ambient precipitation (Fig. 6D). Any effects of temperature or precipitation, as well as their interaction effect, on leaf total nitrogen of *G. spiculosa* were insignificant (Table 3).

Secondary growth of *C. angustifolia*

One-way ANOVA results (Table 4) showed that only above-ground biomass displayed a significant difference with the incubation temperature treatment after one month's secondary growth ($F = 4.096$, $p = 0.049$); no other morphological or photosynthetic indicator showed significant differences (Table 4). Above-ground biomass of secondary growth was greatest in the +4°C manipulation, 1.11 times higher than the ambient temperature (Fig. 7).

Discussion

Interaction effects of warming and precipitation alteration

Boreal wetlands are becoming increasingly vulnerable in a warming world [8]. Measuring responses of dominant plant species to warming and precipitation alteration is fundamental to assess and project the vulnerability of boreal wetlands under

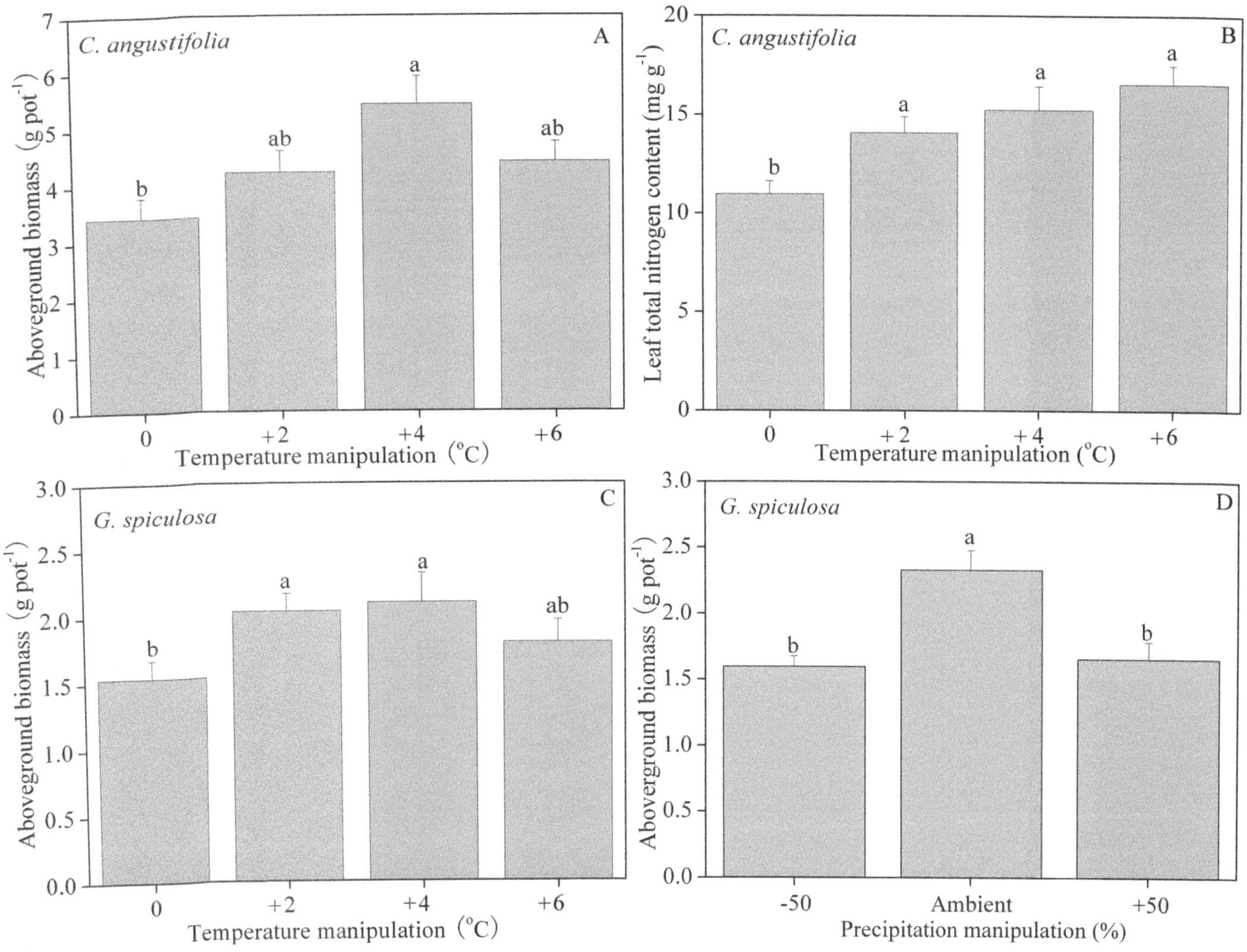

Figure 6. Aboveground biomass (A), leaf total nitrogen (B) of *C. angustifolia*, and aboveground biomass of *G. spiculosa* (C, D) after harvest. Different letters shared by the bars indicate significant differences ($p<0.05$) between temperature or precipitation manipulations. The error bars represent means ± 1 standard error.

climate change. The interactions among multiple climate-change factors may cause responses in wetlands that may not be detected when focusing on a single factor, because climate change will include simultaneous changes in temperature and precipitation regimes [19,20]. The above-ground biomass of *C. angustifolia* and *G. spiculosa* initially showed an increasing trend and then decreased with continued increments in mean monthly temperature. The +4°C temperature treatment produced the highest biomass (Fig. 6A, C), which is consistent with the single factor study by Breeuwer et al. [10] and multi-factor meta-analysis results from Lin et al. [21].

The monthly heights in different warming manipulations show intersecting trend lines (Fig. 1A, 2A), indicating that interaction effects of temperature and precipitation should be considered, especially for the +4°C manipulation. In the decreased precipitation treatment (−50%), greater plant heights were at ambient temperature and in the +4°C manipulation for *C. angustifolia* and *G. spiculosa* (Fig. 1C, 2C, respectively). With the exclusion of injury, height is one of the best indicators of an herbaceous plant's growth and competitiveness in light [22,23]. Our results show that any study on species-specific or interspecific plant response to climate change should also include consideration of the accompanying precipitation regime.

A limited increase of temperature within the optimal range often enhances photosynthesis, depending on the species and its environmental acclimation [24,25]. However, this expected enhancement of Pn can be moderated by concurrent water or drought stress on plants [26]. Our results show that Pn of *C. angustifolia* increased with temperature only in the driest (−50% precipitation) treatment, while ambient precipitation resulted in smaller Pn values in two of the three elevated temperatures (Fig. 4A), indicating that a change in precipitation would counter-intuitively limit the photosynthetic capacity of *C. angustifolia*. The Pn of *G. spiculosa* decreases with increasing precipitation (Fig. 5A), suggesting that soil water availability affects photosynthesis more than temperature variation [27]. Photosynthetic capacity is closely associated with, and approximately proportional to, leaf chlorophyll and nitrogen contents, and all three are favored by increased temperature [28,29,30]. Only the leaf nitrogen content of *C. angustifolia* increased with temperature (Table 3, Fig. 6B), while that of *G. spiculosa* was not significantly affected either by temperature or precipitation (Table 3). Although the leaf chlorophyll contents of two species were altered both by

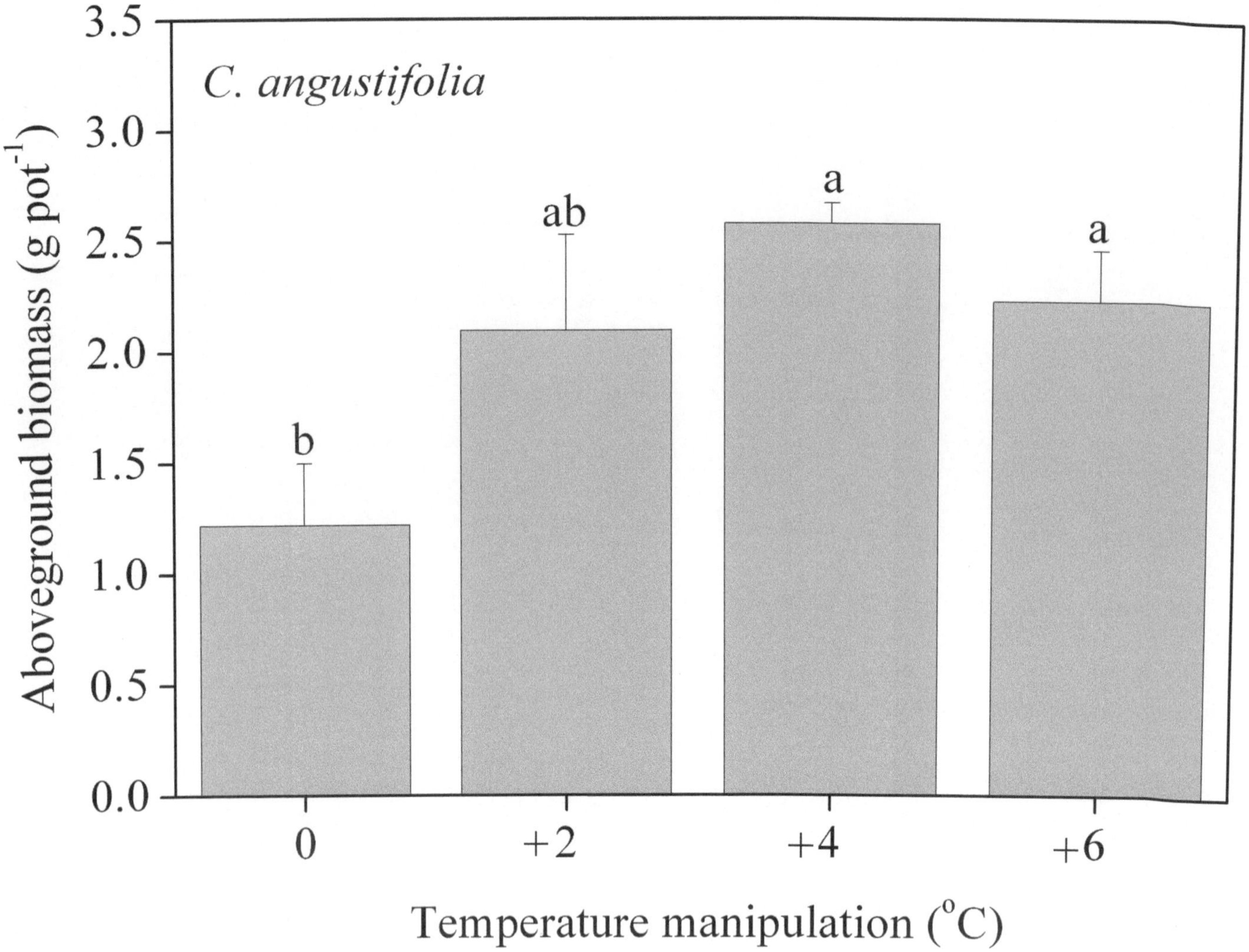

Figure 7. Aboveground biomass of *C. angustifolia* after secondary growth for one month. Different letters shared by the bars indicate significant differences ($p<0.05$) between the temperature manipulations. The error bars represent means ± 1 standard error.

temperature and precipitation (Table 3), the temperature responses under different precipitation treatments differed between species (4D, 5D).

These differential, and even opposite, responses of two species to our manipulated temperature and precipitation treatments are consistent with the widely supported observations and modeling that show species respond individualistically to changes in environmental conditions [5].

Species-specific responses to temperature and precipitation change

The effects of climate change on boreal wetland ecosystems ranges from regional, ecosystem, vegetation type, functional type to species [21,31,32,33]. Our results suggest that within the same functional plant type, species-specific response was a significant determinant in height (Table 3, Fig. 1, 2), leaf area (Fig. 3), photosynthetic indicators, leaf chlorophyll contents (Table 3, Fig. 4, 5), and above-ground biomass (Table 3, Fig. 6).

The most marked species-specific difference was leaf area and above-ground biomass (Table 2, 3). Leaf area is only affected by temperature for *C. angustifolia* (Fig. 3A) and precipitation for *G. spiculosa* (Fig. 3B). Above-ground biomass of *C. angustifolia* is only affected by temperature (Fig. 6A), while both temperature and precipitation influence *G. spiculosa* (Fig. 6C, D). Given the direct relationship with net primary productivity, above-ground biomass is considered as one of the plant characteristics responding strongly to climate change [34]. The +2°C manipulation increased above-ground biomass 23% and 31% compared with ambient conditions for *C. angustifolia* (Fig. 6A) and *G. spiculosa* (Fig. 6C), respectively. According to the linear fitting of local meteorological records in Xingkai Lake lacustrine wetlands, a 2°C warming is likely to take place by the 2080s. Consequently, this species-specific difference in predicted biomass would affect carbon and nutrient budgets in boreal wetland ecosystems responding to future temperature change [31].

Previous species-specific comparisons under simulated climate change are mostly based on only one growth season, and the intergenerational response is not readily apparent [35,36]. The significant difference in above-ground biomass of secondary growth (Fig. 7) with the different temperature treatments indicates that effects of climate change would continue to the next generation, because *C. angustifolia* mainly reproduces clonally by rhizomes [12]. These rhizomes would store a large amount of

Table 4. One-way ANOVA of aboveground biomass, morphological and photosynthetic indicators of *C. angustifolia* after secondary growing for one month.

ANOVA	Height	Basal stem diameter	Leaf area	Pn	Gs	Ts	Aboveground biomass
F	0.281	0.874	1.849	2.963	2.650	2.802	4.096
p	0.838	0.494	0.217	0.097	0.120	0.108	**0.049**

Significant effects ($p<0.05$) are shown in bold.

biomass below-ground, translocated from above-ground when the shoot is harvested. Greater biomass results in better performance during secondary growth [37]. Considering the biomass accumulation of biomass year by year, *C. angustifolia* is more acclimated to temperature and precipitation change. This acclimated capacity supports the continued presence of *C. angustifolia* under climate change and perhaps even expansion, because a competitor is not as evolutionarily successful in adapting to change and hence will decline leaving an ecological niche to be filled [31].

Inter-specific interaction to temperature and precipitation change

Interspecies interactions play a key role in regulating the distribution and composition of plant communities. The impacts of inter-species interactions can be altered by external drivers such as climatic conditions or anthropogenic activities [38]. For well conserved boreal wetlands, climate change may affect wetland plant communities by directly limiting or fostering the performance of particular species or by altering species competition and abundance [10,39].

As the two dominant species in Sanjiang Plain wetlands of Northeast China, *C. angustifolia* competes with *G. spiculosa* both in mesic and hydric habitats. Based on sequentially documented surveys of plant communities in a mesic habitat from 172 sampling plots from 1973 to 2003, the "importance values", as the average of relative density, frequency and coverage which gives an overall estimate of the influence of importance of a plant species in the community, of *C. angustifolia* decreased from 0.55 in the 1970s to 0.50 in the 2000s and that of *G. spiculosa* from 0.30 in the 1970s to 0.13 in the 2000s, accompanied by an increase in mesophytes. Although importance values for both hygrophytes declined, the former decreased by 9%, while the latter decreased 57%. Another sequentially documented *Carex lasiocarpa* community in a hydric habitat showed that the importance value of *G. spiculosa* decreased from 0.19 in the 1970s to 0.13 in the 2000s, while that of *C. angustifolia* increased from 0.021 to 0.023 [40]. Consequently, *C. angustifolia* is more competitive than *G. spiculosa* and *G. spiculosa* is more environmentally sensitive than *C. angustifolia*, which might be partly attributed to the greater acclimated capacity of *C. angustifolia* responding to precipitation and/or temperature change (excluding anthropogenic activities). Firstly, the greatest above-ground biomass of *C. angustifolia* presents in the +4°C treatment (Fig. 6A), while that of *G. spiculosa* is in the +2°C and +4°C manipulations (Fig. 6C), suggesting that *C. angustifolia* might be more acclimated to greater warming, even occasional extreme high temperatures. Secondly, there is no significant difference in the precipitation manipulations for *C. angustifolia*'s leaf area (Table 3) or above-ground biomass (Table 3), while either -50% or +50% of precipitation significantly decreased the leaf area (Fig. 3B) in August and September and above-ground biomass (Fig. 6D) of *G. spiculosa*.

Considering the warming trend of 2.5–5.4°C in the Sanjiang Plain within this century [41], it is predicted that the distribution of *C. angustifolia* will expand while that of *G. spiculosa* will shrink. However, the composition dynamics of these communities will be determined both by the intrinsic biological characteristics of each species and the external environment including climate change [5,42]. The species-specific responses described in this research coupled with inter-species interactions demonstrate the complexity of multi-factor effects. In addition to climate, other factors, e.g. soil nutrient concentrations, will modify or attenuate temperature and precipitation effects [35]. Consequently, minimization of other external (anthropogenic) factors is required in order to assess climate change impacts and this can best be achieved using well-conserved boreal wetlands.

Implications for boreal wetland conservation and restoration

Boreal wetland ecosystems, especially those that are well conserved, have an innate resistance and acclimated self-regulation potential, allowing them to respond to climate change and variation in habitat conditions [7,8]. However, long-term stress caused by high temperature, extreme drought or flooding, or the degradation and eventual loss of dominant species or functional groups will alter fundamental ecosystem properties and processes, because changes in the abundance, production, or distribution of species are often sufficient to alter both structure and function of the ecosystem [43]. Therefore, investigating the thresholds of irreversible climate change effects is both urgent and critical to the conservation and restoration of boreal wetland ecosystems.

Stability and acclimation can by delineated by ecological thresholds. Within these thresholds, the species interactions of freshwater wetland plants tend to shift from competitive to facilitative with increased stress [44]; however, when the environment changes beyond these thresholds, perhaps through a combination of habitat alteration by climate change and direct damage by humans, boreal wetlands may shift to other ecosystems such as grassland or shrubland [45]. This may result in loss of the organic carbon sequestered in these wetlands for hundreds or thousands of years, ultimately reducing the net cooling effect of a carbon sink on global warming [7,45]. To avoid this consequence, conservation policies and restoration practices will need to be revised in the face of potential thresholds for irreversible change to key wetland plant species under climate change. For example, hydrologically damaged wetlands will experience greater vulnerability to climate change effects compared to wetlands with an intact hydrological regime [8]. According to our results (Fig. 6D), the stability of the natural surface water table is critical for the conservation and restoration of the *G. spiculosa* population, which indicates special actions may be required at both the management and policy level if this species is to be safeguarded in the future.

Acknowledgments

The authors express gratitude to the reviewers and editors for their critical comments on an earlier version of the manuscript. We gratefully acknowledge Jiawei Guo and Haiyang Zhao for seedling collection, and Yuxia Zhang and Bokun Lou for sample analyses.

Author Contributions

Conceived and designed the experiments: YZ GW. Performed the experiments: X. Lou. Analyzed the data: XY. Contributed reagents/materials/analysis tools: X. Lu. Contributed to the writing of the manuscript: YZ MG.

References

1. Gorham E (1991) Northern peatlands, role in the carbon cycle and probable responses to climatic warming. Ecol Appl 1:182–195.
2. Aerts R, Callaghan TV, Dorrepaal E, van Logtestijn RSP, Cornelissen JHC (2012) Seasonal climate manipulations have only minor effects on litter decomposition rates and N dynamics but strong effects on litter P dynamics of sub-arctic bog species. Oecologia 170:809–819.
3. Bartlett KB, Harriss RC (1993) Review and assessment of methane emissions from wetlands. Chemosphere 26:261–320.
4. Bridgham SD, Pastor J, Updegraff K, Malterer TJ, Johnson K, et al. (1999) Ecosystem control over temperature and energy flux in northern peatlands. Ecol Appl 9:1345–1358.
5. Weltzin JF, Bridgham SD, Pastor J, Chen J, Harth C (2003) Potential effects of warming and drying on peatland plant community composition. Global Change Biol 9:141–151.
6. Flanagan LB, Syed KH (2011) Stimulation of both photosynthesis and respiration in response to warmer and drier conditions in a boreal peatland ecosystem. Global Change Biol 17:2271–2287.
7. Dise N (2009) Peatland response to global change. Science 326:810–811.
8. Essl F, Dullinger S, Moser D, Rabitsch W, Kleinbauer I (2012) Vulnerability of mires under climate change, implications for nature conservation and climate change adaptation. Biodivers Conserv 21:655–669.
9. Charles H, Dukes JS (2009) Effects of warming and altered precipitation on plant and nutrient dynamics of a New England salt marsh. Ecol Appl 19:1758–1773.
10. Breeuwer A, Heijmans MMPD, Robroek BJM, Berendse F (2008) The effect of temperature on growth and competition between *Sphagnum* species. Oecologia 156:155–167.
11. Thompson RM, Beardall J, Beringer J, Grace M, Sardina P (2013) Means and extremes, building variability into community-level climate change experiments. Ecol Lett 16:799–806.
12. Editorial Committee of Flora Republicae Popularis Sinicae (1987) Flora Republicae Popularis Sinicae 9(3). Beijing, China: Science Press. 221–223 p.
13. Editorial Committee of Flora Republicae Popularis Sinicae (2002) Flora Republicae Popularis Sinicae 9(2). Beijing, China: Science Press. 327–328 p.
14. Ren G, Guo J, Xu M, Chu Z, Zhang L, et al. (2005) Climate changes of China's mainland over the past half century. Acta Meteorol Sin 63:942–956.
15. Guo R, Shi L, Ding X, Hu Y, Tian S, et al. (2010) Effects of saline and alkaline stress on germination, seedling growth, and ion balance in wheat. Agron J 102:1252–1260.
16. Arnon DI (1949) Copper enzymes in isolated chloroplasts Polyphenoxidase in *Beta vulgaris*. Plant Physiol 24:1–15.
17. Porra RJ (2002) The chequered history of the development and use of simultaneous equations for the accurate determination of chlorophylls a and b. Photosynth Res 73:149–156.
18. Zou YC, Jiang M, Yu XF, Lu XG, David JL, et al. (2011) Distribution and biological cycle of iron in freshwater wetlands of Sanjiang Plain, Northeast China. Geoderma 164:238−248.
19. Fenner N, Freeman C, Lock MA, Harmens H, Reynolds B, et al. (2007) Interactions between elevated CO_2 and warming could amplify DOC exports from peatland catchments. Environ Sci Technol 41:3146–3152.
20. Kardol P, Cregger MA, Campany CE, Classen AT (2010) Soil ecosystem functioning under climate change, plant species and community effects. Ecology 91:767–781.
21. Lin DL, Xia JY, Wan SQ (2010) Climate warming and biomass accumulation of terrestrial plants, a meta-analysis. New Phytol 188:187–198.
22. Wilson SD, Tilman D (1993) Plant competition and resource availability in response to disturbance and fertilization. Ecology 74:599–611.
23. Gough L, Gross KL, Cleland EE, Clark CM, Collins SL, et al. (2012) Incorporating clonal growth form clarify the role of plant height in response to nitrogen addition. Oecologia 169:1053–1062.
24. Berry J, Bjorkman O (1980) Photosynthetic response and adaptation to temperature in higher plants. Annu Rev Plant Physiol 31:491–543.
25. Rustad LE, Campbell JL, Marion GM, Norby RJ, Mitchell MJ, et al. (2001) A meta-analysis of the response of soil respiration, net nitrogen mineralization, and aboveground plant growth to experimental ecosystem warming. Oecologia 126: 543–562.
26. De Boeck HJ, Lemmens CMHM, Zavalloni C, Gielen B, Malchair S, et al. (2008) Biomass production in experimental grasslands of different species richness during three years of climate warming. Biogeosciences 5:585–594.
27. Ge Z, Zhou X, Kellomäki S, Zhang C, Peltola H, et al (2012) Acclimation of photosynthesis in a boreal grass (*Phalaris arundinacea* L) under different temperature, CO_2, and soil water regimes. Photosynthetica 50:141–151.
28. Farquhar GD, Sharkey TD (1982) Stomatal conductance and photosynthesis. Annu Rev Plant Physiol 33:317–345.
29. Evans JR (1989) Photosynthesis and nitrogen relationship in leaves of C_3 plants. Oecologia 78:9–19.
30. Mae T (1997) Physiological nitrogen efficiency in rice, nitrogen utilization, photosynthesis, and yield potential. Plant Soil 196:201–210.
31. Weltzin JF, Pastor J, Harth C, Bridgham SD, Updegraff K, et al. (2000) Response of bog and fen plant communities to warming and water-table manipulations. Ecology 81:3464–3478.
32. Erwin KL (2009) Wetlands and global climate change, the role of wetland restoration in a changing world. Wetl Ecol Manage 17:71–84.
33. Liancourt P, Spence LA, Song D, Lkhagva A, Nharkhuu A, et al. (2013) Plant response to climate change varies with topography, interactions with neighbors, and ecotype. Ecology 94:444–453.
34. Van Minnen JG, Onigkeit J, Alcamo J (2002) Critical climate change as an approach to assess climate change impacts in Europe, development and application. Environ Sci Policy 5:335–347.

35. Luo WB, Xie YH (2009) Growth and morphological responses to water level and nutrient supply in three emergent macrophyte species. Hydrobiologia 624:151–160.
36. Gedan KB, Bertness MD (2010) How will warming affect the salt marsh foundation species *Spartina patens* and its ecological role? Oecologia 164:479–487.
37. Qiu ZC, Wang M, Lai WL, He FH, Chen ZH (2011) Plant growth and nutrient removal in constructed monoculture and mixed wetlands related to stubble attributes. Hydrobiologia 661:251–260.
38. Brooker RW (2006) Plant-plant interactions and environmental change *New Phytol*, 171, 271–284.
39. Pastor J, Peckham B, Bridgham S, Weltzin J, Chen J (2002) Plant community dynamics, nutrient cycling and alternative stable equilibria in peatlands. Am Nat 160:553–568.
40. Lou YJ (2008) Species biodiversity spatial patterns of typical wetland plants and their changes within 30 years in the Sanjiang Plain. Graduate University of Chinese Academy of Sciences, Beijing, China.
41. Yin XM (2013) Simulation Study of Climate Change impacts on Wetlands Productivity in Sanjiang Plain. Graduate University of Chinese Academy of Sciences, Beijing, China.
42. Hoeppner SS, Dukes JS (2012) Interactive responses of old-field plant growth and composition to warming and precipitation. Global Change Biol 18:1754–1768.
43. Berry PM, Dawson TP, Harrison PA, Pearson RG (2002) Modelling potential impacts of climate change on the bioclimatic envelope of species in Britain and Ireland. Global Ecol Biogeogr 11:453–462.
44. He Q, Bertness MD, Altieri AH (2013) Global shifts towards positive species interactions with increasing environmental stress. Ecol Lett 16:695–706.
45. Hobbs RJ, Higgs E, Harris JA (2009) Novel ecosystems, implications for conservation and restoration. Trends Ecol Evol 24:599–605.

Transfer of Cadmium from Soil to Vegetable in the Pearl River Delta area, South China

Huihua Zhang[1]*, Junjian Chen[1]*, Li Zhu[2], Guoyi Yang[1], Dingqiang Li[1]*

1 Guangdong Institute of Eco-environmental and Soil Sciences, Guangzhou, China, **2** Management School, Jinan University, Guangzhou, China

Abstract

The purpose of this study was to investigate the regional Cadmium (Cd) concentration levels in soils and in leaf vegetables across the Pearl River Delta (PRD) area; and reveal the transfer characteristics of Cadmium (Cd) from soils to leaf vegetable species on a regional scale. 170 paired vegetables and corresponding surface soil samples in the study area were collected for calculating the transfer factors of Cadmium (Cd) from soils to vegetables. This investigation revealed that in the study area Cd concentration in soils was lower (mean value 0.158 mg kg^{-1}) compared with other countries or regions. The Cd-contaminated areas are mainly located in west areas of the Pearl River Delta. Cd concentrations in all vegetables were lower than the national standard of Safe vegetables (0.2 mg kg^{-1}). 88% of vegetable samples met the standard of No-Polluted vegetables (0.05 mg kg^{-1}). The Cd concentration in vegetables was mainly influenced by the interactions of total Cd concentration in soils, soil pH and vegetable species. The fit lines of soil-to-plant transfer factors and total Cd concentration in soils for various vegetable species were best described by the exponential equation ($y = ax^b$), and these fit lines can be divided into two parts, including the sharply decrease part with a large error range, and the slowly decrease part with a low error range, according to the gradual increasing of total Cd concentrations in soils.

Editor: Manuel Reigosa, University of Vigo, Spain

Funding: Funding was provided by Natural Science Foundation of China (NSFC No. 41171387 and 31270516), the Natural Science Foundation of Guangdong Province, China (No.S2012030006144), the Science and Technology Planning Project of Guangdong Province, China (No.2012A020100003 and 2011B030900013), and Science and Technology Planning Project of Guangzhou (No.2013J2200003). The funders had no role in study design, data collection and analysis, decision to publish, or preparation of the manuscript.

Competing Interests: The authors have declared that no competing interests exist.

* Email: hhzhang@soil.cn (HZ); jjchen@soil.gd.cn (JC); dqli@soil.gd.cn (DL)

Introduction

Cadmium is non-essential element to biota. It is known to be toxic for plants as well as animals to a much higher extent and at lower concentrations than e.g. Zn, Pb, or Cu. It especially affects humans because of their longevity and the accumulation of Cd in their organs by eating Cd-contaminated food [1–3].

Soil Cd is naturally derived from parent materials because of chemical weathering, and as a contaminant in many areas of anthropogenic activities such as mining, smelting, composts, phosphate fertilizer application, waste disposal, and vehicle exhausts [4–7]. The main forms of cadmium in soils and sediments are the exchangeable fraction, followed by the Fe-Mn oxides and residual fractions. Several studies indicated the Cd in soils contaminated by anthropogenic activates such as mining and smelting, seem to be more bioavailable than Cd from unimpacted soils [5]. To protect the safe of soil environment and prevent the soil Cd contamination, various soil quality standards were established in many countries or regions, such as 0.3 mg kg^{-1} in China [8], 0.8 mg kg^{-1} in the Netherlands and Switzerland, 0.5–1 mg kg^{-1} in Austria [9], and 5 mg kg^{-1} in Taiwan [10].

Because soil Cd is easily accumulated by plants, soil-plant-human transfer of Cd has been considered as a major pathway of human exposure to soil Cd [11,12]. Many studies have been conducted in China such as Cd in rice [10,13–15], in orchard [16,17], and in vegetable [18–21]. And the maximum permissible concentrations of Cd of 0.05 mg kg^{-1} for non-environmental pollution vegetable – so called "No-Polluted vegetable" [22] and maximium concentration of 0.2 mg kg^{-1} in leaf vegetables for food security – so called "Safe vegetable" [23] are used in China.

Although undoubtedly soil Cd is the primary source of vegetable Cd, and total soil Cd concentration was commonly used for soil environmental quality estimation, its usefulness to predict soil-to-plant transfer was often questioned since the phytoavailability of Cd in soils [12,21,24]. Yang et al [25] proposed that various vegetable species showed significantly different accumulation capacity for Cd in the same soil sample sites; Soil-to-vegetable transfer factor (TF) of Cd was much higher in a pot experiment than that in a field trial. Previous studies have showed that the TF is decided especially by the soil properties and plant species [26]. Wang et al [12] pointed out that TFs can be considered as a useful index of metals potentially transfer abilities from soil to plant, and the TFs of Cd for leafy vegetables were higher than those for non-leafy vegetables.TF has been widely used in the evaluation of potential health risk of human exposure to metals from soil [3,5,12,15,16].

In the PRD area, the vegetable planting area is about 4830 km^2. The vegetable production was 1129.2×10^4 tons in 2011. The leaf vegetable is the dominant consumption for Guangdong people, the planting area is about 33% of the total planting area [27]. As a report of the Statistics Bureau of Guangdong Province, the export of fresh vegetable was up to 78.3 ×10^4 tons [28], mainly selled to Japan, Korea, Malaysia, Russia, and America. Hence, the quality safety of vegetable is not only

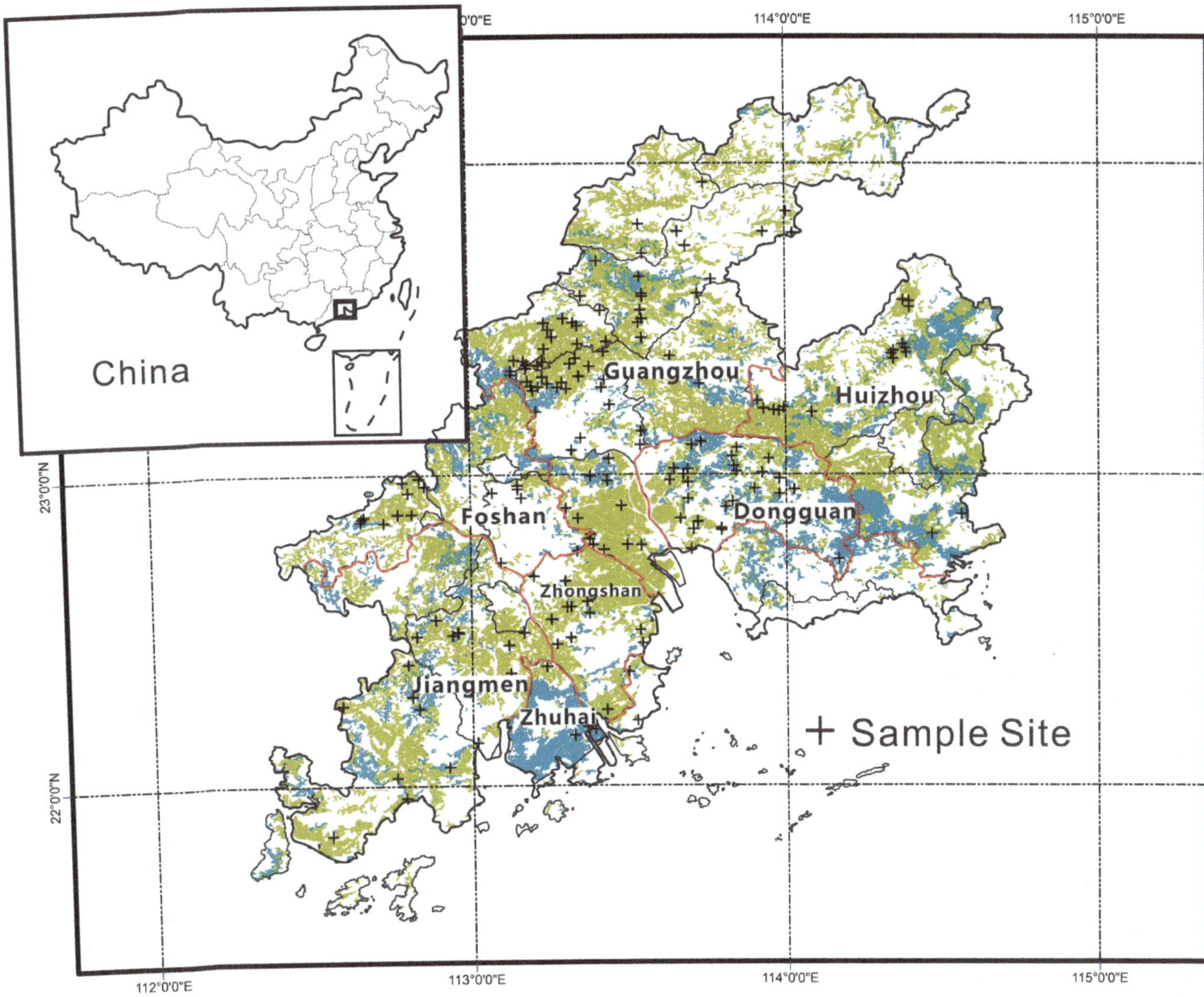

Figure 1. Locations of soil and vegetable samples in the Pearl River Delta area, Guangdong, China. (The yellow and blue areas are the dry lands and the paddy fields, respectively; the red line is the boundary of the city).

concerned by the local government, but it should receive much more attention from the international community for the increasing export and import trade.

In this study, we conducted a systemic environmental quality survey of the vegetable soil and various vegetables in the PRD area. The aim of this study was to: (1) obtain the information on concentrations and spatial patterns of Cd in soils and leaf vegetables across the PRD area; (2) investigate the relationship between soil Cd and vegetable Cd on a regional scale; (3) reveal the transfer characteristic of Cd from soil to various leaf vegetable species.

Materials and Methods

Ethics Statement

This study was carried out on collective-owned lands, and the owners of the lands gave us permission to conduct the study on these sites. The field studies did not involve endangered or protected species.

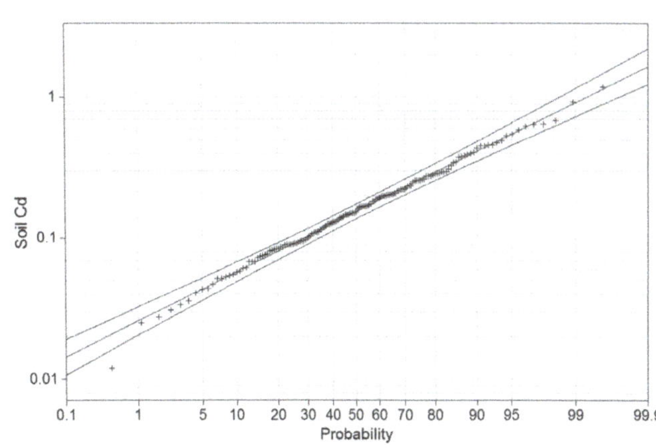

Figure 2. Lognormal probability plot for total Cd concentration in vegetable soils in the Pearl River Delta area, Guangdong, China (mg kg^{-1}, DW).

Table 1. Descriptive statistics of soil Cd concentrations (mg kg^{-1}, DW) and vegetable Cd concentration (mg kg^{-1}, FW) in the PRD area.

Sample Site	n	Soil Cd concentration			Vegetable Cd concentration			pH (H_2O)	OM[e] (%)
		Mean ± SD	Range	C.V. (%)	Mean ± SD	Range	C.V. (%)	Mean ± SD	Mean ± SD
Dongguan	27	0.138±0.079	0.041–0.433	57.2	0.024±0.012	0.004–0.057	48.4	5.76±0.87	2.58±0.70
Foshan	16	0.291±0.200	0.025–0.685	69.0	0.017±0.011	0.002–0.037	62.5	5.84±1.12	2.61±1.01
Guangzhou	73	0.226±0.190	0.027–1.180	84.3	0.027±0.024	0.002–0.082	86.8	6.34±0.96	2.44±0.66
Jiangmen	17	0.188±0.157	0.012–0.644	83.9	0.014±0.016	0.002–0.068	117.8	5.49±0.67	2.28±0.90
Zhuhai	6	0.204±0.160	0.044–0.474	78.5	0.008±0.005	0.001–0.014	58.1	6.11±0.99	2.34±0.63
Zhongshan	13	0.316±0.150	0.083–0.530	47.3	0.022±0.021	0.001–0.067	92.7	6.02±0.97	2.45±0.90
Huizhou	18	0.107±0.052	0.052–0.200	39.6	0.026±0.018	0.010–0.082	69.1	6.69±0.43	2.41±0.62
Total	170	0.208±0.169 0.158 (1.39)[d]	0.012–1.180	81.2	0.024±0.019 0.015 (1.57)[d]	0.002–0.082	84.3	6.12±0.96	2.42±0.74
Threshold		0.3[a]			0.05[b]; 0.2[c]				

[a]The maximum permissible concentrations of Cadmium for agriculture soils (SEPAC, 1995).
[b]the maximum permissible concentrations of Cadmium for non-environmental pollution vegetable (AQSIQ, 2001).
[c]The maximum level of Cadmium for leaf vegetables (MOH, 2012).
[d]Geometric Mean (Geometric standard deviation).
[e]Orgnic matter.

Study area

The PRD area is located in the south of Guangdong province occupying 41698 km^2 of land area (Figure 1). The area has a subtropical – tropical monsoon climate with an average annual temperature of 21–22°C and average annual rainfall of 1600–2000 mm. The main soil types in the PRD area are Ultisol, mostly developed on the granite parent materials in the local hills, and paddy soils developed on the fluvial sediments. In the study area, 10–15 crops of vegetables are planted annually, and the rotation of vegetable and rice is applied in mostly paddy soils.

Over the last 30 yr, rapid urbanization and industrialization has taken place in this area. Heavy metal contents in soils and sediments were elevated compared with historical monitoring results [29]. As a result, heavy metal accumulation in agricultural soils has also become increasingly serious in this area because of increasing reliance on fertilizers and agrochemicals [30].

Field sampling and preparation

Both surface soil samples and vegetable samples used in this study were collected from locations shown in Figure 1. All sample sites were far away at least 100 m from the obviously polluted area such as industries, feedlot, wastewater and highway for avoiding the directly anthropogenic influence. The planting area of each site was larger than 1000 m^2 for.a certain vegetable species. In this study the vegetable species included Pakchoi (*Brassica rapa chinensis*) (n = 31), Chinese flowering cabbage (*Brassica campestris L. ssp. chinensis*) (n = 37), Leaf mustard (*Brassica juncea Coss*) (n = 16), Romaine lettuce (*Lactuca sativa L. var. longifolia*) (n = 31), Chinese lettuce (*Lactuca sativa L. var. asparagina*) (n = 15), Cauliflower (*Brassica oleracea L.var. botrytis L.*) (n = 7), Water spinach (*Ipomaea aquatic Forssk*) (n = 7), Celery (*Apium graveolens*) (n = 6), Chinese chives (*Allium tuberosum*) (n = 5), Spinach (*Spinacia oleracea*) (n = 5), Amaranth (*Amaranthus mangostanus L.*) (n = 5), watercress (*Nasturtium officinale*) (n = 5).

Vegetables and field soils at a soil depth of 0–15 cm which was the root concentrated layer in this study were collected when the vegetable were suitable for harvest. Each vegetable and soil sample consisted of five subsamples, and combined and mixed well. The fresh vegetable samples were put in clean plastic bags and immediately transported to the laboratory for sample treatment.

The vegetable samples were cleaned with tap water and Milli-Q water, and then the edible parts (including leaves and steams) were separated, weighed (Fresh weight). The washed samples were then dried in an oven at 60°C, and their dry weights (DW) were recorded. The dry vegetable samples were ground to pass through a 250 μm sieve in a steel grinder. Soil samples were air-dried at room temperature (25°C) and ground to pass through a 2 mm nylon sieve. The fine vegetable and soil sample powders were stored in polythene zip bags.

Chemical analyses

The sieved soil samples were ground further to pass through a 150 μm nylon sieve. The prepared soil samples were digested to dryness using an acid mixture of 10 ml HF, 5 ml $HClO_4$, 2.5 ml HCl, and 2.5 ml HNO_3. Total Cd concentrations in soils were determined by Inductively Coupled Plasma-Atomic Emission Spectrometry (ICP-AES) (Model PS 1000 AT, USA). The recovered soil Cd concentration of the National Research Center for GeoAnalysis soil standard reference materials (SRM) (ESS-4, Beijing, China; Standard value: 0.083 mg kg^{-1}) was 0.081 mg kg^{-1} (n = 32). The analytic precision was 2.4% for soil Cd.

Seventy five soil samples were analyzed for exchangeable Cd using the first step of Tessier sequential extraction[31]. The extraction procedure is 2.00 g of air-dried soil (<2 mm) were mixed with 16 ml 1 M $MgCl_2$ solution (pH = 7.0), the mixture was shaken for one hour on 25°C. the suspension was filtered immediately after shaking. The Cd concentration in the filtrate solutions were measured by Inductively Coupled Plasma-Atomic Emission Spectrometry (ICP-AES) (Model PS 1000 AT, USA).

The ground vegetable samples were ashed in a muffle furnace for 16 h at 500°C, dissolved in 0.5 M HNO_3, and diluted to 25 mL with deionized water. Cd concentrations in vegetables were determined by Inductively Coupled Plasma-Atomic Emission Spectrometry (ICP-AES) (Model PS 1000 AT, USA). A plant standard reference material (GSV-4, Beijing, China; Standard value:0.057 mg kg^{-1}) was used in order to control the determination quality. The recovered Cd concentration was 0.059 mg kg^{-1} (n = 10). The analytic precision 3.5% for vegetable Cd. The Cd concentration in vegetable samples was expressed on a fresh weight basis.

The pH of soil was measured by taking 10 g of sample into 25 ml of deionized water [32]. The soil organic matter content was measured using potassium bichromate oxidation process [33].

Data analysis

The transfer factors (TFs) of Cd from soil to vegetable (edible part) were calculated using the following equation [12,15–17,20,34]:

$$TF = \frac{C_{vegetable}}{C_{soil}}$$

Where $C_{vegetable}$ is Cd concentration (FW) in the edible parts of vegetables, and C_{soil} is the total Cd concentration (DW) in the soil where the vegetable was grown.

The data statistical analysis was performed using the Minitab 16 statistical software (Minitab Inc., USA). Spatial interpolation was performed using ordinary kriging. For the low-density sampling, the ordinary kriging estimate can be thought of simply as an optimally weighted average of the data [35]. It provides a best linear unbiased prediction of spatial distribution. The spatial interpolation and contour maps displaying the spatial distribution of Cd concentrations in soils and vegetables, and transfer factors were produced based on the geostatistical analysis by using the software of ArcGIS 9.0.

Results and Discussions

Cd concentrations in vegetable soils

Total Cd concentrations in soils range from 0.012 to 1.18 mg kg^{-1} in the PRD area (Table 1). These values fit a log normal distribution (Figure 2), therefore the geometric mean (GM) value of 0.158 mg kg^{-1} and geometric standard deviations (GSD) of 1.39 were used to represent the central tendency and variations of the data. The present GM value was much higher than its background concentration (0.04 mg kg^{-1}) of soil in Guangdong province [35] and significantly lower than Cd mean concentrations of 0.858 mg kg^{-1} in the reclaimed tidal flat soil and 1.4–1.8 mg kg^{-1} in some Cd-contaminated vegetable soils in the PRD area [20,21]. The present results reflect the overall level of soil Cd concentration on the region al scale, not is responsible for local point-source contamination.

According to the maximum permissible concentrations of cadmium for agriculture soils of 0.3 mg kg^{-1} [8], thirty vegetable soils were contaminated with Cd in the study area, mainly located in the Guangzhou, Zhongshan, and Foshan cities. The soil Cd mean concentration was up to 0.316 mg kg^{-1} in the Zhongshan city, followed by 0.291 mg kg-1 in the Foshan city, 0.226 mg kg-1

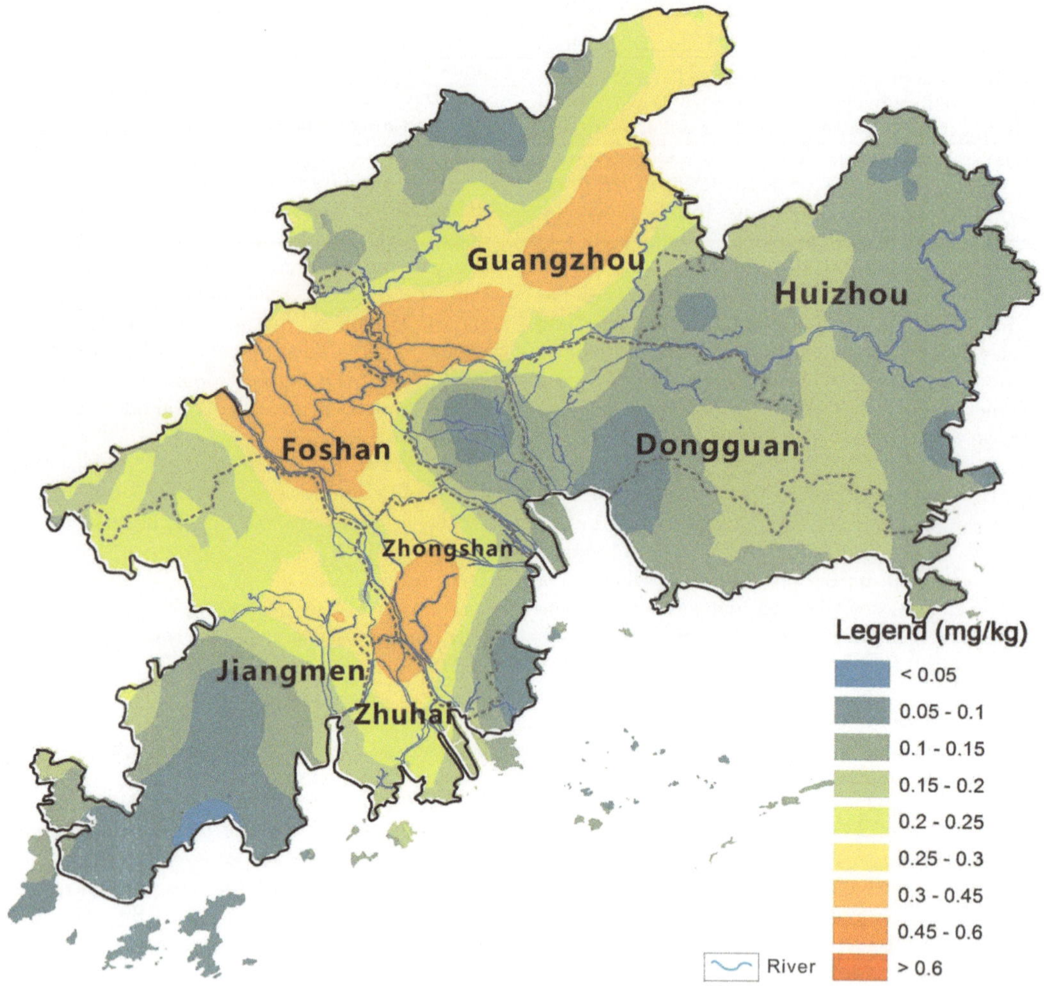

Figure 3. Spatial pattern of total Cd concentration in the soils in the PRD area (mg kg^{-1}, DW).

in the Guangzhou City, and 0.204 mg kg-1 in the Zhuhai City (Table 1). Generally, spatial characteres of soil Cd show that the western areas (including Foshan, Guangzhou, Zhongshan, Zhuhai) influenced by the West River and North River has the higher Cd concentration than the eastern areas (including Dongguan and Huizhou) influenced by the East River (Figure 3). The results coincided with spatial characters of other heavy metal concentrations (Pb, Zn) in sediments in the PRD area, that is, there were higher heavy metal concentrations in sediments of the West River and North River than that of the East River [36]. The spatial correlation between soil Cd cocnetrtiaons and regional rivers suggested that the cadmium produced by the mining and smelting activities in the upper reaches of the West River and North River was the main source of Cd in soils of the PRD area, and the West River and North River was the important transfer approach for Cd entering into surrounding soils through river irrigation.

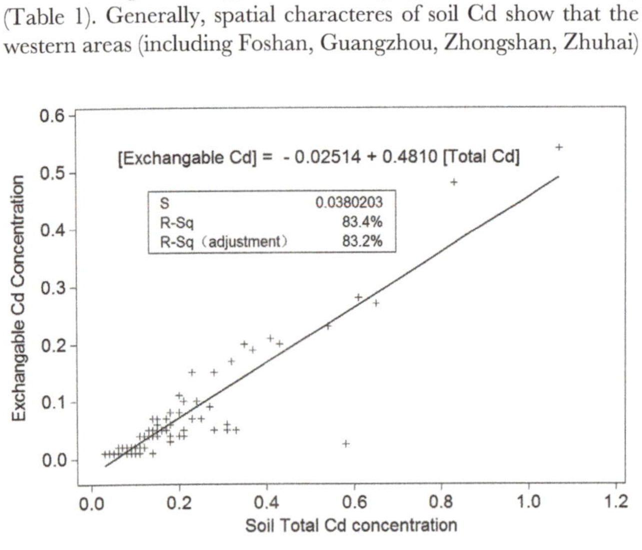

Figure 4. Relationship between exchangeable Cd concentration and total Cd concentration in vegetable soils (mg kg^{-1}, DW).

Although many literatures pointed out that metal availability in soil is of main concern, because the available concentration is an indication of the amount available for plant uptake, and provided some methods to determine plant availability of heavy metals [12,21,24]. Our results reveal that the significant positive correlation between soil total Cd and exchangeable Cd was shown in Figure 4, which can be described by a linear equation:

$$y = 0.4810x - 0.02514 \ (r^2 = 0.834; p < 0.001)$$

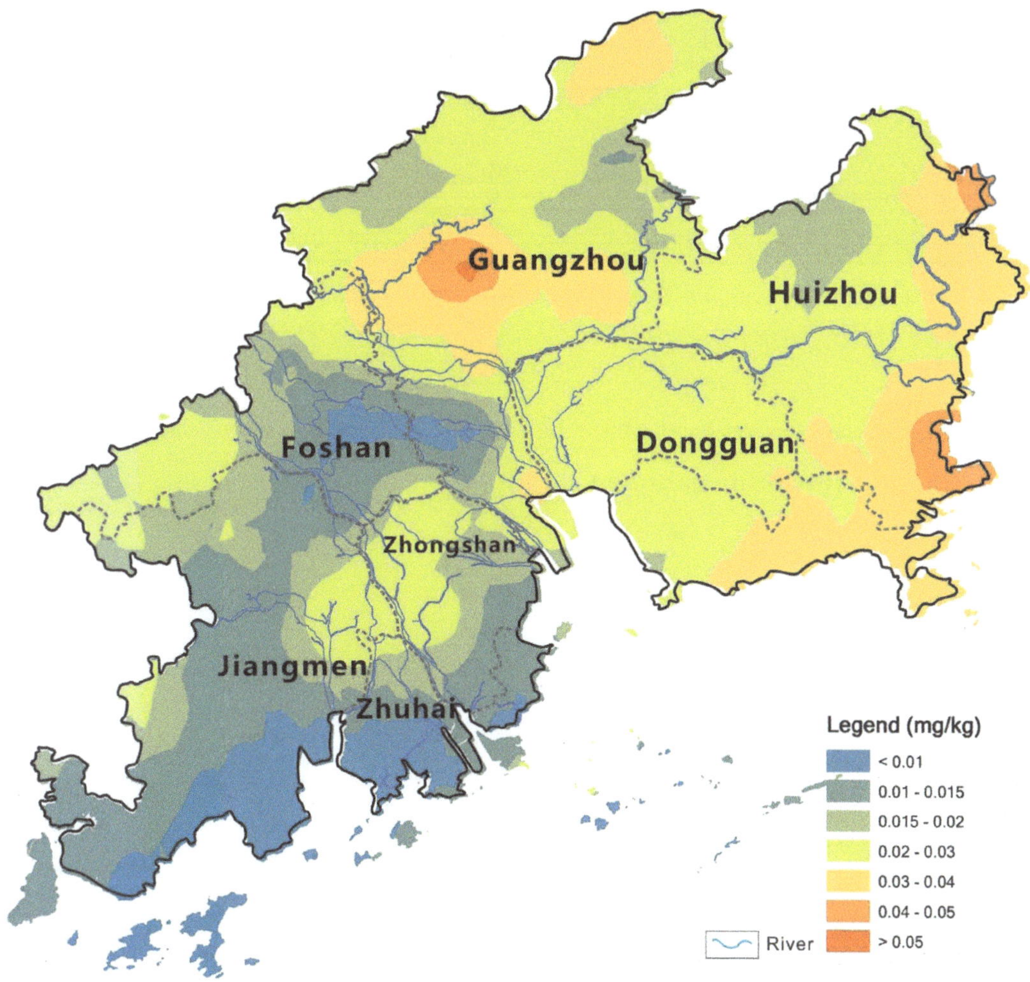

Figure 5. Spatial pattern of Cd concentrations in the vegetables in the PRD area (mg kg^{-1}, FW).

Table 2. Cd concentration (mg kg^{-1}, FW) in different vegetable species and the TF values in the PRD area.

Vegetable species	n	Cd		TF	
		Mean ± SD	Range	Mean ± SD	Range
Pakchoi (*Brassica rapa chinensis*)	31	0.023±0.015	0.002–0.074	0.222±0.220	0.010–0.920
Chinese flowering cabbage (*Brassica campestris L. ssp. chinensis*)	37	0.022±0.015	0.001–0.068	0.160±0.142	0.025–0.774
Leaf mustard (*Brassica juncea Coss*)	16	0.023±0.014	0.001–0.054	0.532±1.087	0.022–4.50
Romaine lettuce (*Lactuca sativa L. var. longifolia*)	31	0.026±0.021	0.003–0.082	0.148±0.111	0.001–0.371
Chinese lettuce (*Lactuca sativa L. var. asparagina*)	15	0.031±0.026	0.005–0.078	0.188±0.219	0.017–0.821
Cauliflower (*Brassica oleracea L.var. botrytis L.*)	7	0.010±0.013	0.002–0.032	0.048±0.042	0.014–0.103
Water spinach (*Ipomaea aquatic Forssk*)	7	0.011±0.019	0.001–0.053	0.045±0.052	0.003–0.117
Celery (*Apium graveolens*)	6	0.043±0.020	0.019–0.064	0.234±0.115	0.110–0.376
Chinese chives (*Allium tuberosum*)	5	0.027±0.032	0.001–0.074	0.229±0.241	0.001–0.563
Spinach (*Spinacia oleracea*)	5	0.037±0.027	0.006–0.067	0.213±0.034	0.182–0.262
Amaranth (*Amaranthus mangostanus L.*)	5	0.060±0.017	0.030–0.082	0.753±0.442	0.258–1.577
watercress (*Nasturtium officinale*)	5	0.021±0.020	0.002–0.039	0.087±0.046	0.003–0.129
Total	170	0.024±0.019	0.001–0.082	0.189±0.211	0.001–4.50

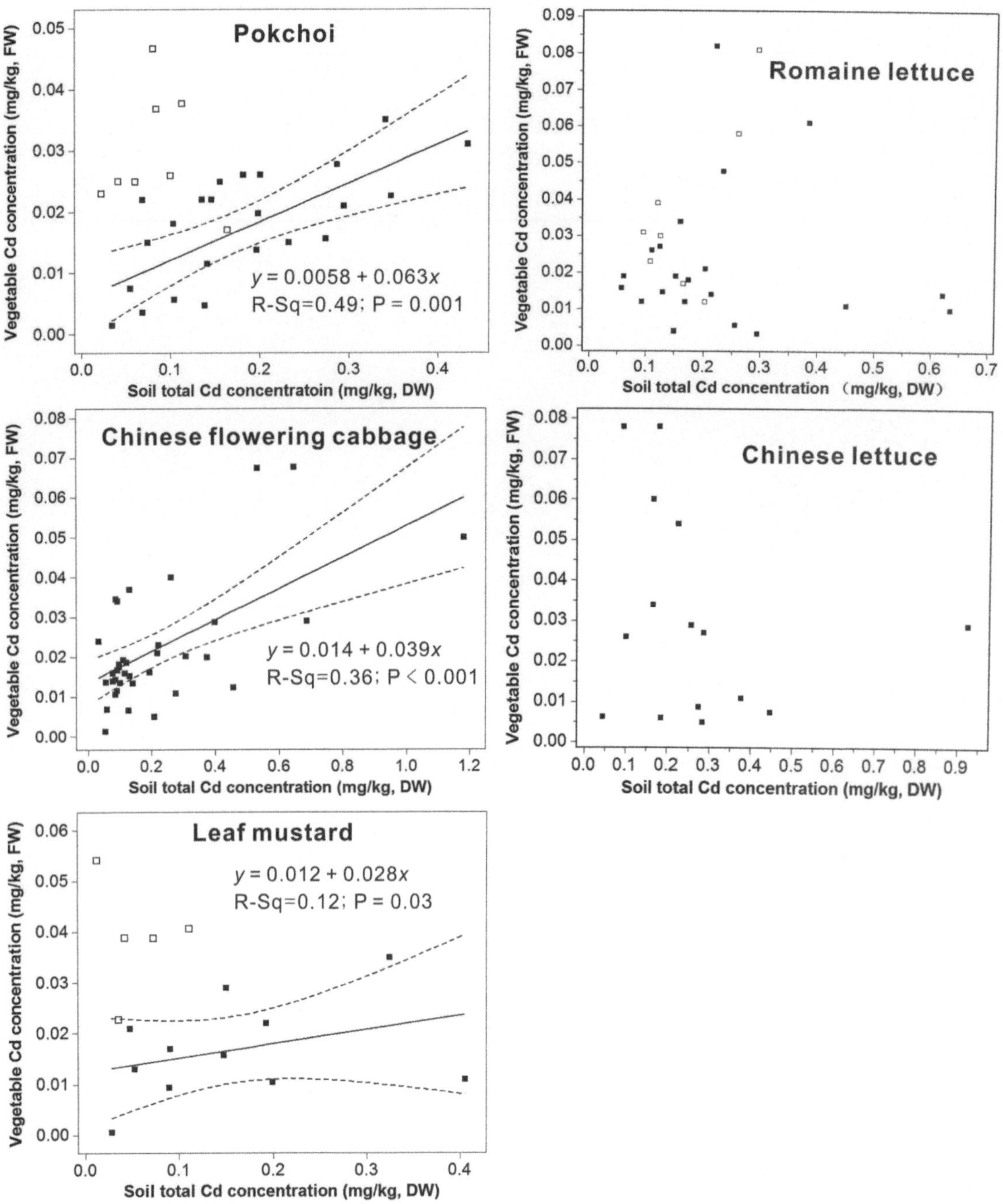

Figure 6. Relationships between soil total Cd concentrations (mg kg^{-1}, DW) and vegetable Cd concentrations (mg kg^{-1}, FW) for five species of main vegetables (soil pH: 3.79–7.72) (hollow diamonds represented samples with lower pH (3.79–5.00) removed for the regression analysis; the solid line is the sample regression line; dotted lines indicated the 95% confidence interval).

Where y is the soil exchangeable Cd concentration, *x* refers to soil total Cd concentration.

This indicated that exchangeable Cd concentration in soils in the PRD area was mainly controlled by the soil total Cd concentration.

Cd concentrations in vegetables

The present results show that in the PRD area the range of Cd concentration in all vegetable samples was 0.002–0.082 mg kg^{-1} with a geometric mean value of 0.015 mg kg^{-1} (Table 1). In this study Cd concentrations in all vegetable samples met the national standard of Safe vegetables (0.2 mg kg^{-1}), 150 vegetable samples met the standard of No-Polluted vegetables (0.05 mg kg^{-1}). Twenty vegetable samples were in between Safe vegetables and No-polluted vegetables, of which14 vegetable sample sites were in the Guangzhou.

According to the different cities, the mean vegetable Cd concentration, was in the order of Guangzhou (0.027 mg kg^{-1})> Huizhou (0.026 mg kg^{-1})> Dongguan (0.024 mg kg^{-1})> Zhongshan (0.022 mg kg^{-1})> Foshan (0.017 mg kg^{-1})> Jiangmen (0.014 mg kg^{-1})> Zhuhai (0.008 mg kg^{-1}) (Table 1). This order is significantly different from Cd concentrations in soils.

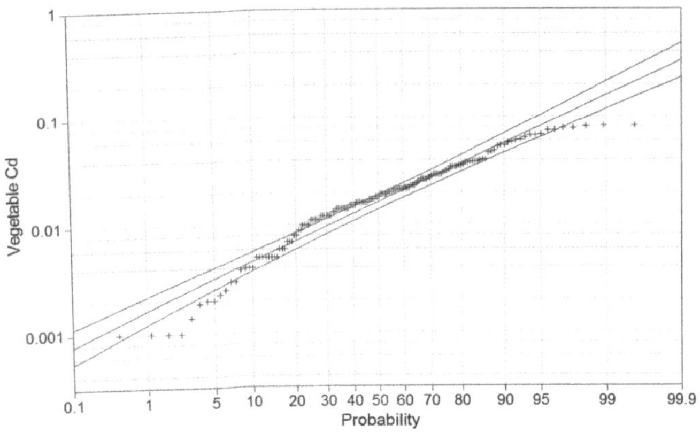

Figure 7. Lognormal probability plot for vegetable Cd concentrations in the Pearl River Delta area, Guangdong, China (mg kg^{-1}, FW).

The spatial pattern of Cd concentrations in vegetables showed that no spatial correlation with that of total Cd cocentrtaions in soils on the regional scale (Figure 3 and Figure 5).

The mean and range of Cd concentrations in various vegetable species were listed in Table 2. In the study area the dominant vegetable species are the pakchoi, Chinese flowering cabbage, leaf mustard, Romaine lettuce and Chinese lettuce for the local residents. The mean Cd concentrations in these vegetables range from 0.022 mg kg^{-1} to 0.031 mg kg^{-1} (Table 2). The amaranth vegetables had the highest Cd concentration of 0.082 mg kg^{-1} with mean value of 0.06 mg kg^{-1}. This value was closed to 0.078 mg kg^{-1} in amaranth on the reclaimed tidal flat soil reported by Li et al [20]. Compared with the previous researches on the Cd-contaminated soil [20,21], the present results revealed that there were low Cd concentrations of various vegetables in the study area.

Relationships between Cd in soils and in vegetables

Many studies have shown that the concentration of heavy metals in vegetables is influenced by many factors, such as total concentration of heavy metals in soils, vegetable species, soil pH value, soil organic matter, climate, atmospheric depositions and temperature [19,21,37–39]. McBride [40] suggested that combination of soil pH and soil total Cd concentration was reasonably predictive of Cd concentration in the above-ground plant tissue.

In this study, we discussed the correlation between Cd concentrations in five species of dominant leaf vegetables and Cd concentrations in soils. The results revealed that in the condition of soil pH>5, Cd concentrations in both pakchoi and Chinese flowering vegetables had obvious positive correlations with Cd concentrations in soils; and Cd concentrations in leaf mustard vegetables and Cd concentrations in soils had a weak positive correlation (Figure 6). These can be described by linear equations, respectively:

$$y=0.063x+0.0058\ (\textit{Pakchoi}: r^2=0.49; p=0.001)$$

$$y=0.039x+0.014\ (\textit{Chinese flowering cabbage}: r^2=0.36; p<0.001)$$

$$y=0.028x+0.012\ (\textit{Leaf mustard}: r^2=0.12; p=0.03)$$

Where y is the Cd concentration in the various vegetable, x is the soil total Cd concentration.

The results of the line regression indicated that in the condition of soil pH>5, soil total Cd concentrations explained 49%, 36% and 12% of the variability of Cd concentration in Pakchoi and Chinese flowering cabbage and leaf mustard, respectively. But Cd concentrations in the Romaine lettuce and Chinese lettuce had no correlations with total Cd concentrations in soils (Figure 6). This phenomenon should be related with various vegetable families. Pakchoi, Chinese flowering cabbage and leaf mustard belong to the brassica of cruciferous plants; Romaine lettuce and Chinese lettuce belong to the Lactuca of Compositae plants. The difference of correlations between Cd concentrations in vegetables and total Cd concentrations in soils should be the result of different absorption capacity of various vegetable families to Cd. And the result showed that the influence of soil total Cd concentrations on the Cd in various vegetable species is obviously different [39,41].

When soils pH is less than 5, total Cd concentrations in soils were low because the environmental capacity of soil Cd was small in acid soils due to the strong eluviation, but Cd concentrations in the pakchoi and leaf mustard vegetables were much higher than the reference line (Figure 6). Obviously, low soil pH can raise the available Cd content for pakchoi and leaf mustard vegetables even in the soils with low Cd concentrations.

The analysis of Pearson correlations among soil Cd, vegetable Cd, soil pH, organic matter and transfer factor showed that there was no obvious correlation among soil Cd, vegetable Cd, organic matter and soil pH; significant correlations were be found between transfer factor and soil Cd ($r^2=-0.295$) and between transfer factor and vegetable Cd ($r^2=0.435$) (Table 3). Our results confirmed that soil pH played important role to Cd absorption of vegetables, but there was no linear correlation between soil pH and Cd concentration in vegetables in this study. Moreover, The lognormal probability distribution graph shows that some vegetable samples of upper tail had less high value than would be

Table 3. Pearson correlations of soil Cd, vegetable Cd, soil pH, organic matter and Transfer factor.

	soil Cd	vegetable Cd	pH	Organic matter
vegetable Cd	0.205(0.088)			
pH	0.033(0.784)	0.137(0.258)		
Organic matter	0.065(0.590)	−0.007(0.952)	0.076(0.532)	
TF	−0.295(0.013)	0.435(0.000)	−0.143(0.239)	−0.103(0.396)

Cell content: Pearson correlation (p-value).

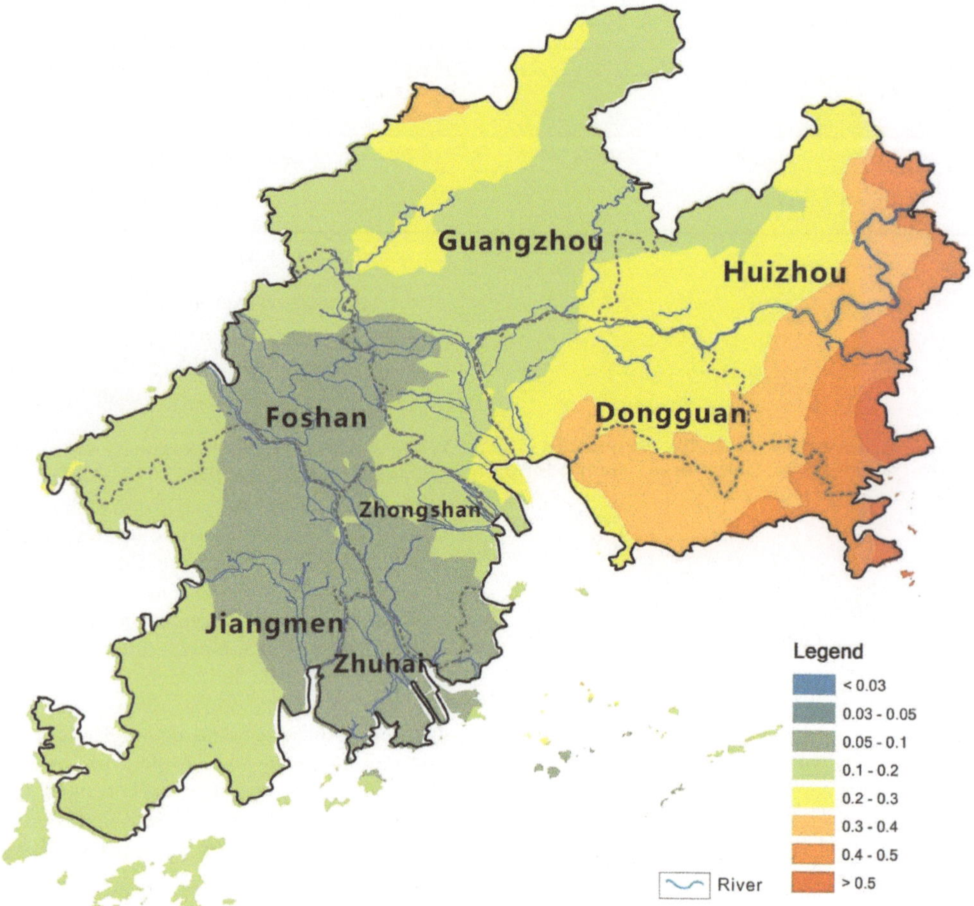

Figure 8. Spatial pattern of soil-to-vegetable transfer factor of Cd in the PRD area.

expected (Figure 7), suggesting the vegetable's ability to absorb Cd would be decreasing with the increasing the Cd content in the vegetables.

Transfer factors of Cd from soil to vegetable

The soil-to-vegetable transfer factors (TFs) reflected the ability of vegetables to take up soil metals. TFs varied significantly with plant species [12,19,21,38,39], and were commonly viewed as a "constant" for a given plant species and a given metal. TFs were used as an important character index for establishing the soil environmental quality criteria or assessing the health risk of soil contamination.

In this study, the TFs of Cd from soil to vegetable for the 12 vegetable species varied from 0.045 in Water spinach to 0.753 in Amaranth (Table 2). The TF mean values in the main five species of leaf vegetables in the study area were 0.532 for leaf mustard, 0.222 for Pakchoi, 0.188 for Chinese lettuce, 0.160 for Chinese flowering cabbage, and 0.148 for Romaine lettuce. Compared with TFs of Cd reported by Wang et al [12], and Hu et al [21], these values were much higher because of the lower Cd concentration in soils. The spatial pattern of TFs also showed that the sample sites with high TFs were always corresponding to the lower Cd concentrations in soils in the study area (Figure 3 and Figure 8).

The regression equations between the TF values and total Cd concentrations in soils can be described by power model ($y = ax^b$) for five kinds of main vegetables:

$$y = 0.0233x^{-0.883}\ (Pakchoi: r^2 = 0.433; p < 0.001)$$

$$y = 0.0392x^{-0.582}\ (Chinese\ flowering\ cabbage: r^2 = 0.354; p < 0.001)$$

$$y = 0.0214x^{-0.915}\ (Leaf\ mustard: r^2 = 0.406; p = 0.008)$$

$$y = 0.0172x^{-1.047}\ (Romiane\ lettuce: r^2 = 0.358; p < 0.001)$$

$$y = 0.0173x^{-1.116}\ (Chinese\ lettuce: r^2 = 0.403; p = 0.011)$$

Where y is the soil-to-vegetable transfer factor of Cd, x is the total Cd concentration in the soils.

For main leaf vegetables, the transfer factors of Cd decreased with increasing total Cd concentrations (Figure 9), suggesting that the ability of vegetable to take up soil Cd decreased with the total soil Cd increasing. Figure 9 also illustrated that when soil Cd concentration was low (about 0.1–0.2 mg kg^{-1}), the TFs decreased sharply; with the gradual increasing of soil total Cd concentration, TFs decreased only slowly, till to a stable value.

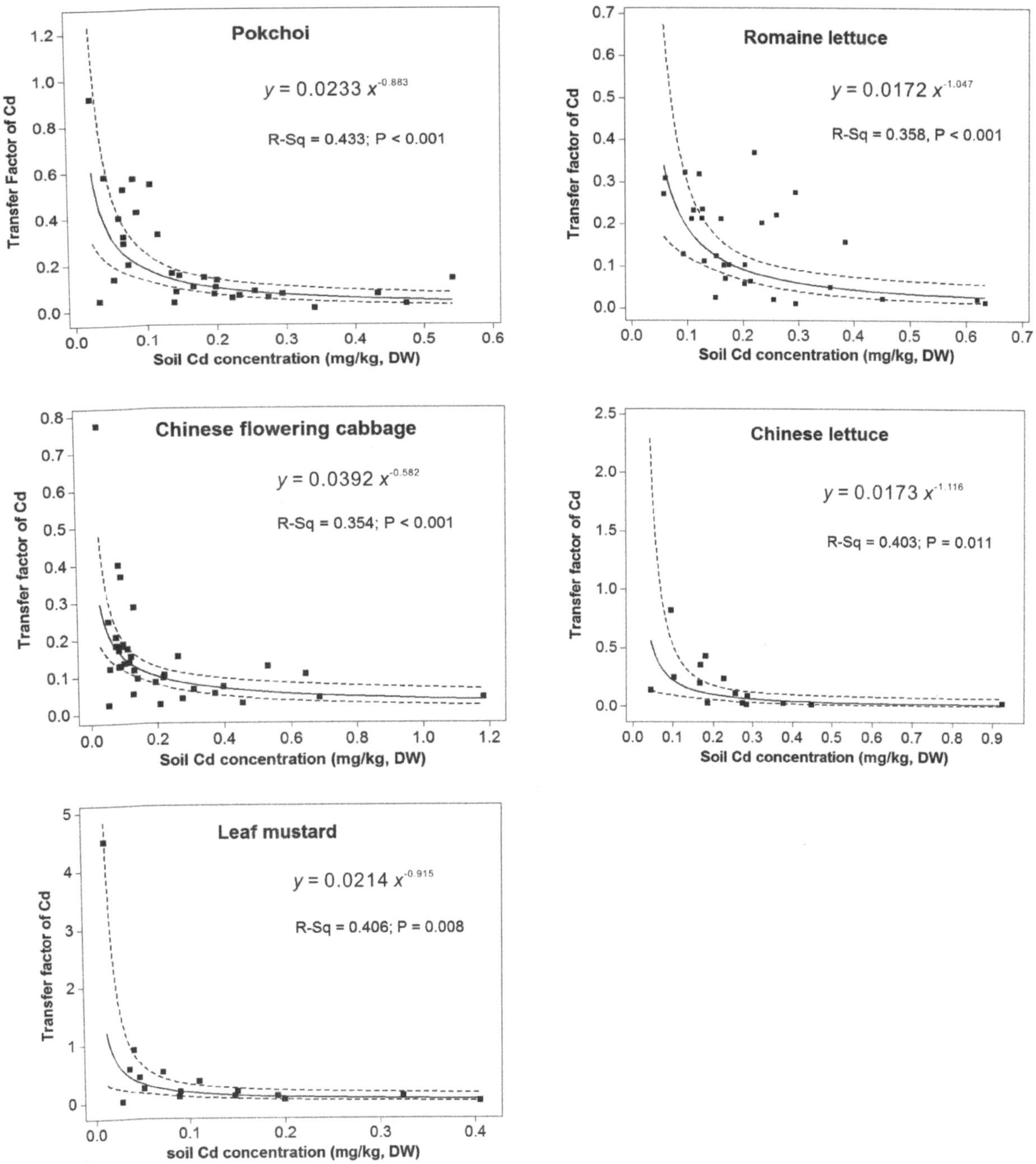

Figure 9. Relationships between soil-to-vegetable transfer factors of Cd and total Cd concentrations in soils for five species of main vegetables (the solid line is the sample regression curve; dotted lines indicated the 95% confidence interval).

Moreover, the 95% confidence interval of regression lines showed that there were a larger error range at the low Cd concentration interval of soils (0–0.1 or 0.2 mg kg^{-1}) for the various leaf vegetable species, and when soil Cd concentrations were greater than 0.1 or 0.2 mg kg^{-1}, the TFs of various vegetables would be reasonable with a lower error range for the assessment of environmental risk (Figure 9). As a result, for carrying out rational environmental health assessment in different Cd concentrations of soils, the present results indicated that the fit line of TFs for vegetable species should be discussed to divide into two parts, including the sharply decrease part and the slowly decrease part, according to the gradual increasing of soil total Cd concentration. This work is agree with the suggestion proposed by Wang et al [12] that the transfer factors of a given crop-metal system estimated from the regression model between the TF values and the corresponding soil metal concentration, are much more reasonable than using the arithmetic means or geometric means at a given soil metal concentration.

Conclusions

The present results showed that the total Cd concentration in soils is safe for planting leaf vegetables in the study area. All vegetable samples met the national standard of safe vegetables, and 88% of which met the standard of no-polluted vegetables. Total Cd concentrations in soils and soil pH played an important role to Cd absorption of vegetables, but there were no obviously linear correlations. There was no spatial correlation between total Cd concentrations in soils and Cd concentrations in vegetables. For a givenl vegetable species the regression model between the TF value and the corresponding total Cd concentrations in soils could be used to estimate the transfer abilities of Cd from soil to vegetable at a given total Cd concentration in soils.

Acknowledgments

We thank four reviewers and editors for their very useful comments that improved the quality of this manuscript.

Author Contributions

Performed the experiments: HZ JC. Analyzed the data: LZ GY. Contributed reagents/materials/analysis tools: HZ DL. Wrote the paper: HZ.

References

1. Singh BR, McLaughlin MJ (1999) Cadmium in Soils and Plants. In: Mclaughlin M., Singh BR, editors. Cadmium in soils and plants. Boston: Kluwer Academic. 357–367.
2. Tudoreanu L, Phillips CJC (2004) Modeling cadmium uptake and accumulation in plants. Adv Agron 84: 121–157. doi:10.1016/S0065-2113(04)84003-3.
3. Singh A, Sharma RK, Agrawal M, Marshall FM (2010) Risk assessment of heavy metal toxicity through contaminated vegetables from waste water irrigated area of Varanasi, India. Trop Ecol 51: 375–387.
4. WHO (1992) Cadmium: Environmental Health Criteria, 134. Geneva.
5. Chlopecka A, Bacon JR, Wilson MJ, Kay J (1996) Forms of Cadmium, Lead, and Zinc in Contaminated Soils from Southwest Poland. J Environ Qual 25: 69. doi:10.2134/jeq1996.00472425002500010009x.
6. Kirkham MB (2006) Cadmium in plants on polluted soils: Effects of soil factors, hyperaccumulation, and amendments. Geoderma 137: 19–32. doi:10.1016/j.geoderma.2006.08.024.
7. Wu L, Tan C, Liu L, Zhu P, Peng C, et al. (2012) Cadmium bioavailability in surface soils receiving long-term applications of inorganic fertilizers and pig manure. Geoderma 173-174: 224–230. doi:10.1016/j.geoderma.2011.12.003.
8. SEPAC (1995) Environmental quality standard for soil (GB 15618–1995). State Environmental Protection Administration of China, Beijing, China
9. Desaules A (2012) Critical evaluation of soil contamination assessment methods for trace metals. Sci Total Environ 426: 120–131. doi:10.1016/j.scitotenv.2012.03.035.
10. Römkens PFAM, Guo HY, Chu CL, Liu TS, Chiang CF, et al. (2009) Prediction of Cadmium uptake by brown rice and derivation of soil-plant transfer models to improve soil protection guidelines. Environ Pollut 157: 2435–2444. doi:10.1016/j.envpol.2009.03.009.
11. Sun W, Sang L, Jiang B (2012) Trace metals in sediments and aquatic plants from the Xiangjiang River, China. J Soils Sediments 12: 1649–1657. doi:10.1007/s11368-012-0596-8.
12. Wang G, Su M, Chen Y, Lin F, Luo D, et al. (2006) Transfer characteristics of cadmium and lead from soil to the edible parts of six vegetable species in southeastern China. Environ Pollut 144: 127–135. doi:10.1016/j.envpol.2005.12.023.
13. Wang Q, Dong Y, Cui Y, Liu X (2001) Instances of soil and crop heavy metal contamination in China. Soil Sediment Contam 10: 497–510.
14. Xiong X, Allinson G, Stagnitti F, Li P, Wang X, et al. (2004) Cadmium contamination of soils of the Shenyang Zhangshi Irrigation Area, China: an historical perspective. Bull Environ Contam Toxicol 73: 270–275. doi:10.1007/s00128-004-0423-z.
15. Williams PN, Lei M, Sun G, Huang Q, Lu Y, et al. (2009) Occurrence and partitioning of cadmium, arsenic and lead in mine impacted paddy rice: Hunan, China. Environ Sci Technol 43: 637–642.
16. Li JT, Qiu JW, Wang XW, Zhong Y, Lan CY, et al. (2006) Cadmium contamination in orchard soils and fruit trees and its potential health risk in Guangzhou, China. Environ Pollut 143: 159–165. doi:10.1016/j.envpol.2005.10.016.
17. Bi X, Ren L, Gong M, He Y, Wang L, et al. (2010) Transfer of cadmium and lead from soil to mangoes in an uncontaminated area, Hainan Island, China. Geoderma 155: 115–120. doi:10.1016/j.geoderma.2009.12.004.
18. Zhou Z., Fan Y., Wang M (2000) Heavy metal contamination in vegetables and their control in China. Food Rev Int 16: 239–255. doi:10.1081/FRI-100100288.
19. Cui Y, Zhu Y, Zhai R, Chen D, Huang Y, et al. (2004) Transfer of metals from soil to vegetables in an area near a smelter in Nanning, China. Environ Int 30: 785–791. doi:10.1016/j.envint.2004.01.003.
20. Li Q, Chen Y, Fu H, Cui Z, Shi L, et al. (2012) Health risk of heavy metals in food crops grown on reclaimed tidal flat soil in the Pearl River Estuary, China. J Hazard Mater 227–228: 148–154. doi:10.1016/j.jhazmat.2012.05.023.
21. Hu J, Wu F, Wu S, Sun X, Lin X, et al. (2013) Phytoavailability and phytovariety codetermine the bioaccumulation risk of heavy metal from soils, focusing on Cd-contaminated vegetable farms around the Pearl River Delta, China. Ecotoxicol Environ Saf 91: 18–24. doi:10.1016/j.ecoenv.2013.01.001.
22. AQSIQ (2001) Safety qualification for agricultural product – Safety requirements for non-environmental pollution vegetable (GB 18406.1–2001). General Administration of Quality Supervision, Inspection and quarantine of China, Beijing, China.
23. MOH (2012) Maximum levels of contaminants in Foods (GB 2762–2012). Ministry of Health of the People's Republic of China.
24. Hart JJ, Welch RM, Norvell WA, Kochian LV (2002) Transport interactions between cadmium and zinc in roots of bread and durum wheat seedlings. Physiol Plant 116: 73–78. doi:10.1034/j.1399-3054.2002.1160109.x.
25. Yang Y, Zhang F-S, Li H-F, Jiang R-F (2009) Accumulation of cadmium in the edible parts of six vegetable species grown in Cd-contaminated soils. J Environ Manage 90: 1117–1122. doi:10.1016/j.jenvman.2008.05.004.
26. McLaughlin M, Parker D, Clarke J (1999) Metals and micronutrients – food safety issues. F Crop Res 60: 143–163. doi:10.1016/S0378-4290(98)00137-3.
27. Wan Z (2013) The development report of Guangdong vegetables industry in 2012. Institute of Agricultural Economics and Rural Development.
28. SBG (2012) Guangdong Statistical Yearbook –2012. Beijing: Statistics Bureau of Guangdong Province, China Statistic Press.
29. Li X, Wai OWH, Li YS, Coles BJ, Ramsey MH, et al. (2000) Heavy metal distribution in sediment profiles of the Pearl River estuary, South China. Appl Geochemistry 15: 567–581. doi:10.1016/S0883-2927(99)00072-4.
30. Wong S, Liu X, Zhang G, Qi S, Min Y (2002) Heavy metals in agricultural soils of the Pearl River Delta, South China. Environ Pollut 119: 33–44. doi:10.1016/S0269-7491(01)00325-6.
31. Tessier A, Campbell PGC, Bisson M (1979) Sequential extraction procedure for the speciation of particulate trace metals. Anal Chem 51: 844–851. doi:10.1021/ac50043a017.
32. Chinese National Standard Agency (1988) Determination of pH value in forest soil. GB 7859–87, UDC 634.0.114:631.422: 171–173.
33. Yu T, Wang Z (1988) Soil Analytical Chemistry. Beijing: Science Press.
34. Chojnacka K, Chojnacki A, Górecka H, Górecki H (2005) Bioavailability of heavy metals from polluted soils to plants. Sci Total Environ 337: 175–182. doi:10.1016/j.scitotenv.2004.06.009.
35. Zhang HH, Chen JJ, Zhu L, Li FB, Wu ZF, et al. (2011) Spatial patterns and variation of soil cadmium in Guangdong Province, China. J Geochemical Explor 109: 86–91. doi:10.1016/j.gexplo.2010.10.014.
36. Zhang C, Wang L (2001) Multi-element geochemistry of sediments from the Pearl River system, China. Appl Geochemistry 16: 1251–1259. doi:10.1016/S0883-2927(01)00007-5.
37. Voutsa D, Grimanis A, Samara C (1996) Trace elements in vegetables grown in an industrial area in relation to soil and air particulate matter. Environ Pollut 94: 325–335. doi:10.1016/S0269-7491(96)00088-7.
38. Peris M, Micó C, Recatalá L, Sánchez R, Sánchez J (2007) Heavy metal contents in horticultural crops of a representative area of the European Mediterranean region. Sci Total Environ 378: 42–48. doi:10.1016/j.scitotenv.2007.01.030.
39. Li Y, Li L, Zhang Q, Yang Y, Wang H, et al. (2013) Influence of temperature on the heavy metals accumulation of five vegetable species in semiarid area of northwest China. Chem Ecol 29: 353–365. doi:10.1080/02757540.2013.769970.
40. McBride M, Murray B (2002) Cadmium uptake by crops estimated from soil total Cd and pH. Soil Sci 167.
41. Hooda PS, McNulty D, Alloway BJ, Aitken MN (1997) Plant Availability of Heavy Metals in Soils Previously Amended with Heavy Applications of Sewage Sludge. J Sci Food Agric 73: 446–454. doi:10.1002/(SICI)1097-0010(199704)73:4<446::AID-JSFA749>3.0.CO;2-2.

16

Controls on the Entrainment of Juvenile Chinook Salmon (*Oncorhynchus tshawytscha*) into Large Water Diversions and Estimates of Population-Level Loss

Steven C. Zeug*, Bradley J. Cavallo

Cramer Fish Sciences, Auburn, California, United States of America

Abstract

Diversion of freshwater can cause significant changes in hydrologic dynamics and this can have negative consequences for fish populations. Additionally, fishes can be directly entrained into diversion infrastructure (e.g. canals, reservoirs, pumps) where they may become lost to the population. However, the effect of diversion losses on fish population dynamics remains unclear. We used 15 years of release and recovery data from coded-wire-tagged juvenile Chinook Salmon (*Oncorhynchus tshawytscha*) to model the physical, hydrological and biological predictors of salvage at two large water diversions in the San Francisco Estuary. Additionally, entrainment rates were combined with estimates of mortality during migration to quantify the proportion of total mortality that could be attributed to diversions. Statistical modeling revealed a strong positive relationship between diversion rate and fish entrainment at both diversions and all release locations. Other significant relationships were specific to the rivers where the fish were released, and the specific diversion facility. Although significant relationships were identified in statistical models, entrainment loss and the mean contribution of entrainment to total migration mortality were low. The greatest entrainment mortality occurred for fish released along routes that passed closest to the diversions and certain runs of Chinook Salmon released in the Sacramento River suffered greater mortality but only at the highest diversion rates observed during the study. These results suggest losses at diversions should be put into a population context in order to best inform effective management of Chinook Salmon populations.

Editor: João Miguel Dias, University of Aveiro, Portugal

Funding: This study was funded by the California Department of Water Resources. The funders had no role in study design, data analysis, decision to publish, or preparation of the manuscript. Staff from the funding agency participated in data collection in coordination with other state and federal resource agencies.

Competing Interests: The authors have declared that no competing interests exist.

* Email: stevez@fishsciences.net

Introduction

Diversion of freshwater for urban, industrial and agricultural use is a common practice around the world and is likely to become more frequent as demand increases [1]. There are numerous changes that take place in aquatic ecosystems as a result of flow reduction that can negatively affect fish including: alteration of sediment budgets, reduction or elimination of floodplain connectivity and altered cues for migration and reproduction [2,3]. Additionally, fish may be lost through direct impingement on intake screens or entrainment into water storage facilities and canals [4]. Although many studies have documented responses of fish populations to altered flow regimes, ecological correlates of the entrainment process and population effects of direct loss at diversions are insufficiently documented and poorly understood [5].

Impingement and entrainment of large numbers of fishes have been reported in water diversions from rivers [6,7], lakes [8] and estuaries [9,10]. Most entrained fish are early life stages (age 0+) and species composition generally reflects habitat adjacent to the diversion [10]. Estimation of population impacts of diversion losses have been more difficult to quantify, although some such studies have been performed [11,12]. Migratory fish species are unique in that their exposure to entrainment is primarily during periods of migration between habitats whereas resident species may be susceptible to entrainment until they leave the diversion vicinity or attain a less susceptible size [13].

Entrainment of juvenile anadromous salmonids (*Oncorhynchus spp.*) into two large water diversions in the San Francisco Estuary, California, USA has frequently been implicated in the decline of these species [14]. A portion of entrained salmon are salvaged and returned to the estuary however, mortality associated with the diversions is thought to impact these populations [15]. Loss densities (fish loss•volume of water diverted^{-1}) at these diversions are currently used as triggers to restrict the volume of water diverted in an effort to protect endangered winter Chinook Salmon (*O. tshawytscha*), spring Chinook Salmon, and steelhead trout (*O. mykiss*). Loss density triggers can be problematic because they are not scaled for population abundance. Thus, triggers may be reached due to abundance fluctuations that do not represent an increase in the proportion of the population lost. In general, the physical and hydrological conditions associated with entrainment remain unclear and population-level effects of fish loss at the diversions are not known.

Our goals for this study were to elucidate these physical, biological and hydrologic conditions and to put entrainment losses in a population context. We assumed that salvage (the metric that can be measured) is proportional to total entrainment at the two diversions. To accomplish these goals we constructed statistical models of salvage and estimated total loss using 15 years of release and recovery data for coded wire tagged salmon raised at hatcheries throughout the Central Valley of California. The use of coded wire tagged fish is important relative to previous work because it allows loss to be scaled by the number of fish released; comparable analyses of raw salvage would be confounded by uncertainty in stock identity and population abundance. The results provide essential information for resource managers charged with recovering salmon stocks and implications for diversion losses in river systems worldwide.

Methods

All data used in this study had previously been collected by state and federal resource agencies. The authors had no role in the handling of organisms.

Study site

The Sacramento and San Joaquin Rivers drain approximately 40% of California' surface area including most of the western slope of the Sierra Nevada Mountains, the eastern slope of the Coast Range and portions of the southern Cascades. The two rivers converge in a tidal freshwater estuary known as the Sacramento-San Joaquin Delta (hereafter referred to as the "Delta") before entering San Francisco Bay (Figure 1). Both rivers have been subjected to intense water development beginning in the late 19th century associated with urban and agricultural development in the Central Valley of California. Dam construction, channelization, levee construction and pollution have been prominent in both systems. Water diverted from these rivers provides water for millions of Californians and supports economically valuable agriculture throughout the Central Valley. Both rivers supported robust populations of Chinook Salmon in the past. However, 48% of historic habitat has been lost [16] and drastic reductions in the number of returning adults have triggered restrictions and even total closures of commercial and recreational fisheries in some years.

Freshwater is extracted in the tidal Delta at two large diversions that divert up to 60% of total flow in some years. Both diversions contain facilities where fish are salvaged and then released in the western Delta, away from the pumps. Fish entering the salvage facilities are subsampled at regular intervals (10–20 minutes•h^{-1}) and total salvage is estimated based on the volume diverted and time since the previous sub-sample. Although salvage occurs at both diversions, there are significant differences in facility design that may affect the number of fish collected. The Central Valley Project (CVP) diverts water directly from a tidal channel in the Delta and fish are directed by a series of louvers into the salvage facility (Figure 2). The State Water Project (SWP) diverts water from a forebay filled by operable gates located on a tidal channel of the Delta (Figure 2). Thus, fish salvaged at the SWP have first been drawn into the forebay where they are exposed to resident predators before they are directed by louvers into the salvage facility as water is pumped out of the forebay. Additionally, the origin of salmon collected at the diversions is likely to have an influence on salvage. Fish released in the San Joaquin River are likely to first encounter the CVP whereas fish released in the Sacramento are likely to encounter the SWP first.

Salmon releases

Chinook Salmon are raised at several hatcheries in the Sacramento-San Joaquin system and released at various locations as mitigation for habitat lost through dam construction, and as part of studies conducted by state and federal resource agencies. A portion of these hatchery fish have coded wire tags (CWTs) inserted for identification when recaptured. These tags are short lengths of steel wire with a numeric code that identifies a specific release group. Fish receiving a CWT also have their adipose fin clipped so tagged fish can be visually identified at capture. The tagging rate and number of fish released can vary considerably among runs. All hatchery winter run and late-fall run are tagged whereas the percentage of fall run tagged and released has varied between years. Spring run are raised at one hatchery; however, only 16 spring run release groups were identified within the study area and few of these fish ever arrived at salvage. Thus, spring run were not included in the analysis. Fish released into tributaries of the Sacramento River including: Battle Creek and Feather River are hereafter referred to as Sacramento River releases. Similarly, fish released into tributaries of the San Joaquin River including the Stanislaus River, Tuolumne River, and Merced River are hereafter referred to as San Joaquin River releases.

Juvenile salmon with an adipose clip collected at the diversions are retained, the coded wire tag is read, and the number of fish salvaged from that release group is estimated. Juvenile salmon exiting the Delta downstream of the diversions are sampled by a mid-water trawl at Chipps Island operated by the United States Fish and Wildlife Service (Figure 1). Trawling effort is variable among and within years and capture probability is low; however, some trawling occurs during all months of the year. Tagged salmon also are recovered from the commercial and recreational ocean fishery for several years after release.

Release data for juvenile salmon were obtained from the Regional Mark Processing Center coded wire tag database maintained by the Pacific States Marine Fisheries Commission (http://www.rmpc.org/). Data from release years 1993–2007 were queried from the database. We chose these years to represent current water management in the Delta which changed in the mid-1990's in response to the Bay-Delta Accord (California State Water Resources Control Board Ruling D-1641). Additionally, we excluded releases under 1000 individuals and releases made downstream of the last entrance to the interior Delta from the Sacramento River at Threemile Slough (Figure 1). The data queried included: release site, release size, date of release, mean fork length at release and age specific recoveries in the ocean. The number of salmon recovered in the ocean was expanded prior to analysis using the method described in Zeug and Cavallo [17]. Ocean recovery information was limited for later release years because the ocean fishery was restricted in 2007 and closed in 2008 in response to the collapse of the fall run. Recovery information was obtained from the United States Fish and Wildlife Service Chipps Island Survival table (http://www.fws.gov/stockton/jfmp/datamanagement.asp). These data included: number of tagged salmon recovered in the Chipps Island trawl, the expanded number of tagged salmon collected at the CVP and SWP salvage facilities, and the range of dates over which fish from each release group were captured in the trawl.

Environmental data

Juvenile salmon are released in the Sacramento and San Joaquin rivers and tributaries (from <30 to >600 km from the diversions); however, they are not vulnerable to entrainment until they enter the tidal Delta. A study of salmon migration with acoustic telemetry indicated juvenile salmon migrated through the Delta in 6.4 days on average [18]. To capture the conditions experienced during Delta migration, hydrologic variables were averaged over 7 days after salmon entered the Delta. To estimate the date when each release group arrived at the Delta, we calculated the median date between the first and last capture in the

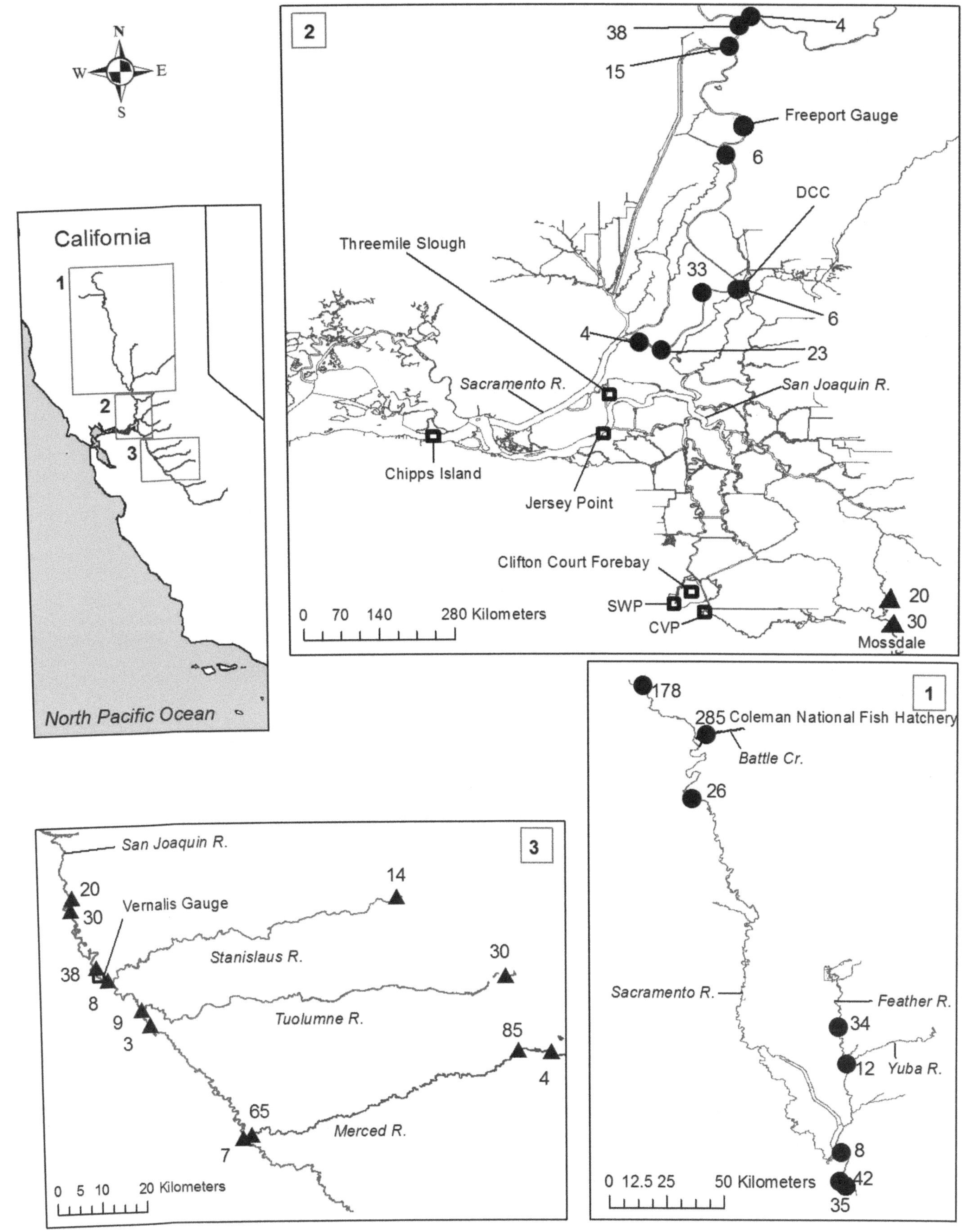
N
W
E
S
California
1
2
3
North Pacific Ocean
2
38
4
15
Freeport Gauge
6
DCC
Threemile Slough
33
6
4
23
Sacramento R.
San Joaquin R.
Chipps Island
Jersey Point
Clifton Court Forebay
SWP
CVP
20
30
Mossdale
0 70 140 280 Kilometers
1
178
285 Coleman National Fish Hatchery
Battle Cr.
26
Sacramento R.
Feather R.
34
12
Yuba R.
8
42
35
0 12.5 25 50 Kilometers
3
San Joaquin R.
20
30
Vernalis Gauge
14
Stanislaus R.
30
38
8
9
3
Tuolumne R.
85
4
65
Merced R.
7
0 5 10 20 Kilometers

Figure 1. Map depicting the location of the study region within California and relevant locations within the study region. Release locations in the Sacramento River are indicated by closed circles and release locations in the San Joaquin River are indicated by closed triangles. The number of releases that occurred at each that location appears next to the marker. Abbreviations: SWP = State Water Project, CVP = Central Valley Project.

Chipps Island trawl at the exit of the Delta. The 7 days prior to the median capture date was the time period over which hydrologic conditions were averaged.

Mean daily flow (hereafter "flow") for the Sacramento River was obtained from the United States Geological Survey (USGS) gauge 11447650 at Freeport, California (Figure 1). San Joaquin River flow was obtained from USGS gauge 11303500 at Vernalis, California (Figure 1). Daily water diversion rates from the CVP and SWP were obtained from the DAYFLOW online data archive maintained by the California Department of Water Resources. An additional variable in the Sacramento River was the position of the Delta Cross Channel (DCC). The DCC is a large gate that diverts water from the main stem Sacramento River into interior portions of the Delta (Figure 1). When the gate is open, there is a greater probability of fish migrating down the Sacramento River will enter routes leading to the diversions [19].

Data analysis

The response variable in all statistical models was the number of fish salvaged. The number of fish released was included as an offset variable to account for differences in release group size. Models were constructed separately for each diversion facility to determine if different independent variables affected salvage at diversions that extract water directly from a tidal channel (CVP) vs. a forebay (SWP). Models were also constructed separately for releases in the Sacramento and San Joaquin Rivers.

Independent variables in statistical models were selected based on hypothesized relationships with salvage. These variables could potentially affect the process of salvage or the exposure of fish to salvage. For example, zero salvage could occur because most fish were not exposed to entrainment or died prior to entering the Delta. Hypothesized relationships between independent variables and salvage are summarized in Table 1. To account for mortality prior to salvage, fork length at release and the shortest distance from release site to the nearest salvage facility were included. We expected survival would be negatively associated with distance [20] and positively associated with mean fork length [21]. For fish in the Delta we hypothesized that salvage would increase as flows decreased and as diversion increased. Previous analyses of fish entrainment have utilized a ratio of diversion to flow as a predictor of entrainment risk instead of using these variables as separate independent predictors. However, analyses of survival in the Delta have suggested diversion rate, and flow alone may have similar predictive capability without conflating these two variables [22]. To determine if a diversion-to-flow ratio was superior to modeling these effects separately, statistical models were constructed using

Figure 2. Aerial view of the layout of the two water diversion and fish salvage facilities. The State Water Project (SWP) diverts from Clifton Court Forebay that is filled from radial gates located on Old River; a distributary of the San Joaquin River. The Central Valley Project (CVP) diverts water directly from Old River. Image was downloaded from The National Map: http://nationalmap.gov/.

Table 1. Predicted relationships between independent variables and salvage (count model) and independent variables and zero salvage (zero-inflation model).

Parameter	Count model	Zero-inflation model
Flow	−	+
Water diversion	+	−
DCC position	+	−
Fork length	−	+
Distance from salvage facilities	−	+
Chipps Island recoveries	−	+
Ocean recoveries	−	+

both methods and then compared using Akaike's corrected information criteria (AIC_c). When the difference between AIC_c values (ΔAIC_c) for a pair of models was greater than 2.0, the model with the lower AIC_c values was considered to have the best support in the data.

To account for fish that survived the Delta and avoided salvage, catch-per-unit effort in the Chipps Island trawl (number•min^{-1}), and expanded ocean recoveries for each release were included in each model. We predicted that salvage would be negatively associated with recoveries at Chipps and in the ocean (i.e. when fewer fish are entrained at the diversion, more are available to be caught later in the trawl). Sacramento models also included a dummy variable for the position of the DCC where 1 = open and 0 = closed. All continuous variables were transformed into z-scores so results could be interpreted in units of standard deviations. A correlation analysis was performed to determine if multicollinearity existed among independent variables however, no strong correlations were identified.

Screening of the response variable indicated that many releases in both rivers resulted in zero salvage. Thus, zero-inflated negative binomial regression was employed. These models are composed of two parts; a count model that explains salvage as a function of covariates and a zero-inflation model that accounts for the processes that result in zero salvage as a function of covariates. The predicted sign of coefficients in the count model are listed in Table 1. Coefficients for the zero-inflation model would be predicted to have a sign that is opposite of the count model. Zero-inflated Poisson regression was explored but model diagnostics indicated over-dispersion. To determine if a zero-inflated model was necessary, a negative binomial regression model was constructed with the same independent variables and a Vuong non-nested hypothesis test was performed to determine if the zero-inflated model provided an improved representation of the data [23]. Once a model was identified, overall model fit was determined with a likelihood ratio test comparing an intercept-only model with the model containing independent variables. All modeling was performed with the R statistical program and the packages "pscl" and "MASS" [24].

To estimate a population-level effect of fish loss at the diversions, the contribution of loss relative to the total mortality rate during migration (hereafter referred to as relative loss) was estimated [11,12]. Loss is defined as the fish that were entrained into the diversion and did not survive to release after salvage and trucking. To estimate loss for each release group, we first estimated the number of fish that encountered the louvers at each facility:

$$F_L = \frac{S}{0.90}$$

where F_L is the number of fish that encountered the louvers, S is estimated salvage and louver efficiency is assumed to be 90% [15]. Total entrainment was then estimated as:

$$E = \frac{F_L}{S_{PL}}$$

where E is total entrainment and S_{PL} is the pre-louver survival. The pre-louver survival at the SWP was assumed to be 15%; the mean rate reported in a study by [25]. No data on pre-screen survival is available for the CVP so we assumed 85% following the methods of [15]. Total loss at each facility was then estimated as:

$$L = E - (S \times 0.96)$$

where L is total loss and survival during trucking and handling was 96% [15]. Loss estimates were summed for each facility and divided by the release group size to estimate the proportion of fish from each release group lost at the diversion.

To bracket the range of relative loss at the diversions, the highest and lowest observed mortality values during migration were used. Because published mortality estimates were not available for all release locations, only releases (n = 285) from Coleman National Fish Hatchery (CNFH) and directly into the tidal Delta (Sacramento River = 129, San Joaquin River = 88) were used. Separate estimates were calculated for each run. Though winter run were released upstream of the CNFH, we assumed that migration mortality of this run was similar to fish released directly from CNFH. Mortality estimates of Sacramento River releases during migration through the Delta were obtained from acoustic tagging experiments [19]. The highest through-Delta (Freeport to Chipps Island) mortality estimate from this study was 64.9% and the lowest was 45.7%. A single mean mortality estimate during migration from CNFH to Chipps Island (88%) was obtained from Michel [18] and San Joaquin estimates were obtained from Newman [26] and Buchanan et al. [27] where the highest through-Delta (Mossdale to Chipps Island) mortality estimate was 95.0% and the lowest was 79% (Mossdale to Jersey Point). The relative loss for each release was estimated as:

$$M_L = \left(\frac{R_L}{M_T}\right) \times 100$$

where M_L is relative loss at the diversions, R_L is the proportion of each release group lost at the diversions and M_T is the total mortality during migration.

To quantify how uncertainty in the estimates of louver efficiency and pre-louver survival at both diversions influenced relative loss, a sensitivity analysis for the estimates of M_L was performed using Monte Carlo methods. A distribution was constructed for each of the three estimates and they were allowed to vary one at a time while the other two were held constant. One hundred resamples were drawn, M_L calculated, and the mean and standard deviation of the 100 resamples was estimated. The mean and standard deviation for pre-louver mortality at SWP [25] was used to inform a normal distribution. No data are available to inform a normal distribution for CVP pre-louver mortality so a uniform distribution was used where mortality ranged from 5–70%. A uniform distribution was also used for louver efficiency where values ranged from 50–95%.

There is considerable interest by resource managers in understanding how losses of salmon are related to diversion rate thus, the M_L value for each release group was plotted against water diversion rate for each run of Chinook Salmon released from CNFH, and directly into the tidal Delta. These M_L estimates were calculated with the lowest M_T estimates to provide an estimate of the maximum mortality that could be accounted for by loss at the diversions. Additionally, relative loss was calculated assuming that no entrained fish were salvaged to estimate the effect of salvage facilities on loss estimates.

Results

Salvage of Sacramento River releases

A total of 749 releases comprised of $>28{\bullet}10^6$ fish were used to model salvage of Sacramento River Chinook Salmon. Fall run accounted for 419 releases, winter run 178 releases and late-fall run 152 releases. Only 16 release groups for tagged spring run Chinook Salmon were available and very few of these fish ever arrived at salvage; spring run Chinook Salmon were not included in further analyses. Across all Sacramento River releases an estimated 19281 CWT salmon were salvaged which represented 0.068% of the tagged fish released. Among the three runs of Chinook Salmon released, late-fall run fish were salvaged more frequently (0.2%) than winter and fall run (0.05 and 0.01% respectively). Average total loss (expanded for louver efficiency and pre-louver mortality) was greatest for late-fall run releases (0.84%) and lowest for fall run (0.03%) with an intermediate value for winter run (0.2%, Table 2).

A zero-inflated negative binomial model was a superior fit to the CVP salvage data for Sacramento River releases ($V = 8.11$, $P < 0.001$), and the model fit the data well (likelihood ratio test, $P < 0.001$). Similarly, salvage of Sacramento River releases at the SWP also was best described by a zero-inflated model ($V = 7.66$, $P < 0.001$) and it was a good fit to the data (likelihood ratio test, $P < 0.001$). The models that included flow and diversion as separate variables were a better fit to the CVP and SWP data than models using a ratio of diversion to flow with ΔAIC_c values of 23.4 and 76.9 respectively. The count models at both diversions revealed that there was a significant increase in salvage as diversion rate increased (Table 3). Contrary to expectations, salvage increased at both diversions when the DCC was closed. The DCC was only open for 48 of the 749 releases (6%) and given the large number of zeros in the data set, there was a lower probability of a large salvage event occurring when the DCC was open. Other significant relationships were specific to each facility. There was a significant negative relationship between flow and salvage, and a positive relationship between distance and salvage at the CVP facility. Fork length and Chipps Island recoveries had significant positive relationships with salvage at the SWP. There was also a significant negative relationship between ocean recoveries and salvage at the SWP (Table 3).

The zero-inflation part of the analysis produced coefficients to estimate when salvage is zero versus any non-zero number. The zero-inflation models for Sacramento releases revealed consistent patterns between SWP and CVP. Specifically, there was a significantly greater likelihood of zero salvage when flows were higher, when water diversion was lower and when fish were released at a smaller size (Table 3). There were no significant relationships with DCC position in this portion of the model.

Salvage of San Joaquin River releases

In the San Joaquin River there were 313 releases comprised of $>7{\bullet}10^6$ juvenile Chinook Salmon (Table 2). Only fall run were released in the San Joaquin River. A greater percentage of salmon released in the San Joaquin Basin were salvaged (0.6%) relative to any run of Sacramento River-origin fish. Mean total loss was also greater for releases in the San Joaquin River (1.4%, Table 2) relative to any run released in the Sacramento River. Similar to the Sacramento River releases, models that used diversion and flow as separate predictors were superior to models that used the diversion-flow ratio for the CVP and SWP ($\Delta AIC_c = 57.5$ and 82.5 respectively).

A Vuong test indicated that a zero-inflated negative binomial model was the best description of San Joaquin releases salvaged at the CVP facility ($V = 7.72$, $P<0.001$). This model was a good fit to the data (likelihood ratio test, $P<0.001$). Additionally, a zero-inflated negative binomial model best represented the SWP salvage data ($V = 6.22$, $P<0.001$) and fit the data well (likelihood ratio test, $P<0.001$). The count models at both facilities yielded a significant increase in salvage as diversion rate increased (Table 4). The only other significant relationship in the count models at either facility was a positive coefficient for ocean recoveries in the SWP model.

The zero-inflation models for both facilities yielded significant negative relationships between the probability of zero salvage and diversion rate and ocean recoveries (Table 4). At the CVP, there was also a significant negative relationship between zero salvage and flow, and a significant positive relationship with recoveries at Chipps Island. For the SWP, fork length was found to have a significant positive relationship with zero salvage.

Contribution to total migration mortality

Relative loss at the diversions was low for Sacramento River fish released at CNFH and directly into the Delta (Table 5). For CNFH releases, relative loss was greater at the SWP facility relative to the CVP facility although both values were <0.4%. A similar pattern was observed for Sacramento River fish released directly into the tidal Delta regardless of the migration mortality estimate. However, relative loss at the CVP was similar for fish released at CNFH and in the Delta whereas relative loss at the SWP was greater for fish released in the Delta. Mean relative loss of San Joaquin River releases was more than double that of Sacramento River releases at both facilities (Table 5). The pattern among the facilities was similar where relative loss was greater at the SWP relative to the CVP.

The sensitivity analysis indicated that incorporating uncertainty in louver efficiency resulted in higher estimates of relative loss. Uncertainty in pre-louver survival at the CVP resulted in lower estimates relative to the baseline and uncertainty in pre-louver survival at the SWP produced similar estimates. The largest

Table 2. Means and coefficients of variation for variables used in models of salvage.

	Sacramento River			San Joaquin River
Parameter	**Late-fall**	**Winter**	**Fall**	**Fall**
Release size	58,365 (0.38)	6,354 (1.02)	43,936 (2.17)	24,917 (0.21)
Total salvage•release^{-1}	118.5 (1.80)	2.7 (2.12)	1.9 (2.89)	149.6 (1.66)
Proportion salvaged	0.002 (1.64)	0.0005 (2.42)	0.0001 (2.96)	0.0058 (1.64)
Proportional loss	0.008 (1.65)	0.002 (2.57)	0.0003 (3.23)	0.014 (2.05)
Distance from salvage (km)	452 (0.47)	623 (0)	348 (0.61)	152 (0.56)
Flow m^3•s^{-1}	941 (0.83)	782 (0.55)	919 (0.69)	255 (0.90)
Water diversion m^3•s^{-1}	213 (0.50)	226 (0.30)	121 (0.74)	74 (0.56)
Salvage at CVP	44.4 (1.91)	1.1 (3.54)	0.5 (5.29)	102.5 (1.66)
Water diversion from CVP m^3•s^{-1}	97 (0.45)	105 (0.19)	54 (0.68)	40 (0.57)
Salvage at SWP	74.1 (1.84)	1.6 (2.34)	1.5 (3.26)	48.2 (2.14)
Water diversion from SWP m^3•s^{-1}	115 (0.59)	116 (0.31)	60 (0.93)	33 (0.75)
Length at release (mm)	128 (0.09)	78 (0.09)	67 (0.19)	82 (0.06)
Expanded ocean recoveries	593 (0.94)	16 (1.74)	365 (1.46)	85 (1.62)
Chipps trawl cpue	9.64 (0.98)	0.49 (1.40)	3.20 (1.57)	7.42 (1.38)

Variables were separated by run for Sacramento releases. Currently the San Joaquin only supports fall run Chinook Salmon.

difference resulting from incorporation of parameter uncertainty was for San Joaquin River-origin fall run where mean estimates incorporating uncertainty in louver efficiency were 2.9% relative to the baseline value of 1.7%. All other differences were <1% (Figure 3).

For Sacramento River fall run Chinook Salmon, combined loss at the diversions (CVP + SWP) was always less than 1% of total migration mortality (relative loss) regardless of the diversion rate or release location (Figure 4). A small percentage of relative loss was observed for late-fall run released from CNFH until the diversion rate exceeded approximately 275 m^3•s^{-1}. Once this level of water diversion was reached, relative loss increased, although the variation also increased (Figure 4). Most late-fall Chinook Salmon released into the Delta experienced relative losses less than 2.5%. However, nine releases had relative losses that ranged between 3.0% and 10.5%. Seven of these releases occurred within days of each other in 2007 when the diversion rate was approximately 187 m^3•s^{-1}. Relative losses of winter run releases were variable throughout the range of observed diversion levels but were less than 2% for most releases and never exceeded 5.5% (Figure 4). Fall run Chinook Salmon released into the San Joaquin River experienced a greater relative loss at the diversions than any run released in the Sacramento River (Figure 4). Water diversion was less than 100 m^3•s^{-1} during most San Joaquin River releases and although relative loss was less than 5% for most releases; this value ranged as high as 17.5%. Three releases occurred when the diversion rate was greater than 100 m^3•s^{-1} and relative loss was less than 1% of total mortality for all three (Figure 4).

Salvage reduced relative loss by 19% for late-fall and winter run Chinook Salmon and 15% for fall run released at CNFH (Table 6). Salmon released in the Sacramento River received the greatest benefit from salvage with a reduction in relative loss of 42% and 41% for fall run and late-fall run respectively. Relative loss of fall run Chinook Salmon released in the San Joaquin River was reduced by 24% due to the presence of salvage facilities.

Discussion

During the study period, over 1000 releases of >35•10^6 juvenile Chinook Salmon were performed. For releases in both rivers and among both diversions, there was a significant positive relationship between water diversion rate and salvage. The salvage facilities at these diversions have been likened to giant sampling devices [13]. Thus, it is not surprising that more fish are encountered as more water is sampled per-unit-time. Kimmerer [15] also found strong effects of water diversion on entrainment of salmon in this system and positive relationships between diversion volume and fish entrainment have been reported in other systems [7–9]. In contrast, the relationship between salvage and flow could not be generalized among rivers. For Sacramento River releases, there was a significant increase in salvage at the CVP facilities with decreasing flow. Supporting the same trend, greater flows significantly increased the probability of zero salvage of Sacramento River releases at both facilities. The lack of strong relationships between salvage and flow and the consistent strong relationships with salvage and diversion rate likely explain why using a ratio of diversion rate to flow where these two variables are conflated was a poor predictor relative to modeling these variables separately.

Perry [28] found that when discharge is low in the Sacramento River, flow changes direction with the tide at the junction of routes leading to the diversions and upstream flow increases the probability of salmon entering these junctions. Several releases of late-fall run in the Delta that were released within days of each other experienced unusually high rates of salvage. The timing of fish arrival at junctions leading to the diversions and tides were unmeasured here but may be important predictors of salvage and may have influenced these anomalous points in the late-fall Delta releases. For San Joaquin River releases, relationships with flow were less clear and the only significant relationship was between zero salvage at the CVP and flow. Salvage of San Joaquin River releases may only occur when fish are abundant near the diversion regardless of flow conditions. Other studies of fish entrainment at

Table 3. Parameter estimates for zero-inflated negative binomial regression describing salvage of coded wire tagged juvenile salmon at the Central Valley Project and State Water Project facilities.

	Central Valley Project				State Water Project			
	Count model		Zero-inflation model		Count model		Zero-inflation model	
Parameter	estimate	*p*-value	estimate	*p*-value	estimate	*p*-value	estimate	*p*-value
Flow $m^3 \cdot s^{-1}$	**−0.651**	**<0.001**	**0.427**	**0.026**	**−0.208**	**0.227**	**0.521**	**<0.001**
Diversion $m^3 \cdot s^{-1}$	**0.663**	**<0.001**	**−0.443**	**0.004**	**0.29**	**0.005**	**−1.296**	**<0.001**
DCC open	**−0.800**	**0.001**	−0.134	0.751	**−0.917**	**<0.001**	−0.293	0.634
Fork length (mm)	0.008	0.944	**−1.497**	**<0.001**	**0.614**	**<0.001**	**−0.977**	**<0.001**
Distance from salvage facilities (km)	**0.331**	**0.001**	−0.246	0.116	0.149	0.085	0.144	0.270
Chipps Island recoveries ($number \cdot min^{-1}$)	0.194	0.067	−0.014	0.928	**0.345**	**0.011**	0.206	0.243
Ocean recoveries	0.020	0.827	−0.186	0.190	**−0.297**	**0.032**	−0.187	0.233

The count model describes the salvage process whereas the zero-inflated model describes the process resulting in zero salvage. All releases were in the Sacramento River.

Table 4. Parameter estimates for zero-inflated negative binomial regression describing salvage of coded wire tagged juvenile salmon at the Central Valley Project and State Water project facilities.

	Central Valley Project				State Water Project			
	Count model		Zero-inflation model		Count model		Zero-inflation model	
Parameter	estimate	*p*-value	estimate	*p*-value	estimate	*p*-value	estimate	*p*-value
Flow $m^3 \cdot s^{-1}$	0.098	0.247	−0.781	0.002	−0.019	0.874	−0.017	0.919
Diversion $m^3 \cdot s^{-1}$	0.453	<0.001	−0.791	<0.001	1.000	<0.001	−0.517	0.003
Fork length (mm)	0.030	0.692	−0.127	0.392	0.104	0.285	0.468	0.006
Distance from salvage facilities (km)	−0.060	0.505	0.091	0.567	−0.158	0.075	0.019	0.896
Chipps Island recoveries ($number \cdot min^{-1}$)	0.161	0.153	0.651	0.042	−0.295	0.054	0.171	0.503
Ocean recoveries	0.084	0.361	−1.996	<0.001	0.666	<0.001	−1.252	<0.001

The count model describes the salvage process whereas the zero-inflated model describes the process resulting in zero salvage. All releases were in the San Joaquin River.

Table 5. Estimates of the % of total migration mortality accounted for by loss at each diversion (relative loss) for releases in the Sacramento and San Joaquin Rivers.

River	Release location	Facility	Migration mortality estimate (%)	Relative loss (%)	Confidence interval
Sacramento	CNFH	CVP	88.0	0.013	0.008–0.017
Sacramento	CNFH	SWP	88.0	0.372	0.328–0.416
Sacramento	Delta	CVP	64.9	0.009	0.004–0.013
Sacramento	Delta	CVP	45.7	0.012	0.006–0.018
Sacramento	Delta	SWP	64.9	0.449	0.237–0.661
Sacramento	Delta	SWP	45.7	0.614	0.324–0.905
San Joaquin	Delta	CVP	95.0	0.091	0.050–0.131
San Joaquin	Delta	CVP	79.0	0.109	0.060–0.157
San Joaquin	Delta	SWP	95.0	1.334	0.739–1.930
San Joaquin	Delta	SWP	79.0	1.596	0.884–2.309

Estimates were generated for Sacramento River fish released at Coleman National Fish Hatchery (CNFH) and directly into the tidal Delta. San Joaquin River estimates were only made for fish released into the tidal Delta.

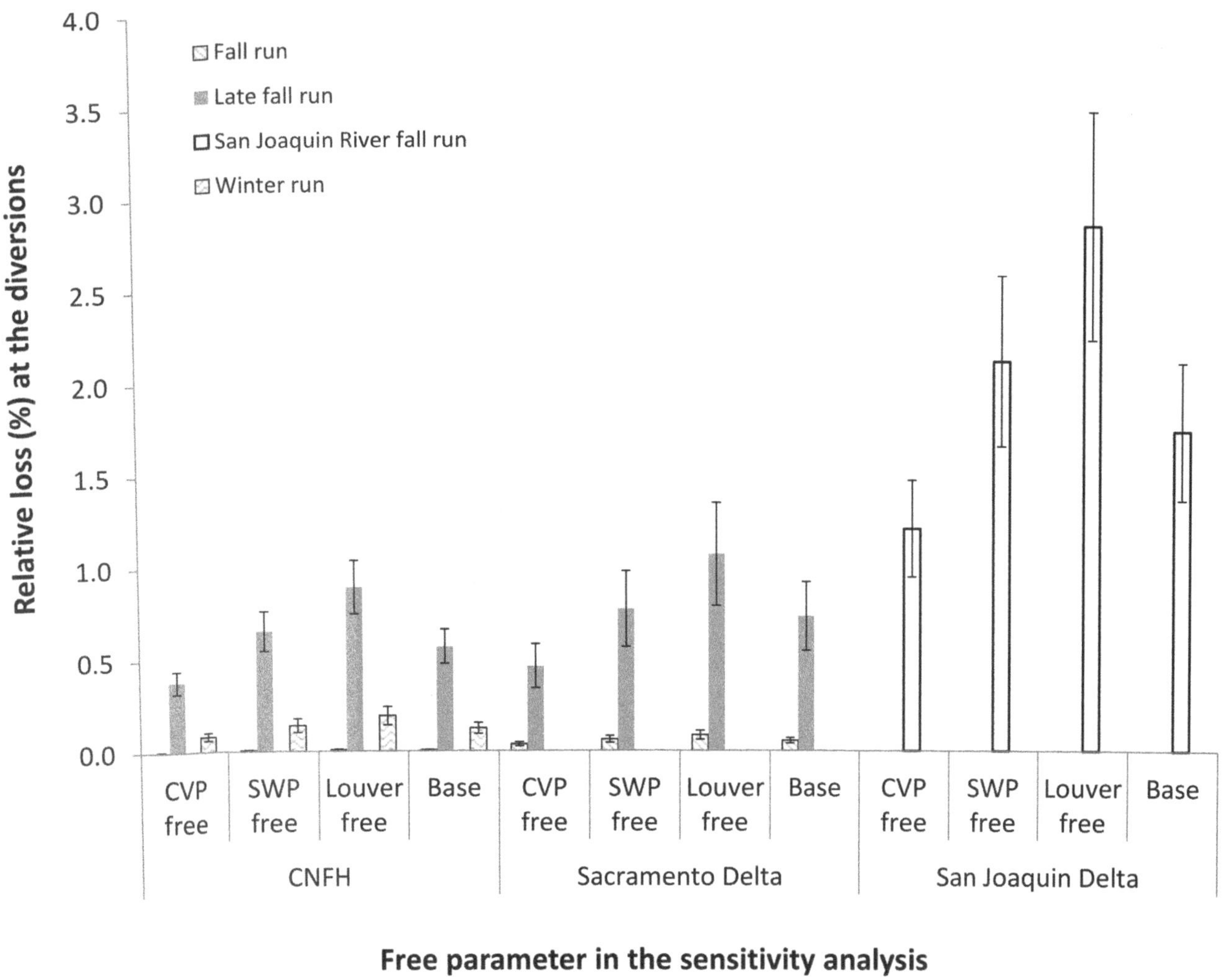

Figure 3. Results of the sensitivity analysis for the calculation of relative loss at the diversions as a function of uncertainty in the parameters used for the calculation. Three parameters (CVP pre-louver mortality, SWP pre-louver mortality and louver efficiency) were modeled as distributions and one parameter at a time was allowed to vary while the others were held constant. 100 re-samples were performed and the means and standard deviations are reported here.

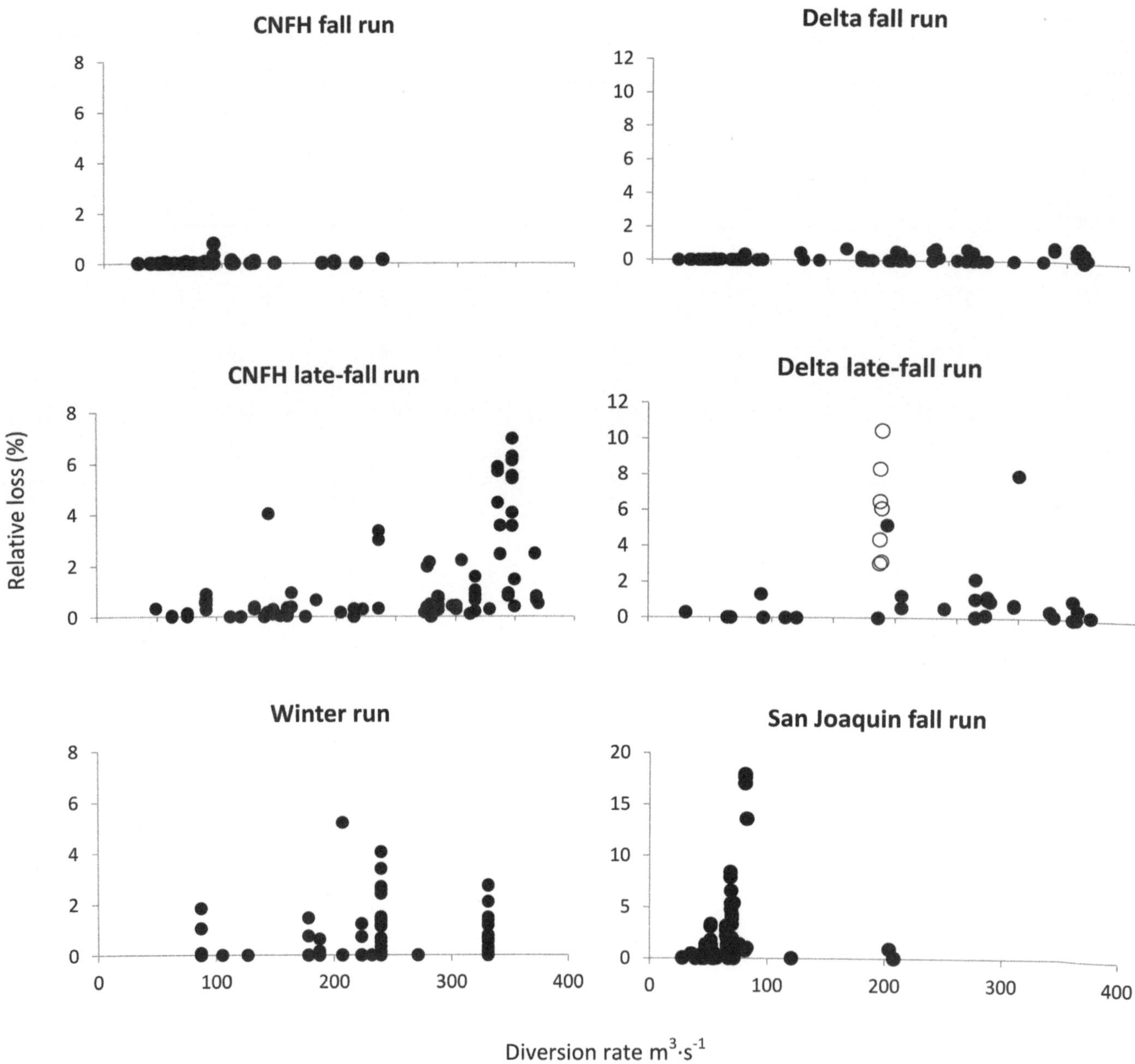

Figure 4. Plot of the percentage of migration mortality accounted for by loss at the two diversions (relative loss) as a function of diversion rate for three runs of Chinook Salmon released from the Coleman National Fish Hatchery (CNFH) or directly into the Delta. Open circles in the Delta late-fall run plot represent a set of releases that occurred within days of each other in 2007 and experienced unusually high loss. Note that the range of the y-axis changes among release locations.

Table 6. Percent of migration mortality accounted for by loss at the diversions (relative loss) with and without accounting for salvage.

		Coleman Releases	Sacramento Delta Releases	San Joaquin Delta Releases
Fall	Salvage	0.017 (0.007)	0.076 (0.024)	1.704 (0.373)
	No salvage	0.020 (0.008)	0.132 (0.029)	2.242 (0.475)
Late-fall	Salvage	0.953 (0.153)	1.339 (0.430)	N/A
	No salvage	1.178 (0.189)	2.279 (0.535)	N/A
Winter	Salvage	0.222 (0.043)	N/A	N/A
	No salvage	0.273 (0.051)	N/A	N/A

water diversions have found that catch is largely proportional to abundance in the channel being diverted from [6,11].

Prior to constructing salvage models, we hypothesized that fish size at release and the distance of release sites from the diversions would influence how many fish would be susceptible to salvage through the effect of these variables on survival [20,21]. However, both fish size and distance from the diversions were only strong predictors of zero salvage for releases in the Sacramento River. Zero salvage was more likely when fish were released at a smaller size and salvage of larger fish was greater at the SWP. Most potential predators of salmon smolts are gape-limited fishes such as striped bass (*Morone saxatilis*) and larger size may confer a survival advantage especially at the SWP where fish are exposed to high predation rates in the forebay prior to salvage [25]. Size effects were mostly insignificant for San Joaquin releases (although zero salvage was positively related to size at the SWP); however, fish were released at a wider range of sizes in the Sacramento River relative to the San Joaquin River. In particular, late-fall run Chinook Salmon were salvaged more frequently than any other run from the Sacramento River and this run was released at larger sizes than any other run.

The distance of release sites from the diversions also was important only for Sacramento River releases however the relationship was opposite of our expectation (positive coefficient). Large, late-fall run Chinook Salmon were salvaged more than any other run and these fish were primarily released at CNFH that was located 577 rkm from the closest salvage facility. There were no significant relationships between distance and salvage of San Joaquin River releases. The maximum distance of a release site in the San Joaquin was 262 rkm, which was less than half of the maximum distance of Sacramento releases (624 rkm) and may have masked distance effects in the San Joaquin River.

Relationships between salvage and recoveries downstream of the facilities were inconsistent and conflicting without any clear patterns. Previous studies in the Delta have attempted to link recovery rates at downstream locations with the water diversion rate from the Delta and have not produced strong evidence of an effect [26,29,30]. Additionally, Zeug and Cavallo [17] failed to find a relationship between salvage at these facilities and recovery rates in the ocean. Unlike many water diversions, these allow for a fraction of entrained fish to be returned to the channel alive and our results suggest that salvage reduced migration mortality due to entrainment by 15–42%.

Although several strong relationships were identified between salvage and predictor variables, total loss and the contribution of juvenile salmon loss at the diversions to total mortality (relative loss) during migration was low. This may partially explain the poor and inconsistent relationships between salvage and recovery of tagged fish in Chipps trawl and the ocean. Although diversion-related entrainment is frequently invoked as a threat to fish populations, few studies have evaluated population-level effects of fish loss at diversions [5]. Several studies of entrainment loss relative to population mortality have reported relatively small contributions of entrainment similar to the estimates reported here [9,11,12]. The location of the diversions may also contribute to low relative loss. Both diversions are located on a distributary of the San Joaquin River thus, only salmon migrating through that route are susceptible to entrainment. In general, less than half of the juvenile salmon migrating down the San Joaquin River are likely to enter channels leading to the diversions [27] and even fewer Sacramento River-origin salmon are likely to enter this channel [19].

Although the results presented here suggest the total effect of loss at these diversions on migrating juvenile salmon is small, caution should be used when applying these results to other systems or even to the Sacramento-San Joaquin Delta generally. First, these diversions include salvage facilities that allow some fish to be returned alive. In many systems, fish that are entrained in diversions cannot return and are lost. Although it is largely unknown how these losses affect populations, Roberts and Rahel [4] suggested these diversions can function as sink habitats. Second, there are a large number of small diversions in the Delta that do not contain fish screens or salvage mechanisms and the aggregate effect of these diversions could be significant [31]. The calculation of entrainment loss includes estimates for louver efficiency and pre-louver survival that have low certainty and better quantitative estimates for these parameters could reveal greater estimates of total loss. Finally, all fish in this study (and in the acoustic studies used to calculate migration mortality) were hatchery reared fish. Thus, we are making the assumption that the behavior and survival of hatchery Chinook Salmon is similar to naturally produced fish in both rivers.

The data presented here indicated that a variety of hydrologic (diversion rate and flow), physical (distance from facilities) and biological (fish size) factors influence the salvage of CWT juvenile Chinook Salmon at two large water diversions. However, the relative importance of these factors varied among the two river systems where fish were released and among the two diversions which differed in the configuration of water diversion. Attempts to increase survival of juvenile salmon migrating through the Delta have largely focused on restriction of water diversion [14]. Yet the total contribution of loss at these facilities was small relative to total migration mortality and relationships between salvage and recoveries downstream were inconsistent and conflicting. The ability of fish to be salvaged and the physical location of the pumps off of the main stem rivers likely reduced the total population-level effect of these diversions. As water development continues worldwide, the inclusion of effective salvage facilities in diversion designs, and careful selection of diversion locations could help mitigate fish losses.

Acknowledgments

We thank Annie Brodsky for assisting in the preparation of data for analysis and creating maps of the study area. Steve Cramer, David Delaney and Tommy Garrison provided insights that significantly improved the statistical analyses. Staff from the California Department of Water Resources and the United States Fish and Wildlife Service staff provided necessary data.

Author Contributions

Conceived and designed the experiments: SZ BC. Analyzed the data: SZ. Wrote the paper: SZ BC.

References

1. Dudgeon D, Arthington AH, Gessner MO, Kawabata ZI, Knowler DJ, et al. (2006) Freshwater biodiversity: importance, threats, status and conservation challenges. Biol Rev 81: 163–182.
2. Poff NL, Allan JD, Bain MB, Karr JR, Prestegaard KL, et al. (1997) The natural flow regime. BioScience 47: 769783.
3. Bunn SE, Arthington AH (2002) Basic principles and ecological consequences of altered flow regimes for aquatic biodiversity. Environ Manage 30: 492–507.
4. Roberts JJ, Rahel FJ (2008) Irrigation canals as sink habitat for trout and other fishes in Wyoming drainage. T Am Fish Soc 137: 951–961.
5. Moyle PB, Israel JA (2005) Untested assumptions: effectiveness of screening diversions for conservation of fish populations. Fisheries 30: 20–28.

6. Carter KL, Reader JP (2000) Patterns of drift and power station entrainment of 0+ fish in the River Trent, England. Fish Manage and Ecol 7: 447–464.
7. Baumgartner LJ, Reynoldson NK, Cameron L, Stanger JG (2009) Effects of irrigation pumps on riverine fish. Fisheries Manag Ecol 16: 429–437.
8. Kelso JRM, Milburn GS (1979) Entrainment and impingement of fish by power plants in the Great Lakes which use the once-through cooling process. J Great Lakes Res 5: 182–194.
9. Hadderingh RH, Jager Z (2002) Comparison of fish impingement by a thermal power station with fish populations in the Ems Estuary. J Fish Biol 61: 105–124.
10. Nobriga ML, Matica Z, Hymanson ZP (2004) Evaluating entrainment vulnerability to agricultural irrigation diversions: a comparison among open-water fishes. In Feyrer F, Brown LR, Brown RL and Orsi JJ, editors. Early Life History of Fishes in the San Francisco Estuary and Watershed. American Fisheries Society Symposium 39: 281–295.
11. Post JR, Van Poorten BT, Rhodes T, Askey P, Paul A (2006) Fish entrainment into irrigation canals: an analytical approach and application to the Bow River, Alberta, Canada. N Am J Fish Manage 26: 875–887.
12. Carlson AJ, Rahel FJ (2007) A basinwide perspective on entrainment of fish in irrigation canals. T Am Fish Soc 136: 1335–1343.
13. Grimaldo LF, Sommer T, Van Ark N, Jones G, Holland E, et al. (2009) Factors affecting fish entrainment into massive water diversions in a tidal freshwater estuary: can fish losses be managed? N Am J Fish Manage 29: 1253–1270.
14. National Marine Fisheries Service (2009) NMFS Final 2009 Biological Opinion on the Long-Term Central Valley Project and State Water Project Operation, Criteria, and Plan. June 4, 2009. Available: http://swr.nmfs.noaa.gov/sac/myweb8/BiOpFiles/2009/Draft_OCAP_Opinion.pdf/. Accessed 2012 Mar.
15. Kimmerer WJ (2008) Losses of Sacramento River Chinook salmon and Delta smelt to entrainment in water diversions in the Sacramento-San Joaquin Delta. San Francisco Estuary and Watershed Science 6(2).Available: http://www.escholarship.org/uc/item/7v92h6fs/. Accessed 2010 Aug.
16. Yoshiyama RM, Gerstung ER, Fisher FW, Moyle PB (2001) Historical and present distribution of Chinook salmon in the Central Valley of California. Calif Depart Fish Game Bull 179: 71–176.
17. Zeug SC, Cavallo BJ (2013) Influence of estuary conditions on the recovery rate of coded-wire-tagged Chinook salmon (*Oncorhynchus tshawytscha*) in an ocean fishery. Ecol Freshw Fish 22: 157–168.
18. Michel CJ (2010) River and estuarine survival and migration of yearling Sacramento River Chinook salmon (*Oncorhynchus tshawytscha*) smolts and the influence of environment. Master's thesis, University of California, Santa Cruz.
19. Perry RW, Skalski JR, Brandes PL, Sandstrom PT, Klimley, et al. (2010) Estimating survival and migration route probabilities of juvenile Chinook salmon in the Sacramento-San Joaquin River Delta. N Am J Fish Manage 30: 142–156.
20. Zabel RW, Faulkner J, Smith SG, Anderson JJ, Van Holmes C, et al. (2008) Comprehensive passage (COMPASS) model: a model of downstream migration and survival of juvenile salmonids through a hydropower system. Hydrobiologia 609: 289–300.
21. Zabel RW, Achord S (2004) Relating size of juveniles to survival within and among populations of Chinook salmon. Ecology 85: 795–806.
22. Newman KB (2003) Modelling paired release-recovery data in the presence of survival and capture heterogeneity with application to marked juvenile salmon. Stat Model 3: 145–177.
23. Vuong QH (1989) Likelihood ratio tests for model selection and non-nested hypotheses. Econometrica 57: 307–333.
24. R Core Team (2012) R: a language and environment for statistical computing. R foundation for statistical computing, Vienna, Austria. Available: http://www.R-project.org/. Accessed 2012 July.
25. Gingras M (1997) Mark-recapture experiments at Clifton Court Forebay to estimate pre-screening losses to juvenile fishes 1976–1993. Interagency Ecological Program Technical Report 55, 26 p. Available: http://science.calwater.ca.gov/pdf/workshops/SP_workshop_predation_M.Gingra.pdf/. Accessed 2012 July.
26. Newman KB (2008) An evaluation of four Sacramento-San Joaquin river Delta juvenile salmon survival studies. United States Fish and Wildlife Service, Stockton, California. 182p. Available: http://www.science.calwater.ca.gov/pdf/psp/PSP_2004_final/PSP_CalFed_FWS_salmon_studies_final_033108.pdf/. Accessed 2010 June.
27. Buchanan RA, Skalski JR, Brandes PL, Fuller A (2013) Route use and survival of juvenile Chinook salmon through the San Joaquin River Delta. N Am J Fish Manage 33: 216–229.
28. Perry RW (2010) Survival and migration dynamics of juvenile Chinook salmon (*Oncorhynchus tshawytscha*) in the Sacramento-San Joaquin river Delta. Doctoral dissertation, University of Washington.
29. Newman KB, Rice J (2002) Modeling the survival of Chinook salmon smolts outmigrating through the lower Sacramento River system. J Am Stat Assoc 97: 983–993.
30. Newman KB, Brandes PL (2010) Hierarchical modeling of juvenile Chinook salmon survival as a function of Sacramento-San Joaquin Delta water exports. N Am J Fish Manage 30: 157–169.
31. Walters AK, Holzer DM, Faulkner JR, Warren CD, Murphy PD, et al. (2012) Quantifying cumulative entrainment effects for Chinook salmon in a heavily irrigated watershed. T Am Fish Soc 141: 1180–1190.

17

Investigating Avian Influenza Infection Hotspots in Old-World Shorebirds

Nicolas Gaidet[1]*, Ahmed B. Ould El Mamy[2], Julien Cappelle[1], Alexandre Caron[1], Graeme S. Cumming[4], Vladimir Grosbois[1], Patricia Gil[3], Saliha Hammoumi[3], Renata Servan de Almeida[3], Sasan R. Fereidouni[5], Giovanni Cattoli[6], Celia Abolnik[7], Josphine Mundava[8], Bouba Fofana[9], Mduduzi Ndlovu[4], Yelli Diawara[10], Renata Hurtado[11,12], Scott H. Newman[13], Tim Dodman[14], Gilles Balança[1]

1 CIRAD-ES, UPR AGIRS, Montpellier, France, **2** Centre National d'Elevage et de Recherches Vétérinaires, Nouakchott, Islamic Republic of Mauritania, **3** CIRAD-BIOS, UMR CIRAD/INRA CMAEE, Montpellier, France, **4** Percy FitzPatrick Institute, DST/NRF Center of Excellence, University of Cape Town, Cape Town, South Africa, **5** Friedrich-Loeffler-Institut, Insel Riems, Germany, **6** OIE/FAO Reference Laboratory for Avian Influenza and Newcastle Disease, Istituto Zooprofilattico Sper.le delle Venezie, Legnaro, Italy, **7** ARC-Onderstepoort Veterinary Institute, and Department of Production Animal Studies, Faculty of Veterinary Science, University of Pretoria, Onderstepoort, South Africa, **8** National University of Science and Technology, Bulawayo, Zimbabwe, **9** Direction Nationale des Eaux et Forêts du Mali, Bamako, Mali, **10** Parc National du Banc d'Arguin, Nouakchott, Islamic Republic of Mauritania, **11** Department of Microbiology, Institute of Biomedical Science, University of São Paulo, São Paulo, Brazil, **12** Department of Preventive Medicine and Animal Health, Faculty of Veterinary Medicine University of São Paulo, São Paulo, Brazil, **13** Food & Agriculture Organisation of the United Nations, Wildlife Health & Ecology Unit, EMPRES Animal Health, Rome, Italy, **14** Wetlands International, Wageningen, The Netherlands

Abstract

Heterogeneity in the transmission rates of pathogens across hosts or environments may produce disease hotspots, which are defined as specific sites, times or species associations in which the infection rate is consistently elevated. Hotspots for avian influenza virus (AIV) in wild birds are largely unstudied and poorly understood. A striking feature is the existence of a unique but consistent AIV hotspot in shorebirds (Charadriiformes) associated with a single species at a specific location and time (ruddy turnstone *Arenaria interpres* at Delaware Bay, USA, in May). This unique case, though a valuable reference, limits our capacity to explore and understand the general properties of AIV hotspots in shorebirds. Unfortunately, relatively few shorebirds have been sampled outside Delaware Bay and they belong to only a few shorebird families; there also has been a lack of consistent oropharyngeal sampling as a complement to cloacal sampling. In this study we looked for AIV hotspots associated with other shorebird species and/or with some of the larger congregation sites of shorebirds in the old world. We assembled and analysed a regionally extensive dataset of AIV prevalence from 69 shorebird species sampled in 25 countries across Africa and Western Eurasia. Despite this diverse and extensive coverage we did not detect any new shorebird AIV hotspots. Neither large shorebird congregation sites nor the ruddy turnstone were consistently associated with AIV hotspots. We did, however, find a low but widespread circulation of AIV in shorebirds that contrast with the absence of AIV previously reported in shorebirds in Europe. A very high AIV antibody prevalence coupled to a low infection rate was found in both first-year and adult birds of two migratory sandpiper species, suggesting the potential existence of an AIV hotspot along their migratory flyway that is yet to be discovered.

Editor: Justin David Brown, University of Georgia, United States of America

Funding: This survey was conducted under four international programmes including the GRIPAVI project sponsored by grants from the French Ministry of Foreign Affairs, the Technical Cooperation Programme of the Food and Agricultural Organization of the United Nations (UN-FAO) through additional financial resources from the government of France and Sweden, the USAID - and Wildlife Conservation Society - sponsored Global Avian Influenza Network for Surveillance (GAINS), and the EU-funded New Flubird project. The funders had no role in study design, data collection and analysis, decision to publish, or preparation of the manuscript.

Competing Interests: The authors have declared that no competing interests exist.

* E-mail: nicolas.gaidet@cirad.fr

Introduction

Heterogeneity in the transmission rates among host species and across geographical ranges is a major determinant of the dynamic of infectious diseases [1]. Particular seasons, environments, or species associations can generate disease "hotspots" in which pathogen prevalence is consistently higher than elsewhere. These hotpots play a major role in the dynamics of infectious diseases: for instance, seasonal peaks in infection rate produce a rapid increase in the level of the population immunity, affecting the long-term maintenance of a pathogen in the host population; elevated pathogen prevalence may facilitate reassortment between hetero-subtypic pathogens; and hotspots may constitute a source of pathogen spillovers to less susceptible or less exposed species, environments or geographical areas that are connected to the hotspot by host movements. Identifying the occurrence of hotspots is therefore of particular importance for the control and prevention of infectious diseases.

Low pathogenic avian influenza viruses (AIV) have been extensively studied in wild birds in recent years in response to the emergence and dispersion of highly pathogenic AIV responsible for major health and economic threat [2]. Shorebirds (Charadriiformes) are classically recognised, together with ducks, geese and swans (Anseriformes), as the major natural reservoir of

AIV [3,4]. Globally and locally, the typical prevalence of AIV infection in shorebird species sampled worldwide is low (c. 1%) [3–6] as compared with prevalence in ducks (c. 10% globally with seasonal peaks of 20–60%) [3,4]. There is, however, one notable exception: a high AIV prevalence (>10%) has been consistently reported in the ruddy turnstone (*Arenaria interpres*) sampled in May during spring migration at Delaware Bay, USA [5,7,8]. This particular species, season and site combination represent the only known shorebird-AIV hotspot worldwide at which the infection rate is consistently higher than elsewhere in the world [5,6,8]. Delaware Bay is one of the world's largest congregation sites of shorebirds, supporting an estimated 1 million shorebirds stopping during spring migration. Surprisingly, all of the other shorebird species that mingle with the ruddy turnstones at Delaware Bay in May show a very low AIV prevalence (<2%) [5,9]. A low prevalence has also been found in shorebirds (including ruddy turnstone) stopping at Delaware Bay during autumn migration [7,10].

Shorebirds form the most abundant and the most species-rich group of waterbirds. Shorebirds are divided taxonomically into three major clades: the Scolopaci (sandpipers, snipes, phalaropes, jacanas), Lari (gulls, terns, auks, pratincoles, skuas) and Charadrii (plovers, oystercatchers, stilts) [11]. Most species share a set of ecological characteristics that are favorable to the transmission and dispersion of AIV: (i) they are generally highly gregarious (at least outside the breeding season), congregating at very high densities at key staging sites along migratory flyways, where they form multispecies foraging or roosting flocks; (ii) most species are very long-distance migrants, including some that undertake non-stop flights of up to 11,000 km [12]; and (iii) many shorebirds breeding in the Northern hemisphere winter in the Southern hemisphere; in this way they connect the northern hemisphere waterbird fauna to the regions south of the equator that are not reached by migratory ducks.

Characterisation of AIVs isolated from shorebirds suggests that they play a key role in the perpetuation of AIV in wild bird communities. The majority of inter-continental exchange and reassortment of AIV genes between Nearctic and Palaearctic regions has, although uncommon, been reported in shorebirds [13,14]. A larger variety of hemagglutinin and neuraminidase combination subtypes has been found in shorebirds than in ducks [8,10] suggesting that shorebirds may maintain a wider spectrum of viruses than ducks [3]. Furthermore, phylogenetic studies consistently indicate the existence of a large overlap between the gene pools of viruses that originate from both ducks and shorebirds [13,15], suggesting that AIVs are commonly transmitted between those two groups of species.

It is not currently known whether Delaware Bay constitutes the world's only shorebird AIV hotspot or whether other hotspots exist, in particular at other large shorebird congregation sites [5,6]. Published studies have limitations in their geographic scope, numbers of sampled birds, taxonomic range, and ability to detect AIVs. So far there have been relatively few intensive studies of AIV infection in shorebirds outside Delaware Bay (see [6] for a review) and it is unclear whether the number of birds sampled at other locations and times has been adequate to detect hotspots. Most of the shorebirds that have been tested belong to only two (Scolopacidae and Laridae) of the 19 families of Charadriiformes. In addition, almost all AIV infection studies in shorebirds have relied only on cloacal or fresh fecal samples [6,8,16,17], whilst oropharyngeal samples have been tested on very few occasions [8,18]. Several recent studies of Anseriformes provide evidence of the importance of the respiratory tract for the replication of AIV [2,19] and demonstrate the complementary nature of cloacal and oropharyngeal samples, since birds are rarely found concurrently infected from both types of samples [8,20]. Results from experimental infection studies also suggest that oropharyngeal excretion may be predominant for some AIV subtypes in gulls [21,22]. As a result of these various limitations, some hotspots of AIV infection in shorebirds may have remained undetected.

As this summary implies, the significance of interspecific variation in AIV prevalence among shorebirds, and in particular the high prevalence in the ruddy turnstone, remains unclear. Ruddy turnstones have rarely been tested for AIV infection outside Delaware Bay (only 9%, see supporting information Table S1). It remains to be determined whether this species is associated with AIV hotspots in other sites or if other species are involved.

In this study we investigated the possibility that other AIV hotspots may be associated with large shorebird congregation sites or with alternative species. First, we conducted a survey of AIV infection in shorebirds, including ruddy turnstone, at the Banc d'Arguin in Mauritania (Table 1, Figure 1). This site constitutes one of the largest wintering sites for shorebirds in the world (c. 2.3 million birds) and supports the greatest number of ruddy turnstones (c. 9,000 birds) across the old world [23]. Second, we extended our analysis to a large-scale dataset of AIV prevalence in 69 shorebird species that we sampled in 25 countries at some of the most important shorebirds sites in Africa and Western Eurasia (Figure 1, and in supporting information Table S2) during various international surveillance programs [20,24,25]. We explored these data for the presence of potential hotspots of AIV infection in relation to the number of birds sampled at a given time and location. We also examined whether variations in prevalence could be linked to ecological factors that affect (i) the local persistence and transmission potential of AIV (such as the environment and local abundance of shorebirds) and (ii) the exposure of host species to AIV (such as their geographical range and foraging behaviour). Third, we conducted a serologic survey in some shorebird species to compare differences in previous AIV exposure between species throughout their annual range and identify the species that might potentially be involved in AIV transmission at other hotspots.

Methods

Ethics statement

Birds used in this study were sampled from Botswana, Burkina Faso, Chad, Egypt, Ethiopia, Iran, Kenya, Mali, Mauritania, Morocco, Niger, Nigeria, Malawi, Mozambique, Ukraine, Senegal, Republic of Sudan, South Sudan, Tanzania, Tunisia, Turkey, Zambia and Zimbabwe (Figure 1). Most birds were captured and released in the wild using conventional techniques (mist-nets and baited walk-in traps) covered in the Ornithological Council "Guidelines to the Use of Wild Birds in Research". Procedures for capture, handling, and sampling were approved by the Centre de Recherches par le Baguage des Populations d'Oiseaux (CRBPO, Natural History Museum Paris - French National Reference Bird Ringing Center). We conducted a serological survey on a subset of birds captured at two sampling sites (Banc d'Arguin, Mauritania; Inner Niger River Delta, Mali) (see below for details). A capture and sampling permit at the Banc d'Arguin was obtained from the Conseil Scientifique du Banc d'Arguin. Capture permits were similarly obtained from the relevant government authority in each country where field studies were conducted. All sampling activities were conducted in the presence of a representative from the animal health and veterinary national services and a representative from the environment national services.

Table 1. Prevalence of AIV infection detected by rRT-PCR in shorebirds sampled at the Banc d'Arguin, Mauritania.

Species	No. bird positive/total						% pos. (±95% CI)
	Feb 2006	Dec 2006	Apr 2008	Nov 2009	Mar 2010	Total	
Ruddy turnstone *Arenaria interpres*		1/28	0/98	0/5	0/27	1/158	0.6 (0.1–3.5)
Sanderling *Calidris alba*		0/19		0/6	0/22	0/47	0 (0–7.6)
Dunlin *C. alpina*		3/186	0/121	0/269	1/160	4/736	0.5 (0.2–1.4)
Red knot *C. canutus*		3/128		0/88	3/131	6/347	1.7 (0.8–3.7)
Lesser black-backed gull *Larus fuscus*	1/129		0/1	0/63	0/30	1/223	0.4 (0–2.4)
Slender-billed gull *Chroicocephalus genei*			1/141	0/26	0/99	1/266	0.4 (0–2.1)
Terns[a]	2/150					2/150	1.3 (0.4–4.7)
other species[b]		1/35	0/4	0/41	0/11	1/92	1.1 (0.1–5.9)
Total	3/279	8/396	1/365	0/498	4/480	16/2018	0.8 (0.5–1.3)

[a]. Caspian tern *Sterna caspia, Royal tern S. maxima, Sandwich tern S. sandvicensis.*

[b]. Curlew sandpiper *Calidris ferruginea*, Little stint *C. minuta*, Common ringed plover *Charadrius hiaticula*, Eurasian oystercatcher *Haematopus ostralegus*, Bar-tailed godwit *Limosa lapponica*, Grey plover *Pluvialis squatarola*, Common redshank *Tringa totanus*, Grey-headed gull *Chroicocephalus cirrocephalus*, Black-headed gull *C. ridibundus*.

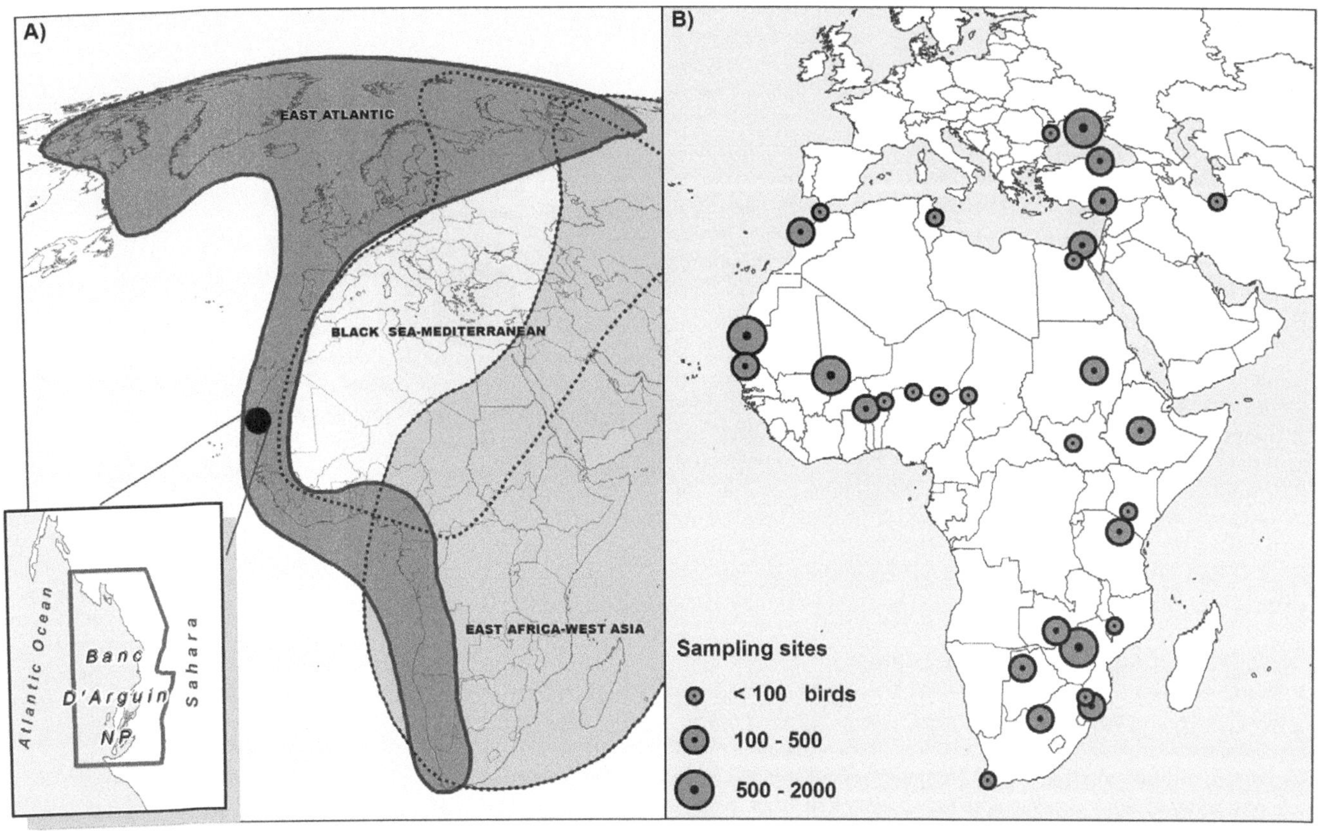

Figure 1. Location of the study sites. (A) The Banc d'Arguin (Mauritania) and the main shorebirds migratory flyways across Western Eurasia and Africa. (B) All shorebird sampling sites considered in our study (list of sites ranked by latitude: Ukraine - Eastern Sivash, Romania-Danube Delta, Turkey - Kizilirmak Delta and Yumurtalik Lagoons, Iran - Fereydoon Kenar marshes, Morocco - Marais du Bas Loukkos and Sidi Moussa-Oualidia Lagoon, Tunisia - Thyna salt pans, Egypt - Nile River Delta and Lake Qarun, Mauritania - Banc d'Arguin National Park, Senegal/Mauritania - Senegal River Delta, Republic of Sudan - El Saggay Island, Mali - Inner Niger Delta, Niger - Kurfunkura pond and Gaya, Chad - Lake Chad, Nigeria - Hadejia-Nguru wetlands, Burkina Faso - Lake Kompienga, Ethiopia - Lake Debre Zeit, South Sudan - Bargel wetland, Kenya - Lakes around Nairobi, Tanzania - Lake Manyara, Malawi - Lake Chilwa, Zambia - Kafue Flats, Zimbabwe - Lakes Manyame-Chivero, Botswana - Lake Ngami, Mozambique - Massingir Dam and Lake Chuali, South Africa - Barberspan wetland and Strandfontein). A detailed list of sampling sites is provided in the supporting information Table S2 (map by M. Gély ©Cirad).

On some occasions birds (8% of the total sample) were provided by hunters. Samples were collected from traditional or commercial safari hunting activities. Birds that had been killed by hunters had been hunted with appropriate permits from each relevant local authority and were obtained with the hunters' permission. None of the collected birds were protected. Safari operators conduct their activities through authorisations delivered by: Morocco: Haut Commissariat aux Eaux et Forêts et à la Lutte Contre la Désertification, Direction de la Lutte contre la Désertification et de la Protection de la Nature; Burkina Faso: Direction de la Faune et des Chasses; Chad: Direction de la Conservation de la Faune et des Aires Protégées. In two countries (Mali, Malawi) birds were acquired through local bird hunters who harvest birds for subsistence and to supplement and sustain the food resources of the local population. In Mali birds were acquired from local bird hunters through an official hunting and sampling authorisation delivered by the Direction National de la Conservation de la Nature (Félix Dakouo, National Director of the DNDN). In Malawi birds were acquired through registered bird hunters who belong to bird hunting clubs that form part of the Lake Chilwa Bird Hunters Association which was formed with the aim of sustainably managing the utilisation of sedentary and migratory waterbirds at Lake Chilwa; sampling authorisation was provided by the Wildlife Conservation Department, Ministry of Agriculture and Rural Development. On two sampling occasions (Niger and Burkina Faso in 2006) birds were shot through special permits for sample collection in the framework of a national emergency surveillance operation implemented after the notifications of avian influenza highly pathogenic H5N1 outbreaks. These birds (n = 163) represent only 2% of the entire birds sampled in this study. Hunting authorisations were obtained from the Direction de la Conservation de la Nature for Niger (Ali Harouna, Director of the Direction de la Conservation de la Nature) and the Direction de la Faune et des Chasses for Burkina Faso (Urbain Belemsobgo, Director of the Direction de la Faune et des Chasses).

The Banc d'Arguin

The Banc d'Arguin National Park is located near the 20th parallel and extends >180 km along the Atlantic coast of Mauritania, bordered by the Sahara desert (Figure 1). This area of about 500 km^2 of very shallow intertidal flats is flooded by upwelling of cold water rich in nutrients. This exceptional ecosystem is the wintering site with the highest density of shorebirds along the East Atlantic Flyway [26]. It also represents a crossroads for migratory shorebirds breeding in the Palearctic, Nearctic and Afro-tropical regions and wintering along the African coast.

Sampling and AIV detection procedures

Shorebirds were sampled at the Banc d'Arguin on five occasions between 2006 and 2010 (Table 1). Samples were collected during the same period at 30 other sites in 25 countries (Figure 1, and supporting information Table S2). Sampling was conducted on a different number of occasions according to sites and years. Three types of samples were collected that we distinguished in our subsequent analyses: a single cloacal or fecal swab, a single oropharyngeal swab, or both cloacal and oropharyngeal swabs tested separately. All samples were collected using cotton swabs and stored in cryovials containing a viral transport medium. Cryovials were stored in liquid nitrogen directly in the field and every effort was made to maintain the cold chain using cryopacks with dry ice during international shipment to the laboratory. Samples were analysed in different laboratories using a similar standard diagnostic procedure based on RNA extraction and real-time RT-PCR virus detection (rRT-PCR) targeting the matrix gene specific for influenza A viruses. Positive samples were also inoculated into embryonated SPF chicken eggs for virus isolation and typed according to standard procedures. Further details on sampling and diagnostic procedures can be found in a previous publication [20]. We computed the observed prevalence for each species for each sampling occasion as the percentage of individuals found positive for AIV compared with the total number of birds tested.

Presence of hotspots of AIV infection

We defined a hotspot as a specific location and time at which the AIV infection rate is consistently elevated and about an order of magnitude greater than in other sites, using 10% as a minimum prevalence threshold according to long-term prevalence measures in ruddy turnstone at Delaware Bay [5,8]. We compared the number of AIV-positive birds detected on each sampling occasion (considering birds from all species altogether or from each species separately) with a threshold number of positive birds for a sample of the same size below which the prevalence is unlikely (probability <0.05) to be greater than 10%. This threshold was defined as the 0.05 quantile of a binomial distribution B(n = sample size, p = 0.10) (*stats* R package, *qbinom* procedure). If the observed number of AIV-positive birds on a sampling occasion was below the threshold defined for a sample of the same size, we concluded that prevalence in the population from which the sample had been drawn was most likely lower than 10%. We restricted our analysis to sampling units that had at least 28 birds sampled (28 being the minimum number of individuals required to be 95% confident of detecting at least one infected bird when prevalence is ≥10%).

Variations in AIV prevalence in shorebirds across Eurasian and Afro-tropical regions

Explanatory variables tested in this analysis are summarized in Table 2. Species behavioral and ecological traits were taken from literature (body mass, main foraging technique: [27,28]; geographic range: [23]). We restricted our analysis to species that had at least 20 individuals sampled. Sampling sites were classified according to four abundance classes of shorebird populations estimated from counts compiled in [23].

We explored the relationships between AIV prevalence and explanatory variables using Generalized Linear Mixed Models (GLMM) with a binomial error distribution and a logit link function (*lme4* R package, *glmer* procedure). We tested the independence among categorical variables (phi coefficient) to identify potential collinearity issues. Samples had usually been collected on several occasions at the same site or during the same year; to avoid pseudo-replication we included a year and a site random effect in models. The potential aggregation of infected birds within a given sampling occasion was also accounted for by incorporating the sampling occasion as a random effect nested within year and site. Finally, we included a random laboratory effect to account for a potential difference in diagnostic sensitivity among laboratories.

Our analysis consisted of two steps. First, we tested for environmental, seasonal and species variation in prevalence, accounting for differences in the types of sample tested. The two variables related to the sampling site (abundance and environment) were considered associated (phi coefficient >0.28) and were tested alternatively in models. We constructed a full model including four explanatory variables (species, season, sampling procedure and abundance/or environment,) as fixed effects and four random factors (year, site, sampling occasion and laboratory). The goodness-of-fit of the model was assessed using the Pearson

Table 2. List of the explanatory variables tested to explain geographical, seasonal and species variations in AIV prevalence in shorebirds across Eurasian and Afro-tropical regions (Figure 1).

Explanatory variables		Eco-epidemiological predictions	Definition
Site	Shorebirds abundance	Aggregation of birds may enhance inter-individual transmission through contact rate	Four abundance classes ([<5], [5–50], [50–500], [>500]×10^3 birds)
	Environment	High salinity, wind and solar radiation exposure (low vegetation cover) and tidal washing may reduce virus persistence in coastal habitat	Marine-saline vs inland-freshwater habitats
Season		Seasonal patterns of migration and reproduction may influence the turnover of susceptible birds	Four trimester periods
Sampling procedure		AIV may replicated preferentially in the respiratory or the digestive tract; the type of sample tested may influence the probability of detecting an infection	Single cloacal, single oropharyngeal or both types of samples
Species	Foraging behaviour[a]	Higher AIV exposure in tactile-foraging (probing) than in visual-foraging (pecking) species	Tactile-foraging vs visual-foraging species
	Geographic range[a]	Lower AIV exposure in high arctic/coastal than boreal-temperate/freshwater species, and in Afro-tropical resident than in boreal-temperate migratory species	High arctic/coastal vs boreal-temperate/ freshwater vs tropical/freshwater species
	Body mass[a]	Demographic rates associated with body mass may influence the turnover of susceptible birds	Mean species body mass

[a]. Only tested for species from the Scolopaci and Charadrii clades.

Chi-Square statistic. The initial full model was simplified by a stepwise backwards elimination of non-significant variables. We used likelihood ratio tests (LRT) to test the significance of each variable, computing the $\chi 2$ of the LRT between the model retaining and the model excluding the variable. We first explore the random part of the model, iteratively removing from the model the random factors with the lower estimated variance components. We then explored the fixed part, iteratively removing variables that had the lowest explanatory power (highest P values, t-test) to identify the minimal adequate model.

In a second step we tested whether species variation in prevalence was related to species traits that may affect exposure and immunity to re-infection (Table 2). In most ecosystems, cohabiting shorebird species are separated in space and time through dietary preferences and foraging techniques leading to potential differences in exposure to infection. Shorebirds of the Scolpaci (sandpipers and allies) and Charadrii (plovers and allies) clades use two main sensory mechanisms to detect their prey: species relying on tactile sensation that forage mainly by probing prey continuously beneath the substrates may have a higher AIV exposure by comparison to species that use visual mechanisms and forage by pecking prey from the surface. Moreover, differences in foraging strategy lead to variations in vigilance; tactile-foraging species tend to form larger foraging flocks than visual-foraging species [29] and hence may be subject to higher contact rates and inter-individual AIV transmission. Species heterogeneity in transmission rate may be caused by latitudinal and habitat differences in AIV exposure. Environmental persistence of AIV is reduced by salinity. A lower AIV exposure is expected in the high Arctic regions where shorebirds breed at low density, forage mainly on terrestrial invertebrates in moist or dry habitat and where the dabbling ducks of the *Anas* genus (presumably the main AIV maintenance hosts [3]) are largely absent. High Arctic breeding shorebird species winter predominantly in coastal-saline environments [30] and therefore remain year-round in AIV-poor environments, whereas species breeding in sub-Arctic to temperate regions winter mainly in inland-freshwater wetlands where the potential for AIV infection is higher (freshwater, cohabitation with dabbling ducks of the *Anas* genus). Eurasian migratory species wintering in Africa move between vastly separated areas with distinct waterbird communities and hence are likely to be exposed to a higher diversity of AIVs than Afro-tropical species that reside year-round within sub-Saharan regions where AIVs generally circulate at a lower level than in temperate or boreal wetlands [20].

In the second step we followed the same model selection procedure as in the first step, starting from the previous minimum adequate model but substituting the variable "species" with variables describing species traits. Since demographic turnover rate in birds is related to body size at the inter-specific level, we included among the potential explanatory variables body mass (log transformed) as a proxy for turnover rate of susceptible birds in the population. In order to account for non-independence between species owing to phylogeny we included a hierarchical taxonomic (Clade–Family–Genus–Species) nested error structure as random effects in the model. The foraging and range classes defined for the Scolopaci and Charadrii were not appropriate for species belonging to the Lari clade (most gull and tern species that we sampled remained year-round in coastal habitats and forage using either specialized techniques, such as aerial-plunging in terns, or opportunism in gulls); they were removed from this second analysis.

Serological survey of antibodies to AIV

We conducted a serological survey on a subset of birds captured at two sampling sites (Banc d'Arguin, Mauritania; Inner Niger Delta, Mali) during three sampling occasions at each site. Birds were ringed and aged (first year, ≥second year, undetermined) on the basis of plumage characteristics, and a blood sample was collected from the jugular vein. We tested individual serum samples for the presence of antibodies specific to the AIV nucleoprotein by using a commercial blocking enzyme-linked immunosorbent assay (bELISA) kit (FlockCheck AI MultiS-Screen nucleoprotein antibody test kit, IDEXX Laboratories) following the manufacturer's instructions. All samples were run in combination with supplied positive and negative controls for validation. Optical density (OD) values were read at 650 nm and sample with S/N values (ratio of the sample OD to the kit negative control mean OD)<0.50 were considered positive for antibodies to AIV.

We also tested a subset of samples from different species at both sites using a second commercially competitive ELISA kit (ID screen, influenza A nucleoprotein antibody competition, ID.VET) following the manufacturer's instructions. We evaluated the reproducibility of diagnostic results using i) the McNemar's test (exact binomial test for correlated proportions) to assess if proportions of positive results differed between the two Elisa assays, and ii) the Cohen's Kappa statistic to measure the level of agreement between the two tests beyond chance. Seroprevalence of antibodies to AIV was computed for each species and sampling occasion from results of the first diagnostic assay.

We investigated the variations in seroprevalence using a GLMM approach, as described above. We tested for effects of age and species as fixed factors, accounting for potential pseudo-replication issues by including year, site and sampling occasions as random factors. We also restricted this analysis to species that had at least 20 individuals sampled.

Results

AIV circulation at PNBA, Mauritania

A total of 2018 shorebirds were sampled at the Banc d'Arguin. AIV was detected on four out of the five sampling occasions and in almost all species, but at a consistently low prevalence (<2%, Table 1). Only one AIV-positive ruddy turnstone was found, and the proportion of birds found infected with AIV in this species (0.6%, n = 158) was not significantly different from that in other shorebird species (0.8%, n = 1860) ($\chi2 = 0.06$, $p>0.5$). We also found no difference in infection rate for all shorebird species between sampling occasions conducted at the beginning (Nov–Dec: 0.9%, n = 894) and the end (Feb–Apr: 0.7%, n = 1124) of the wintering period ($\chi2 = 0.21$, $p>0.5$).

Presence of AIV hotspots in shorebirds across Eurasian and Afro-tropical regions

Altogether, a total of 7715 birds belonging to 69 species from 10 shorebird families were tested for AIV infection during 86 sampling occasions at 31 sites (in supporting information Table S3). The global prevalence was low: AIV infection was detected in only 1.4% (n = 107) of birds tested. Two AIVs were isolated, an H1N1 virus in dunlin (*Calidris alpina*) in Turkey and an H10N7 virus in Spotted Redshank (*Tringa erythropus*) in Romania.

Our analysis for the presence of potential AIV hotspots in the shorebirds revealed that on most sampling occasions, the number of AIV-positive birds detected was below the threshold number of infected individuals expected for a prevalence of at least 10% (Figure 2). For about 90% of sampling occasions when at least 28 birds were tested (42 out of 47 occasions) we can be 95% confident that prevalence was lower than 10%. We found similar results when we considered each shorebird species separately (supporting information Figure S1).

On only five sampling occasions (Table 3) the number of positive birds was too large for rejecting the hypothesis that prevalence could be at least 10%. For these, we could not exclude the presence of a hotspot of AIV infection. Interestingly, three of these sampling occasions originated from the same site (Lakes Manyame-Chivero, Zimbabwe) and from the same bird species (mainly African jacana *Actophilornis africana* and Blacksmith lapwing *Vanellus armatus*), although from three distinct sampling seasons (Sep 2007, Jan 2008 and Nov 2008). The two other sampling

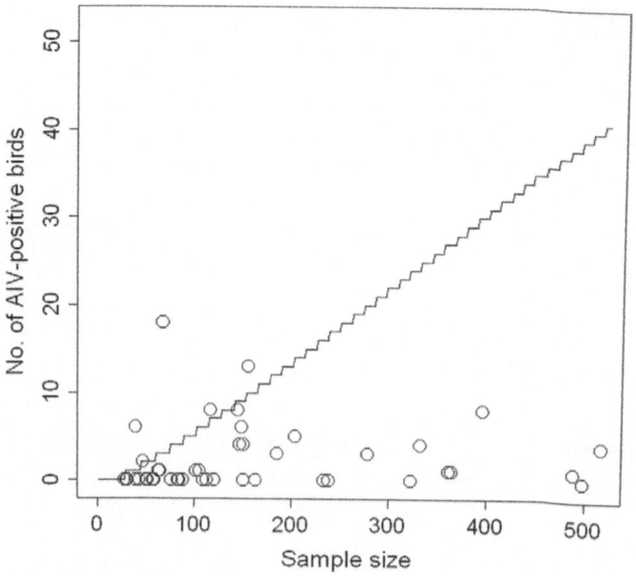

Figure 2. Detection of potential hotspots of AIV infection in shorebirds sampled at various sites across Eurasian and Afro-tropical regions (Figure 1-B; and supporting information Table S2). The number of AIV-positive birds detected in relation to the number of birds sampled per sampling occasion is here compared to the threshold number of positive birds (solid line) below which the prevalence is unlikely (probability <0.05) to be greater than 10% for a sample of the same size. Points on or above the line represent potential AIV hotspots, i.e. sampling occasions (n = 5) for which the number of positive birds was too large for rejecting the hypothesis that prevalence could be >10% (see Table 3). Only sampling occasions (n = 47) that had at least 28 birds sampled were considered in this analysis.

occasions were in the Senegal River Delta and at the Thyna Salt Pans (Gulf of Gabès) in Tunisia. Sampling conducted on these same three sites at the same time-point and on the same species on successive years did not, however, confirm the recurrent presence of a hotspot of AIV infection at these specific locations and times of the year (Table 3).

Variations in AIV prevalence in shorebirds across Eurasian and Afro-tropical regions

Although AIV prevalence was generally low, AIV infection was relatively ubiquitous in shorebirds across Eurasian and Afro-tropical regions. AIV positive individuals were found in a large number of bird species (n = 26) of both Eurasian and Afro-tropical origin, belonging to various families including Scolopacidae (13 species), Charadriidae (8), Laridae (7), Jacanidae (1) and Glareolidae (1) (in supporting information Table S3). Infection with AIV was reported for the first time in about two thirds (n = 18) of these species. AIVs were also detected in birds sampled throughout the year, in all study regions, and in both inland-freshwater and coastal-saline environments.

In our first analysis of the environmental, seasonal and species variations in prevalence, we found that the sampling occasion accounted for most of the variance in the random part (random effect variance estimation = 4.47). The inclusion of each of the other random factors (laboratory, site and year) did not improve the model fit (LRT, p>0.5). We therefore included sampling occasion as a random effect in all subsequent models. During our selection procedure among explanatory variables we found that only sampling procedure and species had a significant effect on

Table 3. List of potential hotspots of AIV infection detected in our study.

Country	Site	Occasion	Main species[a]	No. bird pos./total	% pos. (±95% CI)
Potential AIV hotspots					
Zimbabwe	Lakes Manyame-Chivero	Sep. 2007	African jacana, Kittlitz's plover, Little stint	6/40	15.0 (7.1–29.1)
		Jan. 2008	African jacana, Blacksmith lapwing	18/68	26.5 (17.4–38.0)
		Nov. 2008	African jacana, Blacksmith lapwing, Kittlitz's plover	8/117	6.8 (3.5–12.9)
Mauritania-Senegal	Senegal River Delta	Mar. 2006	Slender-billed gull	13/156	8.3 (4.9–13.7)
Tunisia	Thyna salt pans	May 2006	Curlew sandpiper	2/48	4.2 (1.1–14.0)
Follow-up sampling at the same site and month in the successive years					
Zimbabwe	Lakes Manyame-Chivero	Sep. 2008	African jacana, Kittlitz's plover	0/77	0 (0–4.8)
		Jan. 2009	African jacana, Blacksmith lapwing	0/84	0 (0–4.4)
		Nov. 2009	African jacana, Wood sandpiper, Blacksmith lapwing	1/102	1.0 (0.0–5.3)
		Fev. 2010	African jacana	1/64	1.6 (0.1–8.3)
Mauritania-Senegal	Senegal River Delta	Mar. 2010	Slender-billed gull	0/36	0 (0–9.6)
Tunisia	Thyna salt pans	Apr. 2007	Little stint, Curlew sandpiper	0/44	0 (0–8.0)

These sites correspond to sampling occasions (top) at which the number of AIV-positive birds was above the threshold number of birds for which the hypothesis that prevalence is lower than 10% could not be rejected (see Figure 2). Sampling conducted in different years at the same sites during the same months and on the same species (below) detected a low number of AIV-positive birds.

[a]. African jacana *Actophilornis africana*, Kittlitz's plover *Charadrius pecuarius*, Little stint *Calidris minuta*, Blacksmith lapwing *Vanellus armatus*, Slender-billed gull *Chroicocephalus genei*, Curlew sandpiper Calidris ferruginea, Wood sandpiper *Tringa glareola*.

variation in prevalence (Table 4). Prevalence was significantly higher for birds tested concurrently for both cloacal and oropharyngeal samples than in birds tested for a single cloacal sample (Z value = 2.29, p = 0.022). However, prevalence was similar between birds tested for a single cloacal or a single oropharyngeal sample (Z value = 0.93, p = 0.35). We detect a significant variation in prevalence between species but no species had prevalence significantly different from the prevalence estimated for the ruddy turnstone. We did not detect a statistical difference in prevalence between seasons, the type of environment or the abundance classes of shorebirds at the sampling site (Table 4).

The variable species and sampling procedures were also significant when we restricted our analysis to Scolopaci and Charadrii. None of the variables related to species traits (geographic range, foraging behaviour and body mass), however, was a significant predictor of AIV prevalence and only the variable sampling procedure was retained in the selection process (Table 4). None of the nested taxonomic levels (Clade–Family–Genus–Species) included as a random effect received statistical support from the data and sampling occasion accounted for most of the variance of the random part (variance estimation = 4.89).

Seroprevalence of antibodies to AIV

The presence of antibodies specific to AIV was investigated in a total of 930 birds belonging to 23 species (in supporting information Table S4). The overall seroprevalence was 17.6% and was highly variable between species, ranging from 0 to 77%. The reproducibility of diagnostic results, evaluated by testing a subset of samples (n = 258) using a second ELISA kit, was high: we found no significant difference in the proportion of positive samples between the two ELISA assays (McNemar's chi-squared = 5.79, df = 1, p-value = 0.016) with an almost perfect agreement (Kappa test = 0.87, p<0.001).

Our GLMM analysis indicated that seroprevalence was unrelated to age ($\chi2 = 0.79$, df = 12, $p = 0.67$) but varied significantly between species ($\chi2 = 281.31$, df = 10, $p<0.001$). Age also had no significant effect when we excluded birds of unknown age ($\chi2 = 2.61$, df = 12, $p = 0.11$). Seroprevalence was significantly higher in the red knot *Calidris canutus* (77.5, 95% CI: 70.2–83.4; Z value = 3.93, p<0.001) as compared to the ruddy turnstone (47.1, 95% CI: 36.8–57.5) and significantly lower in the dunlin (1.4, 95% CI: 0.6–3.3; Z value = −7.17, p<0.001) and the sanderling *Calidris alba* (4.8, 95% CI: 0.2–22.7; Z value = −2.53, p<0.05) sampled concurrently at the Banc d'Arguin (Figure 3, supporting information Table S5). No AIV antibodies were found in species sampled in the Inner Niger Delta, in either Eurasian migrants (ruff *Philomachus pugnax*, wood sandpiper *Tringa glareola*) or Afrotropical species (African jacana, painted snipe *Rostratula benghalensis*, spur-winged lapwing *Vanellus spinosus*) (in supporting information Table S4). As in the analysis of AIV infection rate we found that the sampling occasion accounted for most of the variance of the random part (variance estimation = 0.41) while the inclusion of each of the other random factors (site and year) did not improve the model fit (LRT, p>0.5).

The limited number of species in the data set precluded tests of the relationship between species traits and variation in seroprevalence in our modelling analysis. We note, however, that a high seroprevalence was found in both one mainly tactile (red knot) and one mainly visual-foraging species (ruddy turnstone) at the Banc d'Arguin and that globally, no difference was found between species using these two distinct foraging strategies ($\chi2 = 0.45$, df = 1, $p = 0.50$). Among species originating from Arctic breeding ground and wintering in the coastal environment of the Banc

Table 4. Results of the model selection procedure relating variations in AIV prevalence in shorebirds across Eurasia and Africa to species ecological traits, sampling procedure, period and ecological characteristics of the sampling site.

Explanatory variables	Coefficient ± S.E.		χ2	df	p
Full model: all species					
Sample type	Cloacal+oropharyngeal	1.59±0.70*	9.04	2	0.011
	Single oropharyngeal	0.75±0.80			
Species		NS[a]	66.1	34	<0.001
Abundance			3.06	3	0.38
Environment			0.05	1	0.82
Season			1.65	3	0.65
Species traits: only Scolopaci and Charadrii species					
Sample type	Cloacal+oropharyngeal	1.58±0.77*	8.03	4	0.018
	Single oropharyngeal	0.90±0.84			
Geographic range			1.90	6	0.39
Foraging behaviour			0.47	5	0.49
Body mass			0.10	5	0.75

All models were fitted as generalized mixed effects models, with sampling occasion fitted as random intercept terms to control for pseudo-replication and other explanatory variables as fixed effect. The initial full model was simplified by backwards elimination of non-significant variables. Coefficient estimates are given only for variables interpreted as statistically significant (t-test,* p<0·05) and included in the minimal adequate model (NS - no coefficient of individual species was statistically different from zero). The test statistics refer to a likelihood ratio test between the model in which the variable is retained and in which it is excluded. Statistics for variables not included in the final model correspond to the values when added to the minimal adequate model. Variables related to species traits were included in substitution to the variable species.

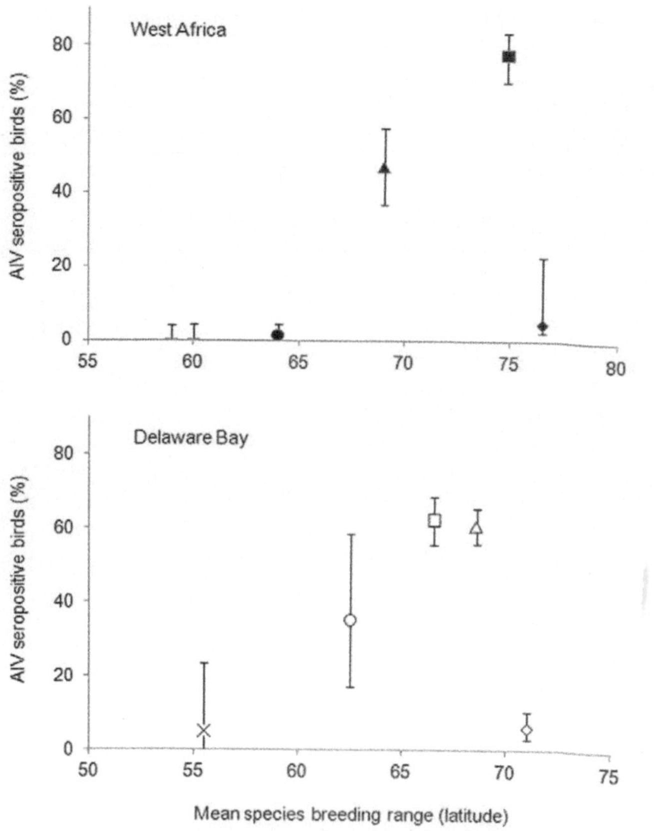

Figure 3. Mean seroprevalence of AIV antibodies among closely related shorebird species in relation to the mean latitude of their breeding range. Seroprevalence were measured in West Africa (the Banc d'Arguin, Mauritania and the Inner Niger Delta, Mali) and the Delaware Bay, USA (from [8,31]) using the same commercial bELISA kit (see Methods). The mean species breeding latitude was computed from the northern and southern limits of the breeding distribution of the populations present at each site using distribution maps from [22,32]. Species include ruddy turnstone (Δ), red knot (□), dunlin (o), sanderling (◇), short-billed dowitcher (×), ruff and wood sandpiper (−). Error bars represent the binomial exact 95% confidence interval.

d'Arguin we found both species with high (red knot, ruddy turnstone) and low seroprevalence (dunlin, sanderling).

Discussion

Despite an unprecedentedly large geographic and taxonomic coverage we did not detect any hotspots of AIV infection in shorebirds that matched the criterion that infection rate should be consistently elevated and about an order of magnitude greater than in other sites (≥10%). We did find a relatively high infection rate at one of our sampling sites - Lakes Manyame-Chivero, Zimbabwe - on three distinct occasions, representing different seasons (Sep. and Nov.- late dry season, Jan.- wet season) and years (2007 and 2008). These two adjacent lakes, though of relatively small area (65 and 185 km2, respectively) represent a key site for shorebirds in Zimbabwe [23] and AIV has been detected continuously in the wild bird community present at these lakes [33]. The relatively low number of individuals sampled per species and sampling occasions limit our ability to identify the species associated with this potential hotspot. The African jacana was the shorebird species that was most commonly sampled and the most frequently found infected at the Lakes Manyame-Chivero, but surprisingly no African jacana was found infected at any other sites across Africa despite a relatively large number of individuals sampled (n = 312 birds). Follow-up sampling studies conducted at Lakes Manyame-Chivero in following years during the same season and on the same species consistently detected AIV-positive birds but at a lower infection rate, making us unable to confirm the existence of a recurrent AIV hotspot at this site. Peaks in AIV prevalence may be associated with very narrow seasonal windows (e.g., few weeks in May at the Delaware Bay [6,9]). However the timing of these seasonal windows may be more variable in tropical than in temperate ecosystems. The high variability of seasonal rainfall in the tropics and the related fluctuations in the timing of reproduction and congregation of waterbirds may produce different seasonal dynamics of AIV infection between years. The inter-annual difference in prevalence measured at Lakes Manyame-Chivero may result from a difference in lake level and the related difference in the local density of waterbirds [33].

The overall absence of shorebird-AIV hotspots in our study is notable for several reasons. First, we looked for hotspots in a remarkably large number of species (n = 69) from ten shorebird families, including species and families that had never been tested for AIV infection before. This wide exploration identified many new AIV host species. Second, the majority of birds were tested for AIV infection from both cloacal and oropharyngeal samples. As predicted, a significantly higher prevalence was found in birds tested using both types of sample than those using one type of sample only. Among birds (n = 4711) tested for both cloacal and oropharyngeal samples, few birds were found positive concurrently for both types of sample (n = 3), while birds were found positive as frequently from cloacal samples only (n = 34) as from oropharyngeal samples only (n = 36) (McNemar's chi-squared = 0.01, df = 1, p-value = 0.90). Moreover, the two viruses isolated in this study originated from both an oropharyngeal (dunlin/H1N1) and a cloacal sample (spotted redshank/H10N7). Though these results confirm the importance of the respiratory tract for the replication of AIV in shorebirds as previously shown for wild ducks [2,19,20], neither the inclusion of new species nor the collection and testing of oropharyngeal swabs revealed new AIV hotspots. Major attention was given to the preservation of the cold chain from the field to the laboratory, although in some instances logistical constrains in remote field locations or unexpected international shipment delays may account for the low virus isolation rate obtained. However detection of AIV by rRT-PCR is insensitive to differences in cold-chain conditions and freeze-thaw cycles [34] therefore differences in storage conditions should not have affected our results.

Third, we found a consistently low AIV prevalence in shorebirds in several sites that are, like Delaware Bay, large seasonal congregation sites of shorebirds. Prevalence was also not related to the abundance classes of shorebirds across all our sampling sites. With two million wintering shorebirds, about one-third of the entire East Atlantic Flyway population, the Banc d'Arguin clearly qualifies as one of the larger congregation sites of shorebirds in the world. Shorebirds forage on the intertidal flats of the Banc d'Arguin at a density four times as high as the average density recorded in the other major wintering sites along the East-Atlantic coast [26]. Despite those ecological characteristics, we found a consistently low AIV prevalence in shorebirds, including in ruddy turnstone, sampled in various seasons and years at the Banc d'Arguin. We found a similarly low prevalence at other major shorebird congregation sites, including the Sivash (Crimea peninsula, Ukraine) and the Senegal River delta (Senegal-Mauritania). In Figure 4 we present the distribution of the world's largest congregation sites of waders (i.e. shorebirds of the Scolpaci (sandpipers and allies) and Charadrii (plovers and allies) clades): AIV detection studies in these birds have been conducted for seven of these sites (including two from this study) but a high infection rate has been reported only at Delaware Bay (in supporting information Table S6). Therefore, a large congregation of birds appears to be an insufficient condition for the existence of a hotspot of AIV infection.

Another characteristic of Delaware Bay is the exceptionally high number of ruddy turnstones which represent the second most abundant shorebird species at this site (annual maximum counts range from c. 32,000 to 105,000 birds [35]). No other site worldwide supports such a concentration of ruddy turnstones. By comparison, the highest abundance of this species over Western Eurasia and Africa is found at the Banc d'Arguin in winter, with maximum annual counts fluctuating between c. 4,100 and 10,300 individuals [23]. During spring migration, shorebirds gather at Delaware Bay to feed on the eggs of horseshoe crabs (*Limulus polyphemus*) spawning in the bay. Variation in spawning activity and wave action creates patches of egg concentration where shorebirds forage in densities as high as 210 birds/m^2 [36]. Much lower shorebird densities are found at other sites where resources are more uniformly distributed over the intertidal zone (e.g. 7×10^{-3} birds/m^2 on average at the Banc d'Arguin [26]). The exceptionally high abundance of ruddy turnstones, the unusually high foraging density of shorebirds and the potential variation in their feeding strategy associated with a unique food source (horseshoe crab eggs) may explain the specific AIV hotspot at the Delaware Bay.

Our repeated sampling revealed a low but continuous circulation of AIVs in shorebirds at the Banc d'Arguin throughout the wintering period of Eurasian migrants. This coastal tropical site combines several constraints to virus persistence in the environment: high temperatures, solar radiations and wind exposure, salinity, little precipitations, and tidal washing of the tidal flats. The main maintenance hosts of AIVs (the dabbling ducks of the *Anas* genus) are also absent. The perpetuation of AIV throughout the wintering period, therefore, likely results from a continuous inter-individual transmission among shorebirds facilitated by the aggregation of birds into large flocks at a few high tide roosts. Besides its importance as a wintering site, the Banc d'Arguin is also a major staging site for migratory shorebirds: it represents the tip of a flyway funnel draining migratory birds breeding across the extensive Arctic tundra of North America and Eurasia through the Atlantic coast from Western Europe down to southern Africa (Figure 1). Continuous AIV circulation at such a migration crossroad creates the potential for reassortment between AIV strains originating from different geographic areas as well as the potential for geographically extensive dispersal of new viruses. Globally, our results reveal a low but widespread circulation of AIV in shorebirds across Eurasian and Afro-tropical regions. AIV-infected shorebirds were detected in a large number of species, in all study regions and in both inland-freshwater and coastal-saline environments. This finding contrast with the absence of AIVs previously reported in studies of shorebirds in northern Europe [3,4] and suggest that these birds play a role, as in the Americas, in the epidemiology of AIV in the old word.

Results from our study also reveal that the ruddy turnstone is not consistently associated with AIV hotspots. Prevalence measured in this species was low and not significantly different from the prevalence measured in cohabiting species. Our multivariate analyses revealed the existence of significant species variations in AIV prevalence and seroprevalence. Contrary to our predictions, prevalence and seroprevalence were no lower in visual-foraging species (feeding mostly by pecking in small foraging flocks) than in tactile-foraging species (mainly probing in large foraging flocks). Nor were they lower in Arctic-coastal species that remain year-round in AIV-poor environments than in boreal-freshwater species (see methods). Conversely, no AIV antibodies were found in freshwater Eurasian or African shorebird species. This absence is striking, given that most of these freshwater shorebirds cohabit, sometimes in close proximity and high numbers, with potentially infectious wild ducks in boreal, temperate, and/or Afro-tropical wetlands.

AIV-specific antibodies acquired after a natural infection have been reported to persist generally for not more than a year in captive mallards [37] and in free-ranging migratory geese [38]. Similarly in this study we found no difference in seroprevalence between first-year and adults birds. Though the potential existence of long-lasting antibodies in shorebirds cannot be completely ruled out, the very high AIV antibody prevalence but low infection rate that we found at the Banc d'Arguin in red knot (c.80%) and ruddy turnstone (c. 50%) suggest that these species experienced a prior exposure to a relatively high AIV infection rate at other staging sites along their annual range. A similar pattern of high antibody prevalence (c. 50–90%, [8,31]) and low infection rate (<1%, [5,9]) was found in red knot at Delaware Bay, also suggesting a prior high AIV infection rate at other sites. These three bird populations have non-overlapping breeding ranges: the sub-species of red knot that winter or stop at the Banc d'Arguin (*canutus*) breed on the Taimyr Peninsula, western Siberia, whereas the sub-species that

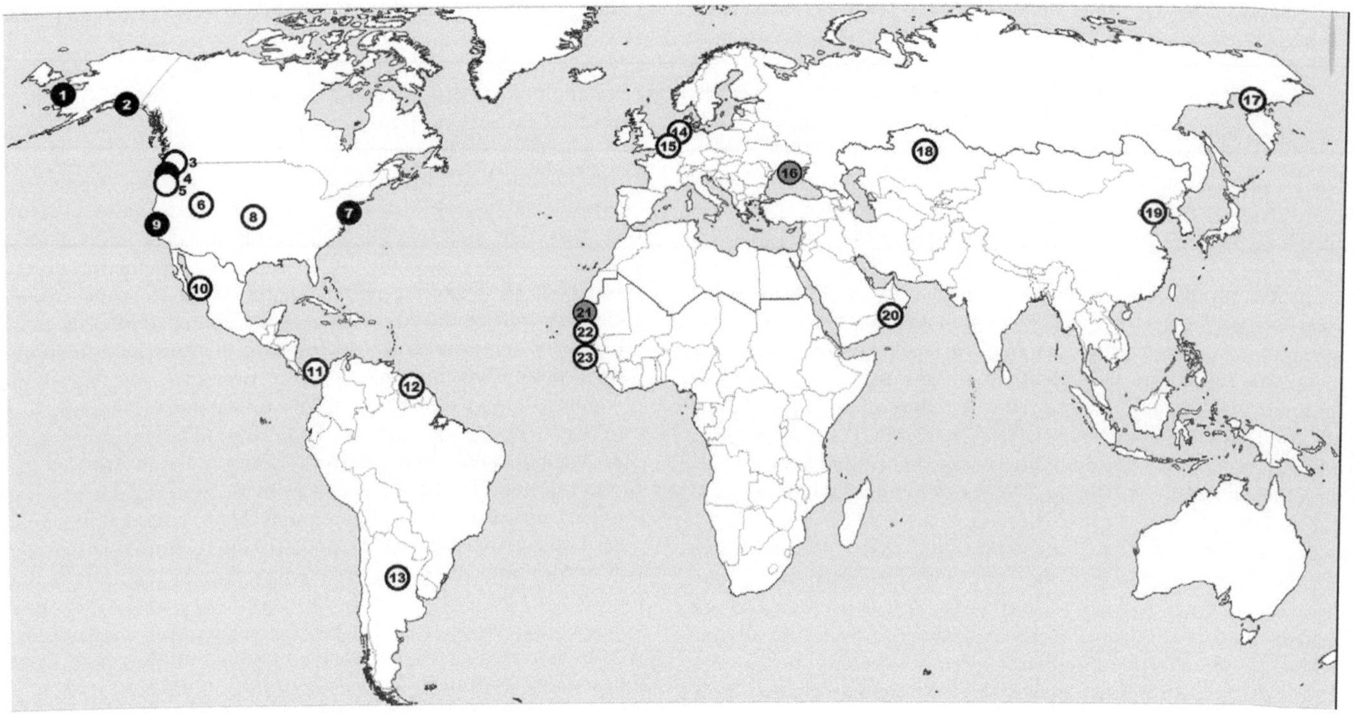

Figure 4. Location of the world's largest congregation sites of waders (sandpipers, plovers and allies), i.e. sites where at least 500,000 birds congregate annually. Among these sites where birds have been tested for AIV infection (black symbols - from the literature; grey symbols - this study) an AIV hotspot has been reported only at the Delaware Bay (no. 7). See Table S6 (supporting information) for detailed information on each site. 1-Yukon-Kuskokwim Delta, 2-Copper River Delta, 3-Fraser' River Estuary, 4-Gray's harbour estuary, 5-Bay of Fundy, 6-Great Salt Lake, 7-Delaware Bay, 8-Cheyenne Bottoms, 9-San Francisco Bay, 10-Bahia de Santa Maria, 11-Upper Bay of Panama, 12-Suriname coast, 13-Laguna Mar Chiquita, 14-Wadden Sea, 15-Rhine-Maas-Schelde Delta, 16-Azov Sea, 17-Sea of Okhotsk, 18-Tengiz-Korgalzhyn Lakes, 19-Yellow Sea coast, 20-Arabian Sea off Oman, 21-Banc d'Arguin, 22-Senegal River Delta, 23-Bijagos Archipelago (map by M. Gély ©Cirad).

stop at Delaware Bay (*rufa*) breed in the central Canadian Arctic [30]; the ruddy turnstone of the Banc d'Arguin breed in both the north-eastern Canada and the Fennoscandia-west Russia regions [23]. Very little information is available on the infection status of these populations along their migratory flyway, and hence the site and season where potentially high infection rates occur – i.e. potential AIV hotspots - remain to be discovered.

Highly contrasting seroprevalence values were found at the Banc d'Arguin among four species - red knot, ruddy turnstone, dunlin and sanderling - that are ecologically and phylogenetically highly related. A very similar pattern in seroprevalence variation was found among the same species sampled at Delaware Bay [8,9,31] (Figure 3). The two species found with a high seroprevalence at both sites - red knot, ruddy turnstone – both breed in the high-Arctic, while closely related species with a lower seroprevalence have a more southerly breeding distribution (except the sanderling, Figure 3). This finding is in contradiction with the adjustment to disease pressure hypothesis that predicts a lower exposure to pathogens and a lower investment in immune system in species breeding at higher latitudes as an adaptation to pathogen-poor environment and a trade-off to compensate the higher energetic costs associated with long-distance migration and breeding in climatically adverse conditions [30,39]. Studies conducted in the high Arctic regions (Svalbard [38], Chukchi Peninsula [40], Northern Alaska [41], Taimyr Peninsula [42]) have consistently reported no or very low AIV infection rate in wild birds. The low seroprevalence found in the sanderling – one of the more northerly breeding shorebird species – in both the Banc d'Arguin and Delaware Bay (Figure 3) also suggests that the high Arctic is not the region where high AIV infection occurs.

Differences in seroprevalence among these closely related shorebird species do not appear to be readily explained by environmental variables. They may instead result from an intrinsic difference between species in their receptivity to AIV infection, and/or in their ability to mount and maintain an acquired antibody-mediated (humoral) immune response. At Delaware Bay the low infection rate in several shorebird species commingling with ruddy turnstone at very high density to forage on the same egg resource also suggest the existence of species-level constraints to interspecies AIV transmission. The spectrum and the distribution of sialic acid receptors of AIV on host epithelial tissues varies substantially among closely related bird species [22,43] and may lead to variations in permissiveness for infection and limit transmission between cohabiting species. Species-specific differences in acquired immune responses have also been found among closely related shorebird species after an experimental infection with the same antigens, with ruddy turnstone showing higher antibody responses than sanderling, ruff and red knot [44]. The lower immune response of red knot compared to ruddy turnstone does not, however, fit with the high AIV antibody prevalence found in these two species.

On few occasions the serological status of birds individually identified from their ring number could be controlled on consecutive sampling occasions at the Banc d'Arguin NP. One dunlin seroconverted between November 2009 and March 2010, as a result of an infection that probably occurred at the Banc d'Arguin, since this site constitutes the larger southernmost staging site of dunlin of its migration flyway [23]. AIV was also detected by rRT-PCR in a red knot sampled in March 2010 that had been previously found seropositive in November 2009; this bird was still seropositive in March 2010. This suggests that acquired immunity is only partial and that a prior exposure does not fully protect against a subsequent AIV infection.

In summary, our study reveals that, when considered separately, the individual features associated with a disease hotspot do not systematically produce a locally and temporally high transmission rate in other contexts. Outside Delaware Bay, the ruddy turnstone has not been found infected at a higher prevalence than other shorebird species (in supporting information Table S1). In addition, no AIV-hotspot has been found at any of the other world's largest shorebird congregation sites investigated so far (Figure 4, supporting information Table S6). Different constituents should be combined to generate an exceptionally high transmission rate. To what extent the constituents (species, environment, and season) of AIV hotspots are identical and temporally stable, hence predictable, remains to be elucidated [1]. More generally, we suggest that interpreting existing hotspots in light of data from other ecosystems and pathogens should help to understand and work towards a more general model of hotspots.

Supporting Information

Figure S1 Number of AIV-positive birds detected per species for a given sampling occasion compared to the threshold number of positive birds (solid line) below which the prevalence is unlikely (probability <0.05) to be greater than 10% for a sample of the same size. Points on or above the line represent potential species-hotspots, i.e. species for a given sampling occasion (n = 11) for which the number of positive birds was too large for rejecting the hypothesis that prevalence could be >10%. Only species sampling occasions (n = 89) that had at least 28 birds sampled were considered in this analysis.

Table S1 Summary of the worldwide investigation of AIV infection in ruddy turnstone.

Table S2 List of sampling sites ranked by latitude and sampling details.

Table S3 Overview of the shorebird species tested for AIV infection († species reported infected with AIV for the first time).

Table S4 Seroprevalence of AIV antibodies in shorebird species sampled at the Banc d'Arguin (Mauritania) and the Inner Niger Delta (Mali).

Table S5 Results of the model selection procedure relating variations in seroprevalence of AIV antibodies to species ecological traits in shorebirds sampled at two West African sites (Banc d'Arguin, Mauritania; Inner Niger Delta, Mali).

Table S6 List of the world's largest wader congregation sites ranked by latitude on all continents. These sites correspond to locations that support at least 500,000 waders annually selected from Internationally Important Sites databases (main sources: America - Western Hemisphere Shorebird Network; Africa and Western Eurasia - Delany et al., 2009; Australasia - Bamford et al., 2008). Sites where AIV infection studies in waders have been conducted are presented in bold with details about AIV detection. Waders include all species from the Scolopaci (sandpipers, snipes, phalaropes, jacanas) and Charadrii clades (plovers, oystercatchers, stilts).

Acknowledgments

The survey at the Banc d'Arguin was conducted under the GRIPAVI project. We thank the Parc National du Banc d'Arguin for permission to work in the park. We also thank the collaborators of the national park at Iwik for their support and Mohamed Camara for his assistance in field activities. Data from the regional dataset were collected during four international programmes: the GRIPAVI project, the TCP of FAO, the GAINS and the New Flubird projects. We are grateful to the numerous ornithologists and veterinarians who participated in field operations. We also thank Ritesh Kumar and Pranav J. Pandya (Wetlands International) who provided useful information on shorebirds congregation sites in the Asia-Pacific.

Author Contributions

Conceived and designed the experiments: NG GB. Performed the experiments: NG GB ABOEM JC AC GSC PG SH RSA SRF GC CA JM BF MN YD RH. Analyzed the data: NG VG. Wrote the paper: NG. Co-coordinated the TCP-FAO project: TD SHN. Coordinated the USAID-GAINS project: GSC.

References

1. Paull SHL, Song S, McClure KM, Sackett LC, Kilpatrick AM, et al. (2012) From superspreaders to disease hotspots: linking transmission across hosts and space. Front Ecol Environ 10: 75–82.
2. Hoye BJ, Munster VJ, Nishiura H, Klaassen M, Fouchier RAM (2010) Surveillance of wild birds for avian influenza virus. Emerg Infect Dis 16: 1827–1834.
3. Olsen B, Munster VJ, Wallensten A, Waldenstrom J, Osterhaus AD, et al. (2006) Global patterns of influenza a virus in wild birds. Science 312: 384–388.
4. Munster VJ, Baas C, Lexmond P, Waldenstrom J, Wallensten A, et al. (2007) Spatial, temporal, and species variation in prevalence of influenza A viruses in wild migratory birds. PLoS Pathog 3: e61.
5. Hanson BA, Luttrell MP, Goekjian VH, Niles L, Swayne DE, et al. (2008) Is the occurrence of avian influenza virus in Charadriiformes species and location dependent? J Wildl Dis 44: 351–361.
6. Krauss S, Stallknecht DE, Negovetich NJ, Niles LJ, Webby RJ, et al. (2010) Coincident ruddy turnstone migration and horseshoe crab spawning creates an ecological 'hot spot' for influenza viruses. Proc R Soc B 277: 3373–3379.
7. Kawaoka Y, Chambers TM, Sladen WL, Webster RG (1988) Is the gene pool of influenza viruses in shorebirds and gulls different from that in wild ducks? Virology 163: 247–250.
8. Stallknecht DE, Luttrell MP, Poulson R, Goekjian V, Niles L, et al. (2012) Detection of avian influenza virsues from shorebirds: evaluation of surveillance and testing approaches. J Wildl Dis 48: 382–393.
9. Maxted AM, Luttrell MP, Goekjian VH, Brown JD, Niles LJ, et al. (2012) Avian Influenza Infection Dynamics in Shorebird hosts. J Wildl Dis 48: 322–334.
10. Krauss S, Walker D, Pryor SP, Niles L, Chenghong L, et al. (2004) Influenza A viruses of migrating wild aquatic birds in North America. Vector Borne Zoonotic Dis 4: 177–189.
11. Baker AJ, Pereira SL, Paton TA (2007) Phylogenetic relationships and divergence times of Charadriiformes genera: Multigene evidence for the Cretaceous origin of at least 14 clades of shorebirds. Biol Lett 3: 205–209.
12. Gill RE, Tibbitts TL, Douglas DC, Handel CM, Mulcahy DM, et al. (2009) Extreme endurance flights by land birds crossing the Pacific Ocean: ecological corridor rather than barrier? Proc R Soc B 276: 447–457.

13. Widjaja L, Krauss SL, Webby RJ, Xie T, Webster RG (2004) Matrix gene of influenza A viruses isolated from wild aquatic birds: ecology and emergence of influenza a viruses. J Virol 78: 8771–8779.
14. Dugan VG, Chen R, Spiro DJ, Sengamalay N, Zaborsky J, et al. (2008) The evolutionary genetics and emergence of avian influenza viruses in wild birds. PLoS Pathog 4: e1000076.
15. Chen R, Holmes EC (2009) Frequent inter-species transmission and geographic subdivision in avian influenza viruses from wild birds. Virology 383: 156–161.
16. Ghersi BM, Blazes DL, Icochea E, Gonzalez RI, Kochel T, et al. (2009) Avian influenza in wild birds, central coast of Peru. Emerg Infect Dis 15: 935–938.
17. Hansbro PM, Warner S, Tracey JP, Arzey KE, Selleck P, et al. (2010) Surveillance and analysis of avian influenza viruses, Australia. Emerg Infect Dis 16: 1896–1904.
18. Hlinak A, Mühle RU, Werner O, Globig A, Starick E, et al. (2006) A virological survey in migrating waders and other waterfowl in one of the most important resting sites of Germany. J Vet Med B 53: 105–110.
19. Jindal N, de Abin M, Primus AE, Raju S, Chander Y, et al. (2010) Comparison of Cloacal and Oropharyngeal Samples for the Detection of Avian Influenza Virus in Wild Birds. Avian Dis 54: 115–119.
20. Gaidet N, Caron A, Cappelle J, Cumming GS, Balança G, et al. (2012) Understanding the ecological drivers of avian influenza virus infection in wildfowl: a continental-scale study across Africa. Proc R Soc B 279: 1131–1141.
21. Costa TP, Brown JD, Howerth EW, Stallknecht DE (2011) Variation in viral shedding patterns between different wild bird species infected experimentally with low-pathogenicity avian influenza viruses that originated from wild birds. Avian Pathol 40: 119–24.
22. Jourdain E, van Riel D, Munster VJ, Kuiken T, Waldenström J, et al. (2011) The Pattern of Influenza Virus Attachment Varies among Wild Bird Species. PLoS ONE 6: e24155.
23. Delany S, Scott D, Dodman T, Stroud D (2009) An Atlas of Wader Populations in Africa and Western Eurasia. Wageningen: Wetlands International. 524 p.
24. Gaidet N, Dodman T, Caron A, Balança G, Desvaux S, et al. (2007) Influenza surveillance in wild birds in Eastern Europe, the Middle East, and Africa: Preliminary results from an ongoing FAO-led survey. J Wildl Dis 43: S22–S28.
25. Cumming GS, Caron A, Abolnik C, Cattoli C, Bruinzeel LW, et al. (2011) The ecology of influenza A viruses in wild birds in southern Africa. Ecohealth 8: 4–13.
26. Zwarts L, Blomert AM, Bruno JE, Hupkes R, Van Spanje TM (1990) Why do waders reach high feeding densities on the intertidal flats of the Banc d'Arguin, Mauritania? Ardea 78: 39–52.
27. Brown LH, Urban EK, Newman K (1982) The Birds of Africa, Vol 1. London: Academic Press. 521 p.
28. Cramp S, Perrins CM (1983–1985) The Birds of the Western Palearctic, vol. III–IV. Oxford: Oxford University Press.
29. Barbosa A (1995) Foraging strategies and their influence on scanning and flocking behaviour of waders. J Avian Biology 26: 182–186.
30. Piersma T (2007) Using the power of comparison to explain habitat use and migration strategies of shorebirds worldwide. J Ornith 148: S45–S59.
31. Brown D, Luttrell MP, Berghaus RD, Kistler W, Shamus P, et al. (2010) Prevalence of antibodies to type a influenza virus in wild avian species using two serologic assays. J Wildl Dis 46: 896–911.
32. Poole A (2005) The Birds of North America Online website. Available: http://bna.birds.cornell.edu/BNA/. Accessed 2012 July12. Cornell Laboratory of Ornithology, Ithaca NY.
33. Caron A, Abolnik C, Mundava J, Gaidet G, Burger CE, et al. (2011) Persistence of Low Pathogenic Avian Influenza Virus in Waterfowl in a Southern African Ecosystem. EcoHealth 8:109–115.
34. Munster VJ, Baas C, Lexmond P, Bestebroer TM, Guldemeester J, et al. (2009) Practical considerations for high-throughput influenza A virus surveillance studies of wild birds by use of molecular diagnostic tests. J Clin Microbiol 47: 666–673.
35. Clark K E, Niles LJ, Burger J (1993) Abundance and distribution of migrant shorebirds in Delaware Bay. Condor 95: 694–705.
36. Gillings S, Atkinson PW, Bardsley SL, Clark NA, Love SE, et al. (2007) Shorebird predation of horseshoe crab eggs in Delaware Bay: species contrasts and availability constraints. J Anim Ecol 76: 503–514.
37. Fereidouni SR, Grund C, Hauslaigner R, Lange E, Wilking H, et al. (2010) Dynamics of specific antibody responses induced in mallards after infection by or immunization with low pathogenicity avian influenza viruses. Avian Dis 54: 79–85.
38. Hoye BJ, Munster VJ, Nishiura H, Fouchier RAM, Madsen J, et al. (2010) Reconstructing an annual cycle of interaction: natural infection and antibody dynamics to avian influenza along a migratory flyway. Oikos 120: 748–755.
39. Hasselquist D (2007) Comparative immunoecology in birds: hypotheses and tests. J Ornithol 148: S571–S582.
40. Wahlgren J, Waldenstrom J, Sahlin S, Haemig PD, Fouchier RAM, et al. (2008) Gene segment reassortment between American and Asian lineages of avian influenza virus from waterfowl in the Beringia area. Vector Borne Zoonotic Dis 8: 783–790.
41. Ip HS, Flint PL, Franson JC, Dusek RJ, Derksen DV, et al. (2008) Prevalence of Influenza A viruses in wild migratory birds in Alaska: Patterns of variation in detection at a crossroads of intercontinental flyways. Virology J 5: 71–80.
42. Sivay MV, Silko NY, Sharshov KA, Prokudin AV, Li L, et al. (2011) The role of wild goose (Anser) populations of Russia and the Tibet Plateau in the spread of the avian influenza virus. Chinese Birds 2: 140–146.
43. Matrosovich MN, Stech J, Klenk HD (2009) Influenza receptors, polymerase and host range. Rev Sci Tech 28: 203–217.
44. Mendes L, Piersma T, Hasselquist D, Matson KD, Ricklefs RE (2006) Variation in the innate and acquired arms of the immune system among five shorebird species. J Exp Biol 209: 284–291.

18

Inferring Species Richness and Turnover by Statistical Multiresolution Texture Analysis of Satellite Imagery

Matteo Convertino[1,2,3*], **Rami S. Mangoubi**[4*], **Igor Linkov**[1,5], **Nathan C. Lowry**[4], **Mukund Desai**[4]

1 Risk and Decision Science Team, Environmental Laboratory, Engineering Research and Development Center, United States Army Corps of Engineers, Concord, Massachusetts, United States of America, **2** Department of Agricultural and Biological Engineering - Institute of Food and Agricultural Sciences, University of Florida, Gainesville, Florida, United States of America, **3** Florida Climate Institute, University of Florida-Florida State University, Gainesville, Florida, United States of America, **4** Algorithms and Software, Charles Stark Draper Laboratory, Inc., Cambridge, Massachusetts, United States of America, **5** Department of Engineering and Public Policy, Carnegie Mellon University, Pittsburgh, Pennsylvania, United States of America

Abstract

Background: The quantification of species-richness and species-turnover is essential to effective monitoring of ecosystems. Wetland ecosystems are particularly in need of such monitoring due to their sensitivity to rainfall, water management and other external factors that affect hydrology, soil, and species patterns. A key challenge for environmental scientists is determining the linkage between natural and human stressors, and the effect of that linkage at the species level in space and time. We propose pixel intensity based Shannon entropy for estimating species-richness, and introduce a method based on statistical wavelet multiresolution texture analysis to quantitatively assess interseasonal and interannual species turnover.

Methodology/Principal Findings: We model satellite images of regions of interest as textures. We define a texture in an image as a spatial domain where the variations in pixel intensity across the image are both stochastic and multiscale. To compare two textures quantitatively, we first obtain a multiresolution wavelet decomposition of each. Either an appropriate probability density function (pdf) model for the coefficients at each subband is selected, and its parameters estimated, or, a non-parametric approach using histograms is adopted. We choose the former, where the wavelet coefficients of the multiresolution decomposition at each subband are modeled as samples from the generalized Gaussian pdf. We then obtain the joint pdf for the coefficients for all subbands, assuming independence across subbands; an approximation that simplifies the computational burden significantly without sacrificing the ability to statistically distinguish textures. We measure the difference between two textures' representative pdf's via the Kullback-Leibler divergence (KL). Species turnover, or β diversity, is estimated using both this KL divergence and the difference in Shannon entropy. Additionally, we predict species richness, or α diversity, based on the Shannon entropy of pixel intensity.To test our approach, we specifically use the green band of Landsat images for a water conservation area in the Florida Everglades. We validate our predictions against data of species occurrences for a twenty-eight years long period for both wet and dry seasons. Our method correctly predicts 73% of species richness. For species turnover, the newly proposed KL divergence prediction performance is near 100% accurate. This represents a significant improvement over the more conventional Shannon entropy difference, which provides 85% accuracy. Furthermore, we find that changes in soil and water patterns, as measured by fluctuations of the Shannon entropy for the red and blue bands respectively, are positively correlated with changes in vegetation. The fluctuations are smaller in the wet season when compared to the dry season.

Conclusions/Significance: Texture-based statistical multiresolution image analysis is a promising method for quantifying interseasonal differences and, consequently, the degree to which vegetation, soil, and water patterns vary. The proposed automated method for quantifying species richness and turnover can also provide analysis at higher spatial and temporal resolution than is currently obtainable from expensive monitoring campaigns, thus enabling more prompt, more cost effective inference and decision making support regarding anomalous variations in biodiversity. Additionally, a matrix-based visualization of the statistical multiresolution analysis is presented to facilitate both insight and quick recognition of anomalous data.

Editor: Kimberly Patraw Van Niel, University of Western Australia, Australia

Funding: The authors have no funding or support to report.

Competing Interests: The authors have declared that no competing interests exist.

* E-mail: mconvertino@ufl.edu (MC); mangoubi@draper.com (RSM)

Introduction

Background

Recent decades witnessed a considerable number of alien species being brought into wetland with significant impacts on local ecosystem structure and biodiversity richness across taxa [1]. In particular, variation in hydrological regimes that occur due to natural and anthropic factors strongly affect water-dependent ecosystems [2,3,4]. Wetlands are particularly fragile due to host species' high sensitivity to variations in water regime fluctuations. In fact, wetlands respond to nutrient enrichment of associated waters in typical fashion [5]: some shift in plant community composition occurs first after nutrient levels in soil increase, followed by changes in both aquatic and wetland-dependent animal communities [6]. In oligotrophic wetlands, such as peatlands and ombrotrophic bogs, responses to eutrophication may be more rapid, more dramatic, and longer lasting, naturally implying an intrinsic strong correlation between water, soils, and vegetation [7,5]. Particularly not well understood is the correlation between species richness and rainfall [8,2]. Yet, an understanding of linkages between soil properties in wetlands and above-ground landscape patterns is critical to developing quantitative soil-landscape models that will aid in detecting, localizing, and characterizing changes in species composition.

Currently, assessment of biodiversity at local and regional scales often relies on fieldwork-based data collection [9]. Species assessment in relatively large or weakly accessible areas is a longtime challenging task for ecologists. As stressed in [10], a key factors need to be determined before a sampling design is ready for implementation, such as: (i) the number of sampling units; (ii) the spatial placement of the sampling units; (iii) clear definition of the statistically meaningful species of concern; and, (iv) an operational definition of a species community. Moreover, field-based approaches are typically labor intensive and costly, and only a small fraction of a study area may be sampled [10]. The same holds true for hydrological and geological fieldworks that aim to better understand the dynamics of water and soils.

The above discussion attests to a crying need for methods that extract information and accurately analyze remote sensing images nowadays available. High-resolution satellite imagery provides detailed spatial characteristics over large areas of ecosystems and offers a promising potential for accurate vegetation mapping [11,12,13,14,15]. However, most multispectral image classification techniques more commonly focus on spectral discrimination of ground objects for single-species detection [16,17], and may overlook pertinent information extractable through analysis of spatial, indeed spatiotemporal, pixel intensity variation residing in high resolution images [18], not to mention the multiscale nature of these variations. Noteworthy, the work in [18] emphasizes the need for methods that rapidly and objectively forecast species diversity thru spatiotemporal data analysis. The work in [18] also demonstrates how previous techniques produced poor assessment of species-richness, or α diversity, and particularly pairwise species dissimilarity, or β diversity. While the work in [19] focuses attention on β diversity as an important measure of species dissimilarity between communities, the spatial analysis presented do not consider the time varying aspects of species dissimilarity. Besides the pairwise species dissimilarity, species-turnover, or β diversity, reflects the change of environmental variations, such as of rainfall and soil, as dictated by either natural events, anthropic events, or both [2]. Species turnover is a major motivation for the work in [20,21], and a significant number of research efforts analyze α diversity at different spatial scales (e.g. [22,23,24,8,25,26]). Yet, apart from [10], time variations in species composition is often overlooked.

Greater Everglades Ecosystem Restoration

In this study we consider the Water Conservation Area 1 (WCA 1) in the Greater Everglades Restoration Area, also known as the Arthur R. Marshall Loxahatchee National Wildlife Refuge. This is a constructed tropical wetland (Figure 1). We demonstrate our approach for 28 year wet and dry seasons using Landsat observations. Among wetlands worldwide, the Greater Everglades Ecosystem Restoration area (GEER) has undergone the most considerable changes in ecohydrological patterns. The changes are due to the constructions of a set of levees and canals aimed to control water flow in the area. Thus, GEER is a reference wetland among ecohydrologists and environmental scientists at large [7,1,27,4,3]. The work in [4] analyzes the effect of rainfall variation induced by climate change for the whole Everglades National Park. This study evidenced the importance of rainfall for the Everglades, an ecosystem defined as strongly rainfall-dominated. Within GEER, WCA 1 is one area that, in comparison to others, underwent the smallest ecohydrological changes; rainfall thus controls the majority of its water. Thus, WCA 1 is a unique area to test the correlation between climatological and ecohydrological patterns. The choice of GEER to demonstrate our proposed methods is opportune, as there is currently a serious ongoing debate regarding the restoration of Everglades according to the original predrainage patterns; the main concern regards effects that such restoration may bring to the ecosystem [1,27]. Prior to undertaking such restoration it is very important to develop analytical methods and tools capable of characterizing and localizing multiscale spatiotemporal changes, such as in vegetation patterns, as a function of changes that derive from previous interventions and/or natural changes of climate. Additionally, the analysis of vegetation, plant species-richness in space and time in particular, is also crucial to the development of models that are capable of predicting scenarios of different management alternatives. Management alternatives for WCA 1 are different configurations of the levee system that regulates the ecohydrology in the area. Generally, ecological models are calibrated and validated on observed data, and accurate parameter estimates are therefore critical to enabling these models. Regarding WCA 1, we are only aware of the effort of [28] for evaluating the restoration of the Everglades using satellite imagery. However, [28] did not consider any species richness indicator.

Approach

In a new approach to this problem, we aim to extract species richness in space and time from multispectral satellite imagery, using statistical multiresolution wavelet texture analysis. We adapt the approach in [29], used for retrieval, to texture classification. We demonstrate our analysis of WCA 1's ecohydrological patterns by considering interseasonal and interannual species dissimilarity - which is better known as "species-turnover" [21,20] - and species richness for twenty-eight years of Landsat observations. The approach consists of applying wavelet decomposition, and statistically modeling then comparing these coefficients either thru a non-parametric histogram, or thru an appropriate probability density function. We elect a model that embraces a wide spectrum of central and tail behavior that can be represented by varying just two parameters at each subband. Specifically, the coefficients from each scale, even each subband of the decomposition are considered samples from a generalized Gaussian (GG) probability density function (pdf). The coefficients provide estimate of the characterizing GG pdf parameters. We next assume independence

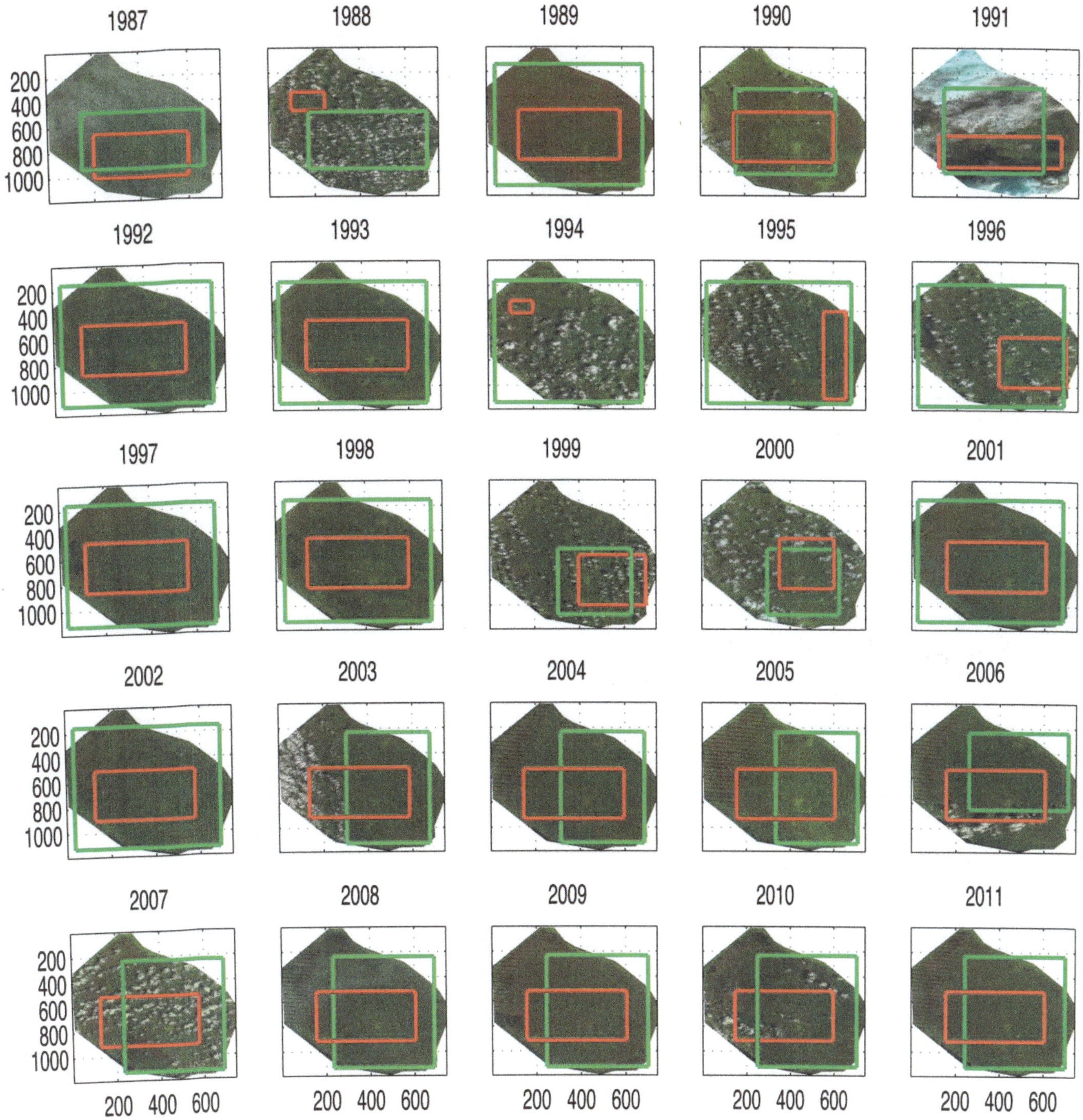

Figure 1. Remote-sensed images for the Arthur R. Marshall Loxahatchee National Wildlife Refuge (WCA 1) during the dry season for the period 1987–2011. The first three years (1984–1986) images are not represented. The representative region in which the texture analysis is performed is delineated in red for each image. The red regions are characterized by a cloud cover lower than 20%. The green regions identify where the data of species are available. Figure S1 reports the images for the wet season.

across scales and across subbands within scales and obtain a joint pdf for each textured image region. The independence assumption is only an approximation, but experience has shown that it provides accurate results when the objective is to quantitatively examine differences across regions and time points. There are many ways to quantitatively assess differences between pdf's; we choose the relative entropy, known as the Kullback-Leibler divergence (KL). Other methods, such as likelihood ratios, and various metrics, could also be examined. We use the KL divergence and the Shannon entropy to assess species dissimilarity in time (β diversity), and species richness (α diversity) respectively. Entropy is widely reported as a proxy of species richness [30], while spectral heterogeneity measured by reflectivity is a measure to quantify the entropy [31,32].

We validate our predictions on independent data from the Global Biodiversity Information Facility [33] that are composed by field-data. We do not compare directly our method with other multispectral analysis methods (e.g. see [18] for a review) as we prefer to provide a novel theoretical and methodological framework - tested against long duration data records - for analyzing multispectral images.

Previous Research

To the best of our knowledge this is the first time that a statistical model based multiresolution texture analysis is applied to satellite imagery for quantifying species-turnover, a multiscale spatiotemporal phenomenon. Previously, the work in [19], [34] and recently [35] used the KL divergence for estimating β diversity in space as a theoretical construct without any model. In [36], texture analysis predicts avian biodiversity richness based on satellite imagery. Species turnover, however, was not considered. Other previous texture-based calculation used the intensity of imagery for estimating biodiversity variables. Texture-based analysis have investigated species habitat relationships [37], successional attributes of forest species [38], classification of species [39], species richness as a function of spatial and spectral resolution [15], intercorrelations of vegetation biodiversity and soil dynamics at both local and regional scales [40], and different band combinations of images [41].

None of the above authors, however, specifically account for multiple scale variations in both space and time. Statistical multiresolution texture analysis proved successful in other applications; it is previously applied to classify stem cell colonies [42,43,44], thus enabling non-invasive analysis of these cells. Non-invasive analysis of cells is a disruptive technology that carries the potential of supplementing, even replacing invasive and at times destructive chemical biomarkers. As such, in addition to lowering cost, it provides two advantages: (1) it provides a quantitative statistical assessment of these cells; and (2) enables the preservation of these cells for their intended use, such as drug testing and tissue formation. For the ecological study of concern here, the analysis will have similar advantages: it will reduce the need for costly field fieldwork-based data collection [9]. It is also non-invasive, while fieldwork data may also perturb the ecosystem and the reliability of the collected data may be low because of the generated disturbance or because of the limitedness of the sampled data [45,1].

Summary of Contributions

The overall objective of this study is to investigate the spatiotemporal relationships between soil, water, and vegetation patterns derived from remote sensing data in a tropical wetland. We adapt and apply a multiresolution statistical texture analysis to remote sensing imagery. The methodology developed aims to account for both the statistical and multiscale spatiotemporal nature of the above mentioned relationships, thus enabling more parsimonious inference at higher spatial and temporal resolution. We also investigate the accuracy of the KL divergence as a measure of β diversity in time versus conventional approaches such as the difference of the Shannon entropy. We detect: (i) the hydrological footprints in term of correlation between water, vegetation, and soil changes; (ii) the interseasonal variation of species-richness in which dry and wet seasons are compared for each year (Figure 2); and, (ii) the interannual variation of species-richness for the same season between consecutive and non-consecutive years (Figure 2). We verify that the use of the KL divergence estimated on texture performs better than conventional methods. This analysis is performed successfully on low-medium resolution data (Landsat images). Moreover, we illustrate the results with a novel color based visualization that provides insight into the ecological variations, and enables a quick recognition of anomalous data.

Materials and Methods

Satellite Imagery

The satellite images of the Water Conservation Area 1, described in Figure 1 and Text S1, are derived from the Landsat database of [46] and [47]. Figure 1 and Figure S1 show WCA 1 for the 1987–2011 period in the dry and wet season respectively. The resolution of these images is 30 m. The images from 1984 to 1998 are from the L4–5 TM dataset, from 1999 to 2003 from the L7 ETM+ with SLC-on (1999–2003), and from 2003 to present from the L7 ETM+ with SLC-off. The Scan Line Corrector (SLC) stopped working on May 31, 2003 and that caused striping of remote sensed images [48]. Thus, we do not consider in our analysis the striped parts of these images. Each Landsat image is cropped along the boundaries of the Water Conservation Area 1 provided by the South Florida Water Management District. The representative area of analysis for each image is selected in oder to guarantee a cloud cover lower than 20% on average for each selected area and a minimum area requirement for the texture method (Section). At the same time, we strived to maximize the extension of the selected area and its representativeness of the whole landscape. For each year two images are collected, one for the dry season, December to April, and one for the wet season, May to November. Images were selected in January and August, the most representative months for the dry and wet season respectively. In Figures 1 and S1, the red squares show the regions selected for texture analysis. We analyze the satellite images for each visible band defined as:

We analyze the satellite images for each visible band defined as:

- Band 1: (Blue; electromagnetic wavelength: 0.45–0.52 μm), which is useful for mapping *water*, differentiating between soil and plants, and identifying manmade objects such as roads and buildings;
- Band 2: (Green; electromagnetic wavelength: 0.52–0.60 μm), which spans the region between the blue and red chlorophyll absorption bands, and shows the green reflectance of *healthy vegetation* (thus vegetation that changes across seasons and years). It is useful for differentiating between types of plants, determining the health of plants, and identifying manmade objects;
- Band 3: (Red; electromagnetic wavelength: 0.63–0.69 μm), which is one of the most important bands for discriminating among different kinds of vegetation. It is also useful for mapping soil type boundaries and geological formation boundaries. It is considered as a proxy of *soil* types.

Species Occurrences

The species richness for WCA 1 is compiled by assembling together data from fieldworks of [45], the Global Biodiversity Information Facility database [33], and the Comprehensive Everglades Restoration Project database [49]. In order to compare the predicted species richness from the analysis of the green band of Landsat images, we build a species richness matrix at 30 m resolution from the aformentioned sources. For more information about WCA 1 we refer the reader to Text S1.

The "base" local species-richness (or α diversity) of plant communities is built from the Global Biodiversity Information

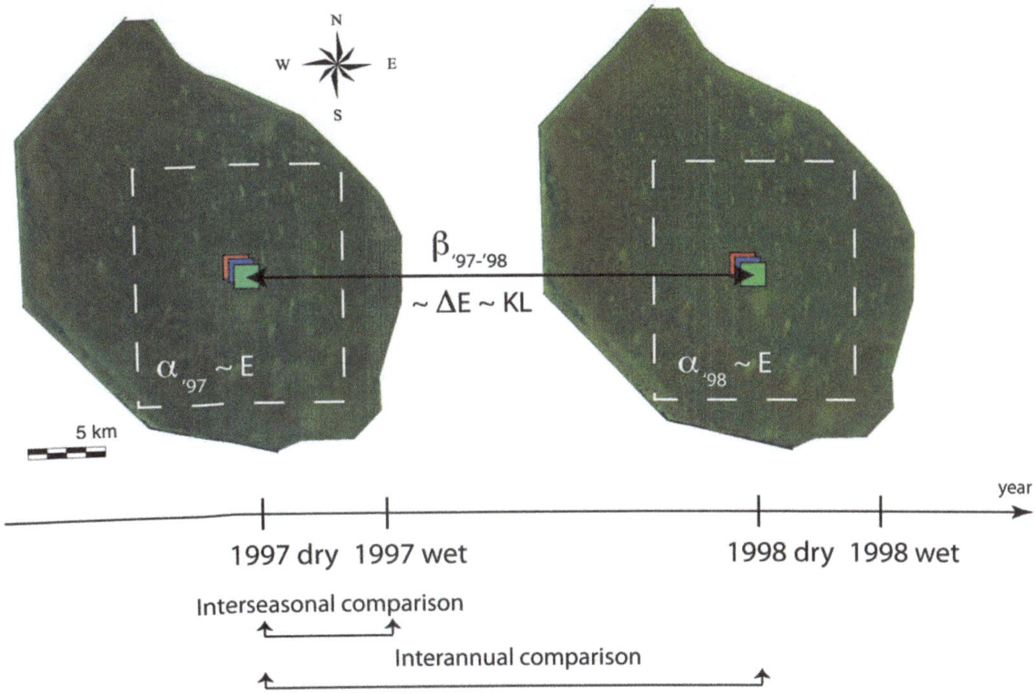

Figure 2. Graphical explanation of the analysis performed for the WCA 1. Landsat images (from 1984 to 2011) are acquired for the dry and the wet season. The example is reported for the years 1997 and 1998. The changes in vegetation composition are analyzed using α and β diversity among seasons and among years. For the interseasonal analysis the time-scale is on average six months between seasons of the same year, while for the interannual analysis the time-scale is about a year between the same season of different years. α and β diversity from data are compared to the estimates of the Shannon entropy (Equation 5) and of the KL divergence (Equation 4) for the green band of the images respectively.

Facility [33] data that provide occurrences of species in space and time at 1 km resolution. The green squares in Figure 1 and S1 indicate the regions where the GBIF data are available. We consider data from 1984 to 2011 such as for the Landsat images. For the GBIF data we downscale the information of species richness from 1 km^2 to 30 m^2 using a simple coarse-graining algorithm [50]. A grid of smaller boxes is created and the number of species are counted at the resolution of the smaller grid. The estimation of species richness is also refined with point data of plant species occurrences from CERPzone, that is the database of the Comprehensive Everglades Restoration Project [49]. CERPzone data are available from 2001. In the species richness matrix we also include the data of [45], where a total of 30 plant species are found along the Loxahatchee National Wildlife Refuge transect which goes from the westernmost to the easternmost boundary of WCA 1 at its maximum width. According to [45] the species-richness at a given site never exceeded $8\,m^{-2}$. Of these 30 species, only 11 were found in both 1989 and 1999 (Table 2 in [45]).

The occurrences of these three datasets are compared together after creating a grid over WCA 1 in which the unitary pixel is characterized by an area of 30 m^2. The point occurrences of CERP and [45] are downscaled to 30 m^2 following the approach of [51] that is a shot-noise Cox process method. All the unique occurrences of species from all datasets are assigned to each pixel according to a distance criterion. Each occurrence characterized by geographic coordinates is assigned to the nearest neighbor pixel. The local species-richness or α diversity is calculated by the sum of unique species in each pixel, and the average local species-richness of WCA 1 is the average over all the species-richness values of each pixel. Species turnover or β diversity is calculated on the data of [33] as complementary to the Jaccard Similarity Index (JSI) evaluated in time at resolution of 30 m. JSI is given by the ratio of the number of common species in two pixels considered and the number of all species in both pixels. Specifically, $\beta = \frac{\alpha_{ij}}{\alpha_i + \alpha_j - \alpha_{ij}}$, where α_i and α_j are the numbers of species present in pixel i and j at different seasons or years). We also note that γ diversity is the number of distinct species in the whole domain of the green squares (Figure 1).

Specific Study Hypothesis

Our study is based on the following hypothesis formulated on some previous studies and data.

1. Rainfall is the main driver of vegetation patterns in WCA 1 and in tropical wetlands [1]. This is evidenced by hydrologic studies [4] and by analysis of satellite imagery pre- and post-construction of the levee-canal drainage systems [1] We do not anticipate any relationship between changes in soil, water, and vegetation richness in WCA 1 because non-linearities were evidenced among these variables [5]. However, because of the high seasonality of climate [8,1] we expect variation of species richness within each year for the wet and dry season such as was reported for single species [52].
2. Landsat RGB imagery, despite its low resolution,reveals information about the spatiotemporal structure of ecosystem components [53,54]. As stated by [54] high resolution imagery is not always available for free for all the regions in the world. Moreover, imagery of higher quality does not always guarantee better estimation of biodiversity variables. This is particularly true in tropical ecosystems where the resolution of high quality imagery can be too small compared to the extent of each single species [54]. We believe that the improvement of methodologies for assessing biodiversity variables and other environ-

mental variables of ecosystems such as in this paper, can overcome the low quality of imagery such as the Landsat imagery.

3. The Shannon entropy of the green band can be considered as the spectral heterogeneity (reflectance) of plant species in water-dominated ecosystems [55,56,57,58,59,60,61,62,63, 64,65,66]. The blue and red bands can be considered as the spectral signatures of water and soil heterogeneities in the ecosystem [55]. Different species have different reflectance levels for different bands of the Landsat images. The higher the range of reflectivity, the higher the entropy because there are potentially more plant species with different degree of reflectivity. In ecology there is an large discussion that different reflectance levels do not always characterize different species, but "functional species" [55,58,59,67,53,68]. The issue about species diversity versus functional diversity is a longstanding issue in ecology. Functional diversity, which is different but related to taxonomic diversity (i.e., species richness), is crucial to ecosystems and the services they provide to humans [59]. In this paper we talk about "species" rather than "functional species"; nonetheless, we make clear that similar reflectance levels may characterize functional species or vegetation types (within the same spectral group) rather than the same species. No distinction is made here between invasive and endemic species; however, the variation of species richness can also be attributed to the invasion of non-native species and inform species management. Other sources of variation in surface reflectance, such as directional effects and shadowing, are also sources of texture variations in Landsat scenes. Nonetheless, we believe our algorithm is capable of detecting the average texture generated by species differences.
4. α and β diversity are related to the Shannon entropy and to the KL divergence of the green band respectively. Previous studies investigated these relationships based on data and theoretical constructs [19,35]. We also hypothesize that α and β diversity in time are independent measures of biodiversity richness as suggested by previous studies [69,70,71,72,73,20]. We conjecture that the average α diversity is a lower but close estimate of γ diversity. This is because the distribution of species is quite homogeneous within WCA 1 as verified by data [33] and as reported in [74]. Thus, because the number of species is on average the same in any representative region within the whole ecosystem and for the period considered, scale-invariance of α diversity is hypothesized.
5. The difference between Shannon entropies provides a worse estimate of β diversity in time than the KL divergence. We expected that this occurs because the KL divergence, that is not calculated on the intensity of the Shannon entropy, potentially accounts for the pairwise interactions between vegetation communities in time (i.e. the mutual information in information theory). On the contrary, the difference between Shannon entropies does not capture the community pairwise interactions [19].

Texture Image Analysis

We model the region of interest as textures. Texture models abound in the literature and textbooks (see e.g. [75]). We favor a model that captures what in our view are two essential properties of texture: variations of intensity across pixels that: (i) occur at multiple scale, and (ii) are stochastic in nature. Such a model suggests that a wavelet analysis can decompose the spatial intensity variation into multiple scales, and subbands within scales, after which the coefficients of the decomposition at each scale can be analyzed statistically. The across scale statistical analysis characterizes the texture. This is the approach followed in [29] for texture retrieval. As wavelet analysis decomposes a signal locally according to orientation and scale, it is especially apt for modeling texture, characterized by intensity randomness at multiple scales. Note that perceived image texture is closely related to the degree of random fluctuation in image gray-scale intensity at multiple frequencies. An n-level wavelet decomposition produces $3n$ detail subbands, three per level, whose coefficients convey information about the fluctuation at a particular scale and orientation (one oriented horizontally, one vertically, and one diagonally). We compute textural features from these subbands by modeling the empirical pdf of the coefficients in each subband. Moving from one level to the next the scale increases. The above method was also adapted to texture comparison and classification, and used to non-invasively and quantitatively classify stem cell colonies [42,43], without using chemical markers, thus preserving the colonies for use. The method is also used to classify nuclei in [44,76,77]. We will adopt this method here to compare Landsat images of regions across seasons and years, and demonstrate in the next section that it successfully predicts species turnover. A complete description of this approach is beyond the scope of this paper, byt the approach is thoroughly developed in [29] and [42]. We briefly outline the steps below. For representing a texture by a probability model:

1. Texture window selection. For the texture of interest within the image, select the largest possible representative window where the patch (or region) of the image is analyzed; a minimum of 64×64 pixels is required. The patches have different size from year to year. The size of the patches varies because we aim to select the largest cloud-free regions possible. Thus, size is increased until the cloud coverage in the area is lower than 20%. We used a moving window in space to select each patch. The size of the patches varies from 10% to 96% of the total area of WCA 1 that is 588 km^2 (Figure 1). Thus the area of the patches varies from 58 to 564 km^2 approximately.

2. Wavelet decomposition. Apply a multiscale wavelet decomposition to the selected window. Collect the coefficients at each detail scale of the decomposition. The specific choice of wavelet and the number of scales is a design choice, and more than one choice will yield appropriate representations. Wavelet analysis is a generalization of Fourier analysis that quantifies the degree at which pixel intensity varies at multiple scales or electronic magnifications. As wavelet analysis decomposes a signal locally according to orientation and scale, it is especially apt for modeling texture, characterized by intensity randomness at multiple scales. Note that perceived image texture is closely related to the degree of random fluctuation in image gray-scale intensity at multiple frequencies. An n-level wavelet decomposition produces $3n$ detail subbands, three per scale, whose coefficients convey information about the fluctuation at a particular scale and orientation (one oriented horizontally, one vertically, and one diagonally). We compute textural features from these subbands by modeling the empirical pdf of the coefficients in each subband. Moving from one level to the next the scale increases.

3. Probabilistic modeling at each scale. For each decomposition scale, either represent the coefficients by a non-parametric histogram, or instead select a probability density function (pdf) as a model and use the coefficients to estimate the pdf's parameters. In [76], the first approach, non-parametric histograms, is used, while in [29,43,44,77], the Generalized Gaussian (GG) probability density function was selected as an appropriate model for the wavelet coefficients at each scale. The GG pdf for detail coefficients at scale s is given by:

$$f(x_s;\, v_s,p_s)=\frac{p_s}{2\omega\Gamma(1/p_s)}\, e^{-(|x|/v_s)^{p_s}} . \qquad (1)$$

Here, x_s is the random variable or detail coefficient at scale s, and v_s and p_s are the distribution width factor and shape parameter, respectively, for subband s. Additionally, $\Gamma(z)=\int_0^\infty t^z e^{-t}dz$ indicates the Gamma function. We assume the location parameter (i.e. process mean) to be zero as the detail coefficients are the outputs of a high-pass filters. The GG density may be used to model a wide variety of symmetric, unimodal density functions; special cases include the (standard) Gaussian ($(v,\, p)=(\sqrt{2}\,\sigma,\, 2)$), standard Laplacian ($(v,\, p)=(\sigma/\sqrt{(2)},\, 1)$), and uniform ($p\rightarrow\infty$) densities. σ is the standard deviation, which for a GG Distribution process (GGD) is $\sigma=v\left(\frac{\Gamma(3/p)}{\Gamma(1/p)}\right)^{\frac{1}{2}}$. A variety of techniques exist for estimating parameters v and p, including moment-matching [78] and maximum-likelihood [29,79,76].

4. Texture joint pdf model. Construct the joint pdf for the texture by combining the pdf's for each scale's subbands, assuming independence across scales. We note that such an assumption is not true in reality, but it eases the computational burden while providing good results for classification. For S scales, with $\mathbf{v}=(v_1,\ldots,v_s,\ldots,v_S)$, and $\mathbf{p}=(p_1,\ldots,p_s,\ldots,p_S)$.

$$f(x;\,\mathbf{v},\mathbf{p})=\prod_{s=1}^{S} f(x_s;\, v_s,p_s) \qquad (2)$$

One of the key advantages of the GG distribution model is that a closed-form solution exists for the KL between two GGD processes simplifying computation: $D_{GGD}(f||g)=log\left(\frac{v_2 p_1\Gamma(1/p_2)}{v_1 p_2\Gamma(1/p_1)}\right)+\left(\frac{v_1}{v_2}\right)^{p_2}\frac{\Gamma[(p_2+1)/p_1]}{\Gamma(1/p_1)}-\frac{1}{p_1}$. Textures are compared in a pairwise fashion by obtaining the difference between their respective pdf's. Specifically, to compare two textures with pdf's $f_i(x)=f(x;\,\mathbf{v}_i,\mathbf{p}_i)$, $i=1,2$, respectively, we calculate a measure of the pdf's divergence or difference. There are many ways to express the divergence between pdf's, including likelihood functions, various metrics such as the L1 divergence ($\int|f-g|$), the Bhattacharyya distance, and the KL divergence. We elect the latter as it is inspired by the relative entropy between pdf's. The information divergence is especially convenient as it admits a tractable closed-form solution for GG distributions. Specifically, we have for the Kulback-Leibler divergence,

$$D_{KL}(f,g)=\int_{-\infty}^{\infty} f(x)\,\log\frac{f(x)}{g(x)}\,dx\,, \qquad (3)$$

that theoretically is $-E_f[\log g]-H(f)$ where $E_f[\log g]=\int f\log g$ is the expectation under f of $\log g$, and $H(f)$ is the true entropy in f, i.e. $H(f)=-\int f\log f$. The divergence in Equation 3 is not symmetric, meaning that $D_{KL}(f,g)\neq D_{KL}(g,f)$. To obtain a symmetric version, we define

$$D_{sym}(f,g)=\sum_{i=1}^{3n} k_i\; D_{KL}(f,g)+D_{KL}(f,g), \qquad (4)$$

summing across all $3n$ subbands, where k_i are simply weights assigned to particular subbands $i=1,\ldots,3n$. In general, we set all $k_i=1$, but the weights might be used to emphasize or penalize certain bands according to prior knowledge. In the results and discussion section, we demonstrate that the above symmetrized KL divergence is an effective predictor of species turnover.

Ecological Equivalence

The ecological equivalence of texture analysis is based on the assumption that the entropy and the KL divergence represent the potential species-richness and the dissimilarity in time of species-richness within communities (α and β diversity respectively). In the following we propose a theoretical description of entropy and KL divergence in ecology making analogies to the definition of these quantities in image analysis (Section). Because of our hypothesis on the reflectance of species, the entropy in shades of green means diversity in types of plants, which determines α diversity. KL divergence between two regions at different times means different shades of green for the two regions, which determines β diversity in time. α and β diversities are among the most employed theoretical concepts in ecology and biodiversity conservation. For most ecologists, α diversity traditionally reflects the within-habitat diversity [80], whereas β diversity is the component of "total diversity" that is produced by differences in species composition among the sampling units [81]. The need to partition diversity within and among habitats has both theoretical and applied interests. Species-richness is a diversity index of order zero, Shannon entropy is a diversity index of order one, and all divergence measures are diversity indices of order two [30,18]. Because the Shannon entropy is a first order measure of species diversity [30], the variation of Shannon entropy in time is as well a first order measure of species diversity between the same or different regions.

The Hill order [82] is the order of any information measure that is derived from the generalized expression $\sum_{i=1}^{S} p_i^q$ or limits of such functions as q approaches unity, where p_i is a probability distribution function and S is the number of equally common species that compose a community. The diversity metrics from $(\sum_{i=1}^{S} p_i^q)^{1/(1-q)}$ are often called "Hill numbers", but they are more general than [82]'s derivation suggests [30]. The exponent and superscript q may be called the "order" of the diversity; the true diversity depends only on the value of q and the species frequencies, and not on the functional form of the index. For q that approaches unity the information functions are the species richness, Shannon entropy, all Simpson measures, all Renyi entropies, all HCDT or "Tsallis" entropies, and many others. All values of q less than unity give diversities that disproportionately favor rare species, while all values of q greater than unity disproportionately favor the most common species [83,30].

Generally the higher the metric's order the better the estimation of the ecological metric. With the knowledge of entropy the species-turnover could be potentially calculated as difference of entropies between two different time steps. However, we consider whether the KL divergence is a better measure with respect to the difference of entropies. In fact, the KL divergence captures the pairwise variation of species-richness of local communities and not only their independent variation of species-richness [19]. For example [84] showed how the pairwise interaction is not negligible in grasslands and this has profound implications for the ecosystem functioning. Recently the KL divergence was related to the Shannon entropy and the species-richness in the whole ecosystem analyzed [35]. We consider that the species-richness and dissimilarity estimations are representative of the whole region extending the analysis from the subregions in which the texture analysis is performed. The entropy calculated by our texture analysis method can be assumed to be an estimate of the maximum likelihood estimation of the Shannon's index [19] defined as:

$$E_P(t) = -\sum_{i=1}^{S} f_i(t) \ln f_i(t) \,, \qquad (5)$$

where $f_i(t)$ is the observed sample fraction a local community P (N_i/N, where N can be considered as the number of individuals of species S, and N_i is the number of individuals of the i-th species) at time t. The entropy here introduced has the same meaning of the entropy calculated using the intensity of the green band images. The same entropy can be calculated over an other local community Q for an other function g. The entropy as defined above is one of the most commonly used index of α diversity [19,71], that is the number of species in a subarea of the region considered, in theoretical ecology and ecological modeling. Certainly there are other metrics to estimate α diversity, but the entropy is one of the most profound and useful of all diversity indices. However its value gives the uncertainty rather than the diversity. Thus, the entropy needs to be rescaled to diversity given some data or calculated as in [30]. The entropy is easily derived from the "order generating function" of [81]. The predicted α diversity is tested vs. the estimation of α diversity from the data described in Section at resolution of 30 m.

The KL divergence calculated by our texture analysis method is an estimate of the divergence:

$$\begin{aligned} KL[P(t):Q(t+\Delta t)] &= D_{KL}[P(t):Q(t+\Delta t)] \\ &+ D_{KL}[Q(t):P(t+\Delta t)] = \\ &= \sum_{i=1}^{S} [f_i(t) - g_i(t+\Delta t)] \ln \frac{f_i(t)}{g_i(t+\Delta t)} \\ &= -E_P(t) - E_Q(t+\Delta t) + E_{PQ}(t) \,, \end{aligned} \qquad (6)$$

where D_{KL} is the discriminant information introduced by [85] (Equation 3). Thus, KL is the symmetric version of the divergence 4. Equation 6 is estimated using Equation 4. Lets assume that local communities (or regions) P and Q are composed by the same species (labelled from 1 to S) with probability distributions $P = f_1, f_2, \ldots, f_s$ and $Q = g_1, g_2, \ldots, g_s$ respectively, with $f_i > 0$, $g_i > 0$ and $\sum_i^S f_i = 1$, $g_i^S f_i = 1$. P and Q are evaluated at different time step, t and $t + \Delta t$ where in our analysis Δt is approximatively six-month and a year for the interseasonal and interannual comparison of the species dissimilarity, respectively. E_P and E_Q are the Shannon entropies of communities P and Q calculated as in Equation 5, and E_{PQ} is the "reciprocal information". The reciprocal information can be assessed only after the knowledge of KL (Equation 4) and of E_P and E_Q. The reciprocal information represents the pairwise interactions among vegetation species in different communities in time (or in space) which is not captured by completely by the differences of entropies of the two local communities. The KL divergence gives a measure of how much the distribution of species within the communities differs from one another. This has been traditionally called β diversity in ecology. The β diversity is the most useful metric from which it is possible to assess the number of species and their abundance. We do not use the KL divergence for measuring the dissimilarity of species in space within the same region or between different regions although this is a possible operation that can be performed. Thus, Equation 4 is a proxy of Equation 6 of the landscape analyzed.

Results

The entropy of each band for the dry and wet seasons is shown in Figure 3. The Shannon entropy is calculated on the green band of the Landsat images using Equation 5. In this paper we use the Shannon entropy of the RGB bands as spectral signature of temporal heterogeneity of the landscape for soil, vegetation and water. Thus, we deviate from [11] which uses the spectral heterogeneity as the mean of the pairwise Euclidean distances in the wavebands dimensional space. There is no significant decadal change in plant community patterns as confirmed by [45]. In fact, the entropy of the green band that is a proxy of α diversity is overall constant both for the dry and wet season in the twenty-eight years analyzed. This may be both related to the confinement of species within WCA 1 (the set of surrounding levees block immigration of species) and to the absence of dominating species that do not spread over large regions. Thus, the set of species living on WCA 1 is believed to be the same on average. The comparison of Figure 3 (a) and (b) shows that the fluctuations of the Shannon entropy for each band in the wet season (Figure 3, b) are moderately smaller than in the dry season (Figure 3, a). Moreover, the fluctuations of the Shannon entropy of the red, green, and blue bands are highly correlated within the same season. We note also that the Shannon entropy of the RGB bands for the dry and wet seasons are inversely correlated. This is expected considering the opposite ecohydrological dynamics in South Florida in the dry and in the wet season.

The dry season is characterized by a lower density of vegetation than the wet season which theoretically increases the probability to detect different species. However, the reflectance of the green band is higher in the wet season but it tends to be more homogeneous than in the dry season. The Shannon entropy can be considered as the probability to detect species in a given area. Thus, the higher the Shannon entropy, the higher the probability to find all the species within the area considered. Figure 3 (a, b) shows that this probability increases during the dry season due to an higher Shannon entropy than the wet season. Thus, reflectance and heterogeneity may plays a contrasting role in assessing local species richness. We observed a considerable turnover of species-richness that is associated to the fluctuation of the average annual rainfall (Figure 3, c). The entropy of the red band is higher in the dry period than in the wet period as well as the entropy of the blue band in the same season. Despite the decrease of the average annual rainfall in the last decade (Figure S2 (b) shows this trend from about the year 2003) the average species-richness is observed to be constant (Figure S2, a). Figure S2, S3, S4 (Text S1) report the relationships between the average Shannon entropy and the average rainfall at the year scale. Figure S2 reports the interannual entropy and the average annual rainfall derived from [86]; Figure S3 shows the cross-correlation between these two quantities; and, Figure S4 shows the functional relationships between the Shannon entropy and the average rainfall at the season-scale. The correlation between the interannual entropy and the average annual rainfall is almost equal to one for a lag equal to zero. Thus, the highest/lowest species-richness is observed for the highest/lowest rainfall respectively. This result shows that the driver of vegetation richness is the average annual rainfall. The nutrient concentration depends on the water depth that depends itself on the rainfall. Fires, are local phenomena that shape vegetation mostly locally and their frequency decrease for lower rainfall. We believe that the variation of Shannon entropy is a first-oder measure of dissimilarity of a quantity evaluated in different time periods. Thus, this explain why the average entropy for all the bands is higher in the dry season in which the biggest landscape

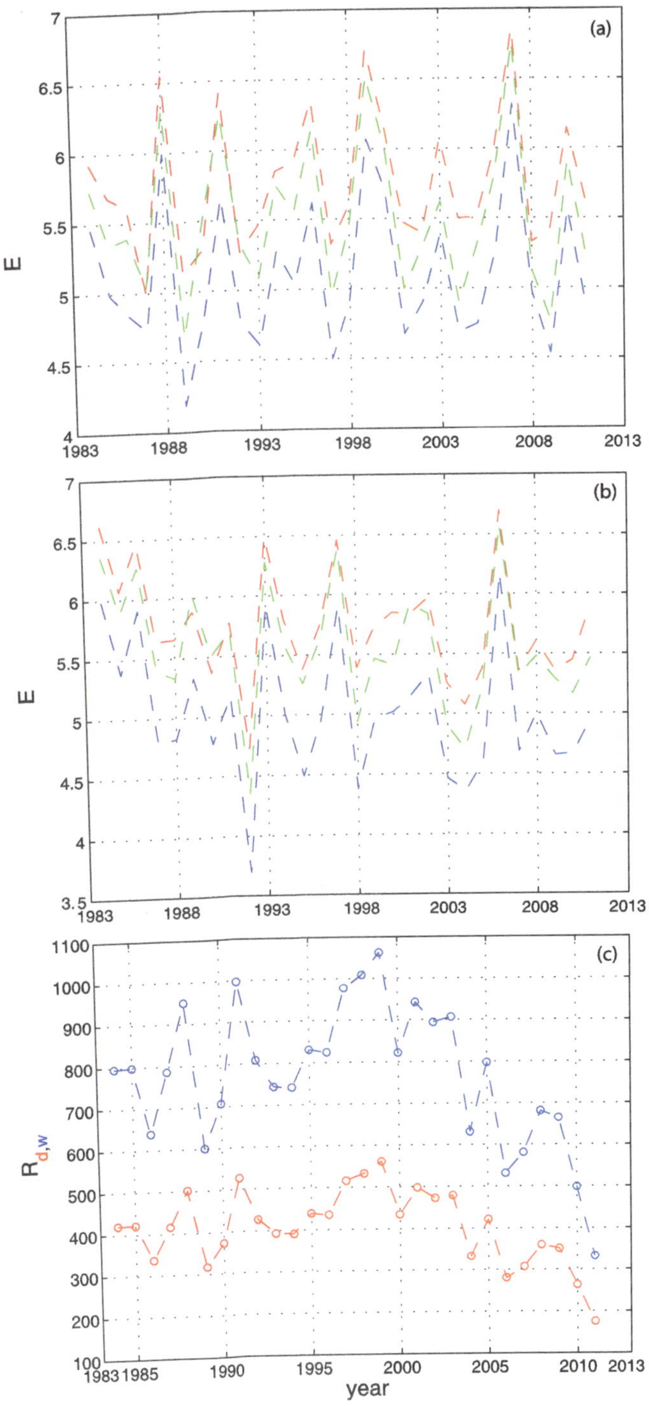

Figure 3. Interannual entropy of WCA 1 Landsat images and rainfall in the period 1984–2011 considered. (a, b) Shannon entropy of the representative regions of WCA 1 in the dry and in the wet season respectively (Figure 1, and Figure S1) for the red, green, and blue bands. The Shannon entropy for the green bad is proportional to the α diversity of plant species. (c) Average annual rainfall (in mm) in the dry (red line) and wet season (blue line).

heterogeneities are observed. Figure S5 confirms the strong correlation between the Shannon entropy of the red, green, and blue band. We consider the Shannon entropy of the green band (E_g) as the dependent variable as a function of water and soil composition that are represented by the Shannon entropy of the blue and red bands respectively. The correlation is very high as evidenced already by the temporal sequences of Figure 3 and Figure S2.

The estimated α diversity on the data (Global Biodiversity Information Facility [33], [45], and the Comprehensive Everglades Restoration Project database [49]) at resolution 30 m after downscaling of the original data is reported in Figure 4. In 1989, 41 species of vegetation were sampled [45]. According to the Global Biodiversity Information Facility [33] the species were 38. The green band Shannon entropy for that year is equal to six approximatively. Thus, if the entropy is taken proportional to 41 species, we observe that the average number of species fluctuated between 29 (in 1992) and 46 species (in 2006). The potential average α diversity is 38 species. In 1999, 30 species of vegetation were sampled [45]. According to the Global Biodiversity Information Facility [33] the species were 33. In Figure 4 (b) the green band Shannon entropy corresponding to the year 1999 is 5.7 that corresponds to 36 species when rescaled to real diversity [30]. Thus, our estimation of α diversity by image analysis is capturing the second observation with an error of three species. Comparing Figure 4 (a) and (b) we observe an high cross-correlation between the measured and estimated α diversity. The inset in Figure 4 (c) shows in fact that the entropy of the green band estimates at least 70% of the measured α diversity. This percentage is 80% and 77% in the dry and wet season respectively that confirms the assumption of easier detectability of species during the dry season. Although we recognize that the Global Biodiversity Information Facility data [33] is affected by some uncertainty in the estimation of species occurrence [87], we believe that the uncertainty for this area is overall lower due to the extensive sampling campaigns occurred in time (we recorded an average of 20,000 occurrences for each year considering together the data from [33] and [49]). Hence, we believe to have captured the local plant species richness and plant species turnover, that is demonstrated by the excellent similarity between predictions and data. We also believe that the texture estimation algorithm on the red squares in Figure 1 and S1 is a very efficient tool to estimate the species richness of the green squares which are representative of the richness of the whole WCA 1. In our study area because the scale of the patch (or region) considered is very large, and the variation of α diversity is very low, the maximum α diversity tends to the average γ diversity. Because the species are evenly distributed γ diversity has not a strong dependency to the scale of analysis. Because the reflectance is capable to distinguish different species, the larger the scale the more the estimated α diversity resembles γ diversity. Gamma diversity is in general greater or equal to the maximum value of α diversity. For ecosystems with very heterogeneous distribution of species this is no more true and local α diversity, average α diversity, and γ diversity are very different [88].

The average change in species-richness in time is expressed by the variation of the image entropy. The calculated difference between Shannon entropies at different time steps was reported as a worse estimator than the KL divergence of the species-turnover [19]. Nonetheless, It was shown that Shannon's diversity (that is the difference between entropies) is the KL divergence between the actual plot and the "average" plot within the region [35]. A plot is defined as the region where the analysis of biodiversity is performed (e.g. the red squares in Figures 1 and S1). If the average plot (e.g. the green squares in Figures 1 and S1) is sufficiently big to contain the plot considered, the pairwise variation of species-richness due to plot interaction is included in the Shannon diversity. However this is not always the case because disturbances (such as clouds and satellite recognition imperfections) limits the

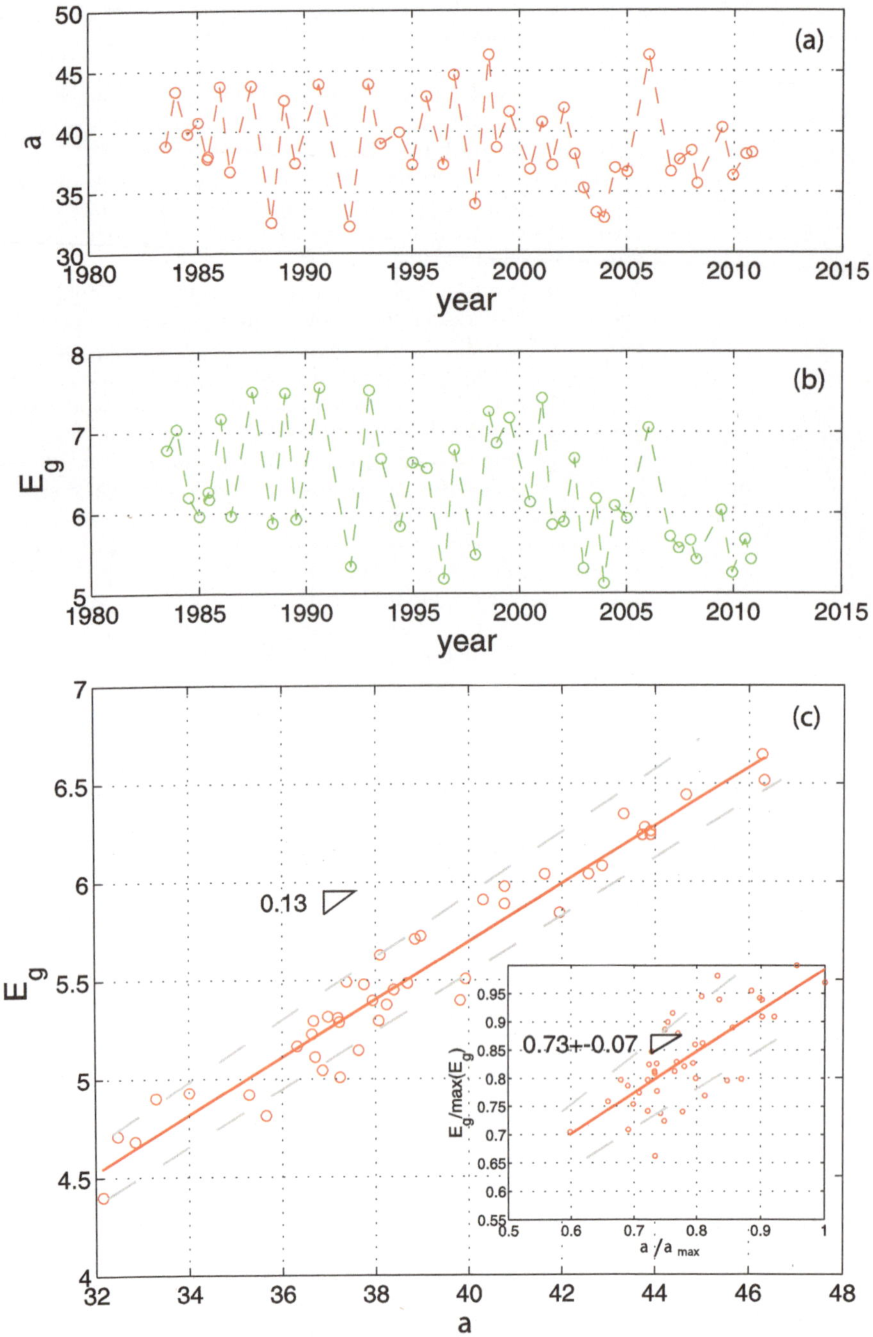

Figure 4. Estimated and measured local species-richness. The local species-richness (α diversity) and the estimated local species-richness (i.e. the Shannon entropy of the green band, E_g using Equation 5) are reported in plot (a) and (b) from 1984 to 2011 respectively. In plot (c) the functional relationship between the Shannon entropy of the green band and the local species-richness for WCA 1 is reported. The inset, which reports the normalized entropy, shows the ability of the Shannon entropy to capture at least 70% of the measured local species-richness. This percentage is 80% and 77% in the dry and wet season respectively. The dashed grey curves are the 95% confidence interval of the linear regression exponent. Variabilities of measured exponents are found by bootstrapping over points and deriving slopes by the linear and the Jackknife models [92].

extent of the representative average plot of the region. Moreover it is not easy to define the representative plot due to the intrinsic heterogeneities of the region.

Figure 5 (a) confirms the independence of β and α diversity as hypothesized and commonly found in literature [69,70,72,71,73,20]. In the specific case of WCA 1, the dissimilarity in species-richness between communities (KL_g) slightly decreases when increasing the local species-richness of communities (E_g). Such trend is more markedly observed when speciation is high and dispersal very limited. In the case of WCA 1 we believe that speciation is fairly high and dispersal is moderately high, however further species sampling are needed to confirm this supposition. This independence between β and α diversity is far to be universal among wetland ecosystems and it is one of the most discussed biodiversity patterns in ecosystems at large. Figure 5 (b) shows that the KL divergence (Equation 4) differs from the difference of Shannon entropies for the green band. In Figure 5 (b) we test the hypothesis of the KL divergence as a better measure of β diversity than the difference of Shannon entropies independently of the season considered. The difference of Shannon entropies captures 85% of the β diversity at the most. Instead the KL divergence represents faithfully the whole β diversity that is estimated on the data. This suggests that interactions and non-linearities in vegetation processes are not negligible. The KL divergence includes the "reciprocal information" of communities in its estimation (Equation 6). Species-turnover can be in fact, lower or higher, than the simple difference of species-richness between the same local communities in time. There may be cases where the disappearance of one species can bring to a large dominance of another species (e.g. this is the case of sawgrass and cattail in many water conservation areas [52]), and cases of "colonization boom" of many alien species that bring the ecosystem to be very species rich (e.g. this is the case of the Everglades National Park [89,90,1]). Considering that the reflectance is characterizing species (or functional species) we observe fluctuations of the species richness in time without anomalous spikes that may be related to a net decrease or increase of species. Thus, we can claim that in WCA 1 there have not been relevant issues with alien species as confirmed in [1]. This is potentially related to the confinement of species within WCA 1 by the set of surrounding levees that block immigration of species from outside. Because we did not perform any species classification our conclusion is about average ecosystem properties (in terms of species richness) rather than certainty about no invasion. Some species whose reflectance is similar to endemic species may be introduced in WCA 1 but their number is certain low.

By considering Figure 5 (b), we see that the difference of Shannon entropies is actually a fairly good estimate of β diversity. This may be related to the quite strong time-invariance of species composition for the representative regions chosen for the analysis (Figure 1). Scale dependence in species-turnover has been recently suggested to reflect variance in species occupancy [91]. Thus, the variability of scale of the red squares analyzed might be a source of variability in the estimation of β diversity. The similarity between the KL divergence, the difference of Shannon entropies, and β diversity appears moderately high for the dry season. This potentially occurs because of the high spatial heterogeneity in species composition and low reflectance of vegetation during the dry season with respect the wet season. During the dry season the difference in entropies is capable to reproduce only 47% of the observed β diversity (Figure S6, b). During the wet season the difference in entropies is capable to reproduce 76% of the observed β diversity (Figure S6, a). The KL divergence is capable to estimate 61% and 82% of the observed β diversity in the dry and in the wet season respectively. Thus, it is advisable to compute the KL divergence rather the difference of Shannon entropies for estimating the β diversity. This is observed to be true both for the interannual and interseasonal species turnover. Overall we verify the superiority of second-oder indices of richness (i.e. the KL divergence) vs first oder indices (i.e. the difference of Shannon entropies) in reproducing β diversity.

In Figure 6 the KL divergence (Equation 4) is represented as a symmetric matrix in which the x and y axis are the consecutive seasons (dry and wet) for each year considered in the analysis. The plots are matrices representing the pairwise textural divergence. Figure S7 and S8 provide the KL divergence for the RGB bands for the dry and wet season respectively. The color scale is proportional to the relative generalized Gaussian pairwise distance (i.e. the KL divergence) between images for different seasons/years. The texture is the set of estimated pdfs for each subband, and the dissimilarity measure between textures is based on the KL divergence which is defined between two pdfs (Equation 4). The dissimilarity between and among years is higher for the blue band (Figure 6, c) than for the red and green bands (Figure 6, a and b respectively). In 1991 a red vertical and red horizontal line occurs in the three plots (a, b, and c). Considering the Landsat image in 1991 (Figure 1) we verify that this occurs because of the high cloud cover in that year that makes the analysis unreliable. Thus, this year should be removed by the time series when analyzing temporal variations of α and β diversity. We include the 1991 image to show that these pairwise textural divergence tables can also be used as a-posteriori indicators to detect images to be excluded in the analysis. The matrices of Figure 6 are useful for comparison of species-richness variation (species-turnover): (i) between seasons of the same or different years (pixels of the upper and lower diagonal along the main diagonal; for example considering the pixels in (a1)); of a selected year with respect to a historical period or with respect to a period in the past (pixels along a column; e.g. considering pixels in (a2)); and, (iii) between periods in time (any group of pixels within a submatrix defined within the matrix; e.g. considering pixels in (a3)). Considering (a1) for example, by taking year Y as a row and year $Y+1$ as a column, we are capable to estimate the difference in species between these two years which is the species turnover. Figure S9 reports the KL divergence for the maximum of the RGB signal in the twenty-eight years period considered. Because the maximum KL divergence can be from the red, green, or blue band in this case the KL divergence is representing the overall maximum ecosystem change (being change in vegetation, soil, or water structure) in the period of observation. Thus, this matrix may be useful to detect the periods of maximum change of the ecosystem.

Discussion

The following points are worth reiterating and discussing.

- The study emphasizes the role of spatial and temporal heterogeneity of the landscape in shaping biodiversity patterns, as a texture method quantifies the strong differences of each season in term of species composition, soil structure and water distribution. Specifically, a positive correlation is found between the Shannon entropy of the red, green, and blue band for the dry and wet seasons and the average annual rainfall of each season. This proves the strong correlation between hydro-geomorphological and ecological changes of wetland ecosystems. The hydro-geomorphological and the ecological changes are represented by the Shannon entropy of the blu-red bands, and of the green band respectively. A negative correlation is found between the RGB band variations of the dry and the wet season as expected because of the opposite ecohydrological dynamics of the two seasons. The texture method quantifies the strong differences of each season in term of species composition, soil structure and water distribution. We believe that our good predictions of species richness from data proves that reflectance is a good metric for assessing taxonomic diversity rather than functional diversity.
- The average species-richness of WCA 1 is found to be constant in the observed period (1984–2011). Despite the decrease in the average annual rainfall from 2000 to present the average species-richness seems to be fairly invariant as reported by [86] and as shown by the data extracted from the Global Biodiversity Information Facility [33]. However, a decrease in species-richness may occur if the drying trend persists in the future. Thus, the monitoring of wetland spectral heterogeneity,

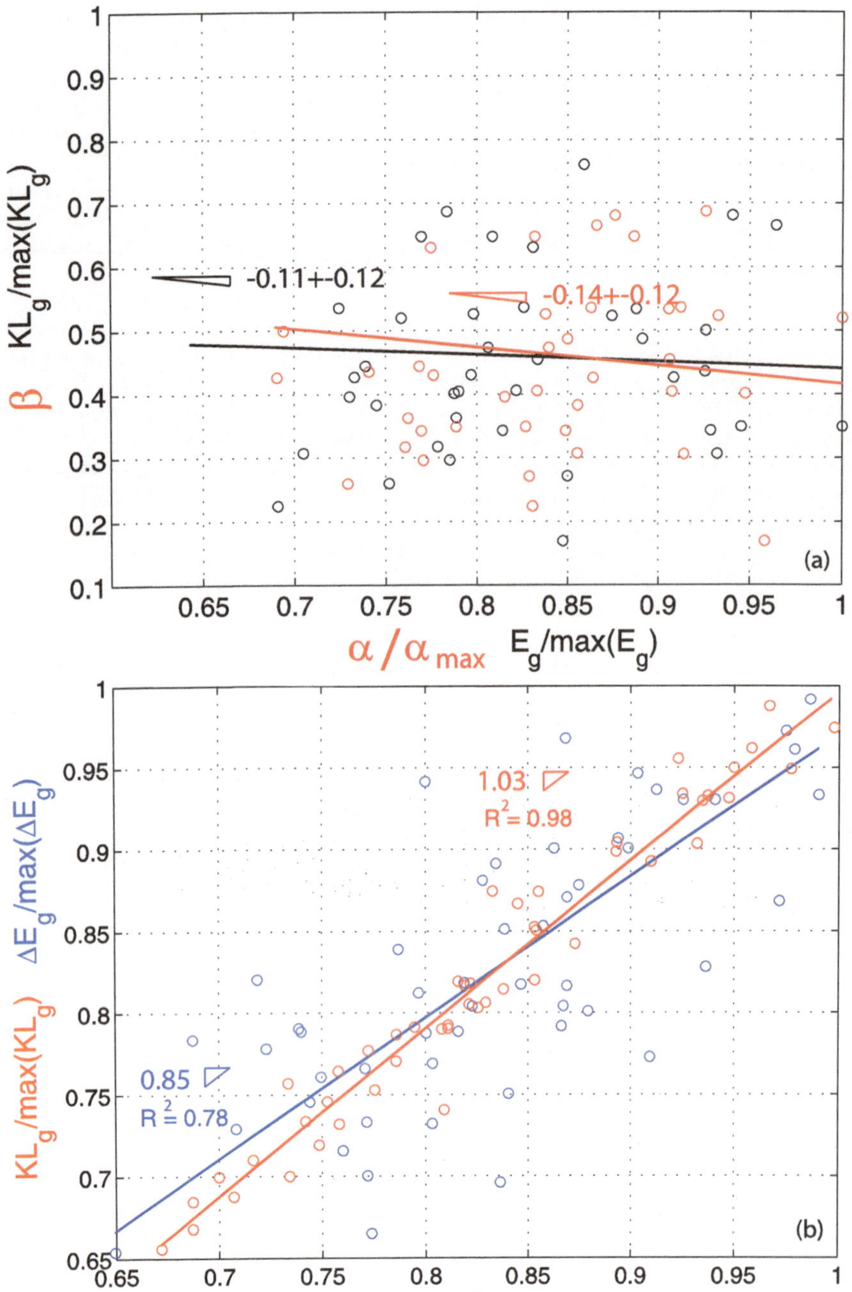

Figure 5. Relationship between α and β diversity, and estimation of β diversity using KL divergence and difference of entropies. (a) Relationship between the interseasonal KL divergence and the green band Shannon entropy for the period 1984–2011, and same relationship estimated on data. The plot shows the independency of β diversity as a function of α diversity. Plot (b) shows how the KL divergence better predicts β diversity than the difference of the Shannon entropy between seasons. The dashed grey curves are the 95% confidence interval of the linear regression exponent. In (a) the dashed grey curves are the 95% confidence interval of the linear regression exponent. In (b) the error in the estimation of the regression coefficient is ± 0.04. The R^2 is the coefficient of determination. Variabilities of measured exponents are found by bootstrapping over points and deriving slopes by the linear and the Jackknife models [92].

performed for example by analyzing image texture, is suggested as a powerful tool in the management of WCA 1 and in general of ecosystems at large. Large-scale restoration requires ecosystem performance measures that can function as rapid quantitative benchmarks of recovery or degradation over time. One of these measures can be the species richness for instance. Our study provides a methodology that can be used for this purpose versus expensive monitoring campaigns post restoration. Our method can also be useful to detect species variation due to species invasion. A novel matrix visualization is proposed to compare the change in species differences (β diversity) between seasons, years, and multiple years. This is potentially useful in the analysis of ecological processes and their coupling with anthropic and natural stressors. We also

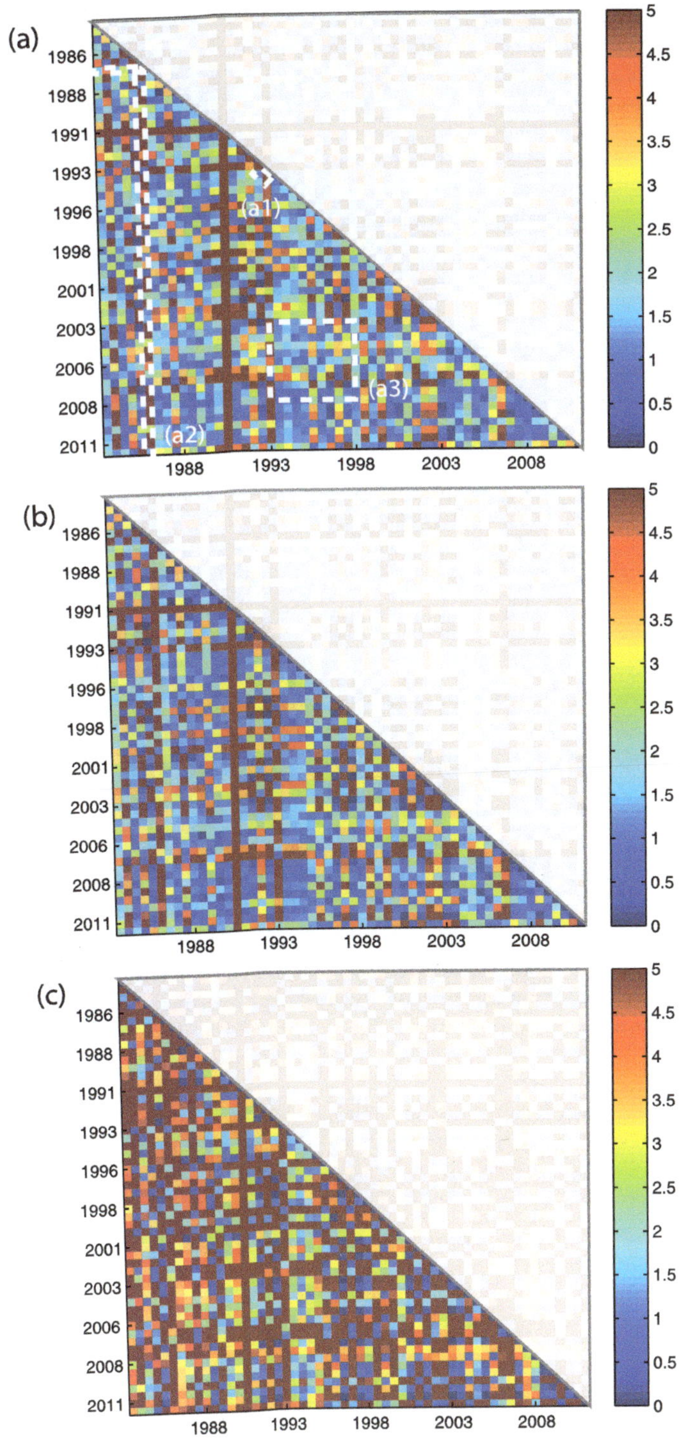

Figure 6. Interseasonal KL divergence matrices for the period 1984–2011. (a), (b), and (c) are the season-season comparisons (56 wet and dry seasons for the period 1984–2011) for the red, green, and blue image band respectively. The more blue/red the pixel the lower/higher the potential β diversity between seasons. Matrices are symmetric and the upper triangular part is made transparent.

emphasize the utility of fast and accurate multispectral image analysis for ecosystem management versus costly field campaigns as suggested by [18].

- The Shannon entropy and the KL divergence of the green band of satellite imagery are verified as good estimators of the species-richness (α diversity) and of the species turnover (β diversity). This is certainly true in ecosystems where spectral heterogeneity is very large. However, the Shannon entropy needs to be calibrated on data in order to assess the number of species as already evidenced by [30]. Our texture analysis is conducted on representative regions of variable extent of the ecosystem analyzed. We found that the predicted α diversity of each representative region is a good proxy the average α diversity for the whole ecosystem. Moreover we realized that the average α diversity is a lower but close estimate of γ diversity for the ecosystem analyzed after validation of predictions with the data of the Global Biodiversity Information Facility. This proves the scalability of the texture-based analysis method in assessing species richness for ecosystems where species is evenly distributed. For WCA 1 the Shannon entropy of the green band reproduces at least 70% of the observed α diversity. The Shannon entropy of the green band has a very high cross-correlation with the average annual rainfall for a lag equal to zero. Thus, every fluctuation of the average annual rainfall (every six-months) implies changes in species-richness and community composition. Therefore we believe that the Shannon entropy of the green band can be considered as the spectral heterogeneity (without considering all the bands as in [11]) for assessing α diversity. We verify the assumption of independence of α and β diversity.
- The KL divergence reproduces β diversity in time (species-turnover) better than the difference of the Shannon entropy of the green band of images between seasons and years. This is because the KL divergence is capable to consider the non-independent variation of species-richness between pairs of local communities which can arise from ecological interactions and feedbacks of local communities. This was proven for β diversity in space by [19]. This is particularly true in the wet season because the higher homogeneity in ecohydrological conditions compared to the dry season makes lower the uncertainty in the estimation of the species-richness. Considering dry and wet seasons together from 1984 to 2011 the difference in Shannon entropy and the KL divergence reproduce 85% and 100% of the observed β diversity. The KL divergence reproduces 61% and 82% of the observed β diversity for the dry and wet season respectively. In general our texture analysis method performs better in predicting species turnover than other indicators used in previous studies, for example in [10]. However studies of species turnover are still lacking. We also show that medium spatial resolution Landsat images can provide extremely good estimates of species richness and turnover without the necessity of hyperspectral images.

Conclusions

We analyze species richness, species turnover, and other spectral heterogeneities derived from satellite imagery for a constructed wetland ecosystem. We use wavelet-based statistical multiresolution texture analysis, a method that, to the best of our knowledge, is new to the field. This method accurately estimates parameters for the marginal distribution of wavelet coefficients using Generalized Gaussian density (GGD) and provides a closed form expression form for the KL divergence between GGD models of different textures. We demonstrate the method on remotely sensed images of the Water Conservation Area 1 in the Greater Everglades Ecosystem Restoration in

South Florida but it can applied to any ecosystem. The results suggest that the method is indeed promising for the analysis of species-richness of any ecosystem with high spatial heterogeneity. The method is shown to provide better estimates of α and β diversity in ecosystems where data-sampling is not feasible (e.g. inaccessible regions) or when it is partially available in time. In particular, statistical wavelet based multiresolution analysis provides a means for estimating divergence between two textures, specifically the Kullback-Leibler divergence between the pair of density functions representing the textures. The KL divergence proved to be a near perfect predictor ($R^2 = 0.98$) of species turnover, or β diversity. Additionally, the visualization representing the KL divergence results between texture pairs provide quick insight into species turnover across many years and seasons. It also enables quick recognition of anomalous data. Texture modeling is also helpful to the theoretical understanding of fundamental ecosystem processes, classification of land-cover, classification of species [39], and as inputs to ecohydrological models that are capable of predicting hydroperiod and runoff of wetlands. We anticipate further effort in using Generalized Gaussian, as well as other non-Gaussian multiresolution texture methods, together with the KL and other divergences for comparing textures, in order to achieve several tasks of importance, more accurately at a higher spatiotemporal resolution. These tasks include detecting single species occurrence and abundance through segmentation analysis citref [76], assessing species-dissimilarity in space, and more generally, analyzing satellite imagery including stereoscopic images. Mathematical features for representing textures other than wavelet coefficients and their distribution could also be of interest, especially if they provide models that are simultaneously simple and parsimonous. Moreover, the application of the presented model to other biological systems (for example for the identification of cell communities [44,76]) and other scales of analysis of ecosystems is a promising research direction. Finally, the methods reduce the need for field-work, while enabling more effective, less costly monitoring, inference, and decision making support.

Supporting Information

Figure S1 Remote-sensed images for the Arthur R. Marshall Loxahatchee National Wildlife Refuge (WCA-1) during the wet-season for the period 1987–2011. The first three years (1984–1986) images are not represented. The representative region in which the texture analysis is performed is delineated in red for each image. The red regions are characterized by a cloud cover lower than 20%. The green regions identify where the data of species are available.

Figure S2 Interseasonal entropy of WCA-1 Landsat images for the red, green, and blue bands. (a, b) are the Shannon entropy and average annual rainfall (m) in the period 1984–2011 respectively.

Figure S3 Cross-correlation between the average annual rainfall and the Shannon entropy of the green-band. A lag is equivalent to a year. For lag = 0 there is an almost perfect correlation ($C(R,E_g)\tilde{1}$) between rainfall and potential α diversity that shows an almost immediate feedback between rainfall and vegetation seasonality.

Figure S4 Predicted α diversity as a function of rainfall. (a) Shannon entropy for the green band vs. average annual rainfall (m). The maximum of the rainfall is about 600 and 1100 mm in the dry and in the wet season respectively. (b) Shannon entropy for the green band in the dry season vs. average annual rainfall (mm) in the dry season; and, (c) Shannon entropy for the green band in the wet season vs. average annual rainfall (mm) in the wet season. Variabilities of measured exponents are found by bootstrapping over points and deriving slopes by the linear and the Jackknife models [92].

Figure S5 Functional relationships between Shannon entropies of ecosystems components (soil, vegetation, and water spectral signatures). (a) Shannon entropy for green band vs. blue band, and (b) Shannon entropy for the green band vs red band. The entropy is calculated for every dry and wet season of each year in the period 1984–2011. These relationships hold also considering separately the entropy for the wet- and for the dry season. The dashed grey curves are the 95% confidence interval of the linear regression exponent. Variabilities of measured exponents are found by bootstrapping over points and deriving slopes by the linear and the Jackknife models [92].

Figure S6 Estimation of the interannual β diversity using KL divergence and the difference of Shannon entropies. Relationship between the interannual KL divergence and the green-band Shannon entropy variation vs. the β diversity for the period 1984–2011 for the dry and wet seasons respectively (a, and b). The KL divergence better predicts β diversity than the difference of the Shannon entropy between years. Variabilities of measured exponents are found by bootstrapping over points and deriving slopes by the linear and the Jackknife models [92].

Figure S7 Interannual KL divergence matrices for the decomposed RGB signal. Plot (a), (b), and (c) is for the red, green, and blue band, respectively, in the dry season from 1984 to 2011. The higher the KL divergence the higher the dissimilarity (β diversity for the green band) between seasons of the same or different years. Matrices are symmetric and the upper triangular part is made transparent.

Figure S8 Interannual KL divergence matrices for the decomposed RGB signal. Plot (a), (b), and (c) is for the red, green, and blue band, respectively, in the wet season from 1984 to 2011. The higher the KL divergence the higher the dissimilarity (β diversity for the green band) between seasons of the same or different years. Matrices are symmetric and the upper triangular part is made transparent.

Figure S9 Interseasonal KL divergence matrix for the maximum of the RGB signal from 1984 to 2011. The maximum value of the KL divergence can be considered as total ecosystem change (in terms of soil, vegetation, and water) among the years considered. The matrix is symmetric and the upper triangular part is made transparent.

Text S1 Supporting Study Area Information.

Acknowledgments

The authors thank the service provided by http://edcsns17.cr.usgs.gov/NewEarthExplorer/http://edcsns17.cr.usgs.gov/NewEarthExplorer/ (Accessed 2012) and http://glovis.usgs.gov/http://glovis.usgs.gov/ (Accessed 2012) for the Landsat images. K.M. Baker at the Risk and Decision Science Team (ERDC - USACE) is gratefully acknowledged for revising the text of the manuscript. Andy LoSchiavo (Jacksonville District, US Army Corps of Engineering) is gratefully acknowledged for providing comments to the paper. Permission is granted by the USACE Chief of Engineers to publish this material. We kindly acknowledge the Editor Dr. Kimberly Patraw Van Niel, an anonymous reviewer, and Dr. J. Alberto Gallardo-Cruz (Departamento de Ecologia y Recursos Naturales, Facultad de Ciencias, Universidad Nacional Autonoma de Mexico, Mexico City, Mexico) for their useful comments to the first version of the manuscript. The views and opinions expressed in this paper are those of the individual authors and not those of the US Army, or other sponsor organizations.

Author Contributions

Conceived and designed the experiments: MC RM IL. Performed the experiments: MC RM NL. Analyzed the data: MC RM. Wrote the paper: MC RM. Provided a review of the paper: MD.

References

1. McVoy C, Said W, Obeysekera J, VanArman J, Dreschel T, editors (2011) Landscape and Hydrology of the Predrainage Everglades. University Press of Florida.
2. Adler PB, Levine JM (2007) Contrasting relationships between precipitation and species richness in space and time. Oikos 116: 221–232.
3. Todd M, Muneepeerakul R, Pumo D, Azaele S, Miralles-Wilhelm F, et al. (2010) Hydrological drivers of wetland vegetation community distribution within everglades national park, orida. Advances in Water Resources 33: 1279–1289.
4. Todd MJ, Muneepeerakul R, Miralles-Wilhelm F, Rinaldo A, Rodriguez-Iturbe I (2011) Possible climate change impacts on the hydrological and vegetative character of everglades national park, orida. Ecohydrology.
5. Cohen MJ, Watts DL, Heffernan JB, Osborne TZ (2011) Reciprocal biotic control on hydrology, nutrient gradients, and landform in the greater everglades. Critical Reviews in Environmental Science and Technology 41: 395–429.
6. Mitsch WJ, Gosselink JG, editors (2000) Wetlands. John Wiley and Sons, New York.
7. Bernhardt C, Debra A (2009) Response of the everglades ridge and slough landscape to climate variability and 20th-century water management. Ecological Applications 19: 17231738.
8. Prigent C, Matthews E, Aires F, Rossow WB (2001) Remote sensing of global wetland dynamics with multiple satellite data sets. Geophysical Research Letters 28: 4631–4634.
9. Ferretti M, Chiarucci A (2003) Design concepts adopted in long-term forest monitoring programs in europe-problems for the future? Science of The Total Environment 310: 171–178.
10. Rocchini D, He K, Oldeland J, Wesuls D, Neteler M (2010a) Spectral variation versus species beta-diversity at different spatial scales: a test in african highland savannas. Journal of environmental monitoring JEM 12: 825–831.
11. Rocchini D, Chiarucci A, Loiselle S (2004) Testing the spectral variation hypothesis by using satellite multispectral images. Acta Oecologica 26: 117–120.
12. Ricotta C, Anand M (2006) Spatial complexity of ecological communities: Bridging the gap between probabilistic and non-probabilistic uncertainty measures. Ecological Modelling 197: 59–66.
13. Rocchini D, Ricotta C, Chiarucci A (2007b) Using satellite imagery to assess plant species richness: The role of multispectral systems. Applied Vegetation Science 10: 325–331.
14. Gillespie TW, Foody GM, Rocchini D, Giorgi AP, Saatchi S (2008) Measuring and modelling biodiversity from space. Progress in Physical Geography 32: 203–221.
15. Nagendra H, Rocchini D, Ghate R, Sharma B, Pareeth S (2010) Assessing Plant Diversity in a Dry Tropical Forest: Comparing the Utility of Landsat and Ikonos Satellite Images. Remote Sensing 2: 478–496.
16. Southworth J, Munroe D, Nagendra H (2004) Land cover change and landscape fragmentation-comparing the utility of continuous and discrete analyses for a western honduras region. Agriculture, Ecosystems & Environment 101: 185–205.
17. Belluco E, Camuffo M, Ferrari S, Modenese L, Silvestri S, et al. (2006) Mapping saltmarsh vegetation by multispectral and hyperspectral remote sensing. Remote Sensing of Environment 105: 54–67.
18. Rocchini D, Balkenhol N, Carter G, Foody G, Gillespie T, et al. (2010b) Remotely sensed spectral heterogeneity as a proxy of species diversity: Recent advances and open challenges. Ecological Informatics 5: 318–329.
19. Ludovisi A, TaticchiM(2006) Investigating beta diversity by kullback-leibler information measures. Ecological Modelling 192: 299–313.
20. Anderson MJ, Crist TO, Chase JM, Vellend M, Inouye BD, et al. (2011) Navigating the multiple meanings of beta diversity: a roadmap for the practicing ecologist. Ecology Letters 14: 19–28.
21. Buckley LB, Jetz W (2008) Linking global turnover of species and environments. Proceedings of the National Academy of Sciences of the United States of America 105: 17836–17841.
22. Lhermitte S, Verbesselt J, Verstraeten W, Coppin P (2011) A comparison of time series similarity measures for classification and change detection of ecosystem dynamics. Remote Sensing of Environment 115: 3129–3152.
23. Latifi H, Fassnacht F, Koch B (2012) Forest structure modeling with combined airborne hyperspectral and lidar data. Remote Sensing of Environment 121: 10–25.
24. Poveda G, Salazar L (2004) Annual and interannual (enso) variability of spatial scaling properties of a vegetation index (ndvi) in amazonia. Remote Sensing of Environment 93: 391–401.
25. Saatchi S, Buermann W, ter Steege H, Mori S, Smith T (2008) Modeling distribution of amazonian tree species and diversity using remote sensing measurements. Remote Sensing of Environment 112: 2000–2017.
26. Tomppo E, Gagliano C, Natale FD, Katila M, McRoberts R (2009) Predicting categorical forest variables using an improved k-nearest neighbour estimator and landsat imagery. Remote Sensing of Environment 113: 500–517.
27. SFWMD (2000) Ecological effects of phosphorus enrichment in the everglades. chapter 3. In: Redfield G, editor, Everglades Consolidated Report., South Florida Water Management District.
28. Cohen M, Grunwald S, Clark M, Reddy K (2007) Rapid assessment of restoration performance measures at multiple scales in the greater everglades using near infrared reectance spectroscopy (nirs). Technical report, Institute of Food and Agricultural Sciences, University of Florida. http://www.nps.gov/ever/naturescience/upload/ASS04-1FinalReportSecure.pdf; Accessed 2012.
29. Do MN, Vetterli M (2002) Wavelet-based texture retrieval using generalized Gaussian density and kullback-leibler distance. IEEE Trans Image Processing 11: 146–158.
30. Jost L (2006) Entropy and diversity. Oikos 113: 363–375.
31. Rocchini D (2009) Algorithmic foundation of spectral rarefaction for measuring satellite imagery heterogeneity at multiple spatial scales. Sensors 9: 303–310.
32. Viedma O, Torres I, Perez B, Moreno J (2012) Modeling plant species richness using reectance and texture data derived from quickbird in a recently burned area of central spain. Remote Sensing of Environment 119: 208–221.
33. GBIF (2012) Global biodiversity information facility. Technical report, Global Biodiversity Information Facility. http://www.gbif.org/; Accessed 2012.
34. McGlinn D, Palmer M (2010) Quantifying the inuence of environmental texture on the rate of species turnover: evidence from two habitats. Plant Ecology: 1–12.
35. Marcon E, Herault B, Baraloto C, Lang G (2012) The decomposition of shannon's entropy and a confidence interval for beta diversity. Oikos 121: 516–522.
36. St-Louis V, Pidgeon A, Clayton M, Locke B, Bash D, et al. (2009) Satellite image texture and a vegetation index predict avian biodiversity in the chihuahuan desert of new mexico. Ecography 32: 468–480.
37. Bellis L, Pidgeon A, Radeloff V, Navarro J, Martella M (2008) Modeling habitat suitability for the endangered greater rhea (rhea americana) in central argentina based on satellite image texture. Ecological Applications 18: 1956–1966.
38. Gallardo-Cruz, Meave JA, González EJ, Lebrija-Trejos EE, Romero-Romero MA, et al. (2012) Predicting tropical dry forest successional attributes from space: Is the key hidden in image texture? PLoS ONE 7.
39. Key T, Warner TA, McGraw JB, Fajvan MA (2001) A comparison of multispectral and multitemporal information in high spatial resolution imagery for classification of individual tree species in a temperate hardwood forest. Remote Sensing of Environment 75: 100–112.
40. Townsend AR, Asner GP, Cleveland CC (2008) The biogeochemical heterogeneity of tropical forests. Trends in Ecology & Evolution 23: 424–431.
41. Wulder M (2004) High spatial resolution remotely sensed data for ecosystem characterization. Biogeoscience 54: 511–521.
42. Jeffreys C (2004) Support vector machine and parametric wavelet-based texture classification of stem cell images – phd thesis. Technical report, Massachusetts Institute of Technology. http://dspace.mit.edu/bitstream/handle/1721.1/16651/56472941. pdf?sequence = 1; Accessed 2012.
43. Mangoubi R, Jeffrey C, Copeland A, Desai M, Sammak P (2007) Non-invasive image based support vector machine classification of human embryonic stem cells. International Symposium on Biomedical Imaging: 284–287.
44. Mangoubi R, Desai M, Sammak P (2008) Non-gaussian methods in biomedical imaging. Applied Image Pattern Recognition Workshop 0: 1–6.
45. Childers D, Doren R, Jones R, Noe G, Rugge M, et al. (2003) Decadal change in vegetation and soil phosphorus pattern across the everglades landscape. J Environ Qual 32.
46. GLOVIS (2011) Global visualization viewer. Technical report, USGS. http://glovis. usgs.gov/; Accessed 2012.
47. USGS (2011) Earth explorer. Technical report, USGS. http://edcsns17.cr.usgs.gov/NewEarthExplorer/; Accessed 2012.

48. USGS (2003) Preliminary assessment of the value of landsat 7 etm+ data following scan line corrector malfunction. Technical report, U.S. Geological Survey, EROS Data Center, NASA, and Landsat 7 Science Team. http://landsat.usgs.gov/documents/SLC_off_Scientific_Usability.pdf; Accessed 2012.
49. USACE (2012) Comprehensive everglades restoration project database. Technical report, US Army Corps of Engineers. http://www.cerpzone.org/; Accessed 2012.
50. Convertino M, Muneepeerakul R, Azaele S, Bertuzzo E, Rinaldo A, et al. (2009) On neutral metacommunity patterns of river basins at different scales of aggregation. Water Resources Research 45: 8424.
51. Azaele S, Cornell S, Kunin W (2012) Downscaling species occupancy from coarse spatial scales. Ecological Applications: in press.
52. Lagerwall G, Kiker G, Munoz-Carpena R, Convertino M, James A, et al. (2012) A spatially- distributed, deterministic approach to modeling typha domingensis (cattail) in an everglades water- controlled wetland. Ecological Processes; special issue "Wetland in a Complex World" forthcoming.
53. Cho HJ (2007) Depth-variant spectral characteristics of submersed aquatic vegetation detected by landsat 7 etm+. Int J Remote Sens 28: 1455–1467.
54. Nagendra H, Rocchini D (2008) High resolution satellite imagery for tropical biodiversity studies: the devil is in the detail. Biodiversity and Conservation 17: 3431–3442.
55. Rose P, Rosendahl P (1979) An application of landsat multispectral imagery for the classification of hydrobiological systems, shark river slough, everglades national park, orida. Technical report, South Florida Research Center Report. http://www.nps. gov/ever/naturescience/upload/SecureTRT-544.pdf; Accessed 2012.
56. Bubier JL, Rock BN, Crill PM (1997) Spectral reectance measurements of boreal wetland and forest mosses. Journal of Geophysical Research 102.
57. Rundquist D, Narumalani S, Narayanan R (2001) A review of wetlands remote sensing and defining new considerations. Remote Sensing Reviews 20: 207–226.
58. Ozesmi SL, Bauer ME (2002) Satellite remote sensing of wetlands. Wetlands Ecology and Management 10: 381402.
59. Petchey OL, Gaston KJ (2002) Functional diversity (FD), species richness and community composition. Ecology Letters 5: 402–411.
60. Tilley D, Baldwin A, Jenkins E (2004) Leaf-scale hyperspectral reectance models for determining the nitrogen status of freshwater wetlands. Technical report, Maryland Water Resources Research Center.
61. Rocchini D (2007a) Effects of spatial and spectral resolution in estimating ecosystem α-diversity by satellite imagery. Remote Sensing of Environment 111: 423–434.
62. Wang C, Menenti M, Stoll MP, Belluco E, Marani M (2007) Mapping mixed vegetation communities in salt marshes using airborne spectral data. Remote Sensing of Environment: 559–570.
63. Ajithkumar TT, Thangaradjou T, Kannan L (2008) Spectral reectance properties of mangrove species of the muthupettai mangrove environment, tamil nadu. Journal of Environmental Biology 29: 785–78.
64. Sun Y, Liu XH, Wu Y (2011) Identifying hyperspectral characteristics of wetland species using in situ data. ISPRS Journal of Photogrammetry and Remote Sensing.
65. Miao S, Zou CB (2009) Seasonal variation in seed bank composition and its interaction with nutrient enrichment in the everglades wetlands. Aquatic Botany 90: 157–164.
66. Rocchini D, Neteler M (2012) Spectral rank-abundance for measuring landscape diversity. International Journal of Remote Sensing 33: 4458–4470.
67. Thenkabail PS, Enclona EA, Ashton MS, Meer BVD (2004) Accuracy assessments of hyperspectral waveband performance for vegetation analysis applications. Remote Sensing of Environment 91: 354–376.
68. Muneepeerakul R, Rinaldo A, Levin SA, Rodriguez-Iturbe I (2008) Signatures of vegetational functional diversity in river basins. Water Resources Research 44.
69. Jost L (2007) Partitioning diversity into independent alpha and beta components. Ecology 88: 2427–2439.
70. Baselga A (2010) Multiplicative partition of true diversity yields independent alpha and beta components; additive partition does not. Ecology 91: 1974–1981.
71. Jost L (2010) Independence of alpha and beta diversities. Ecology 91: 1969–1974.
72. Veech JA, Crist TO (2010) Diversity partitioning without statistical independence of alpha and beta. Ecology 91: 1964–1969.
73. Wilsey BJ (2010) An empirical comparison of beta diversity indices in establishing prairies. Ecology 91: 1984–1988.
74. Brandt L, KitchensW(1998) Spatial and temporal changes in tree islands of the arthur r. marshall loxahatchee national wildlife refuge in response to altered hydrologies. Technical report, Florida Cooperative Fish and Wildlife Research Unit, University of Florida.
75. Engler O, Randle V, editors (2009) Introduction to Texture Analysis: Macrotexture, Microtexture, and Orientation Mapping. CRC Press.
76. Lowry N, Desai M, Mangoubi R, Sammak P (2010) Nonparametric segmentation and classification of small size irregularly shaped stem cell nuclei using adjustable windowing. IEEE International Symposium on Biomedical Imaging: From Nano to Macro: 141–144.
77. Erb T, Schneider C, Mucko S, Sanfilippo J, Lowry N, et al. (2011) Paracrine and epigenetic control of trophectoderm differentiation from human embryonic stem cells: the role of bone morphogenic protein 4 and histone deacetylases. Stem Cells Dev 20: 1601–14.
78. Van DeWouwer G, Scheunders P, Van Dyck D (1999) Statistical texture characterization from discrete wavelet representations. IEEE Transactions on Image Processing 8: 592–598.
79. Mangoubi R, Desai M, Sammak P (2008) Performance evaluation of multiresolution texture analysis of stem cell chromatin. IEEE International Symposium on Biomedical Imaging: From Nano to Macro: 380–383.
80. MaCarthur R (1965) Patterns of species diversity. Biological Reviews 40: 510–533.
81. Whittaker RH (1960) Vegetation of the siskiyou mountains, oregon and california. Ecological Monographs 30.
82. Hill MO (1973) Diversity and evenness: a unifying notation and its consequences. Ecology 54: 427–432.
83. Keylock CJ (2005) Simpson diversity and the shannonwiener index as special cases of a generalized entropy. Oikos 109: 203–207.
84. Connolly J, Cadotte M, Brophy C, Dooley A, Finn J, et al. (2011) Phylogenetically diverse grasslands are associated with pairwise interspecific processes that increase biomass. Ecology 92: 13851392.
85. Kullback S, Leibler R (1951) On information and sufficiency. Ann Math Stat 22: 79–86.
86. Childers D, Boyer J, Davis S, Madden C, Rudnick D, et al. (2006) Relating precipitation and water management to nutrient concentrations in the oligotrophic "upside-down" estuaries of the orida everglades. Limnol Oceanogr 51: 602616.
87. Yesson C, Brewer P, Sutton T, Caithness N, Pahwa J, et al. (2007) How global is the global biodiversity information facility? PLoS ONE 2.
88. Jurasinski G, Retzer V, Beierkuhnlein C (2009) Inventory, differentiation, and proportional diversity: a consistent terminology for quantifying species diversity. Oecologia 159: 15–26.
89. Daehler CC (2003) Performance comparisons of Co-Occurring native and alien invasive plants: Implications for conservation and restoration. Annual Review of Ecological and Evolutionary Systematics 34.
90. Zedler JB, Kercher S (2004) Causes and consequences of invasive plants in wetlands: Opportunities, opportunists, and outcomes. Critical Reviews in Plant Sciences 23: 431–452.
91. McGlinn D, Hurlbert A (2012) Scale dependence in species turnover reects variance in species occupancy. Ecology.
92. Warton DI, Wright IJ, Falster DS, Westoby M (2006) Bivariate line-fitting methods for allometry. Biological Reviews 81: 259–291.

Linking Stoichiometric Homeostasis of Microorganisms with Soil Phosphorus Dynamics in Wetlands Subjected to Microcosm Warming

Hang Wang[1], HongYi Li[1], ZhiJian Zhang[1]*, Jeffrey D. Muehlbauer[2], Qiang He[3], XinHua Xu[1], ChunLei Yue[4], DaQian Jiang[5]

1 College of Natural Resource and Environmental Sciences, China Academy of West Development, Zhejiang University, Hangzhou, Zhejiang Province, China, **2** Curriculum for the Environment & Ecology, University of North Carolina at Chapel Hill, Chapel Hill, North Carolina, United States of America, **3** Department of Civil & Environmental Engineering, University of Tennessee, Knoxville, Tennessee, United States of America, **4** Institute of Ecology, Zhejiang Forestry Academy, Hangzhou, China, **5** Department of Earth and Environmental Engineering, Henry Krumb School of Mines, Columbia University, New York, New York, United States of America

Abstract

Soil biogeochemical processes and the ecological stability of wetland ecosystems under global warming scenarios have gained increasing attention worldwide. Changes in the capacity of microorganisms to maintain stoichiometric homeostasis, or relatively stable internal concentrations of elements, may serve as an indicator of alterations to soil biogeochemical processes and their associated ecological feedbacks. In this study, an outdoor computerized microcosm was set up to simulate a warmed (+5°C) climate scenario, using novel, minute-scale temperature manipulation technology. The principle of stoichiometric homeostasis was adopted to illustrate phosphorus (P) biogeochemical cycling coupled with carbon (C) dynamics within the soil-microorganism complex. We hypothesized that enhancing the flux of P from soil to water under warming scenarios is tightly coupled with a decrease in homeostatic regulation ability in wetland ecosystems. Results indicate that experimental warming impaired the ability of stoichiometric homeostasis (*H*) to regulate biogeochemical processes, enhancing the ecological role of wetland soil as an ecological source for both P and C. The potential P flux from soil to water ranged from 0.11 to 34.51 mg m^{-2} d^{-1} in the control and 0.07 to 61.26 mg m^{-2} d^{-1} in the warmed treatment. The synergistic function of C-P acquisition is an important mechanism underlying C:P stoichiometric balance for soil microorganisms under warming. For both treatment groups, strongly significant ($p<0.001$) relationships fitting a negative allometric power model with a fractional exponent were found between n-$H_{C:P}$ (the specialized homeostatic regulation ability as a ratio of soil highly labile organic carbon to dissolved reactive phosphorus in porewater) and potential P flux. Although many factors may affect soil P dynamics, the n-$H_{C:P}$ term fundamentally reflects the stoichiometric balance or interactions between the energy landscape (i.e., C) and flow of resources (e.g., N and P), and can be a useful ecological tool for assessing potential P flux in ecosystems.

Editor: Roeland M. H. Merks, Centrum Wiskunde & Informatica (CWI) & Netherlands Institute for Systems Biology, Netherlands

Funding: The authors wish to thank the National Natural Science Foundation of China (41373073) and Zhejiang Science and Technology Program (2011F20025) for providing the financial support for this project. The funders had no role in study design, data collection and analysis, decision to publish, or preparation of the manuscript.

Competing Interests: The authors have declared that no competing interests exist.

* E-mail: zhangzhijian@zju.edu.cn

Introduction

The principle of stoichiometric homeostasis states that organisms maintain relatively stable levels of biologically-relevant elements (e.g., carbon, nitrogen, and phosphorous) over time. This concept is based on ecological stoichiometry and is primarily applied to trophic interactions [1]. By modulating organism responses to key environmental drivers (e.g., nutrient fertilization), stoichiometric homeostasis is also a major mechanism responsible for the structure, function, and stability of ecosystems. Therefore, deviations from stoichiometric homeostasis can serve as indicators of environmental fluctuations that may be useful in ecological evaluation. The homeostatic regulation coefficient (*H*), which may be accessed through a regression model (equation 1, in the Materials and Methods) is a measure of homoeostatic regulation ability and is commonly used as an indicator of these ecological changes [2].

The average global surface temperature has increased by 0.74°C since 1850 and is likely to increase an additional 1.1–6.4°C by the end of this century [3,4]. Temperature is one of the primary determinants affecting the metabolic rates of organisms, from cells to global-scale ecosystems, based on the metabolic theory of ecology [5,6]. Even a slight increase in temperature can alter energy flow and resource cycles in ecosystems and may affect environmental quality [3,7]. For instance, global soil respiration increased by 0.1 Pg yr^{-1} between 1989 and 2008 due to a high temperature anomaly [8]. Understanding the response of stoichiometric homeostasis to global warming and its associated ecological feedback to the biosphere is gaining an increasing degree of attention worldwide.

Studies of stoichiometric homeostasis have strongly emphasized physiological variation in elemental composition in macroscopic aquatic species [9]. To a lesser extent, some studies have also quantified the degree of stoichiometric homeostasis in terrestrial plants [2,10,11]. Stoichiometric homeostasis investigations in soil microorganisms subjected to environmental changes, however, are rare. Biogeochemical processes at the soil-water interface play a fundamental role in overall soil development, and they are the primary driving force behind key ecosystem functions such as community productivity and water quality [12,13]. Although stoichiometric homeostasis reflects the net outcome of many underlying physiological and biochemical adjustments as organisms respond to their surroundings [14], to our knowledge, few studies have linked any change in stoichiometric homeostasis to biogeochemical processes in a soil ecosystem affected by global warming.

Wetlands are one of the most productive and biologically diverse ecosystem types on the planet, where the chemical and, in particular, nutrient composition of shallow groundwater and surface water can be altered by a range of biogeochemical processes [15]. Many of the elements and nutrients that cycle through wetlands have ecosystem and global-scale effects; for example, phosphorous (P), is the key element affecting eutrophication in aquatic ecosystems [16], limiting coastal ecosystem processes [17], and in regulating allometry in soil food webs [18]. However, global warming may impact the inherent biogeochemical balance of P and other major elements. Diffusive models predict that elemental fluxes of P and nitrogen (N) to receiving waters are more than 2 times higher during summer (i.e., when temperatures are warmer) than in other seasons [19,20]. Further, an 18-month microcosm investigation using a minute and seasonal scale temperature manipulation verified that warming induced substantial mobilization of P from wetland sediment to water [21,22]. In other studies, excessive P loading under high temperature accelerated the risk of eutrophication in receiving waters [15,20,23] and altered carbon (C) assimilation and N accumulation in aquatic plants [24]. These studies suggest that warming or increasing temperature disturbs P equilibrium between sediment and water. However, the link between stoichiometric homeostasis and P dynamics in wetland soil subjected to global warming is poorly understood, which impedes the development of ecological regulatory strategies for coping with global warming scenarios. Additionally, wetlands play a vital role in the global C cycle and may respond strongly to climate change [25]. This is important because the flow of C energy and material also controls the biogeochemical cycling of many other elements in Earth's ecosystems [5,26,27,28]. Thus, integrating C:P stoichiometry is vital to understanding the connection between soil P dynamics and the regulation of stoichiometric homeostasis in wetland ecosystems.

The objective of our study was to shed new light on P dynamics in response to global warming, from the perspective of C:P stoichiometric homeostasis in six subtropical wetlands located in the Yangtze River delta in southeast China. We simulated global warming using a novel, in situ deployment of a microcosm device that mimics a warming scenario of 5°C above ambient temperature. The link between potential P flux from soil to water and the homeostatic regulation ability of soil microorganisms was also examined. We hypothesized that enhancing P flux from soil to water under warming scenarios is tightly coupled with a decrease in homeostatic regulation ability in wetland ecosystems.

Materials and Methods

This work is unrelated to any ethics issues. No specific permit was required for the described field study because the sampling locations were not located in protected areas or private land. The experimental field studies did not involve endangered or protected species.

Study sites

The study sites (120°41′31″E, 30°53′55″N to 120°33′32″E, 30°01′58″N) were located in the southern region of the Taihu Lake Basin and the NingShao Plain within the Yangtze River delta in southeast China. The climate in this area is subtropical monsoon with an annual average rainfall of 1350 mm and an annual average temperature of 26°C in summer and 4°C in winter.

The Supporting Information contains details about the six sites related to geographical position, hydrological parameters, and dominant vegetation during sampling (Table S1), as well as the physico-chemical properties of the six collected soils (Table S2). In brief, YaTang riverine wetland (YT) is in an advanced state of eutrophication and is classified as eutrophic, with the highest P and C total soil content among the six study sites. Soils sampled from XiaZhuhu wetland (XZ), XiXi National Wetland Park (XX), BaoYang riverine wetland (BY), and JinHu wetland (JH) are in a meso-eutrophic state, while the lowest soil C and P content among the sites was found in ShiQiu multipond wetland (SQ). Because these selected study sites represented the range of typical soil C and P conditions across subtropical wetland ecosystems in southeast China, we did not design any additional C or P treatments, which would have required artificial manipulations of the soil concentrations of these elements.

Microcosm configuration and sampling regime

A custom-built, novel microcosm (Fig. S1) simulating climate warming was developed under both present-day ambient temperature conditions (Control) and simulated warming conditions of 5°C above ambient temperature (Warmed). Details on the configuration of this microcosm system and its operation were either described previously [21] or can be found in Text S1. Compared to fixed-temperature laboratory incubations, this novel microcosm offers higher temporal resolution temperature control (on a minute scale), which simulates more realistic warming conditions.

Details on the establishment of wetland columns and water-soil samplings during the microcosm investigation are provided in Text S2. Briefly, each wetland column consisted of a prefabricated PVC pipe assembly (45.0 cm in height and 10.0 cm in internal diameter) that was designed to hold 20 cm of fresh soil and 20 cm of corresponding overlying water. Columns were installed in May 2008, and three replicates were placed inside each of the two incubation boxes. In this study, three samplings of water (overlying water and porewater) and 0–5 cm of topsoil were carried out in July and November 2010 and March 2011, which were used to illustrate the link between soil microorganism stoichiometric homeostasis and soil P dynamics in wetlands subjected to experimental warming.

Water and soil analysis

All water and soil samples were frozen at −15°C prior to analysis. After thawing, water was filtered through a 0.45-μm filter. Phosphorous in filtered samples was measured using a continuous flow analyzer (Autoanalyzer III, Bran+Luebbe, Germany) with a spectrophotometer set at 880 nm (Murphy and Riley, 1962). This is the absorbance of dissolved reactive phosphorus (DRP), which represents the bioavailable P fraction for microorganisms and/or algae in surface water. Porewater samples for dissolved organic carbon (DOC) analysis were first acidified (10% HCl) and purged with inert gas to remove any inorganic carbon, then analyzed using a Shimadzu TOC 5000 analyzer (Shimadzu Scientific Instruments, Columbia, Maryland, USA).

After thawing, soil samples were analyzed for C and P in microbial biomass (Table 1) using the fumigation-extraction method described by Wu et al. [29] and Brookes et al. [30]. In brief, soil C and P in fumigated and non-fumigated soil samples was extracted using solutions of 0.5 mol L^{-1} K_2SO_4 and 0.5 mol L^{-1} $NaHCO_3$. The differences in extractable C and P between fumigated and non-fumigated soil were assumed to be released from lysed soil microbes, i.e., microbial biomass C and P (MB-C, MB-P). By utilizing the susceptibility of organic C to $KMnO_4$ oxidation [31], three fractions of labile organic C in these wetland columns; namely, highly labile organic C (HLOC), mid-labile organic C (MLOC) and labile organic C (LOC), were determined using 33 mmol L^{-1}, 167 mmol L^{-1}, and 333 mmol L^{-1} of $KMnO_4$, respectively.

Calculation of homeostatic regulation ability of soil microorganisms

The homeostatic regulation ability of soil microorganisms was estimated by calculating the homeostatic regulation coefficient (*H*) according to the following equation [1]:

$$y = c \cdot x^{(1/H)} \quad (1)$$

where y is the microbial C or P concentration (mg g^{-1} or mg kg^{-1} of dry soil weight) or the C:P molar ratio of soil microorganisms; x is the organic C concentration (including HLOC, MLOC, and LOC) in soil, DRP or DOC concentration in porewater, or the C:P molar ratio among the soil organic C and porewater DRP; and c is a constant. H is defined as the homeostatic regulation ability of soil microorganisms; as the value of H increases, the ecosystem trends toward a higher degree of ecological stability and resilience. For example, higher values of H_N and H_P in grasses receiving field N and P fertilizations were significantly associated with higher species dominance and stability [2]. According to ecological allometry [5,32], H may be classified as an equilibrium constant with respect to functions of organism metabolic and stoichiometric processes.

The first derivative of regression equation (1), reflecting the response rate of y to x, is given as follows:

$$\frac{dy}{dx} = c \cdot (1/H) \cdot \left(x^{(1/H-1)}\right) = \left(c \cdot x^{1/H}\right) \cdot (1/x) \cdot (1/H) \quad (2)$$

By substituting in variables from equation (1), equation (2) can be transformed as:

$$\frac{dy}{dx} = y \cdot (1/\mathrm{x}) \cdot (1/\mathrm{H}) = (\mathrm{y}/\mathrm{x}) \cdot (1/H) \quad (3)$$

Here, the value of (y/x) is defined as the specific variable for $1/H$, which may be transformed into (x/y) responding to H. Therefore, we define n-H_i as the specialized homeostatic regulation ability for each of the tested wetland soil columns:

$$n - H_i = (x_i / y_i) \cdot H_i \quad (4)$$

where x_i is the measured value of the organic C concentration (including HLOC, MLOC, and LOC) in soil, DRP or DOC concentration in porewater, or the C:P molar ratio among the soil organic C and porewater DRP for each soil sample; y_i is the microbial C or P concentration (mg g^{-1} or mg kg^{-1} of dry soil weight) or the C:P molar ratio of soil microorganisms for each soil sample; i is the type of parameter (form of HLOC, MLOC, LOC, DRP, DOC, or ratio of C:P for soil or soil microbes); and H refers to the homeostatic regulation coefficient calculated by equation (1) according to the respective for each of the two treatments (Control and Warmed) in terms of the i parameter.

Potential P flux (F_P, mg m^{-2} d^{-1}) was used to evaluate the potential P transfer out of porewater under experimental warming according to the following equation:

$$F_p = (C_{in} - C_{out}) \times V/(S \times T) \quad (5)$$

where $C_{in}C_{in}$ is the DRP concentration in the porewater (mg L^{-1}), C_{out} is the P concentration in the overlying water (mg L^{-1}), V is the volume of overlying water (mL), S is the area of the wetland column (m^2) and T is the interval between two sampling dates (d). Due to the relatively static hydrological conditions in the wetlands (Table S1), DRP was the most important form for P exchange between soil and water in the study. Soil microorganisms are also more sensitive to DRP than to TP due to its bioavailability. Therefore, values of F_P (and n-H_i) were evaluated by the form of DRP rather than TP. The potential P flux defined here provides a good indicator for assessing P concentration gradients between overlying water and porewater, with high values suggesting a high risk of P transfer from soil to overlying water.

In this study, aquatic macrophytes (e.g., *Phragmites communis*) were temporarily excluded from soil columns, but small benthic organisms (e.g., *Margarya melanoide*) and plankton (e.g., *Spirogyra*) were preserved. Data collected from water and soil samples were expressed as the mean plus standard error. Paired Student's *t*-tests were used to compare the effects of warming on soil microorganism and C fractions, P concentrations in porewater and overlying water, and potential P flux. Two-way analysis of variance (repeated ANOVA) tests were carried out using SPSS software (version 15.0) to examine the effects of experimental warming, sampling time, and their interaction on soil biochemical traits and potential P flux. Statistical test results were considered significant at the $p<0.05$ level.

Results

Soil carbon and phosphorus potential flux in response to experimental warming

After >2.5 y of incubation (in Nov 2010), experimental warming increased soil microbial biomass, soil labile C, and porewater DOC (Table 1); however, warming decreased concentrations of total soil C at all wetlands except site SQ. Meanwhile, although fluctuations in DRP concentrations were found in the overlying water, DRP in porewater increased in the warmed treatment by 52% to 137% compared to the ambient treatment, except in the SQ wetland (Table 2). On average, potential P flux in the six wetlands ranged from 0.11 to 34.51 mg m^{-2} d^{-1} in the control and 0.07 to 61.26 mg m^{-2} d^{-1} in the warmed treatment (Table 2). Two-way analysis of variance indicated that experimental warming significantly increased potential P flux, except in the SQ wetland. There were also significant differences in soil C pools and potential P flux between Warmed and Control treatments (Table 3). A significant interaction between treatment and season (sampling date) was found for soil C pools and potential

Table 1. Concentrations of soil microbial biomass (MB; MB-C: microbial carbon; MB-P: microbial phosphorus), soil carbon (TOC: total organic carbon; HLOC: highly labile organic carbon; MLOC: mid-labile organic carbon; and LOC: labile organic carbon), and carbon in porewater (DOC: dissolved organic carbon) in wetland columns in the microcosm experiment (Control: ambient temperature; Warmed: ambient temperature +5°C).

Treatment		Soil microbial biomass				Soil carbon				DOC in porewater (mg kg^{-1})
		Total mass (mg g^{-1})	MB-C (mg g^{-1})	MB-P (mg kg^{-1})	Microbial C:P stoichiometric ratio	TOC (g kg^{-1})	HLOC (mg g^{-1})	MLOC (mg g^{-1})	LOC (mg g^{-1})	
JH	Control	0.402±0.023	0.386±0.022	15.0±2.3	68.6:1	25.6±1.4	2.28±0.09	2.75±0.66	3.75±0.71	13.2±2.3
	Warmed	0.493±0.010	0.477±0.008	16.1±2.1	79.0:1	22.3±2.1	2.31±0.37	3.16±0.70	5.48±1.31	16.6±3.1
p		0.064	**0.050**	0.568		**0.028**	0.899	0.498	0.113	0.182
XZ	Control	1.12±0.02	1.08±0.02	25.9±2.4	111:1	54.8±5.9	5.78±0.31	8.55±0.29	9.93±0.43	17.3±0.9
	Warmed	1.47±0.03	1.42±0.04	33.4±2.3	113:1	49.2±2.6	6.35±0.06	8.67±0.13	9.93±0.32	21.6±2.2
p		**<0.001**	**0.009**	**0.005**		0.054	0.055	0.575	0.990	**0.038**
YT	Control	1.42±0.01	1.37±0.01	36.4±2.9	100:1	133±10	9.13±0.23	13.0±1.0	18.9±4.2	20.2±3.3
	Warmed	1.73±0.06	1.66±0.04	46.1±5.7	96.0:1	117±3	10.9±0.1	18.6±0.9	20.5±4.3	22.7±9.9
p		**0.001**	**0.035**	**0.045**		**0.024**	**<0.001**	**0.002**	0.681	0.701
XX	Control	1.01±0.03	0.964±0.024	33.4±4.4	77.0:1	35.6±1.7	2.14±0.07	3.12±0.23	3.70±0.43	16.2±3.0
	Warmed	1.31±0.08	1.25±0.11	42.0±2.3	79.4:1	31.8±1.3	3.39±0.06	3.65±0.19	4.32±0.83	24.3±8.7
p		**0.003**	0.067	**0.028**		**0.011**	**<0.001**	**0.026**	**0.039**	0.208
BY	Control	1.07±0.06	1.04±0.06	34.5±2.4	80.4:1	37.6±1.9	2.90±0.12	3.98±0.05	4.50±0.09	18.6±1.5
	Warmed	1.13±0.07	1.10±0.06	36.6±2.6	80.1:1	32.4±2.1	3.59±0.36	5.91±0.16	7.06±0.33	21.8±1.0
p		0.299	0.317	0.435		**0.043**	**0.033**	**<0.001**	**<0.001**	**0.027**
SQ	Control	0.183±0.018	0.179±0.017	4.25±0.87	112:1	14.2±2.3	0.556±0.025	0.889±0.673	1.73±0.33	8.14±0.96
	Warmed	0.469±0.011	0.456±0.013	13.5±2.7	90.1:1	15.3±1.7	2.61±0.05	3.74±0.15	4.23±0.16	12.3±1.2
p		**<0.001**	**<0.000**	**0.005**		0.443	**<0.001**	**0.002**	**<0.001**	**0.006**

p-values refer to significant differences between two treatments. Sampling date is November 2010.

Table 2. Phosphorus (P) concentrations in overlying water and porewater, and the potential flux of dissolved reactive P in wetland columns in the microcosm experiment (Control: ambient temperature; Warmed: ambient temperature +5°C).

Wetland column	Treatments	Overlying water (mg L^{-1})	Porewater (mg L^{-1})	Potential P flux (mg m^{-2} d^{-1})
	July 2010			
JH	Control	0.066±0.038	0.063±0.005	0.21±0.07
	Warmed	0.017±0.009	0.095±0.001	0.32±0.09
p		**0.022**	0.004	**0.018**
XZ	Control	0.070±0.026	0.266±0.003	0.90±0.03
	Warmed	0.024±0.024	0.612±0.063	2.50±0.38
p		**0.012**	**0.039**	**0.003**
YT	Control	1.507±1.22	2.334±0.703	15.30±3.56
	Warmed	0.909±0.51	4.537±1.510	35.26±4.61
p		0.304	**0.002**	**<0.000**
XX	Control	0.106±0.076	0.330±0.015	2.14±0.17
	Warmed	0.038±0.031	0.570±0.033	3.65±0.10
p		0.091	**0.023**	**0.004**
BY	Control	0.063±0.061	0.104±0.001	0.61±0.02
	Warmed	0.051±0.079	0.231±0.001	0.98±0.24
p		0.783	**<0.000**	**0.017**
SQ	Control	0.016±0.008	0.088±0.015	0.11±0.02
	Warmed	0.012±0.010	0.021±0.009	0.08±0.01
p		0.431	0.025	0.412
	March 2011			
JH	Control	0.018±0.000	0.182±0.102	0.63±0.37
	Warmed	0.016±0.000	0.354±0.048	1.07±0.23
p		0.467	**0.016**	**0.004**
XZ	Control	0.029±0.014	0.720±0.286	2.10±0.13
	Warmed	0.018±0.000	1.704±0.189	4.36±0.78
p		0.187	**0.013**	**0.006**
YT	Control	0.042±0.008	5.83±1.23	34.51±10.36
	Warmed	0.047±0.012	8.34±0.41	61.26±6.12
p		0.572	**0.011**	**<0.000**
XX	Control	0.016±0.006	1.17±0.18	3.38±0.71
	Warmed	0.013±0.004	2.53±0.34	11.72±2.10
p		0.320	**0.004**	**<0.000**
BY	Control	0.016±0.000	0.390±0.067	1.06±0.12
	Warmed	0.022±0.000	0.653±0.182	2.38±0.53
p		**0.012**	**0.039**	**0.034**
SQ	Control	0.012±0.001	0.068±0.019	0.16±0.09
	Warmed	0.016±0.000	0.044±0.040	0.07±0.03
p		0.490	0.193	0.387

p-values refer to significant differences between two treatments.

P flux as well, although there were no significant differences in these variables by season alone.

Microorganism stoichiometric homeostasis in response to experimental warming

The relationships between soil microorganism biomass and three forms of soil carbon concentrations (HLOC, MLOC, and LOC) sampled in Mar. 2011 were rigorously and significantly described by the stoichiometric homeostasis model (Fig. 1, $p<0.01$). The values of H in the warmed treatment were lower than those in the control, with H values of 1.22 and 1.73 for HLOC, 1.48 and 1.75 for MLOC, and 1.09 and 2.20 for LOC in warmed and control treatments, respectively (Fig. 1). Results from the other two samplings (data not shown) were fundamentally similar to the Mar. 2011 sampling. There was no consistent pattern in the homeostasis model fit between DOC and microorganism C among the three samplings (Table 4). However, strongly significant and positive correlations between DRP in porewater and soil microorganism biomass P fit a power model for

Table 3. Results of two-way analysis of variance (repeated ANOVA) showing the *p*-values for soil microorganisms and soil carbon and potential phosphorus (P) flux under experimental warming and sampling time. MB-C: microbial carbon, MB-P: microbial phosphorus, TOC: total organic carbon, HLOC: highly labile organic carbon, MLOC: mid-labile organic carbon, and LOC: labile organic carbon.

Factor	Soil microorganisms			Soil carbon				Potential P flux ($mg\ m^{-2}\ d^{-1}$)
	Total ($mg\ g^{-1}$)	MB-C ($mg\ g^{-1}$)	MB-P ($mg\ kg^{-1}$)	TOC ($g\ kg^{-1}$)	HLOC ($mg\ kg^{-1}$)	MLOC ($mg\ kg^{-1}$)	LOC ($mg\ kg^{-1}$)	
Treatment	**0.007**	**0.005**	**0.008**	**0.013**	**0.011**	**0.038**	**0.007**	**0.033**
Time	0.285	0.321	0.308	0.079	0.563	0.439	0.407	0.185
Treatment×Time	**0.017**	**0.016**	**0.009**	**0.006**	**0.019**	0.076	**0.016**	**0.047**

p-values smaller than 0.05 are in bold to indicate statistical significance.

these three consecutive samplings (Fig. 2), with lower values of H under the warmed treatment (1.79 to 2.49) compared to the control (3.08 to 4.24). Although no significant difference between these two treatments was found for the C:P stoichiometric ratios of soil microorganisms (paired *t*-test $p = 0.607$, Table 1), strongly significant correlations existed between soil C:P ratios of HLOC to DRP and soil microorganism C:P for these three samplings under the two treatments (Fig. 3). More importantly, the C:P ratio-related H values (i.e., $H_{C:P}$) obtained from the warmed group were consistently lower than those from the control group.

No significant relationship was found between porewater ratios of DOC:DRP and soil microbial C:P (Table 4). Relationships were observed for other comparisons of soil LOC to porewater DRP and soil MLOC to porewater DRP vs soil microorganism C:P ratios (Table 4); however, among the three samplings, corresponding $H_{C:P}$ values showed no consistent trends between the two treatment groups. This is in contrast to soil HLOC to porewater DRP results, in which a consistent increase in $H_{C:P}$ was observed from control to warmed treatments across all three samplings (Fig. 3). Specific n-$H_{C:P}$ values for each wetland column and soil treatment (Warmed or Control) were subsequently calculated using the $H_{C:P}$ values derived from these HLOC:DRP regressions, using C:P ratios (HLOC:DRP) in porewater as the 'x_i' variable and soil microbial C:P ratios as the 'y_i' variable in equation (4). Again, strongly significant ($p<0.001$) linkages that fit a power model between n-$H_{C:P}$ and potential P flux (i.e., allometric relationships [5]) were found for both treatment groups (Fig. 4). Notably, these two fitted curves strongly overlapped, and the regression equation for the two datasets combined ($y = 5.261 \times^{-1.7509}$, $R^2 = 0.5893$) was almost identical to the equations derived from each dataset individually. The data for potential P flux and n-$H_{C:P}$ values were concentrated in ranges less than 20 $mg\ m^{-2}\ d^{-1}$ and 3.0, respectively (Fig. 4).

Discussion

Does homeostasis relate to the ecological role of soil in C and P cycling?

The tight linkages in this study between different pools of soil labile C and soil microbial C (Fig. 1), and porewater DRP and soil microbial P (Fig. 2) fit the stoichiometric homeostasis model. This supports the view that stoichiometric homeostasis is an important mechanism underpinning the dynamics of C and P resource availability at the soil-microbe interphase in wetland ecosystems. It confirms that the composition of C and P in microbial assemblages is closely related to resource availability in ecosystems [33,34], as soil microorganism C and P increases were related to the increases in soil C and P availability observed in the warmed treatment in this study (Table 1 & 2). Based on the principle of homeostatic regulation ability [1] and its prior application to a grassland ecosystem [2], decreased H values for soil labile C and DRP in warmed columns indicate that experimental warming impaired the ability of soil organisms to regulate C and P biogeochemical turnover in the studied wetland ecosystems. This loss of homoeostatic regulation ability is consistent with the observed decrease in soil TOC (Table 1) and the increase in potential P flux (Table 2). Comparison of the data for the decrease in soil TOC and the increase in both soil labile C and microorganism C (Table 1) further indicates that losses of total organic C in warmed soils were mainly derived from non-active C pools; i.e., the recalcitrant organic C. Recalcitrant organic fractions generally represent a large proportion of the total C pool and have relatively slow turnover rates [35,36], and generation of recalcitrant organic matter is regarded as a vital process for C sequestration [36]. Excluding P that is assimilated into microbial biomass (Table 1), warming subjects the remaining P pool to increases in enzymatic biodegradation of organic phosphorus [21], P desorption [20,37], and inorganic P solubilization [38], which could enhance the strength of potential P flux (Table 2) as well as increase DRP in porewater. The ecological role of wetland soil as a source for C and P, coupled with lower H values observed in the warming treatment compared to the control, suggests that H may yield valuable insight into the ecological role of soil in C and P biogeochemical turnover and its response to global warming.

Does homeostasis relate to stoichiometric balance?

Although experimental warming did affect the soil microbial C:P stoichiometric ratios for particular samples to some degree (Table 1), no overall significant difference was found between the two (Warmed and Control) groups. However, C:P stoichiometric ratios in the soil microbial community did seem to vary according to geographical (biotic and abiotic) features [39] across the six wetlands (Table 1 and Fig. 3). Because macronutrients are coupled with various biochemical and cellular constituents, the ratios of these elements remained relatively constant compared to the potentially decoupled resource stoichiometry changes in the surrounding environment that were induced by warming [1,39].

In contrast, tight and significant associations between ratios of soil C:P and microorganism C:P were obtained with the fitted curves of the power model (Fig. 3), which seems contradictory to the results described above. This may be explained by soil microorganisms regulating the equilibrium between the elemental

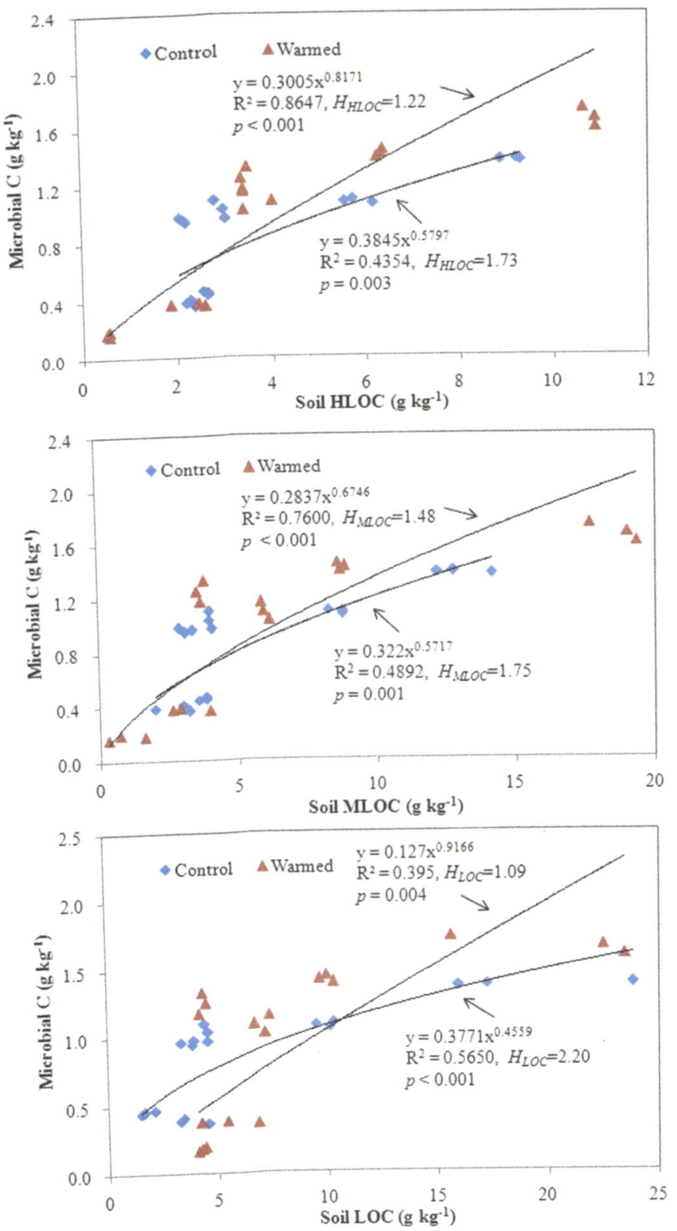

Figure 1. Relationships between soil microbial carbon (MB-C) and soil organic carbon (in forms of HLOC: highly labile organic carbon, MLOC: mid-labile organic carbon, and LOC: labile organic carbon). Control and Warmed represent treatments of ambient temperature and ambient temperature +5°C, respectively. Sampling date is March 2011. Coefficients fitting the data to the equation for stoichiometric homeostasis are given along with an estimate of H_C.

composition of their biomass and the efficiencies of microbial nutrient assimilation and growth through ecoenzymatic interactions [26]. Further, simultaneous N and P enrichment (by photosynthetic biota) produces strongly positive, synergistic responses in ecosystems [16]. This synergistic effect can aid in maintaining "Redfield-like" ratios in soil microorganisms [39], and may also have served to regulate the stoichiometric balance of soil microbial C and P acquisition for the warmed wetland soils in this study. This is also in line with general ecological theory on the elemental stoichiometries of microbial biomass and environmental availability relative to microbial nutrient assimilation and growth [1,32]. For example, increases in N concentration in biomass are usually accompanied by increases in P uptake for vascular plants [11]. This suggests that the synergistic function of C-P acquisition is an important mechanism underlying the C:P stoichiometric balance for soil microorganisms in wetland ecosystems subjected to warming.

Field N and P fertilization in another study showed that higher values of $H_{N:P}$ in grasses were significantly associated with higher species dominance and stability [2]. In this study, the homeostatic regulation coefficient for C:P stoichiometry in the soil microorganism community vs soil HLOC:DRP ratios (i.e., $H_{C:P}$) was lower under the warmed treatment than the ambient treatment (Fig. 3). In ecoenzymatic stoichiometry theory [26,32], the relative availability of hydrolyzed P via ecoenzymes (such as phosphatase) in soils relative to C in the microbial community composition was higher under experimental warming than in the control treatments. Ecosystem function and stability may have been impacted in the warmed soil columns in this study, with a negative allometric relationship to soil microorganism $H_{C:P}$. Our previous research has shown that a significant increase in soil microorganism biomass measured by total phospholipid fatty acids (PLFAs) occurred under warming [21], which is line with the data in Table 1; however, bacterial abundance in the soil tended to decrease along with an increase in the ratio of fungi:bacteria [21,40,41]. Furthermore, relatively high abundances of fungi in soil ecosystems increase the secretion of P-hydrolyzing enzymes in soils [21,42,43]. Meanwhile, functional gene array data verified that soil P utilization genes (such as polyphosphate kinase and exopolyphosphate) were also enhanced by warming [27], leading to increases in soil microbial P pools (Table 1) and speeding up nutrient cycling processes.

In a 13 d laboratory incubation test of YT wetland sediment (Text S3), lower dissolved oxygen (DO) concentrations in overlying water (Fig. S2) and higher ferrous iron (Fe^{2+}) concentrations in sediment (Fig. S3) were found for sediment sampled from the warmed column compared to the control treatment. This indicates that the redox-enhanced sediment under the warmed treatment could increase dissimilatory reduction of P-bound metal oxides and the decomposition of organic matter, liberating P from sediment aggregates. This may explain the relatively high levels of P in porewater and potential P flux under warming (Table 2). In addition to stimulating net primary production, warming could significantly increase both soil respiration and the associated C-degrading microbial genes [27,44], increasing the C flux from terrestrial ecosystems to the atmosphere in the global C cycle [8].

Results from our previous investigation [21], coupled with the data reported here, lend further support to the use of $H_{C:P}$, H_C, and H_P as ecological 'indicators' (Fig. 3, 1, & 2, respectively). In other words, $H_{C:P}$ values in the soil microorganism community in this study were positively correlated with C-P-related ecological function and stability. Because C:P stoichiometric homeostasis is associated with synergistic C-P acquisitions (as discussed above), a decreased $H_{C:P}$ value under warming conditions (Fig. 4) means that C biodegradation and P mineralization exceeded microorganism utilization of these elements, leading to increases in both C loss and P export. This has a "double-negative" effect on climate change and water quality.

We did not find a consistent relationship between $H_{C:P}$ values in warmed vs. control treatments in terms of LOC:DRP and MLOC:DRP ratios (Table 4). Soil C losses in this study were traced to the recalcitrant fraction, while pools of labile C were not depleted (Table 1). Carbon availability from HLOC for the soil microorganism assemblage and metabolism is generally higher

Table 4. Homeostatic regulation coefficient (*H*) in soils collected from wetland columns in the microcosm experiment (Control: ambient temperature; Warmed: ambient temperature +5°C).

Variable	Dependent variable	Sampling data	Control group			Warmed group		
			H	R^2	*p*	*H*	R^2	*p*
Porewater DOC	Microbial biomass C	Jul 2010	−5.00	0.084	0.203	4.26	0.085	0.219
		Nov 2010	2.76	0.274	**0.004**	3.00	0.257	**0.008**
		Mar 2011	1.71	0.386	**0.000**	−20.0	0.019	0.452
Porewater DOC:DRP	Microbial C:P	Jul 2010	500	0.014	0.672	100	0.005	0.953
		Nov 2010	250	0.000	–	71.4	0.002	0.976
		Mar 2011	55.6	0.007	0.872	43.4	0.016	0.553
Soil LOC: porewater DRP	Microbial C:P	Jul 2010	6.76	0.303	**0.004**	9.43	0.463	**0.000**
		Nov 2010	12.9	0.275	**0.003**	13.7	0.237	**0.010**
		Mar 2011	12.6	0.154	**0.013**	7.46	0.545	**0.000**
Soil MLOC: porewater DRP	Microbial C:P	Jul 2010	8.26	0.340	**0.003**	6.90	0.391	**0.000**
		Nov 2010	16.1	0.268	**0.008**	22.2	0.071	0.237
		Mar 2011	12.8	0.213	**0.013**	5.18	0.461	**0.000**

p-values refer to significant differences between two treatments. DOC and DRP refer to dissolved organic carbon and dissolved reactive phosphorus in porewater while MLOC and LOC refer to mid-labile organic carbon and labile organic carbon in soil, respectively.

than for the rest of the C sources [31]; therefore the index of HLOC is relatively sensitive to C dynamics for microorganism acquisition. Further mechanistic investigations are needed to establish direct links between the destabilization of soil organics and soil microorganism community dynamics in order to better assess the relationship between the availability of different forms of C to C:P stoichiometric dynamics and their ultimate effects on $H_{C:P}$.

Does homeostasis link to soil phosphorus flux?

Efforts to predict P flux or P dynamics from the soil biosphere to aquatic ecosystems are becoming increasingly important in light of concerns related to eutrophication, climate change, and other anthropogenic impacts [19,20,45,46]. Statistics-based modeling integrates various parameters, such as soil physicochemical properties, regional meteorology, and anthropogenic activities [19,20,23], and can be used to predict P flux under IPCC climate scenarios [4]. However, these models are insufficiently related to the ecological forcing of microorganisms in the soil micro-ecosystem, even though such microorganisms essentially mediate and drive Earth's biogeochemical cycles [13,27]. We found that potential P flux (Table 2) was significantly related to n-$H_{C:P}$ according to a negative allometric model (Fig. 4), verifying the hypothesis that a n-$H_{C:P}$ index may be useful as an ecological tool to predict P flux. Although C:P stoichiometric dynamics alone can provide some insight into P dynamics in an ecosystem [5,39], the n-$H_{C:P}$ term (as outlined in equation 4) comprehensively incorporates both soil C and P availability and soil microorganism nutrient acquisition under the principle of stoichiometric homeostasis [1]. The n-$H_{C:P}$ response mirrors mechanisms related to the metabolic theory of ecology [5,6] in that the soil micro-environment of wetland ecosystems responds positively (albeit via a negative allometric relationship) to temperature increases.

According to the derived models of potential P flux to n-$H_{C:P}$ (Fig. 4), lower n-$H_{C:P}$ values in this study indicate higher P flux, which is fundamentally consistent with the ecological interpretation of H_C (Fig. 1), H_P (Fig. 2), and $H_{C:P}$ (Fig. 3) between the two treatments. Although regression models differed, the ecological role of n-$H_{C:P}$ in this study is essentially identical in its nature to the linear and positive relationships between community *H* and ecological production and stability in a Mongolian grassland [2]. The consistency of these results suggests that although diverse factors may affect P dynamics, n-$H_{C:P}$ may be a useful ecological tool for assessing potential P flux. Notably, the mathematic formulations between n-$H_{C:P}$ and potential P flux were identical between the two experimental treatments (Fig. 4), suggesting that the feasibility of n-$H_{C:P}$ to quantitatively assess potential P flux might be mainly determined by the features of soil C-P stoichiometric homeostasis itself.

It should be noted that both n-$H_{C:P}$ and potential P flux are calculated using one of the same parameters; namely, soil porewater DRP, and are therefore inherently interrelated to some extent. However, n-$H_{C:P}$ (Eq 4) is computed using a multifactor equation linking soil C:P ratio (in form of soil HLOC: soil porewater DRP), soil microorganism biomass C:P ratio, and the associated homeostatic coefficient ($H_{C:P}$). Similarly, changes in potential P flux (Eq 5) were jointly determined by differences in the concentration of DRP in soil porewater vs. overlying water, by the quantity of overlying water and the area of topsoil in the wetland column, and by the overlying water replacement time interval. Moreover, values of these two indices were distinctly influenced by experimental warming (Table 2, Fig. 3). Therefore, the strong, negative allometric relationship observed between these indices (Fig. 4) is indeed an indication of the potential ecological management role of n-$H_{C:P}$ in assessing potential P flux, and is not be expected outright based only on co-variation arising from their common incorporation of a soil porewater DRP parameter.

The stoichiometric parameter $S_{C/P}$ (or $S_{C/N}$) has been defined as a scalar for the relative availability of hydrolyzed P (or N) in relation to microbial community composition [32,47]. Utilizing this $S_{C/P}$ (or $S_{C/N}$) metric, an allometric biogeochemical equilibrium model was developed to predict microbial growth efficiency (GE) from elemental C:N and C:P ratios in biomass ($B_{C/N}$ and $B_{C/P}$, respectively) and environmental substrate sources

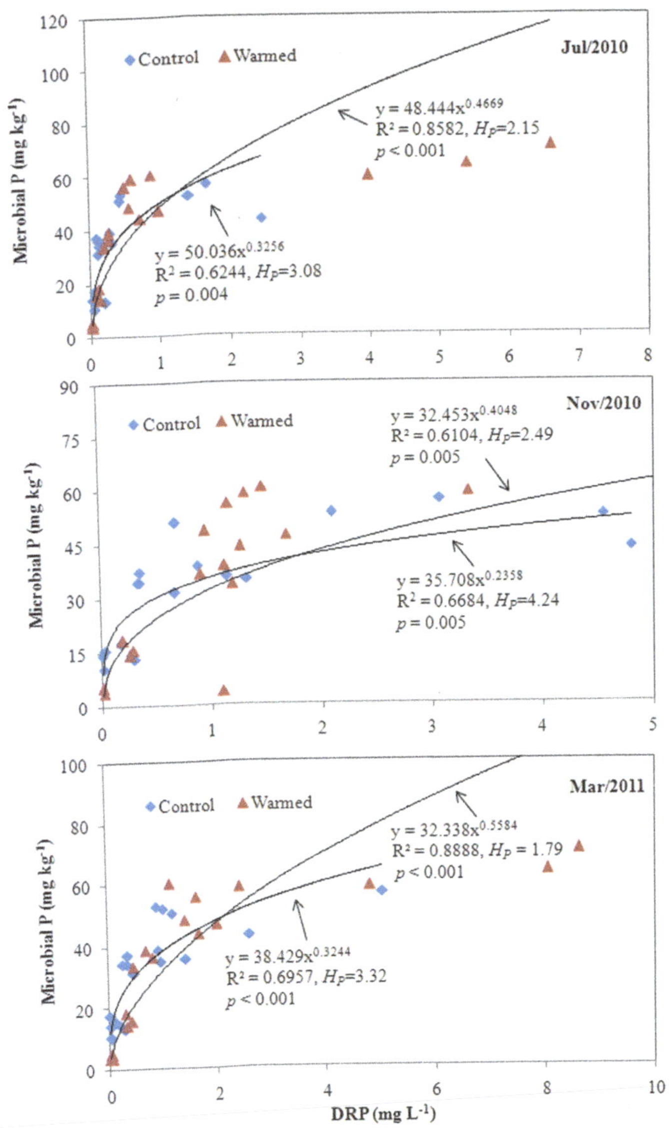

Figure 2. Relationships between soil microbial phosphorus (MB-P) content and dissolved reactive phosphorus (DRP) in porewater. Control and Warmed represent treatments of ambient temperature and ambient temperature +5°C, respectively. Coefficients fitting the data to the equation for stoichiometric homeostasis are given along with an estimate of H_P.

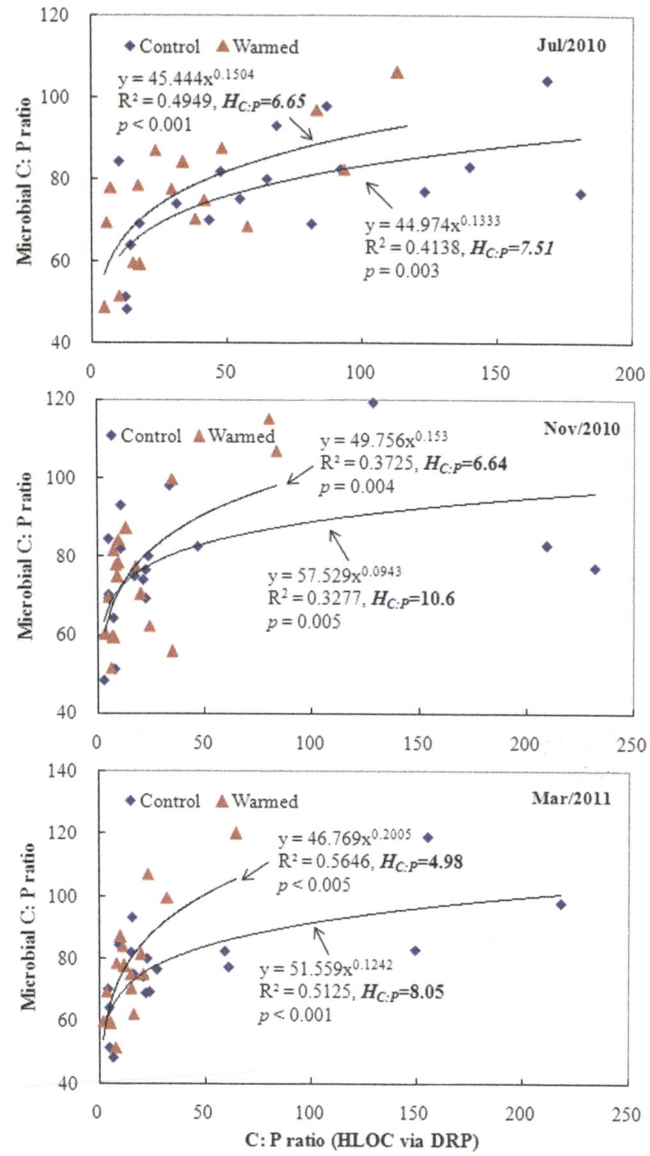

Figure 3. Relationships between the molar ratios of carbon to phosphorus (C:P) in soil microbial biomass and the C:P ratios of soil highly labile organic carbon (HLOC) to dissolved reactive phosphorus (DRP) in porewater. Control and Warmed represent treatments of ambient temperature and ambient temperature +5°C, respectively. Coefficients fitting the data to the equation for stoichiometric homeostasis are given along with an estimate of $H_{C:P}$.

($L_{C/N}$, $L_{C/P}$), integrated with ratios of ecoenzymatic activities (EEAs) that mediate C, N, and P acquisition ($EEA_{C/N}$, $EEA_{C/P}$) [32]. Although the intended prediction outcomes of these approaches differ, the nature of the relationship between microbial n-$H_{C:P}$ and sediment potential P flux is fundamentally similar to the use of the $S_{C/P}$ indictor, because both scalars reflect how stoichiometric balance/interaction is regulated by environmental signals. Further, they can also both be viewed in terms of an energy landscape (i.e., C) that directs the availability and flow of resources (e.g., N and P). Therefore, our experimental data and the measurement of soil n-$H_{C:P}$ might be practical for comparative investigations of soil P dynamics across a variety of wetland ecosystems. This n-$H_{C:P}$ method offers a new alternative for the measurement and understanding of soil P biogeochemical cycling, comparable to current methodologies that utilize biochemistry [37,42], molecular biology [21,27,43,46], isotopic tracing [48], and mechanistic models [19,23,45]. To the best of our knowledge, this is the first study of integrated C-P stoichiometric homeostasis to clearly demonstrate the linkage between soil n-$H_{C:P}$ and potential P flux in a freshwater wetland ecosystem.

Undoubtedly, aquatic plants are another primary pool for P biogeochemical cycling, because 12%–85% of P in water may be immobilized by wetland plants [48,49]. However, addressing the influence of aquatic plants on n-$H_{C:P}$ and other parameters was beyond the purpose of this study, and thus they were temporarily excluded from the soil columns during the course of the experiment. The involvement of aquatic plants in influencing soil

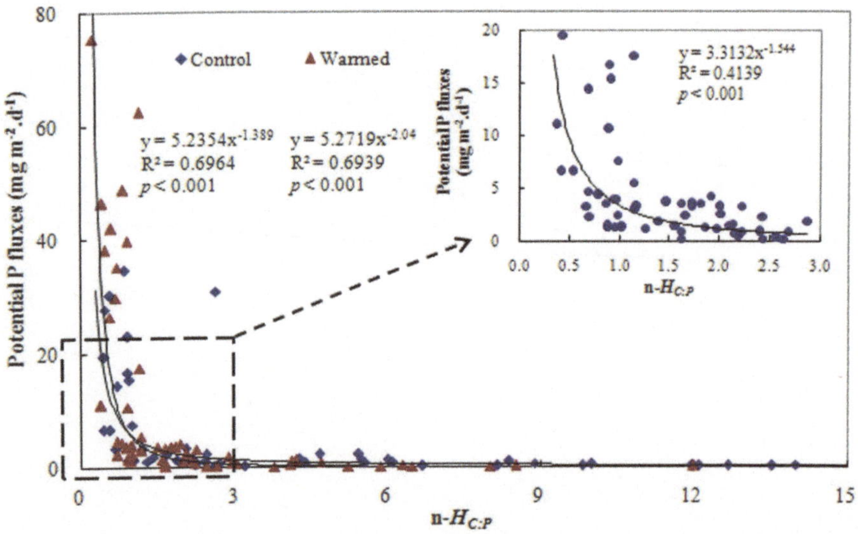

Figure 4. Relationship between potential phosphorus (P) flux and the specialized ability of stoichiometric homeostasis (n-$H_{C:P}$) as a molar ratio of soil highly labial organic carbon (HLOC) to dissolved reactive phosphorus (DRP) in each of the wetland soil columns. Control and Warmed represent treatments of ambient temperature and ambient temperature +5°C, respectively. The data for P flux and n-$H_{C:P}$ less than 20 mg m^{-2} d^{-1} and 3.0, respectively, are plotted in the figure inset.

C:P stoichiometric homeostasis and associated variations in n-$H_{C:P}$ thus requires further investigation.

Although soil respiration at the global scale responds positively to air temperature, in the long term warming may also mobilize older stored carbon, potentially resulting in higher carbon inputs to the soil biosphere [8]. If so, labile P forms would be 'returned' to the soil-microorganism complex under warming according to principles of ecoenzymatic stoichiometry [26] and mechanisms of synergistic C-P acquisition. Such an outcome would be somewhat inconsistent with the near-term decrease in soil C stock (Table 1) and increase in potential P flux (Table 2) we observed in the current microcosm warming experiment. Thus, it will also be necessary to further explore the linkage between n-$H_{C:P}$ and P flux under long-term warming.

In summary, microcosm warming impaired the stoichiometric homeostatic ability (H) of soil microorganisms to regulate P and C biogeochemical processes. This resulted in a "double negative" effect for wetland ecosystem services: increasing their potential for P export and enhancing their role as a recalcitrant organic C source. Moreover, the specialized homeostatic regulation ability (n-$H_{C:P}$, as a function of the C:P ratio in the form of HLOC to DRP in soil porewater relative to the C:P ratio in soil microbial biomass) of soil microorganisms was inversely linked to potential P flux at the soil-water interface. Based on our results, we advocate further use of homeostatic regulation indices (H) as novel ecological tools for assessing potential P flux and other biogeochemical dynamics in the face of global climate change.

Supporting Information

Figure S1 The design of the experimental wetland microcosm system setup by using independently monitored water bath jackets under the current climate condition (Left: Ambient temperature, Control) and the warming climate condition (Right: Ambient temperature +5°C, Warmed treatment).

Figure S2 Dynamics of dissolved oxygen (DO) concentration in overlying water during 12 d of sediment incubation. Sediment samples were collected in July 2010 from YaTang riverine wetland (YT) under control (ambient temperature) and warmed (ambient temperature +5°C) treatments in microcosm experiment. Error bars show ± SD. The differences between control and warmed treatments were tested by Student's t-test for each sampling point, indicated by * $p<0.05$, ** $p<0.01$.

Figure S3 Dynamics of ferric iron (Fe^{3+}) and ferrous iron (Fe^{2+}) concentration in sediment measured during a 13-d laboratory incubation for YaTang riverine wetland (YT) sediment samples under control (ambient temperature) and warmed (ambient temperature +5°C) treatments. Error bars show ± SD. The differences between control and warmed treatments were tested by Student's t-test for each sampling point, indicated by * $p<0.05$, ** $p<0.01$.

Table S1 Details of the six selected wetland sites used in the study.

Table S2 Select physico-chemical parameters of 20-cm depth soil samples collected from the JinHu wetland (JH), XiaZhuhu wetland (XZ), YaTang riverine wetland (YT), XiXi national wetland park (XX), BaoYang riverine wetland (BY), and ShiQiuyang multipond wetland (SQ) in May 2008.

Text S1 Microcosm configuration.

Text S2 Wetland column preparation and field sampling.

Text S3 Laboratory incubation for sediment oxygen demand and reducing capability.

References

1. Sterner RW, Elser JJ (2002) Ecological Stoichiometry: The Biology of Elements From Molecules to the Biosphere. Princeton University Press, Princeton.
2. Yu Q, Chen QS, Elser JJ, He NP, Wu HH, et al. (2010) Linking stoichiometric homoeostasis with ecosystem structure, functioning and stability. Ecology Letters 13: 1390–1399.
3. Hansen J, Sato M, Ruedy R, Lo K, Lea DW, et al. (2006) Global temperature change. Proceedings of the National Academy of Sciences of the United States of America 103: 14288–14293.
4. IPCC (2001) Climate Change 2001:The Scientific Basis. Intergovernmental Panel on Climate Change. Cambridge University Press, Cambridge.
5. Allen AP, Gillooly JF (2009) Towards an integration of ecological stoichiometry and the metabolic theory of ecology to better understand nutrient cycling. Ecology Letters 12: 369–384.
6. Gillooly JF, Brown JH, West GB, Savage VM, Charnov EL (2001) Effects of size and temperature on metabolic rate. Science 293: 2248–2251.
7. Englund G, Ohlund G, Hein CL, Diehl S (2011) Temperature dependence of the functional response. Ecology Letters 14: 914–921.
8. Bond-Lamberty B, Thomson A (2010) Temperature-associated increases in the global soil respiration record. Nature 464: 579-U132.
9. Rhee GY (1978) Effects of N: P atomic ratios and nitrate limitation on algal growth, cell compostion and nitrate uptake. Limnology and Oceanography 23: 10–25.
10. Elser JJ, Fagan WF, Kerkhoff AJ, Swenson NG, Enquist BJ (2010) Biological stoichiometry of plant production: metabolism, scaling and ecological response to global change. New Phytologist 186: 593–608.
11. Yu QA, Elser JJ, He NP, Wu HH, Chen QS, et al. (2011) Stoichiometric homeostasis of vascular plants in the Inner Mongolia grassland. Oecologia 166: 1–10.
12. Totsche KU, Rennert T, Gerzabek MH, Kogel-Knabner I, Smalla K, et al. (2010) Biogeochemical interfaces in soil: The interdisciplinary challenge for soil science. Journal of Plant Nutrition and Soil Science 173: 88–99.
13. Falkowski PG, Fenchel T, Delong EF (2008) The microbial engines that drive Earth's biogeochemical cycles. Science 320: 1034–1039.
14. Hessen DO, Agren GI, Anderson TR, Elser JJ, De Ruiter PC (2004) Carbon, sequestration in ecosystems: The role of stoichiometry. Ecology 85: 1179–1192.
15. Verhoeven JTA, Arheimer B, Yin CQ, Hefting MM (2006) Regional and global concerns over wetlands and water quality. Trends in Ecology & Evolution 21: 96–103.
16. Elser JJ, Bracken MES, Cleland EE, Gruner DS, Harpole WS, et al. (2007) Global analysis of nitrogen and phosphorus limitation of primary producers in freshwater, marine and terrestrial ecosystems. Ecology Letters 10: 1135–1142.
17. Sundareshwar PV, Morris JT, Koepfler EK, Fornwalt B (2003) Phosphorus limitation of coastal ecosystem processes. Science 299: 563–565.
18. Mulder C, Elser JJ (2009) Soil acidity, ecological stoichiometry and allometric scaling in grassland food webs. Global Change Biology 15: 2730–2738.
19. Serpa D, Falcao M, Duarte P, da Fonseca LC, Vale C (2007) Evaluation of ammonium and phosphate release from intertidal and subtidal sediments of a shallow coastal lagoon (Ria Formosa-Portugal): A modelling approach. Biogeochemistry 82: 291–304.
20. Nicholls KH (1999) Effects of temperature and other factors on summer phosphorus in the inner Bay of Quinte, Lake Ontario: implications for climate warming. Journal of Great Lakes Research 25: 250–262.
21. Zhang ZJ, Wang ZD, Holden J, Xu XH, Wang H, et al. (2012) The release of phosphorus from sediment into water in subtropical wetlands: a warming microcosm experiment. Hydrological Processes 26: 15–26.
22. Wang H, Holden J, Spera K, Xu XH, Wang ZD, et al. (2013) Phosphorus fluxes at the sediment-water interface in subtropical wetlands subjected to experimental warming: A microcosm study. Chemosphere 90: 1794–1804.
23. Jeppesen E, Kronvang B, Meerhoff M, Sondergaard M, Hansen KM, et al. (2009) Climate change effects on runoff, catchment phosphorus loading and lake ecological state, and potential Adaptations. Journal of Environmental Quality 38: 1930–1941.
24. Cheng WG, Sakai H, Matsushima M, Yagi K, Hasegawa T (2010) Response of the floating aquatic fern Azolla filiculoides to elevated CO(2), temperature, and phosphorus levels. Hydrobiologia 656: 5–14.
25. Hilbert DW, Roulet N, Moore T (2000) Modelling and analysis of peatlands as dynamical systems. Journal of Ecology 88: 230–242.
26. Sinsabaugh RL, Hill BH, Shah JJF (2009) Ecoenzymatic stoichiometry of microbial organic nutrient acquisition in soil and sediment. Nature 462: 795-U117.

Author Contributions

Conceived and designed the experiments: ZZ JDM XX QH. Performed the experiments: HW CY. Analyzed the data: HW HL DJ. Wrote the paper: HW JDM ZZ. Contributed the experimental materials: ZZ.

27. Zhou JZ, Xue K, Xie JP, Deng Y, Wu LY, et al. (2012) Microbial mediation of carbon-cycle feedbacks to climate warming. Nature Climate Change 2: 106–110.
28. Bradford MA, Fierer N, Reynolds JF (2008) Soil carbon stocks in experimental mesocosms are dependent on the rate of labile carbon, nitrogen and phosphorus inputs to soils. Functional Ecology 22: 964–974.
29. Wu J, Joergensen RG, Pommerening B, Chaussod R, Brookes PC (1990) Measurement of soil microbial biomass C by fumigation extraction - an automated procedure. Soil Biology & Biochemistry 22: 1167–1169.
30. Brookes PC, Powlson DS, Jenkinson DS (1982) Measurement of microbial biomass phosphorus in soil. Soil Biology & Biochemistry 14: 319–329.
31. Logninow W, Wisniewski W, Strony WM (1987) Fractionation of organic carbon based on susceptibility to oxidation Polish. Journal of Soil Science 20.
32. Sinsabaugh RL, Follstad Shah JJF (2012) Ecoenzymatic stoichiometry and ecological theory. In: Futuyma DJ, editor. Annual Review of Ecology, Evolution, and Systematics, Vol 43. Palo Alto: Annual Reviews. pp. 313–343.
33. Small GE, Pringle CM (2010) Deviation from strict homeostasis across multiple trophic levels in an invertebrate consumer assemblage exposed to high chronic phosphorus enrichment in a Neotropical stream. Oecologia 162: 581–590.
34. Marichal R, Mathieu J, Couteaux MM, Mora P, Roy J, et al. (2011) Earthworm and microbe response to litter and soils of tropical forest plantations with contrasting C:N:P stoichiometric ratios. Soil Biology & Biochemistry 43: 1528–1535.
35. Olk DC, Gregorich EG (2006) Overview of the symposium proceedings, "Meaningful pools in determining soil carbon and nitrogen dynamics". Soil Science Society of America Journal 70: 967–974.
36. Schmidt MWI, Torn MS, Abiven S, Dittmar T, Guggenberger G, et al. (2011) Persistence of soil organic matter as an ecosystem property. Nature 478: 49–56.
37. Liikanen A, Murtoniemi T, Tanskanen H, Vaisanen T, Martikainen PJ (2002) Effects of temperature and oxygen availability on greenhouse gas and nutrient dynamics in sediment of a eutrophic mid-boreal lake. Biogeochemistry 59: 269–286.
38. Rinu K, Pandey A (2010) Temperature-dependent phosphate solubilization by cold- and pH-tolerant species of Aspergillus isolated from Himalayan soil. Mycoscience 51: 263–271.
39. Cleveland CC, Liptzin D (2007) C: N: P stoichiometry in soil: is there a "Redfield ratio" for the microbial biomass? Biogeochemistry 85: 235–252.
40. Zhang W, Parker KM, Luo Y, Wan S, Wallace LL, et al. (2005) Soil microbial responses to experimental warming and clipping in a tallgrass prairie. Global Change Biology 11: 266–277.
41. Feng XJ, Simpson MJ (2009) Temperature and substrate controls on microbial phospholipid fatty acid composition during incubation of grassland soils contrasting in organic matter quality. Soil Biology & Biochemistry 41: 804–812.
42. Newman S, McCormick PV, Backus JG (2003) Phosphatase activity as an early warning indicator of wetland eutrophication: problems and prospects. Journal of Applied Phycology 15: 45–59.
43. Allison VJ, Condron LM, Peltzer DA, Richardson SJ, Turner BL (2007) Changes in enzyme activities and soil microbial community composition along carbon and nutrient gradients at the Franz Josef chronosequence, New Zealand. Soil Biology & Biochemistry 39: 1770–1781.
44. Wang H, He ZL, Lu ZM, Zhou JZ, Van Nostrand JD, et al. (2012) Genetic linkage of soil carbon pools and microbial functions in subtropical freshwater wetlands in response to experimental warming. Applied and Environmental Microbiology 78: 7652–7661.
45. Alvarez-Cobelas M, Sanchez-Carrillo S, Angeler DG, Sanchez-Andres R (2009) Phosphorus export from catchments: a global view. Journal of the North American Benthological Society 28: 805–820.
46. He ZL, Deng Y, Van Nostrand JD, Tu QC, Xu MY, et al. (2010) GeoChip 3.0 as a high-throughput tool for analyzing microbial community composition, structure and functional activity. Isme Journal 4: 1167–1179.
47. Sinsabaugh RL, Van Horn DJ, Shah JJF, Findlay S (2010) Ecoenzymatic Stoichiometry in Relation to Productivity for Freshwater Biofilm and Plankton Communities. Microbial Ecology 60: 885–893.
48. Noe GB, Scinto LJ, Taylor J, Childers DL, Jones RD (2003) Phosphorus cycling and partitioning in an oligotrophic Everglades wetland ecosystem: a radioisotope tracing study. Freshwater Biology 48: 1993–2008.
49. White JR, Reddy KR, Majer-Newman J (2006) Hydrologic and vegetation effects on water column phosphorus in wetland mesocosms. Soil Science Society of America Journal 70: 1242–1251.

Has Alberta Oil Sands Development Altered Delivery of Polycyclic Aromatic Compounds to the Peace-Athabasca Delta?

Roland I. Hall[1]*, Brent B. Wolfe[2], Johan A. Wiklund[1], Thomas W. D. Edwards[3], Andrea J. Farwell[1], D. George Dixon[1]

1 Department of Biology, University of Waterloo, Waterloo, Ontario, Canada, **2** Department of Geography and Environmental Studies, Wilfrid Laurier University, Waterloo, Ontario, Canada, **3** Department of Earth & Environmental Sciences, University of Waterloo, Waterloo, Ontario, Canada

Abstract

Background: The extent to which Alberta oil sands mining and upgrading operations have enhanced delivery of bitumen-derived contaminants via the Athabasca River and atmosphere to the Peace-Athabasca Delta (200 km to the north) is a pivotal question that has generated national and international concern. Accounts of rare health disorders in residents of Fort Chipewyan and deformed fish in downstream ecosystems provided impetus for several recent expert-panel assessments regarding the societal and environmental consequences of this multi-billion-dollar industry. Deciphering relative contributions of natural versus industrial processes on downstream supply of polycyclic aromatic compounds (PACs) has been identified as a critical knowledge gap. But, this remains a formidable scientific challenge because loading from natural processes remains unknown. And, industrial activity occurs in the same locations as the natural bitumen deposits, which potentially confounds contemporary upstream-downstream comparisons of contaminant levels.

Methods/Principal Findings: Based on analyses of lake sediment cores, we provide evidence that the Athabasca Delta has been a natural repository of PACs carried by the Athabasca River for at least the past two centuries. We detect no measureable increase in the concentration and proportion of river-transported bitumen-associated indicator PACs in sediments deposited in a flood-prone lake since onset of oil sands development. Results also reveal no evidence that industrial activity has contributed measurably to sedimentary concentration of PACs supplied by atmospheric transport.

Conclusions/Significance: Findings suggest that natural erosion of exposed bitumen in banks of the Athabasca River and its tributaries is a major process delivering PACs to the Athabasca Delta, and the spring freshet is a key period for contaminant mobilization and transport. This baseline environmental information is essential for informed management of natural resources and human-health concerns by provincial and federal regulatory agencies and industry, and for designing effective long-term monitoring programs for the lower Athabasca River watershed.

Editor: Caroline P. Slomp, Utrecht University, The Netherlands

Funding: The research was supported by Suncor Energy Ltd. (http://www.suncor.com), an oil sands mining and production company. Dr. Jon Fennell (WorleyParsons Canada), a respected consultant to industry and government, and a scientific authority on water-related environmental issues in the oil sands industry, helped obtain the funding for the project. During periodic meetings, Dr. Fennell became familiar with the authors' prior hydroecological research in the Peace-Athabasca Delta. And, the authors' discussed with him the concepts and proposals for the project presented in this paper. He was also aware of challenges faced when trying to get the project funded by more traditional research funding routes. Dr. Fennell presented the concept of the research project the authors had developed to a manager at Suncor Energy Ltd., which provided an introduction that led to development of the funding for this project. During the project, Dr. Fennell acted as a liaison for communications between the authors and staff at Suncor Energy Ltd. The funder had no role in study design, data collection and analysis, or preparation of the manuscript. The agreement between the University of Waterloo and the funder allowed the funder an opportunity to delay publication by up to 18 months, which they did not execute.

Competing Interests: The research was supported by Suncor Energy Ltd. (http://www.suncor.com), an oil sands mining and production company. Dr. Jon Fennell (WorleyParsons Canada), a respected consultant to industry and government, and a scientific authority on water-related environmental issues in the oil sands industry, helped the authors in our quest to obtain the funding for the project. During periodic meetings with the authors, Dr. Fennell became familiar with the authors' prior hydroecological research in the Peace-Athabasca Delta. And, the authors discussed with him the concepts and proposals for the project presented in this paper. He was also aware of challenges the authors faced when trying to get the project funded by more traditional research funding routes. Dr. Fennell presented the concept of the research project the authors had developed to a manager at Suncor Energy Ltd., which provided an introduction that led to development of the funding for this project. During the project, Dr. Fennell acted as a liaison for communications between the authors and staff at Suncor Energy Ltd. There are no patents, products in development or marketed products to declare. This does not alter the authors' adherence to the PLOS ONE policies on sharing data and materials.

* E-mail: rihall@uwaterloo.ca

Introduction

Industrial development of oil sands deposits in northern Alberta, Canada, has grown rapidly during the past few decades. Production is projected to reach >3.3 million barrels per day by 2020 [1] and generate $300 billion in tax revenue over the next 25 years [2]. The economic and oil-security benefits are undeniable, but public concerns continue to grow over the potential downstream cumulative environmental effects of the oil sands industry on the lower Athabasca River watershed and residents of its communities. Debate has become polarized in response to the accelerating pace of development. This has led to numerous recent reports by expert review panels that conclude current monitoring programs are inadequate to address the growing concerns [3–5].

Focal points of concern include the downstream community of Fort Chipewyan, the lower Athabasca River, Lake Athabasca and the Peace-Athabasca Delta (PAD; Fig. 1) where worries have been expressed regarding effects of increasing oil sands development on human and ecosystem health [1,6,7]. There, First Nation and Métis residents value a traditional way of life, and many believe that contaminants emitted from the oil sands industry are responsible for higher-than-expected rates of diseases including cancers [e.g., 6] and perceived increases of fish deformities [7]. Of particular concern are polycyclic aromatic compounds (PACs), because many PACs in bitumen are known or suspected human carcinogens [8] and have been associated with toxicological effects in fish [9]. Recent studies have identified that concentrations of some PACs and metals are elevated in snowpack and river water within 50 km of the oil sands development, and concentrations of some metals remain greater in river water at the PAD than upstream of development [10,11]. Yet, other lines of evidence point to the delta as a natural repository for contaminants. For instance, in Lake Athabasca and Richardson Lake, which continuously receive inflow from the Athabasca River, PAC concentrations are high and comparable in sediments deposited pre- and post-development [12]. Black, bitumen-rich sediment in strata deposited during the early 1800s in oxbow lakes of the PAD prone to floods from the Peace River, which passes through unexploited surficial deposits of oil sands, has also been reported [13]. These findings provide evidence that natural fluvial processes deposit contaminants downstream in lakes of the PAD, but relative contributions of contaminants from natural versus industrial causes remain unknown.

Discriminating industrial contributions from natural loads is difficult, and it remains a critical scientific challenge because the Athabasca River and its tributaries traverse surface deposits of the bitumen-rich McMurray Formation in the same areas where surface mining and processing occur. As a consequence, natural sources confound attempts to attribute differences between upstream and downstream levels of contaminants to industrial activity [3]. Moreover, PACs are a diverse group of organic compounds created by natural processes as well as by human activities [8,14].

To date, monitoring programs have been unable to establish baseline reference conditions of natural contaminant loading to the delta, a situation identified as "almost unique in monitoring for toxic compounds" by the Federal Expert Oil Sands Advisory Panel [3]. Based on attempts to analyze PAC data of the Regional Aquatics Monitoring Program (RAMP) from annual collections of bulk river sediment, Timoney and Lee [15] recently suggested that downstream PAC concentrations increased during 1999–2009, which they concluded "may be related to industrial activities" [p. 4282 in 15] based on correlations between PAC concentrations and particulate emissions and bitumen production by oil sands development. However, these results are based on periodic collection of samples of the upper 4–6 cm of river sediment. As

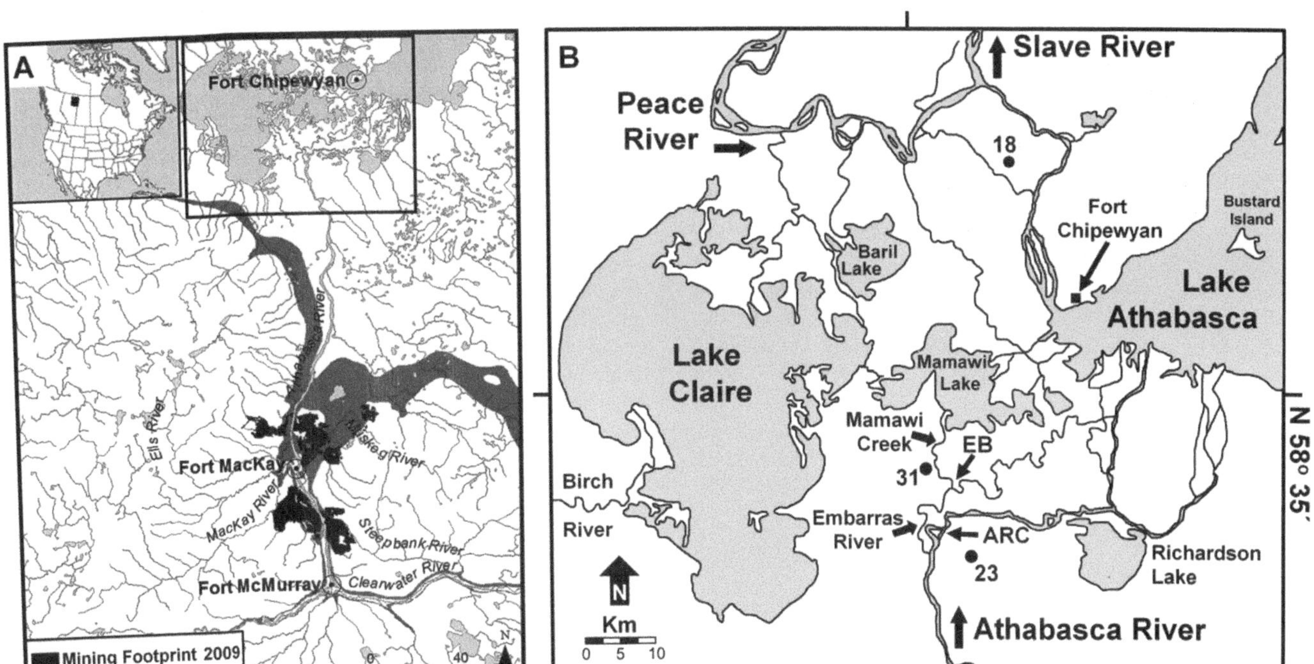

Figure 1. Maps of study area. Panel a): map showing locations of near-surface Athabasca oil sands exposures of the McMurray Formation (light grey), the 2009 mining footprint (dark grey) and the downstream Peace-Athabasca Delta (top center box) in northern Alberta, Canada. Top-center box in **a)** outlines area shown in Panel **b)**: Map identifying locations of the study lakes (solid circles) PAD 18 (N 58° 53.7′, W 111° 21.7′), PAD 23 (N 58° 23.5′, W 111° 26.6′) and PAD 31 (N 58° 29.6′, W 111° 31.0′) within the Peace-Athabasca Delta. Also shown are locations of the Athabasca River Cutoff (ARC) and the Embarras Breakthrough (EB). The ARC, an engineered channel excavation in 1972, straightened and deepened the Athabasca River at the location of a large meander bend where ice-jam floods occurred, and reduced flood frequency at PAD 23. The EB, a natural avulsion that occurred in 1982, has increased flood frequency at PAD 31 because it directs substantial and increasing Athabasca River flow towards PAD 31 via Mamawi Creek. Further details on these geomorphic changes are provided in [27].

acknowledged by the authors, uncertainty regarding the time interval represented by these samples is substantial, with individual samples potentially encompassing time periods that approach the entire decade of sample collection. These data are further complicated by the possibility of sediment remobilization and loss given the dynamic nature of fluvial and fluvio-deltaic depositional environments [16]. In fact, the sampling approach has been identified as problematic [1,3,17], in part because these factors are likely sufficient to mask any meaningful temporal trends. Thus, the suggestion that industrial activity has caused measureable enhancement of pollutant fluxes to the delta remains a testable hypothesis, rather than a scientific certainty.

Unfortunately, RAMP monitoring and other scientific studies began several decades after industrial development of the oil sands, so they cannot readily identify nor distinguish any additional contaminant loading due to industry beyond what is contributed by natural processes [3]. And, they have not adequately characterized contaminant transport during the spring freshet, an erosive annual event when natural contaminant dispersal occurs in northern rivers because ice-scour is substantial and flow velocity is relatively high [18,19]. This is when accumulated industrial-sourced contaminants in the winter snowpack become released as a pulse to the rivers, but its importance for dispersal of industrial contaminants remains a key unknown [1,7,10,11]. Releases of PACs during the spring melt are of particular concern because they coincide with fish spawning [1,20] and when river flow conditions maximize dispersal [19]. Clearly, *without* such knowledge, concerns about contaminant delivery to the delta will inevitably increase as oil sands development expands.

Here, we utilize analyses of precisely-dated lake sediment cores to establish baseline, pre-development, reference conditions for PAC deposition in the PAD. This represents the only method available to directly measure PAC concentrations prior to industrial development of the oil sands, as required to characterize the role of natural riverbank erosion as a vector for downstream deposition of PACs. Indeed, it is an approach recommended by the Federal Expert Oil Sands Advisory Panel [3] and advocated by Environment Canada's Phase 2 monitoring program [21]. We assess whether the Athabasca River and its tributaries have been important vectors for PAC accumulation in the downstream Athabasca Delta over the past ~200 years. We then address the question: Is there evidence to indicate that industry has measurably altered the composition and concentration of PACs in lake sediments in the Athabasca Delta from that provided by natural supplies due to riverbank erosion and subsequent downstream transport by the Athabasca River? We also use our findings to test the hypothesis [10] that oil sands development has altered sedimentary composition and concentration of PACs due to regional atmospheric transport to the delta.

Methods

Ethics Statement

Permission for collection of the sediment samples was provided by Parks Canada Agency (Wood Buffalo National Park) Research and Collection Permit WB-2010-6893.

Study Design

Our study design capitalizes on knowledge of past hydrological conditions that we have gained from more than a decade of field-based studies in the PAD, which is essential for informed interpretation of stratigraphic records of PACs given the hydrological complexity of the delta over space and time [22–32]. Three previously studied and strategically-positioned lakes were selected for stratigraphic analyses of PACs (Fig. 1). Lakes 'PAD 23' and 'PAD 31' (unofficial names) are located in the Athabasca Delta, but have markedly different flood histories during the latter part of the 20th century [27]. PAD 23 (elevation = 213.3 m a.s.l.), one of the most elevated lakes within the Athabasca Delta, was prone to flooding until the engineered Athabasca River Cutoff of 1972 allowed flow to bypass a bend in the river where ice-jams had previously formed and routed floodwaters towards PAD 23 [27]. Thus, sediments deposited before 1972 were designated as flood-prone, whereas sediments deposited afterwards were designated as not flood-prone. The pre-1972 flood-prone period was further divided into two periods, pre- and post-1933. Sediments deposited pre-1933 were characterized by lower organic matter (OM) content, lower cellulose-inferred lake water $\delta^{18}O$ values and markedly lower content of macrofossils from the submerged moss *Drepanocladus* compared to post-1933 sediments, suggesting stronger river influence pre-1933 [22]. In contrast, PAD 31 (elevation ~209.5 m a.s.l.), which lies approximately 40 cm above nearby Mamawi Lake (an open-drainage lake receiving continuous river inflow),has experienced three distinct periods of differing hydrological conditions during the past >200 years. PAD 31 was designated as flood-prone from the base of the core (pre-1800) until 1940 when it was frequently flooded and inundated under a former highstand of Lake Athabasca, as identified by low sediment OM content and abundant open-drainage indicator diatom taxa [22,30]. During 1940 to 1982, PAD 31 was not flood-prone and existed as an isolated closed-drainage lake due to declining Lake Athabasca water levels, as identified by declining relative abundance of open-drainage indicator diatom taxa and a shift to dominance by taxa indicative of closed-drainage conditions in previously analyzed cores from PAD 31, as also occurred in other delta lakes around 1940 (see Johnston et al.'s Fig. 7 [ref. 30]). PAD 31 became flood-prone for a second time after the 1982 Embarras Breakthrough event, a natural avulsion that diverted substantial flow from the Athabasca and Embarras rivers into Cree and Mamawi creeks and towards PAD 31, as identified by a visible contact between darker organic-rich sediment pre-Breakthrough and lighter mineral-rich sediments afterwards, a marked rise in sedimentation rate (Fig. 2), a sharp decline in OM content (Fig. 3) and changes in other sedimentary variables [27]. At present, PAD 31 is one of the most flood-prone of the restricted-drainage lakes in the Athabasca Delta [22,27]. With knowledge of these site-specific hydrological changes, analyses of PACs in lake sediment cores from PAD 23 and PAD 31 provide data to identify the role of the Athabasca River as a source of contaminant loading to the PAD. In contrast, 'PAD 18', an elevated (214.3 m a.s.l.) closed-drainage lake located in the northern Peace Delta, has not flooded for at least the past 100 years [22]. Consequently, all sediment samples were defined as not flood-prone. To assess influence of Athabasca oil sands exploration on atmospheric deposition of PACs, the PAD 18 sediment record was divided into pre- and post-1967 onset of oil sands mining activity [1,4]. Also, we obtained two samples of flood-deposited river sediment from the ground surface at two locations between Mamawi Creek and PAD 31 (approximately mid-way and 50 m from PAD 31) shortly after an ice-jam flood event along the Athabasca River in spring of 2007. These sediments, hereafter referred to as the '2007 Athabasca Delta flood deposit', provided reference materials to assess the composition and concentration of PACs supplied to PAD lakes by Athabasca River floodwaters.

Field Sampling and Laboratory Analyses

In September 2010, four sediment cores, 27–53 cm in length, were retrieved from each of PAD 18, 23, 31 using a gravity corer [33] and transported to the field base in Fort Chipewyan where

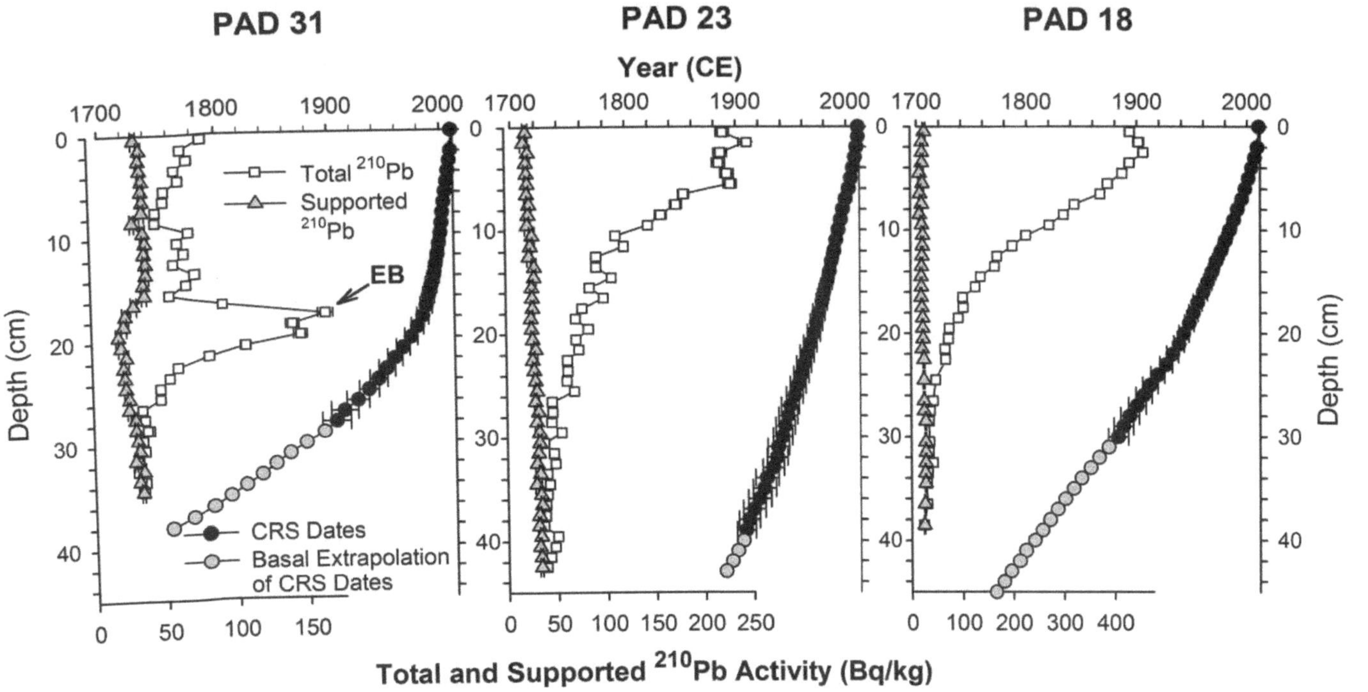

Figure 2. Sediment core dating. ^{210}Pb activity (supported and total (= supported+unsupported)) and depth-age profiles for sediment cores taken from the three study lakes (PAD 31, PAD 23, PAD18) in the Peace-Athabasca Delta, northern Alberta (error bars = ±2 SD). The irregular ^{210}Pb activity profile for PAD 31 is a result of an increase in sedimentation rate associated with increased flooding following the Embarras Breakthrough in 1982 [27].

they were sectioned into 1-cm stratigraphic intervals. Samples were kept refrigerated (4°C) in the dark prior to lab analyses. Sequential loss-on-ignition (LOI) analyses were performed on ~0.5 g subsamples of wet sediment from all core slices to provide profiles of water (post 90°C for 24 hours), OM (post 550°C for 1 hour) and carbonate (post 950°C for 1 hour) contents, following standard methods [34]. The OM content profiles were used to transfer the sediment core chronology developed from a single dated core per lake to a second core that was analyzed for the composition and concentration of PACs (Fig. 2, 3).

^{210}Pb Dating of Sediment Cores

One sediment core per lake was selected for dating by gamma ray spectrometric determination of ^{210}Pb activity at contiguous 1-cm intervals. A measured mass of ~2-4 g of dry sediment was packed into pre-weighed tubes (SARSTEDT product No. 55.524) to a uniform height of 35 mm and a TFA Silicone Septa (Supelco®) was placed overtop the sediment, followed by 1 cm^3 of 2 Ton Clear Epoxy (Devcon® product No. 14310). Three weeks were allowed for ^{222}Rn and its decay products to reach equilibrium with ^{226}Ra (parent isotope) prior to quantifying ^{210}Pb, ^{214}Bi and ^{214}Pb activity. Activities of the radioisotopes were measured in an Ortec co-axial HPGe Digital Gamma Ray Spectrometer (Ortec GWL-120-15) interfaced with Maestro 32 software (version 5.32) at the University of Waterloo's WATER Lab. Sample count time ranged from 23–95 hours per sample, and varied so that net ^{210}Pb counts were >10 × the standard deviation of the net counts of a well-defined machine blank to ensure precision [35]. Calibration of detector energy efficiency for the radioisotopes was previously established for our gamma ray spectrometers using a similarly prepared known standard (3.012 g of IAEA 300, see [36]). Detection efficiency values were determined from the mean of three sample runs performed using the IAEA-300 standard. ^{210}Pb activities were decay-corrected to the coring date for each lake and the ^{210}Pb activity was density-corrected as described by [37].

The Constant Rate of Supply (CRS) model [38,39] was used to develop the sediment core chronologies. The CRS model requires an estimation of supported ^{210}Pb activity, which was determined from activity of ^{214}Bi and ^{214}Pb. Results showed that supported ^{210}Pb activity is variable over time in the sediment cores (Fig. 2). To reduce influence of this temporal variability, a five-point running mean of ^{214}Pb and ^{214}Bi activity was used as an estimate of the supported ^{210}Pb activity for a given interval. The core depth where total ^{210}Pb activity equals supported ^{210}Pb activity was determined using standard methods [40].

The sediment chronology for the ^{210}Pb-dated core was transferred to the core analyzed for PACs based on alignment of OM content profiles in both cores determined by LOI analysis (Fig. 3). Slight depth offsets in the OM content profiles of cores from PAD 23 and PAD 31 were adjusted using a constant depth multiplier (+6% and −11.5%, respectively, for PAD 31 GC-2 and PAD 23 GC-1). This approach preserved sample-to-sample variability in the sedimentation rate determined for the dated core. The CRS chronology was then applied to the modified depth interval of the core analyzed for PACs by a straight linear interpolation of the nearest-neighbour dates (depth-wise) of the ^{210}Pb-dated core (e.g., if 10 and 11 cm-depth in the ^{210}Pb-dated core corresponded to 2000 and 1998.5 respectively, then a modified sample depth of 10.6 cm depth in the core analyzed for PACs equates to 1999.1). Results are consistent with cores we have previously analyzed for these lakes [22,27]. Also, the stratigraphic contact at 18 cm depth marking the 1982 Embarras Breakthrough was used to confirm alignment of cores from PAD 31.

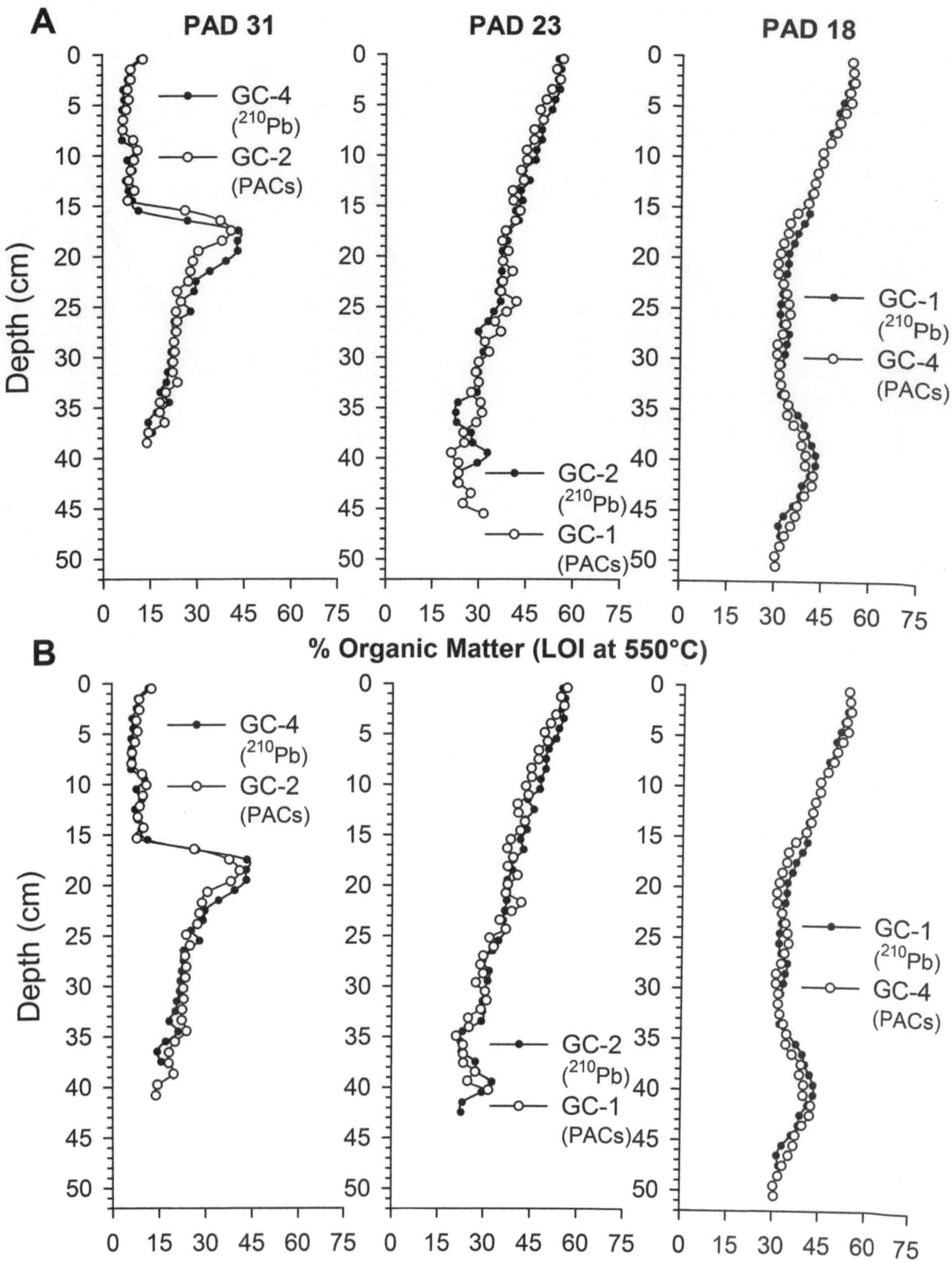
A
PAD 31
PAD 23
PAD 18
Depth (cm)
GC-4 (^{210}Pb)
GC-2 (PACs)
GC-2 (^{210}Pb)
GC-1 (PACs)
GC-1 (^{210}Pb)
GC-4 (PACs)
% Organic Matter (LOI at 550°C)
B
Depth (cm)
GC-4 (^{210}Pb)
GC-2 (PACs)
GC-2 (^{210}Pb)
GC-1 (PACs)
GC-1 (^{210}Pb)
GC-4 (PACs)
% Organic Matter (LOI at 550°C)

Figure 3. Alignment of duplicate cores. Graphs showing stratigraphic profiles of organic matter content in duplicate cores taken from each of the three study lakes in the Peace-Athabasca Delta, northern Alberta. One of the cores from each lake was analyzed for ^{210}Pb dating and the other provided material for analysis of PACs. Upper panel **a)** shows the raw data, whereas the lower panel **b)** shows the data after minor vertical adjustment of the organic matter content profile for the core that was not dated. ^{210}Pb-derived dates from the dated core were transferred to the corresponding undated cores used for PAC analyses. No depth adjustments were required for cores from PAD 18.

PAC Analyses

Samples from one core from each of the study lakes were analyzed for PACs at ALS Canada Ltd. (Edmonton). Using knowledge of the water content derived from LOI analysis, an amount of wet sediment equivalent to ≥2 g dry sediment (mean = 4.6 g dry mass) was placed into amber glass jars with Teflon lid-liners that were pre-washed with acid (5% HCl) and rinsed with solvent (dichloromethane). We combined adjacent sediment intervals for the uppermost water-rich samples in cores from PAD 18 and PAD 23 to obtain sufficient sample mass. Additionally, two samples of the 2007 Athabasca Delta flood deposit were analyzed. These flood deposit samples were kept refrigerated since collection and were similarly prepared and shipped with the other sediment core samples to ALS for analysis of 52 PACs and alkylated PACs using US Environmental Protection Agency method EPA 3540/8270-GC/MS.

Identification and quantification of individual PACs had good repeatability based on analyses of replicate subsamples (mean precision: ±12.5% of PAC concentration averaged across all replicate analyses). Where repeated measurements of a sample were available, the average value was used for data analysis and display. In a few instances, an individual PAC was below detection limit (e.g., <0.040 mg/kg) in one replicate but slightly above the detection limit (e.g., 0.042 mg/kg) in the other replicate for a given sediment interval (e.g., FLPY2 in the sample at 35–36 cm from PAD 18). In these cases, a random number between zero and the detection limit was generated for the first replicate below the detection limit, and then averaged with the value obtained for the second replicate to avoid underweighting the PAC concentration. For situations where all measurements for a given PAC in a sediment interval were below the analytical detection limit, the PAC measurement was treated as having a concentration of zero.

Numerical Analyses

A one-way Analysis of Similarities (ANOSIM) test, a nonparametric and multivariate approximate analogue of analysis of variance (ANOVA) tests [41], was performed on the relative abundances (expressed as proportions of the PAC sum) of the quantified PACs in all sediment core samples from the three study lakes (n = 127) to determine if composition of PACs differs between sediments deposited during flood-prone versus not flood-prone intervals, based on assignment of stratigraphic intervals as described above. ANOSIM tests are commonly, and increasingly, applied to determine differences in biological community composition among site-categories [41], and they are now being used for analysis of contaminant composition [42]. It is common practice to remove rarely encountered species from multivariate analysis and transform the data to up-weight or down-weight uncommon species, depending upon research goals [41]. We opted not to remove rare PACs for this study to avoid the possibility of dismissing historically-rare compounds that could have become recently deposited in the PAD due to anthropogenic activities or other processes. Our intent was to analyze the full PAC dataset with as little manipulation as possible. Consequently, the data were not transformed prior to calculating Bray-Curtis similarity coefficients used by ANOSIM computations. The 2007 Athabasca Delta flood deposit samples were not used in the ANOSIM test or the SIMPER analysis described below. Instead, they served as an independent benchmark to help identify river-transported PACs in the stratigraphic records. The test statistic (global R) computed by the ANOSIM test reflects the observed differences between samples in the two categories (flood-prone versus not flood-prone) contrasted with the differences among replicates within each group, and it ranges from 0 to 1. A value of 0 indicates that the similarity between and within groups is the same on average. A value of 1 indicates that replicates within a group are more similar to each other than to all other replicates of other groups [41]. A p-value was computed by comparing the distribution of within- and across-group rank Bray-Curtis similarities (999,999 computations) to the initial rank similarity, as reported by the global R value [41,43]. For the ANOSIM test, we set alpha = 0.05.

A one-way Similarity of Percentages (SIMPER; [43]) analysis was performed on the full, untransformed relative abundance PAC data to identify which of the 43 detected PACs best discriminated between samples deposited during flood-prone versus not flood-prone intervals. As detailed below, this provided a method to select PACs indicative of a bitumen origin and transport by Athabasca River floodwaters versus those that are more abundant due to other processes. ANOSIM and SIMPER analyses were performed using the software PRIMER, version 6.1.5 [41,43].

For PAD 31, the study lake that was flood-prone for several decades both before and after onset of oil sands development, independent-samples t-tests were performed to determine if the mean of the sum of the river-transported bitumen-associated indicator PACs, identified by use of SIMPER analysis, differed in sediments deposited during flood-prone periods pre- versus post-onset of oil sands development. Independent-samples t-tests were performed on these data when expressed as concentrations and as proportions, because Kolmogorov-Smirnov tests identified that the distributions do not differ significantly from normal (P>0.05). Non-parametric Mann-Whitney tests provided the same conclusions as the independent-samples t-tests. These tests were performed using the software SPSS version 16.0.

Results

PAC Composition

The composition of PACs in the 2007 Athabasca Delta flood deposit (Fig. 4c) closely matches 'reference' compositions reported for oil sands samples within the oil sands region (Fig. 4a; [10]), and for sediments in the Athabasca River and two tributaries downstream of riverbank deposits of bitumen in areas unaffected by industrial development (Fig. 4b; [44]). The oil sands samples and the natural oil sands region river sediments reported by [10,44] originate from sites located ~200 km upstream of the Athabasca Delta. Based on geochemical fingerprinting, the main source of the PACs in the natural oil sands region river sediments is bitumen-rich material in the riverbanks [45]. Similar to the oil sands samples and the reference river sediments, the 2007 Athabasca Delta flood deposit was dominated by alkyl-substituted PACs, including C4 naphthalenes (N4), C2- through C3-fluorenes (F2–F3), C2- through C4-dibenzothiophenes (D2–D4), C1- through C4-phenanthrenes/anthracenes (P1–P4), C2- through

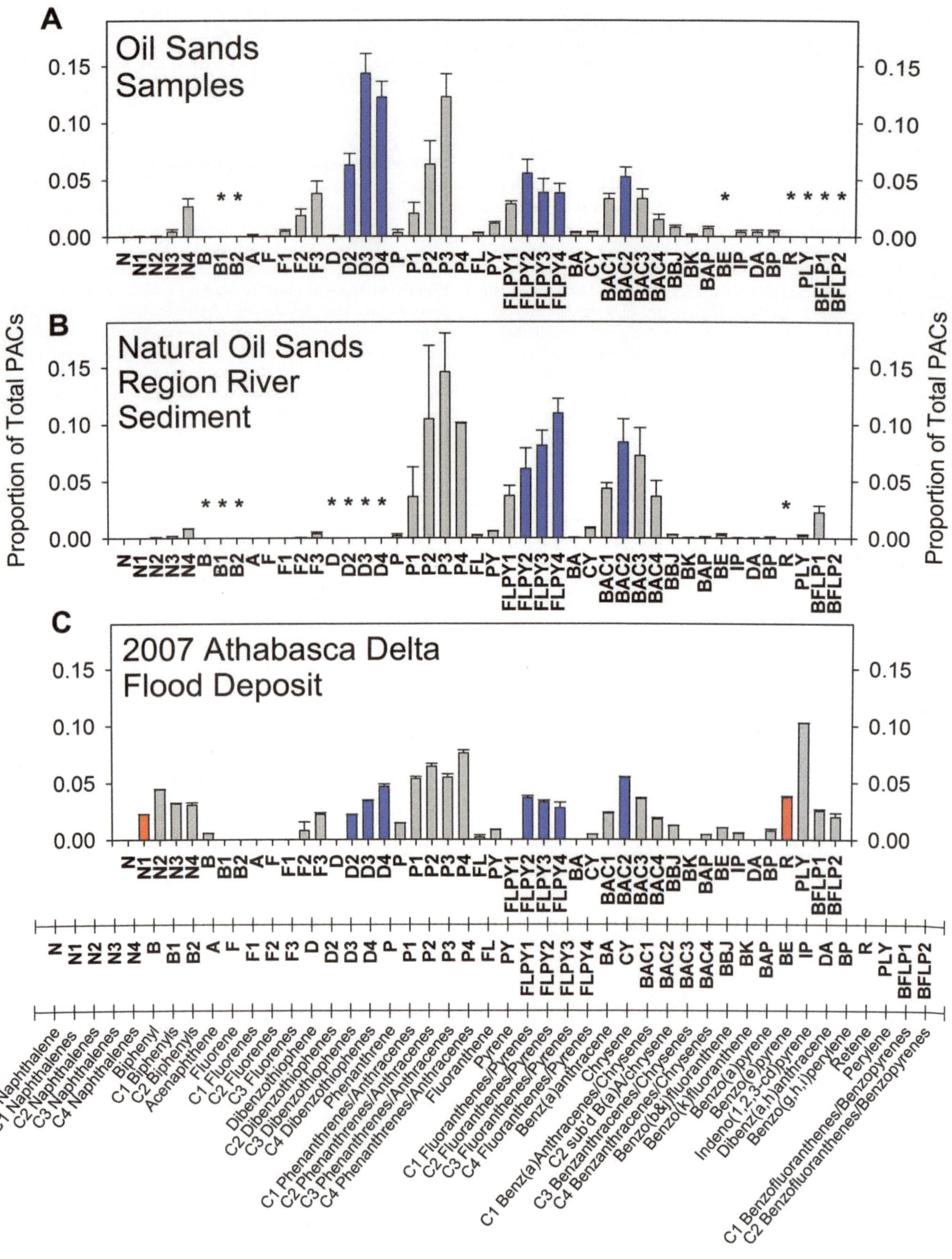

Figure 4. Composition of polycyclic aromatic compounds in oil sands and river sediment downstream of McMurray Formation. Bar graphs showing average composition of polycyclic aromatic compounds (PACs), expressed as relative concentration (proportion; error bars = ±1 SEM) in **a)** oil sands samples from Syncrude, Steepbank River mouth, and east and west banks of the Athabasca River (from [10]), **b)**

natural oil sands region river sediment (from [44]) and **c)** the 2007 Athabasca Delta flood deposit. Names of the PAC compounds and the corresponding codes are presented along the lower expanded horizontal axis. Blue vertical bars are PACs identified by SIMPER analysis as river-transported bitumen-associated indicator PACs deposited during flood-prone intervals (Table 1). Red vertical bars are PACs identified as common in sediments deposited during not flood-prone periods. * identifies PACs that were not measured in one sample but were measured and reported in another sample presented in the figure.

C4-fluoranthenes/pyrenes (FLPY2-FLPY4) and C1- through C4-benzo[a]anthracenes/chrysenes (BAC1–BAC4) (Fig. 4). Compounds C1- through C3-naphthalenes (N1–N3) were also common in the 2007 Athabasca Delta flood deposit. Levels of D2–D4 in the 2007 Athabasca Delta flood deposit could not be compared with the natural bitumen-rich Athabasca River sediments because dibenzothiophenes were not analyzed by [44] (Fig. 4b), but they are a major component of the oil sands samples (Fig. 4a; [10]). Except for perylene, unsubstituted (parent) PACs were absent or rare in the 2007 Athabasca Delta flood deposit. Perylene is formed naturally by metabolic decomposition of plant matter by bacteria and fungi [46,47] and is associated with transport of terrestrial materials from watersheds and with in-situ generation in aquatic systems [47–49], suggesting that the Athabasca River also transports PACs of terrestrial vegetation origin to the Athabasca Delta. However, the predominance of alkyl-substituted PACs over unsubstituted PACs in both the 2007 Athabasca Delta flood deposit and reference oil sands samples and river sediments (Fig. 4; [10,44,45]) identify that a major component of the sediment-associated PACs transported downstream and to the Athabasca Delta by the Athabasca River during floods derives from a common petrogenic origin – namely, bitumen from exposures of the McMurray Formation along the riverbanks in the oil sands area.

The average composition of PACs in sediments deposited in PAD 31 during the two periods when the lake was flood-prone (pre-1940 and post-1982) closely matches that of the 2007 Athabasca Delta flood deposit (Fig. 5a, panels 1, 2, 4), and to that of bitumen-rich natural deposits located upstream (see Fig. 4) [10,44,50]. PACs in these sediments contain high proportions of the alkyl-substituted forms C2- through C4-naphthalenes (N2–N4), C2- through C3-fluorenes (F2–F3), C2- through C4-dibenzothiophenes (D2–D4), C1- through C4-phenanthrenes/anthracenes (P1–P4), C2- through C4-fluoranthenes/pyrenes (FLPY2–FLPY4) and C1- through C4-benzo[a]anthracenes/chrysenes (BAC1–BAC4). Other relatively abundant PACs are retene (R) and perylene (PLY). Retene, a C4-phenanthrene, is associated with combustion of plant material [51,52]. Only minor differences occur between sediments deposited in PAD 31 when it was flood-prone and the 2007 Athabasca Delta flood deposit. Specifically, naphthalene (N) is absent in the 2007 Athabasca Delta flood deposit, but is present in low proportions in PAD 31 sediments deposited post-1982 (<1%) and pre-1940 (<4%). Also, only subtle differences occur in PAD 31 sediments deposited during the two flood-prone periods, with slightly lower proportions of F2–F3, BAC1–BAC3 and R, and a slightly higher proportion of N in sediments deposited pre-1940 compared to post-1982. Clearly, the sediment record from PAD 31 identifies that bitumen-derived PACs carried by the Athabasca River are deposited in this lake when it has received river floodwaters, and that this has occurred both before and since onset of industrial development of the oil sands.

By contrast, the average composition of PACs in sediments from PAD 31 when the lake was not flood-prone (1940–1982: Fig. 5a, panel 3) differed notably from that of the 2007 Athabasca Delta flood deposit, as well as sediments deposited in PAD 31 during the two flood-prone periods (pre-1940, post-1982). Specifically, sedimentary proportions of F3, D2–D4, FLPY2–FLPY4 and BAC1–BAC3 were markedly lower when PAD 31 was not flood-prone (1940–1982), and proportions of N, C1-naphthalenes (N−1), biphenyl (B) and fluorene (F) were more than double. Overall, the data show that the composition of PACs in PAD 31 sediments differs detectably between periods of frequent versus infrequent flooding. During flood-prone periods, PAC composition closely matches that of river-transported sediment originating from bitumen deposits of the McMurray Formation exposed along the riverbanks (Fig. 4, 5). During periods of reduced flood frequency, higher proportions of unsubstituted PACs (notably N, B, F) identify greater influence of hydrocarbons from fire and catchment vegetation [46,52,53].

At PAD 23, the average composition of PACs in sediments deposited post-Cutoff (post-1972), when flood frequency at the lake dropped precipitously [27], is very different from that of the 2007 Athabasca Delta flood deposit (Fig. 5b, panels 1, 2). Many of the alkylated PACs abundant in the flood deposit (D2–D4, FLPY2–FLPY4, BAC1–BAC4) were nearly absent in post-Cutoff sediments at PAD 23. Also, N and N−1 dominated the PACs in post-Cutoff sediments at PAD 23, but these compounds are absent or in low relative abundance, respectively, in the flood deposit. The high proportions of unsubstituted N, F and PLY, and R in PAD 23 sediments during this 'not flood-prone' period suggest the PACs originate from a pyrogenic source (e.g., forest fires) as well as biodegradation products from plant materials [47,50,51,53].

Composition of PACs in sediments deposited at PAD 23 prior to the Cutoff (1972), when the lake was prone to flooding from the Athabasca River [27], is notably different compared to the post-Cutoff period of reduced flood frequency. During the pre-Cutoff flood-prone period, N, N1 and F were relatively less abundant on average, and several of the alkylated PACs common in the 2007 Athabasca Delta flood deposit (D2–D3, FLPY2–FLPY4, BAC2) were more abundant (Fig. 5b, panels 3, 4). Interestingly, this pattern was more strongly apparent during the earliest portion of the flood-prone period (1901–1933), when paleohydrological data identify greater influence of flooding at PAD 23 [27]. Also, PAC composition during the flood-prone periods at PAD 23 shows some similarities with the subsequent period of reduced flood frequency (post-Cutoff), with generally higher relative abundance of N, N1 and P2–P4 than in the 2007 Athabasca Delta flood deposit. This suggests a mixture of PACs from both petrogenic sources supplied by river floodwaters and pyrogenic sources, likely due to the lake's proximity to forests subject to periodic fires.

Based on data obtained previously, PAD 18 has not received floodwaters for at least the past century [22]. The composition of PACs in sediments from this lake, during both pre- and post-oil sands development periods, is markedly different than that of the 2007 Athabasca Delta flood deposit, the natural oil sands region river sediments and sediments deposited during flood-prone periods in the other two study lakes (Fig. 4; Fig. 5c, panels 2, 3). PACs in sediments from PAD 18 are overwhelmingly dominated by N, N1 and R. And, the alkylated PACs characteristic of river-transported bitumen in the 2007 Athabasca Delta flood deposit and during flood-prone periods at PAD 31 and PAD 23 (notably,

F2–F3, D2–D4, FLPY3–FLPY4, BAC1–BAC3) are absent at PAD 18. Only FLPY2 was detected, but at low relative abundance (3–5%) and only in five samples deposited pre-1900. Thus, PACs in sediments at PAD 18 are characteristic of mainly pyrogenic and vegetation sources, with little to no inputs from PACs of petrogenic origin.

Identification of River-transported Bitumen-associated Indicator PACs

Based on results of a one-way analysis of similarities (ANOSIM) test, performed on the relative abundance of the PACs in sediment core samples from the three study lakes (Fig. 5), PAC composition differs significantly between flood-prone versus not flood-prone periods in the study lakes (R-statistic = 0.499, $P<1.0\times10^{-6}$, n = 127). This finding confirms a strong influence of river floodwaters on the sedimentary composition of PACs.

Subsequent similarities percentages (SIMPER) analysis was used to identify the 'indicator' PAC compounds that most strongly contribute to differences in PAC composition between the flood-prone and not flood-prone periods of the study lakes. Ten of the 43 PACs were found to have good indicator qualities: seven for flood-prone sediment intervals (D2–D4, FLPY2–FLPY4, BAC2) and three for not flood-prone sediment intervals (N, N1, R; Table 1). Collectively, these ten PACs accounted for >50% of the total dissimilarity between samples deposited during flood-prone and not flood-prone intervals.

We sought to identify compounds that were consistently more abundant in either the flood-prone or not flood-prone sediment samples and that strongly contribute to the within-group similarity of one sample-type (but not both). This selection criterion was met for the PACs indicative of flood-prone intervals. The seven PAC indicators of flood-prone status (D2–D4, FLPY2–FLPY4, BAC2) had higher relative abundances (proportionally >2x) in sediments deposited during flood-prone intervals and contributed consistently to the between-group (flood-prone versus not flood-prone) dissimilarity (dissimilarity/std. dev. >1, contribute >1% to dissimilarity; Table 1). Individually, they contributed substantially to the within-group similarity of flood-prone samples (>1%), but not to the within-group similarity of the not flood-prone samples (<0.1%). These 'river-transported indicator PACs' (coloured blue in Fig. 4, 5) contributed 21.42% to the total between-group dissimilarity for flood-prone versus not flood-prone samples, and 20.24% to the total within-group similarity of the flood-prone samples (Table 1). They contributed only 0.11% to the total within-group similarity of not flood-prone samples. They averaged 22% of the total PACs present in sediments in the flood-prone samples, but were <1% of the total PACs present in the not flood-prone samples. These river-transported indicator PACs, based on our independent statistical analysis, are in agreement with PACs previously identified as characteristic of bitumen-rich materials associated with the oil sands of the McMurray Formation (Fig. 4) [10,44,45,50]. In fact, these seven river-transported indicator PACs account for 51% of the total PACs in oil sands samples reported by Kelly *et al.* [10]. Thus, we refer to them hereafter as 'river-transported bitumen-associated PACs'.

Using the same criteria, SIMPER analysis identified three PACs indicative of sediments deposited during not flood-prone intervals (N, N1, R; Table 1). These three indicator PACs were consistently found at higher relative abundances in sediments deposited during the not flood-prone intervals (proportionally >2x), contributed consistently to the between-group (flood-prone versus not flood-prone) dissimilarity (dissimilarity/std. dev. ≥0.8), and contributed considerably more to the within-group similarity of the not flood-prone samples than that of the flood-prone samples. These 'not flood-prone indicator PACs' (coloured red in Fig. 4, 5) contributed 29.90% to the total dissimilarity between flood-prone and not flood prone-samples and 45.94% to the total within-group similarity of the not flood-prone samples. They contributed 10.51% to the total within-group similarity of flood-prone samples. On average, they comprised 40% of the total PACs found in the

Table 1. Indicator PACs for flood-prone and not flood-prone sediments.

	Between-group Dissimilarity (Flood-prone *vs* Not flood-prone)		Within-group Similarity (Flood-prone or Not Flood-prone)		Average Proportion of total PACs (0–1)	
	Average Dissimilarity %	**Contribution %**	**Average Similarity %**	**Contribution %**	**Flood-prone**	**Not flood-prone**
Indicator PACs for Flood-prone intervals						
C4 Dibenzothiophenes	2.30	4.53	2.25	3.10	0.05	0.00
C3 Dibenzothiophenes	2.07	4.08	3.42	4.71	0.04	0.00
C2 sub'd B(a)A/chrysene	1.93	3.79	3.03	4.18	0.04	0.00
C2 Fluoranthenes/Pyrenes	1.25	2.46	1.73	2.38	0.03	0.00
C2 Dibenzothiophenes	1.21	2.39	1.59	2.19	0.02	0.00
C3 Fluoranthenes/Pyrenes	1.14	2.25	1.58	2.18	0.02	0.00
C4 Fluoranthenes/Pyrenes	0.97	1.92	1.09	1.50	0.02	0.00
Indicator PACs for Not flood-prone intervals						
Naphthalene	8.47	16.65	14.25	23.90	0.05	0.21
C1 Naphthalenes	4.86	9.56	9.00	15.09	0.03	0.13
Retene	1.92	3.78	4.14	6.95	0.03	0.06

Results of similarity percentages (SIMPER) analysis of PAC composition in sediment core samples from the three study lakes of the Peace-Athabasca Delta, northern Alberta, with samples categorized as flood-prone or not-flood prone, to identify PACs most indicative of river transport versus other vectors and processes.

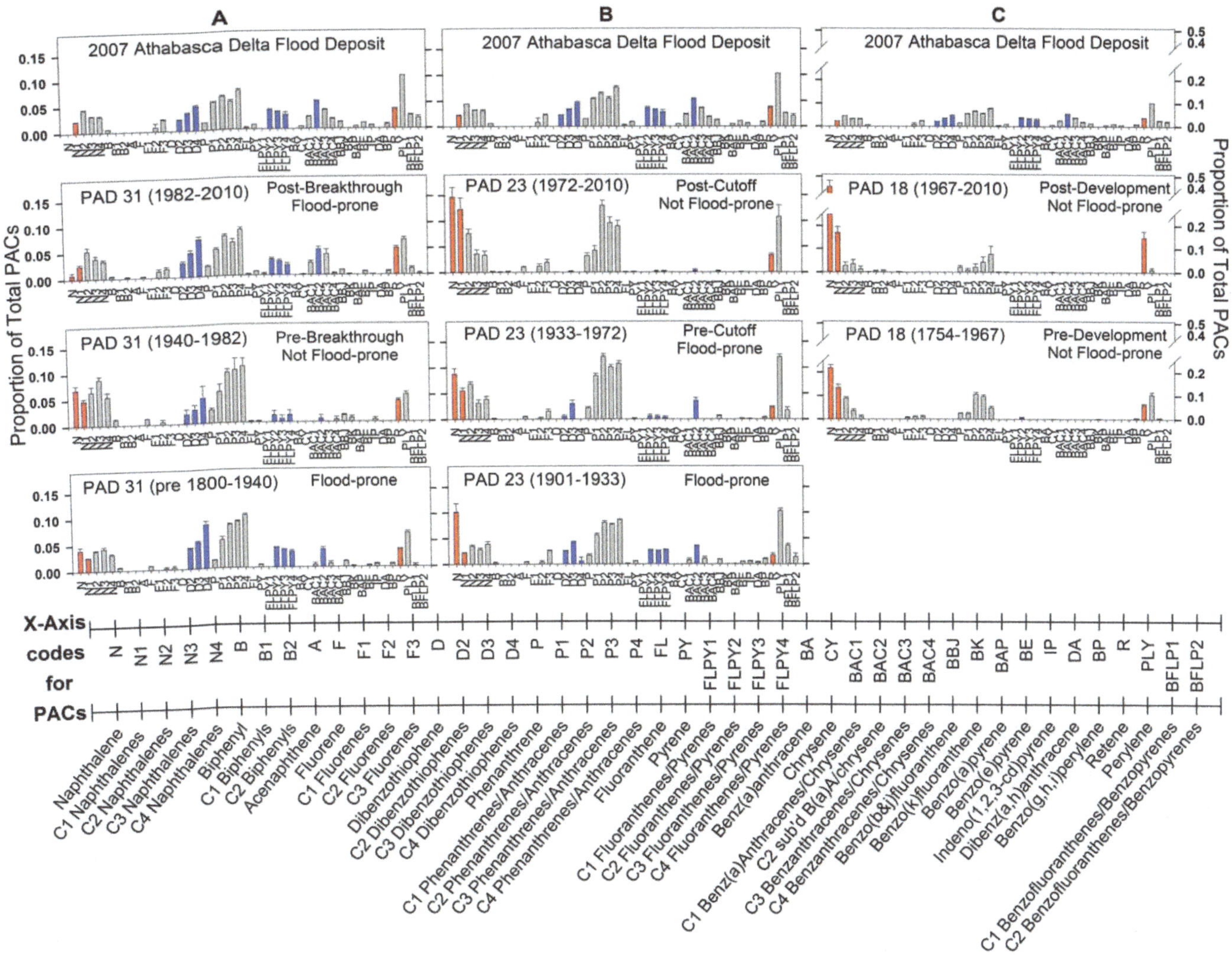

Figure 5. Composition of polycyclic aromatic compounds in river flood sediment and lake sediment cores. Bar graphs showing average composition of polycyclic aromatic compounds (PACs), expressed as relative concentration (proportion; error bars ±1 SEM) in **a)** the 2007 Athabasca Delta flood deposit and three intervals of differing flood regime at PAD 31, **b)** the 2007 Athabasca Delta flood deposit and three intervals of differing flood regime at PAD 23, and **c)** the 2007 Athabasca Delta flood deposit and two intervals that distinguish pre- and post-1967 onset of major commercial oil sands development at not flood-prone PAD 18 (note different scale of vertical axis). Names of the PAC compounds and the corresponding codes are presented along the lower expanded horizontal axis. Blue vertical bars are PACs identified by SIMPER analysis as river-transported bitumen-associated indicator compounds common in sediments deposited during flood-prone intervals (Table 1). Red vertical bars are PACs identified as indicator compounds in sediments deposited during not flood-prone periods.

not flood-prone samples compared with 11% for flood-prone samples. The indicator PACs for sediments deposited during not flood-prone periods are rare or absent in oil sands bitumen [10,44,54] and are often associated with a pyrogenic source such as forest fires [14,51,52]. However, they also occurred in sediments deposited during flood-prone intervals, suggesting that these PACs are most strongly associated with delivery to lakes at the Athabasca Delta by non-river sources and vectors, such as atmospheric transport of PACs produced by fires and from biological degradation of vegetation within the catchment. But, river floodwaters may also acquire these compounds from terrestrial landscapes and deliver them downstream to the delta [55].

Phenanthrene/Anthracene homologues (P1–P4) were not identified by SIMPER analysis as indicator PACs because they were common in all sediment samples (Fig. 4, 5). Indeed, P1–P4 are common in river sediments both upstream and downstream of Athabasca oil sands deposits [44], as well as in river sediments draining unburned and burned forests of northern Alberta in areas outside the Athabasca oil sands region [52].

Relations between River-transported Bitumen-associated Indicator PAC Concentration and Sediment Organic Matter Content

Two distinct trends are evident when comparing river-transported bitumen-associated indicator PAC concentrations in river and lake sediments in the Athabasca Delta versus organic matter content, which further reveal processes responsible for deposition of these PACs (Fig. 6). Data obtained from RAMP [56] show that the organic content of river-bottom sediments is much lower than that of the lake sediments. Also, concentrations of the river-transported bitumen-associated indicator PACs in river sediment are positively correlated with organic matter content of bottom sediments obtained in the Athabasca River and distributaries within the delta ($R^2 = 0.35$,

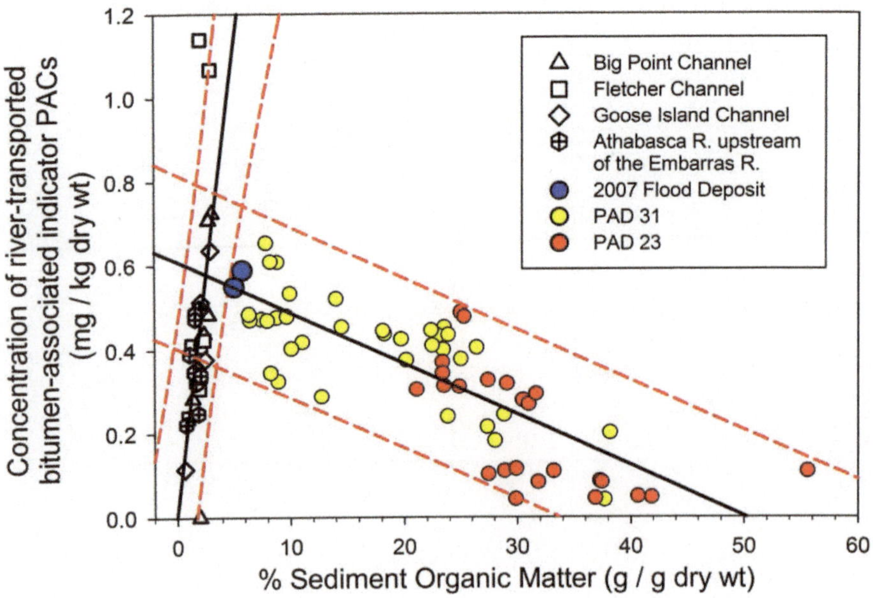

Figure 6. Relations between polycyclic aromatic compound concentration and organic matter content in river versus lake sediment. Scatterplot showing the relations between organic matter (as a percentage of dry sediment mass) and concentration of river-transported bitumen-associated indicator PACs in sediments of lakes PAD 23 (red circles) and PAD 31 (yellow circles) in the Athabasca Delta (data from this study), in bottom sediments of the Athabasca River and its distributaries at four stations within the Athabasca Delta (open symbols; data from [56]), and in sediments deposited on land between Mamawi Creek and PAD 31 by an ice-jam flood in spring of 2007 (blue circles; data from this study). Data include the sum of the concentration of six of the seven river-transported bitumen-associated indicator PACs that were common to both the lake sediment data set (this study) and the river bottom sediment from RAMP [56] (note: FLPY-4 was not reported by [56]). The lines of best fit (solid black line) and 95% prediction intervals (dashed red lines) are presented for linear regressions of sediment organic matter content and river-transported bitumen-associated indicator PAC concentrations for both the river bottom sediments and the lake sediments.

$P = 0.0004$, d.f. = 30; Fig. 6). This positive association probably derives mainly from variations in grain size distribution and composition of river sediment, because the finer-grained fraction possesses higher organic matter content and higher affinity for hydrophobic contaminants, such as PACs [57], than coarser fractions.

In contrast to the sediments of the Athabasca River and distributaries, river-transported bitumen-associated indicator PAC concentrations in the lake sediment records of PAD 23 and PAD 31 are significantly and *negatively* correlated with organic matter content ($R^2 = 0.63$, $P < 0.0001$, d.f. = 60; Fig. 6). This relation reflects periodic inoculation of the lakes with river-borne sediments that are relatively low in organic matter content, but relatively high in river-transported bitumen-associated PAC concentration. Under non-flood conditions, concentrations of the river-transported bitumen-associated PACs in the lake sediments are diluted by in-lake aquatic productivity that increases organic matter content. A similar pattern has been reported for sediments of floodplain lakes in the Mackenzie Delta [55]. The presence of higher concentrations of river-transported bitumen-associated PACs in PAD 31 (and correspondingly lower organic matter content) compared to PAD 23 is consistent with the more proximal location of PAD 31 to river floodwaters. Notably, high river-transported bitumen-associated PAC concentrations in the 2007 flood deposits obtained near PAD 31 lie close to the intersection of the two relations, confirming that flood events deposit river sediment and associated river-transported bitumen-associated PACs directly into PAD 31.

River-transported Bitumen-associated Indicator PAC Profiles in Sediment Cores

Summing the seven PACs identified by the SIMPER analysis as indicative of flood-prone conditions provided an approach to quantify changes over time in the proportion and concentration of bitumen-associated, river-transported PACs to the three study lakes, as described below (Fig. 7).

At PAD 31, the average concentration of the river-transported bitumen-associated indicator PACs does not differ significantly between sediments deposited since the Embarras Breakthrough in 1982 (0.476 mg/kg ±0.165 (1 SD)), when the lake became frequently and increasingly flooded by channelized flow of Athabasca River-sourced water along Mamawi Creek, and sediments deposited pre-1940 when PAD 31 received sediments from the Athabasca River when frequently submerged under a highstand of Lake Athabasca (0.466 mg/kg ±0.065 (1 SD); independent-samples t-test, $P = 0.821$; d.f. = 31; Fig. 7a, top panel). The independent-samples t-test achieved greater than 99% power to detect a mean difference of 0.1 mg/kg (~1 SD), indicating low probability of a Type-2 error. Indeed, concentrations of the summed river-transported bitumen-associated indicator PACs fell close to that of the 2007 Athabasca Delta flood deposit during flood-prone periods both prior to and since onset of oil sands development. Peak concentration occurred in 2007, the year when ice-jam flooding delivered the reference Athabasca Delta flood deposit to the lakeshore. The two other concentration peaks post-1982 corresponded with high open-water flow on the Athabasca River (1990) and ice-jam floods (1996–97; [58,59]). Consistent with this evidence of strong regulation of sediment PACs by river flooding, the sum of the river-transported bitumen-associated indicator PACs (expressed as both concentration and proportion) declined to very low levels during 1940–1982 when PAD 31 was not flood-prone (except for two peaks within this period that correspond to ice-jam flood years of 1963–64 and 1974 [58]). Interestingly, the average proportion of river-transported bitumen-associated indicator PACs during 1983–2010 (27.1% ±7.9 (1 SD))

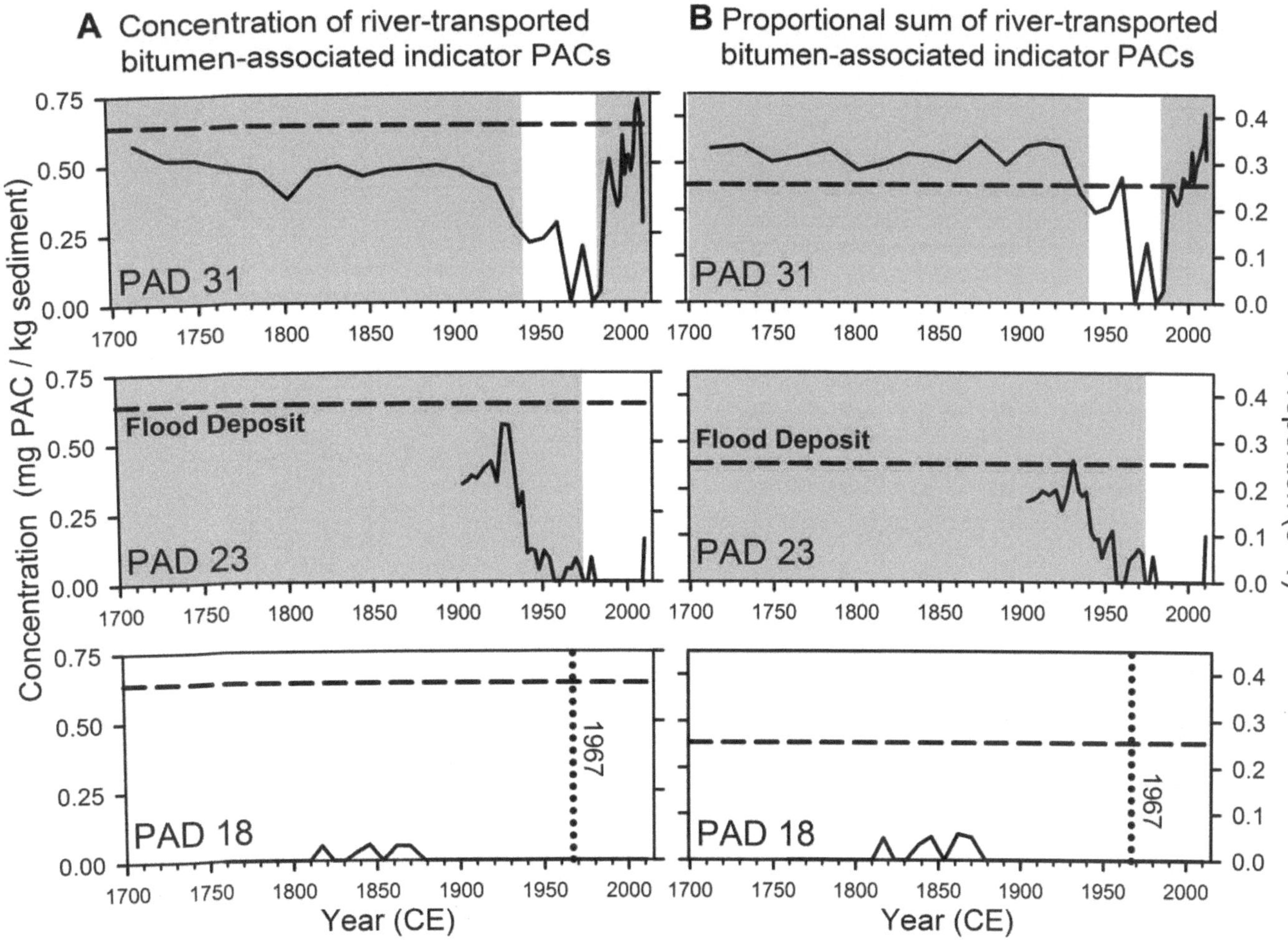

Figure 7. Time trends of river-transported bitumen-associated indicator PACs. Stratigraphic records from sediment cores of the three study lakes (PAD 31, 23, 18) showing **a)** the sum of the concentrations of the seven river-transported bitumen-associated indicator PACs (C2-through C4-dibenzothiophenes, C2- through C4-fluoranthenes/pyrenes, C2-sub'd B(a)A/chrysene), and **b)** the sum of the proportions of the seven river-transported bitumen-associated indicator PACs. Grey shading identifies periods when the lakes were flood-prone. Absence of shading identifies periods when the lakes were not flood-prone. The dotted vertical line in plots for PAD 18 identifies 1967, the year when major commercial operations began [1,4]. Dashed horizontal lines in panels **a)** and **b)** correspond to the average value for the sum of the seven river-transported bitumen-associated indicator PACs in the 2007 Athabasca Delta flood deposit, where **a)** shows the concentration (=0.638 mg/kg) and **b)** shows the proportion (=0.254).

was significantly lower than during the pre-1940 interval (31.5% ±2.8 (1 SD); independent-samples t-test, P = 0.044, d.f. = 31; Fig. 7b, top panel). Overall, these data provide evidence that the proportion and concentration of accumulated river-transported bitumen-associated indicator PACs have not increased measurably at PAD 31 since onset of oil sands development.

At PAD 23, highest amounts of river-transported bitumen-associated indicator PACs were deposited in sediments before ~1940 (middle panels in Fig. 7a,b), when prior research shows strongest flooding [27]. The long-term 20th century decline in proportions and concentrations of river-transported bitumen-associated indicator PACs likely reflects declining susceptibility of this basin to floods [27]. Except for a brief rise in the sample that includes 1974, the year of a known major ice-jam flood event [58], and in the top-most sample, possibly representing a recent unrecorded flood event, river-transported bitumen-associated indicator PACs have been virtually absent in sediments deposited in PAD 23 since the Athabasca River Cutoff (1972). These findings highlight that natural erosion and transport via the Athabasca River are major processes delivering these contaminants.

At PAD 18, where floods have been absent for the past century [22], river-transported bitumen-associated indicator PACs were not detected, except for very low concentrations (0.032–0.057 mg/kg) of FLPY2 during the early- to mid-1800s (bottom graphs in Fig. 7a,b). Clearly, the river-transported bitumen-associated indicator PACs are not supplied in substantial amounts to lakes in the absence of flooding, but their presence in a few samples suggests that PAD 18 may have received a few floods during this time, which was marked by notably high water levels in Lake Athabasca [30].

Total PAC Concentration Profiles in Sediment Cores

Total sedimentary PAC concentrations ranged from 0.053 to 3.008 mg/kg at PAD 23, from 0.837 to 2.234 mg/kg at PAD 31 and from 0.0271 to 1.596 mg/kg at PAD 18 (Fig. 8). Similar to the seven river-transported bitumen-associated indicator PACs, concentrations of total PACs are markedly higher during the flood-

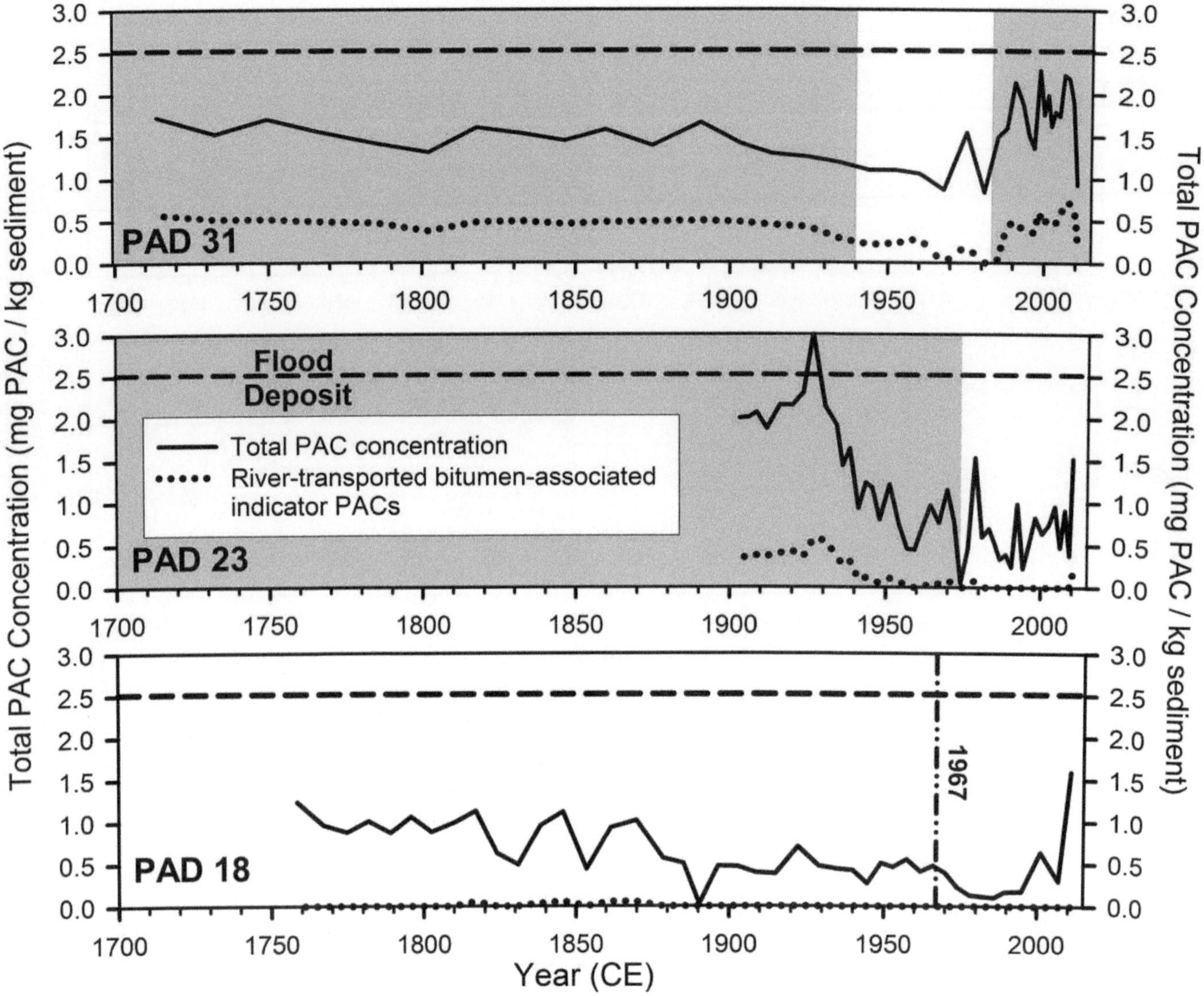

Figure 8. Time trends of total PAC concentration. Stratigraphic records showing the total PAC concentration (solid line) and the concentration of the seven PACs indicative of a bitumen origin and transport by the Athabasca River (dotted line) in the sediment cores from lakes PAD 31, PAD 23 and PAD 18. Grey shading identifies periods when the lakes were flood-prone. Absence of shading identifies periods when the lakes were not flood-prone. The vertical dashed-dotted line in the plot for PAD 18 identifies 1967, the year when major commercial oil sands operations began [1,4]. The dashed horizontal line in all three panels corresponds to the average value for the total PAC concentration in the 2007 Athabasca Delta flood deposits (2.518 mg/kg).

prone versus not flood-prone periods (Fig. 8). Analyses of sediment cores from PAD 31 and PAD 23 identified transient increases in total and river-transported bitumen-associated indicator PAC concentrations in samples dating to 1974, a year of widespread flooding that inundated both lakes despite their not flood-prone status. At PAD 31, three distinct peaks occur in total PAC concentration during the post-1982 flood-prone period (~1990, the late 1990s and 2007), which correspond to major floods of summer 1990, combined ice-jam flood years of 1996 and 1997, and the ice-jam flood of 2007, respectively. At PAD 23, average total PAC concentration during the post-1972 not flood-prone period is less than half that of the pre-1972 average when the lake was flood-prone. High total PAC concentration in the uppermost sample (2010) at PAD 23 is due mainly to increases in retene and C2- through C4 phenanthrenes/anthracenes, which are elevated in river sediments near forest fires in northern Alberta [52] and elsewhere [51], and common in sediments of the Athabasca River and its tributaries both upstream and downstream of oil sands deposits [10,44].

At PAD 18, total PAC concentration shows a gradually declining trend throughout the sediment core, except for an increase in the uppermost sample to values more typical of the late 1700s (Fig. 8). Consistently lowest total PAC concentrations occurred ~1975–1995, a result that indicates oil sands activities are not elevating sedimentary concentrations of atmospherically-deposited PACs in the delta. Although total PAC concentration is elevated in the uppermost sample (2010), PAC composition is dominated by the parent compounds fluorene (F), fluoranthene (Fl), naphthalene (N), phenanthrene (P), perylene (PLY), and other PACs that do not have a unique association with bitumen or oil sands production (C1-biphenyls, C1- to C4 naphthalenes (N1–N4); Fig. 5, 8) [10,44]. Oil sands-associated PACs and retene (a marker of conifer fire; [51,52]) do not rise in the uppermost sample, which suggests oil sands development and fire cannot be the underlying causes. Instead, the rapid down-core decline in concentration of most parent PACs between the uppermost two sediment samples suggests that degradation by biological, chemical or physical processes (i.e., diagenesis) could account for the rise in the uppermost sample [46]. Consistent with this interpretation, several compounds only attain values above the detection limits in the uppermost core sample from PAD 18 (perylene, fluorene, fluoranthene, C1-biphenyls, C4-naphthalenes).

Discussion

Sediment cores of three lakes in the PAD provide distinctly different profiles of PAC deposition over the past >200 years. Prior paleohydrological reconstructions of the study lakes were required to inform accurate interpretations of the PAC stratigraphic records. Without paleohydrological knowledge, entirely conflicting conclusions could readily have been drawn from the PAD 31 and PAD 23 records, whose profiles show increasing and decreasing supply of river-transported bitumen-associated indicator PACs during the past few decades, respectively (Fig. 7). These trends clearly reflect natural and engineered geomorphic changes in the flow of the Athabasca River [27]. Results from these two sites identify that flooding from the Athabasca River is an important vector that naturally supplies bitumen-sourced PACs to the Athabasca Delta.

Predictable responses of sedimentary PAC composition and concentration to changing hydrological conditions confirm that the sources and trends of these contaminants can be determined effectively from analyses of sediment cores from lakes in the PAD, an approach recommended by the Federal Expert Oil Sands Advisory Panel to assess environmental impacts of oil sands development [3] and proposed for the new federal Phase 2 monitoring program [21]. Here, we demonstrate that these methods yield informative results, but that detailed knowledge of hydrological variations is essential because shifting hydrological conditions at the study lakes complicates blind comparison of sedimentary PAC characteristics between stratigraphic intervals representing pre- and post-1967 development of the oil sands. Consequently, the PAC record from PAD 23 cannot be used to evaluate whether additional industrial loading of PACs has occurred, because the lake has received river floodwaters less frequently during the period since oil sands development than it did pre-development. The stratigraphic record at PAD 23 does show, however, that natural river processes, in particular ice-jam floods, have been an important source of PAC delivery to the lake in the recent past. In contrast, the increase in flood frequency at PAD 31 that has occurred since the Embarras Breakthrough in 1982 can be used to evaluate the role of industrial contributions of PACs by comparing results from this stratigraphic interval to the pre-1940 frequently-flooded interval (Fig. 5a, 7). This comparison identifies that natural processes responsible for delivering bitumen from along the banks of the Athabasca River and its tributaries can account for the PACs most associated with a bitumen origin in the post-1982 sediments of PAD 31. Thus, despite rapid growth of oil sands development during the past 25 years [1], the data reveal no measurable increase in concentration or proportion of river-transported bitumen-associated indicator PACs.

Although we fully recognize that our conclusions of no post-industrial increase in the sedimentary composition and concentration of bitumen-associated and river-transported PACs is based on only one lake record, PAD 31 is pivotally located at the upstream front of the Athabasca Delta where river floodwaters spread out over the delta [23,26,27]. PAD 31 is one of the most frequently flooded restricted-drainage lake in the Athabasca Delta [22,27], and it is easily flooded both during high-water events in the spring due to snowmelt and ice-jams and during rain events in summer [27,60]. The spring melt period and rain events have been identified as particularly important times when pulses of contaminants from oil sands development become transported to the Athabasca River and its distributaries and dispersed downstream [10,11,15,20]. Thus, floodplain lake sediments can be expected to capture signals of industrial contamination. For this reason, PAD 31 should be a particularly sensitive recorder of changes in river-borne PAC transport to the numerous restricted-drainage lakes of the Peace-Athabasca Delta, and is well positioned for continued monitoring of bitumen-derived PACs from flooding of the Athabasca River.

Conclusions we have drawn from the sediment record at PAD 31 are consistent with a previous study [12] showing that concentrations of most PACs were similar in sediments deposited during the 1950s (i.e., pre-development) and the late 1990s in Lake Athabasca and Richardson Lake, lakes that continuously receive Athabasca River water. From data at these two time points, Evans *et al.* [12] concluded that there was little evidence of a temporal increase in PAC concentrations due to the oil sands industry. Moreover, concentrations of the compounds we have identified as indicative of transport by the river (D, FLPY, BAC) are similar in sediments from the PAD 31 post-1982 strata (deposited during periodic flood events) and sediments in Richardson Lake from the 1950s and late 1990s [12] (deposited during continuous inflow from the Athabasca River). Agreement of findings at these sites suggests that high-flow (flood) events on the Athabasca River are a major vector for PAC-transport to the delta, as occurs downstream in the lower Mackenzie River basin [18,55]. This conclusion is also supported by equivalent PAC concentrations in flood deposits of PAD 31 and the upper range of concentrations in Athabasca River sediment measured by RAMP and reported in [15] (see Fig. 6). Given the importance of natural processes and industrial sources during river flood stages, monitoring and research programs must begin to sample at locations and times that capture contaminant dispersal during flood events. We propose that a network consisting of a larger number of monitoring sites and broader range of hydrological basin types is critically needed to further evaluate spatial and temporal distributions of PACs in the delta, their sources and their potential toxicity to biota. Such a monitoring network would serve as an early warning signal for any future changes in PAC delivery via the Athabasca River.

Results also reveal no evidence that industrial activity has contributed measurably to the sedimentary concentration of PACs supplied by long-range atmospheric transport and deposition in the vicinity of the PAD, as was also found for key metals of concern [61]. Dominance of PACs supplied by atmospheric transport in sediments from PAD 18 (Fig. 5) reflects pyrogenic-sourced PACs typically produced by forest fires [52,53], and the total PAC profile (Fig. 8) indicates that this has been a decreasing source of PACs since the mid-1700s. Maximum pyrogenic-sourced PACs during the 1700s at PAD 18 may reflect more frequent or larger fires during the regionally arid climatic conditions that prevailed in elevated regions of the PAD during the Little Ice Age [23,26], prior to European settlement and associated forest clearance and fire management. These baseline data can be used for continued monitoring of sediment PAC composition and concentration at PAD 18 to track changes in atmospheric PAC deposition to the delta as oil sands development expands, climate fluctuates and land use changes.

Our findings add new dimensions to conclusions developed by recent studies that have assessed oil sands industrial contributions to downstream PAC fluxes [10,15]. For example, Kelly *et al.* [10] identified that, during the non-breakup period, *dissolved* PAC concentrations at sites in the Athabasca River adjacent to and immediately downstream of the oil sands mining area were significantly greater than concentrations at upstream sites. However, because PACs are strongly hydrophobic, PAC loads to the Athabasca Delta should be dominated by the *particulate* fraction that is archived in the sediments of floodplain lakes. The study by Timoney and Lee [15] did attempt to track changes in particulate PAC loads to the Athabasca Delta via use of total PAC concentration data in river-bottom sediment samples collected by RAMP during 1999–2009. Yet, this sampling design has been identified as problematic [3,17], in part because of the shifting nature of fluvial and fluvio-deltaic sedimentary environments [16]. Our experimental methods, on the other hand, offer an alternative approach to address concerns related to the possible effects of oil sands industrial activities. An important advantage is that analyses of floodplain lake sediment profiles provide the hydrological and temporal context required to define pre-development baseline conditions and decipher trends over time due to natural versus anthropogenic processes.

As we have demonstrated, the unique perspectives offered by the lake sediment core records indicates that the PAD has been a natural repository of PAC-laden sediment carried by the Athabasca River for the past >200 years (and likely millennia). This finding is consistent with an independent estimate of natural hydrocarbon inputs from exposed bitumen-rich sediment along the Athabasca River and its distributaries between Fort McMurray and Embarras [62]. Of the estimated 6.35 Mt of suspended sediment that pass the Embarras gauging station annually, they concluded that at most 3%, and possibly 1% or less, was likely derived from the natural oil sands exposures. Assuming that 1% of the suspended sediment is derived from the natural oil sands exposures (63,500 tonnes/year), which comprise 8–14% bitumen (as reported in [4]), then natural processes may account for ~5,000 to 9,000 tonnes/year of bitumen to the downstream delta and Lake Athabasca. By mass, this is approximately equivalent to an Exxon Valdez crude oil spill [63] deposited every four to eight years by natural processes. Using estimates of the PAC content of the oil sands sediment (0.0095% PAC by mass; from [10]) and sediments from the Athabasca River and two tributaries downstream of riverbank deposits of bitumen in areas unaffected by industrial development (0.0275% PACs by mass; from [44]), this translates to 6 - 17.5 tonnes of total PACs potentially supplied to downstream ecosystems annually by natural erosion and transport of oil sands material by the Athabasca River and tributaries. Although these estimates of natural loads of bitumen and PACs may appear high, they are diluted by 'cleaner' sediment from upstream locations in the watershed [62]. Consequently, the mean PAC content of sediments deposited during flood-prone periods at PAD 23 and PAD 31 within the Athabasca Delta (1.55 mg PACs/kg sediment ±0.51 (1 SD, range: 0.45–3.01 mg/kg)) is lower than that commonly measured in soils from urban environments in the USA [64,65].

As we demonstrate in this and other studies [24], ice-jam floods are associated with peak concentrations and proportions of river-transported bitumen-associated PACs both prior to and since onset of oil sands development. This occurs because moving river-ice and high river flow velocities can mobilize and transport naturally occurring bitumen-rich materials in the riverbanks and in localized depositional accumulations downstream to the delta [62]. Similarly, studies at the Mackenzie Delta have identified that naturally occurring petrogenic PACs eroded from the mainstem of the Mackenzie River are carried with the sediment loads during the spring freshet [18] and that concentrations of PACs in floodplain-lake sediments are positively related to flood frequency [55].

Ironically, flood events, which have long been considered a crucial natural process for maintaining hydrological and ecological integrity of the delta [66] are also a major vector supplying organic contaminants. Consequently, our findings identify that the design of monitoring and research studies must adequately capture contaminant transport and deposition during the influential spring freshet and other periods of river flood stage in order to account for the substantial natural PAC loads to the Athabasca River and its distributaries, as well to capture loading by short-lived hydrological events associated with potential pulses of industrial contaminants [10,11,20]. Additional lakes should be cored and analyzed to verify whether our findings are representative of other parts of the delta and upstream along the Athabasca River floodplain, and to provide further critically-needed baseline information of PAC deposition over space and time. An important future research direction is to examine the toxicity of sediments throughout the delta to aquatic biota.

Limitations

We carefully considered multiple ways to present the sedimentary data generated by our study, including in flux units (e.g., as mass of PACs per unit area per unit time) to directly assess if industrial activity is accelerating the rate of delivery of PACs to lakes in the delta. But, this was not possible based on the experimental design of this study, because at PAD 31 sedimentation rates increased since the lake became more susceptible to flooding following the Embarras Breakthrough (Fig. 2; [27]). As a consequence, it is not possible to distinguish natural hydrological causes of changes in PAC *fluxes* from those potentially due to erosion and other processes by industrial activities because they are confounded in time. Rather, we contend that if industry has substantially enhanced PAC loads to the delta, then the likelihood of doing so with the same sedimentary *concentration* and *composition* of PACs that has occurred naturally due to erosion of the Athabasca River would seem highly unlikely, but remains to be tested at additional sites. We note that Timoney and Lee [15] have concluded that oil sands industrial activity may be related to increasing sedimentary PAC *concentration* in the Athabasca River, but this was not evident in the sediment core records we generated.

The conclusions arising from this study are based on analyses performed at three lakes. While these lakes span a broad range of basin hydrology and river connectivity, they do not cover the full range that exists in the Peace-Athabasca Delta. For example, our study did not include continuously flooded open-drainage lakes due to challenges of obtaining interpretable records from systems with such temporally variable hydrological and sedimentary environments [e.g., 27]. Nor did it include lakes flooded mainly by the Peace River which traverses undeveloped surficial deposits of bitumen. Sediments from the study lakes capture PACs transported by the Athabasca River to the delta during flood stages in spring and summer, but may not record PACs dispersed during lower river stages and by the Peace River. Thus, we recommend that periodic synoptic surveys of surface sediments and water isotope tracers be conducted for a larger suite of delta lakes (40 or more; e.g., [25,32]) that span more complete gradients of basin hydrology (i.e., lakes with open- to closed-drainage hydrology; lakes receiving inflow from the Peace River as well as the Athabasca River) to quantify relationships between connectivity with the rivers and the sedimentary concentrations and composition of PACs, and to further elucidate the potential role of Alberta's oil sands industry on contaminant delivery to the delta. Toxicity tests performed with the surficial sediments could provide important information to examine their effects on aquatic biota in a risk-assessment framework.

Acknowledgments

We thank Wood Buffalo National Park, Wood Buffalo Helicopters and R. Grandjambe for logistical support. M. Elmes assisted with fieldwork. K. Timoney and P. Lee provided the left panel of Figure 1. E. Kelly provided data for Fig. 4a. Comments from two anonymous reviewers and Kevin Timoney led to further improvements of this paper.

Author Contributions

Conceived and designed the experiments: RIH BBW TWDE AJF DGD. Performed the experiments: RIH BBW JAW. Analyzed the data: RIH BBW JAW. Wrote the paper: RIH BBW JAW. Edited the manuscript: TWDE AJF DGD.

References

1. Schindler D (2010) Tar sands need solid science. Nature 468: 499–501.
2. Government of Alberta (2006) Alberta's Oil Sands. Available: http://www.oilsands.alberta.ca/. Accessed 18 Apr 2012.
3. Dowdeswell L, Dillon P, Ghoshal S, Miall A, Rasmussen J, et al. (2010) A foundation for the future: Building an environmental monitoring system for the oil sands. A report submitted to the Minister of Environment. 49 p. Available: http://www.ec.gc.ca/pollution/E9ABC93B-A2F4-4D4B-A06D-BF5E0315C7A8/1359_Oilsands_Advisory_Panel_report_09.pdf. Accessed 18 Apr 2012.
4. Gosselin P, Hrudey S, Naeth A, Plourde A, Therrien R, et al. (2010) Environmental and health impacts of Canada's oil sands industry. Ottawa: The Royal Society of Canada. 414 p. Available: http://www.rsc.ca/documents/RSCreportcompletesecured9Mb_Mar28_11.pdf. Accessed 18 Apr 2012.
5. Auditor General of Canada (2010) 2010 Fall Report of the Commissioner of the Environment and Sustainable Development. Ottawa: ISBN 978-1-100-17245-3. Available: http://www.oag-bvg.gc.ca/internet/docs/parl_cesd_201012_02_e.pdf. Accessed 18 Apr 2012.
6. Timoney KP (2007) A study of water and sediment quality as related to public health issues, Fort Chipewyan, Alberta. Fort Chipewyan, Alberta: on behalf of the Nunee Health Board Society. Available: http://www.tothetarsands.ca/wp-content/uploads/2007/11/fc_final_report1-1.pdf. Accessed 18 Apr 2012.
7. Timoney KP, Lee P (2009) Does the Alberta tar sands industry pollute? The scientific evidence. The Open Conserv Biol J 3: 65–81.
8. Irwin RJ, van Mouwerik M, Stevens L, Seese MD, Basham W (1997) Environmental contaminants encyclopedia. Fort Collins, Colorado: National Park Service, Water Resources Division. Available: http://amser.org/index.php?P=AMSER-ResourceFrame&resourceId=668. Accessed 18 Apr 2012.
9. Colavecchia MV, Hodson PV, Parrott JL (2006) Cyp1a induction and blue sac disease in early life stages of white suckers (*Catostomus commersoni*) exposed to oil sands. J Toxicol Environ Health, Part A 69: 967–994.
10. Kelly EN, Short JW, Schindler DW, Hodson PV, Ma M, et al. (2009) Oil sands development contributes polycyclic aromatic compounds to the Athabasca River and its tributaries. Proc Natl Acad Sci USA 106: 22346–22351. DOI:10.1073/pnas.0912050106.
11. Kelly EN, Schindler DW, Hodson PV, Short JW, Radmanovich R, et al. (2010) Oil sands development contributes elements toxic at low concentration to the Athabasca River and its tributaries. Proc Natl Acad Sci USA 107: 16178–16183. DOI:10.1073/pnas.1008754107.
12. Evans MS, Billeck B, Lockhart L, Bechtold JP, Yunker MB, et al. (2002) PAH sediment studies in Lake Athabasca and the Athabasca River ecosystem related to the Fort McMurray oil sands operations: sources and trends. In: Brebbia CA, editor. Oil and hydrocarbon spills III, modelling, analysis and control. Southampton MA: WIT Press. 365–374.
13. Wolfe BB, Hall RI, Last WM, Edwards TWD, English MC, et al. (2006) Reconstruction of multi-century flood histories from oxbow lake sediments, Peace-Athabasca Delta, Canada. Hydrol Process 20: 4131–4153. DOI:10.1002/hyp.6423.
14. Yunker MB, Macdonald RW (2003) Alkane and PAH depositional history, sources and fluxes in sediments from the Fraser River Basin and Strait of Georgia, Canada. Org Geochem 34: 1429–1454.
15. Timoney KP, Lee P (2011) Polycyclic aromatic hydrocarbons increase in Athabasca River Delta sediment: Temporal trends and environmental correlates. Environ Sci Technol 45: 4278–4284.
16. Garcia-Aragon J, Droppo IG, Krishnappan BG, Trapp B, Jaskot C (2011) Erosion characteristics and floc strength of Athabasca River cohesive sediments: towards managing sediment-related issues. J Soils Sediments 11: 679–689.
17. Ayles GB, Dubé M, Rosenberg D (2004) Oil Sands Regional Aquatic Monitoring Program (RAMP) Scientific Peer Review of the Five Year Report (1997–2001). Winnipeg: Fisheries and Oceans Canada. Available: go.nature.com/tdsuag. Accessed 18 Apr 2012.
18. Yunker MB, Macdonald RW, Cretney WJ, Fowler BR, McLaughlin FA (1993) Alkane, terpene, and polycyclic aromatic hydrocarbon geochemistry of the Mackenzie River and Mackenzie shelf: Riverine contributions to the Beaufort Sea coastal sediment. Geochim Cosmochim Acta 57: 3041–3061.
19. Scrimgeour GJ, Prowse TD, Culp JM, Chambers PA (1994) Ecological effects of river ice break-up: a review and perspective. Freshwat Biol 32: 261–275.
20. Giesy JP, Anderson JC, Wiseman SB (2010) Alberta oil sands development. Proc Natl Acad Sci USA 107: 951–952.
21. Environment Canada (2011) Integrated monitoring program for the oil sands: Expanded geographic extent for water quality and quantity, aquatic biodiversity and effects, and acid-sensitive lake component. Ottawa: Minister of the Environment Cat. No.: En14-49/2011E-PDF. Available: http://www.ec.gc.ca/pollution/EACB8951-1ED0-4CBB-A6C9-84EE3467B211/Integrated%20Oil%20Sands_low_e.pdf. Accessed 18 Apr 2012.
22. Hall RI, Wolfe BB, Edwards TWD (2004) A multi-century flood, climatic, and ecological history of the Peace-Athabasca Delta, northern Alberta, Canada. Final Report. Vancouver: BC Hydro. 163 p + Appendices.
23. Wolfe BB, Karst-Riddoch TL, Vardy SR, Falcone MD, Hall RI, et al. (2005) Impacts of climate and river flooding on the hydro-ecology of a floodplain basin, Peace-Athabasca Delta, Canada since A.D. 1700. Quat Res 64: 147–162. DOI:10.1016/j.yqres.2005.05.001.
24. Wolfe BB, Hall RI, Last WM, Edwards TWD, English MC, et al. (2006) Reconstruction of multi-century flood histories from oxbow lake sediments, Peace-Athabasca Delta, Canada. Hydrol Process 20: 4131–4153. DOI:10.1002/hyp.6423.
25. Wolfe BB, Karst-Riddoch TL, Hall RI, Edwards TWD, English MC, et al. (2007) Classification of hydrological regimes of northern floodplain basins (Peace-Athabasca Delta, Canada) from analysis of stable isotopes ($\delta^{18}O$, $\delta^{2}H$) and water chemistry. Hydrol Process 21: 151–168. DOI:10.1002/hyp.6229.
26. Wolfe BB, Hall RI, Edwards TWD, Jarvis SR, Sinnatamby RN, et al. (2008) Climate-driven shifts in quantity and seasonality of river discharge over the past 1000 years from the hydrographic apex of North America. Geophys Res Lett 35: L24402. DOI:10.1029/2008GL036125.
27. Wolfe BB, Hall RI, Edwards TWD, Vardy SR, Falcone MD, et al. (2008) Hydroecological responses of the Athabasca Delta, Canada, to changes in river flow and climate during the 20th century. Ecohydrology 1: 131–148. DOI:10.1002/eco.13.
28. Wolfe BB, Edwards TWD, Hall RI, Johnston JW (2011) A 5200-year record of freshwater availability for regions in western North America fed by high-elevation runoff. Geophys Res Lett 38: L11404. DOI:10.1029/2011GL047599.
29. Yi Y, Brock BE, Falcone MD, Wolfe BB, Edwards TWD (2008) A coupled isotope tracer method to characterize input water to lakes. J Hydrol 350: 1–13. DOI:10.1016/j.jhydrol.2007.11.008.
30. Johnston JW, Köster D, Wolfe BB, Hall RI, Edwards TWD, et al. (2010) Quantifying Lake Athabasca (Canada) water level during the Little Ice Age highstand from paleolimnological and geophysical analyses of a transgressive barrier-beach complex. The Holocene 20: 801–811. DOI:10.1177/0959683610362816. [Also see Johnston JW, Köster D, Wolfe BB, Hall RI, Edwards TWD, et al. (2010) Erratum. The Holocene 20: 812. DOI:10.1177/0959683610362816.]
31. McGowan S, Leavitt PR, Hall RI, Wolfe BB, Edwards TWD, et al. (2011) Interdecadal declines in flood frequency increase primary production in lakes of a northern river delta. Global Change Biol 17: 1212–1224. DOI:10.1111/j.1365-2486.2010.02304.x.
32. Wolfe BB, Hall RI, Edwards TWD, Johnston JW (2012). Developing temporal hydroecological perspectives to inform stewardship of a northern floodplain landscape subject to multiple stressors: Paleolimnological investigations of the Peace-Athabasca Delta. Environ Rev. DOI:10.1139/A2012-008.
33. Glew JR (1989) A new trigger mechanism for sediment samples. J Paleolimnol 2: 241–243.
34. Dean WE (1974) Determination of carbonate and organic matter in calcareous sediments and sedimentary rocks by loss on ignition: comparison with other methods. J Sed Petrol 44: 242–248.
35. Currie LA (1968) Limits for qualitative detection and quantitative determination: application to radiochemistry. Anal Chem 40: 586–593.
36. Ballestra SJ, Gastaud J, Lopez JJ, Parsi P, Vas D (1994) Intercomparison of radionuclide measurements on marine sediment sample IAEA-300. Monaco Cedex: International Atomic Energy Agency Marine Environment Laboratory B.P. No. 800 MC 98012. 63p.
37. Schelske CL, Peplow A, Brenner M, Spencer C (1994) Low-background gamma counting: applications for ^{210}Pb dating of sediments. J Paleolimnol 10: 115–128.

38. Robbins JA (1978) Geochemical and geophysical applications of radioactive lead. In: Nriagu JO, editor. The biogeochemistry of lead in the environment. Amsterdam: Elsevier. 285–393.
39. Appleby PG (2001) Chronostratigraphic techniques in recent sediments. In: Last WM, Smol JP, editors. Tracking environmental change using lake sediments: Basin analysis, coring, and chronological techniques. Dordrecht: Kluwer Academic Publishers. 171–203.
40. Binford MW (1990) Calculation and uncertainty analysis of ^{210}Pb dates for PIRLA project lake sediment cores. J Paleolimnol 3: 253–267.
41. Clarke KR, Gorley RN (2006) Primer v6: user manual/tutorial. Plymouth, UK: PRIMER-E Ltd. 190 p.
42. Jones R (2010) Environmental contamination associated with a marine landfill ('seafill') beside a coral reef. Mar Poll Bull 60: 1993–2006.
43. Clarke KR, Warwick RM (2001) Change in Marine Communities: An Approach to Statistical Analysis and Interpretation, 2nd Edition. Plymouth, UK: PRIMER-E Ltd. 7-1 to 7-4.
44. Colavecchia MV, Backus SM, Hodson PV, Parrott JL (2004) Toxicity of oil sands to early life stages of fathead minnows (*Pimephales promelas*). Environ Toxicol Chem 23: 1709–1718.
45. Headley JV, Akre C, Conly MF, Peru KM, Dickson LC (2001) Preliminary characterization and source assessment of PAHs in tributary sediments of the Athabasca River, Canada. J Environ Foren 2: 335–345.
46. Wakeham SG, Schaffner C, Giger W (1980) Polycyclic aromatic-hydrocarbons in recent lake-sediments 2. Compounds derived from biogenic precursors during early diagenesis. Geochim Cosmochim Acta 44: 415–429.
47. Grice K, Nabbefeld B, Maslen E (2007) Source and significance of selected polycyclic aromatic hydrocarbons in sediments (Hovea-3 well, Perth Basin, Western Australia) spanning the Permian–Triassic boundary. Org Geochem 38: 1795–1803.
48. Tolosa I, Bayona JM, Albaiges J (1996) Aliphatic and polycyclic aromatic hydrocarbons and sulphur/oxygen derivatives in northwestern Mediterranean sediments: Spatial and temporal variability, fluxes and budgets. Environ Sci Technol 30: 2495–2503.
49. Itoh N, Tamamura S, Kumagai M (2010) Distributions of polycyclic aromatic hydrocarbons in a sediment core from the north basin of Lake Biwa, Japan. Org Geochem 41: 845–852.
50. Yang C, Wang Z, Yang Z, Hollebone B, Brown CE, et al. (2011) Chemical fingerprints of Alberta oil sands and related petroleum products. Environ Foren 12: 173–188.
51. Ramdahl T (1983) Retene – a molecular marker of wood combustion in ambient air. Nature 306: 580–582.
52. Gabos S, Ikonomou MG, Schopflocher D, Fowler BR, White J, et al. (2001) Characteristics of PAHs, PCDD/Fs and PCBs in sediment following forest fires in northern Alberta. Chemosphere 43: 709–719.
53. Vergnoux A, Malleret L, Asia L, Doumenq P, Theraulaz F (2011) Impact of forest fires on PAH level and distribution in soils. Environ Res 111: 193–198. DOI:10.1016/j.envres.2010.01.008.
54. Meyers P, Ishiwatari R (1993) Lacustrine organic chemistry – an overview of indicators of organic matter sources and diagenesis in lake sediments. Org Geochem. 20: 867–900.
55. Headley JV, Marsh P, Akre CJ, Peru KM, Lesack L (2002) Origin of polycyclic aromatic hydrocarbons in lake sediments of the Mackenzie Delta. J Environ Sci Health Part A 37: 1159–1180.
56. Regional Aquatics Monitoring Program (RAMP) (2011) Data available: http://www.ramp-alberta.org/ramp/data.aspx.Accessed 9 Sep 2011.
57. Kersten M, Smedes F (2002) Normalization procedures for sediment contaminants in spatial and temporal trend monitoring. J Environ Monitor 4: 109–115, DOI:10.1039/B108102K.
58. Peters DL, Prowse TD, Pietroniro A, Leconte R (2006) Flood hydrology of the Peace-Athabasca Delta, northern Canada. Hydrol Process 20: 4073–4096. DOI:10.1002/hyp.6420.
59. Schindler DW, Donahue WF (2006) An impending water crisis in Canada's western prairie provinces, Proc Natl Acad Sci USA 103: 7210–7216.
60. Wiklund JA, Hall RI, Wolfe BB (2012) Timescales of hydrolimnological change in floodplain lakes of the Peace-Athabasca Delta, northern Alberta, Canada. Ecohydrology 5: 351–367. DOI:10.1002/eco.226.
61. Wiklund JA, Hall RI, Wolfe BB, Edwards TWD, Farwell AJ, et al. (2012) Has Alberta oil sands development increased far-field delivery of airborne contaminants to the Peace-Athabasca Delta? Sci Total Environ 433: 379–382. DOI:10.1016/j.scitotenv.2012.06.074.
62. Conly FM, Crosley RW, Headley JV (2002) Characterizing sediment sources and natural hydrocarbon inputs in the lower Athabasca River, Canada. J Environ Engineer Sci 1: 187–199. DOI:10.1139/S02-013.
63. Peterson CH, Rice SD, Short JW, Esler D, Bodkin JL, et al. (2003) Long-term ecosystem response to the Exxon Valdez. Science 302: 2082–2086. DOI:10.1126/science.1084282.
64. Bradley LJN, Magee BH, Allen SL (1994) Background levels of polycyclic aromatic hydrocarbons (PAH) and selected metals in New England urban soils. J Soil Contamin 3: 349–361.
65. Mauro DM, DeClercq PJ, Siegener R, Coleman A (2006) Survey of the distribution and sources of PAHs in urban surface soils. Land Contamin Reclam 14: 513–521. DOI:10.2462/09670513.749.
66. Timoney K (2002) A dying delta? A case study of a wetland paradigm. Wetlands 22: 282–300.

Common Distribution Patterns of Marsupials Related to Physiographical Diversity in Venezuela

Jacint Ventura[1]*, Guillem Bagaria[1,2], Maria Assumpció Sans-Fuentes[3,4], Roger Pérez-Hernández[5]

1 Departament de Biologia Animal, Biologia Vegetal i Ecologia, Facultat de Biociències, Universitat Autònoma de Barcelona, Cerdanyola del Vallès, Spain, **2** CREAF, Cerdanyola del Vallès, Spain, **3** School of Geography and Development, University of Arizona, Tucson, Arizona, United States of America, **4** Departament de Fisiologia Animal, Facultat de Farmàcia, Universitat de Barcelona, Barcelona, Spain, **5** Instituto de Zoología y Ecología Tropical, Facultad de Ciencias, Universidad Central de Venezuela, Caracas, Venezuela

Abstract

The aim of this study is to identify significant biotic regions (groups of areas with similar biotas) and biotic elements (groups of taxa with similar distributions) for the marsupial fauna in a part of northern South America using physiographical areas as Operational Geographical Units (OGUs). We considered Venezuela a good model to elucidate this issue because of its high diversity in landscapes and the relatively vast amount of information available on the geographical distribution of marsupial species. Based on the presence-absence of 33 species in 15 physiographical sub-regions (OGUs) we identified Operational Biogeographical Units (OBUs) and chorotypes using a quantitative analysis that tested statistical significance of the resulting groups. Altitudinal and/or climatic trends in the OBUs and chorotypes were studied using a redundancy analysis. The classification method revealed four OBUs. Strong biotic boundaries separated: i) the xerophytic zone of the Continental coast (OBU I); ii) the sub-regions north of the Orinoco River (OBU III and IV); and those south to the river (OBU II). Eleven chorotypes were identified, four of which included a single species with a restricted geographic distribution. As for the other chorotypes, three main common distribution patterns have been inferred: i) species from the Llanos and/or distributed south of the Orinoco River; ii) species exclusively from the Andes; and iii) species that either occur exclusively north of the Orinoco River or that show a wide distribution throughout Venezuela. Mean altitude, evapotranspiration and precipitation of the driest month, and temperature range allowed us to characterize environmentally most of the OBUs and chorotypes obtained.

Editor: Brock Fenton, University of Western Ontario, Canada

Funding: The study was funded by the Spanish Ministerio de Educación y Cultura (Programa de Cooperación Científica con Iberoamérica, A) Proyectos de Investigación Conjunta). The funders had no role in study design, data collection and analysis, decision to publish, or preparation of the manuscript.

Competing Interests: The authors have declared that no competing interests exist.

* E-mail: jacint.ventura.queija@uab.es

Introduction

The current geographic distribution of the fauna in northern South America is strongly linked to dynamic geologic history. In particular, the Andean uplift was a major driver for change in the landscapes and biota in this area and, consequently, was crucial for the evolution of its ecosystems and for the diversification of its species [1,2]. Within northern South America, Venezuela provides a good example of how this tectonic event, posterior geodynamic processes and variations in the surface geology (e.g., [3,4] and references therein) have led to a high physiographical diversity. According to Huber [5,6], the following main landscapes are currently recognized in Venezuela: (A) insular and coastal environments, typical of the Continental coast; (B) lowland plains, found in both the Llanos and the lowland areas along the main Amazonian Rivers; (C) hills and low mountains, which occur in the Lara-Falcón hill system and the Guayana Shield; and (D) high mountains, found either in the ranges or the hills of the Andes (Cordillera de Mérida), Perijá, San Luis, Santa Ana, Central and Eastern Coasts, Copey (in Margarita Island), and Guayana. In turn, these four main landscapes have been divided into different sub-regions (see Fig. 1) identified by geological, climatic and vegetation characteristics [5,6].

Regarding the mammalian fauna, Venezuela is ranked as the eighth most diverse country in the world [7]. As for marsupials, 34 species of Didelphimorphia and one species of Paucituberculata occur in this zone of the Neotropics [8], representing approximately 30% of the total living species of American opossums. In spite of this rich diversity, information on biogeographical associations of these species in Venezuela is very limited and relates exclusively to observations on the mammal fauna, either in the whole country [9,10] or in particular geographic systems [4]. In order to detect general patterns of the geographical distribution of mammal species, several approaches have been undertaken using different biogeographical units, such as bioregions [9], regions established on the basis of their "biogeographic and physical-natural" characteristics [10], and zoogeographical units [11,12]. Nevertheless, in these studies common biogeographical patterns for these species have only been based on comparisons of their distribution areas, without statistical testing. Furthermore, quantitative biogeographical analyses have never been applied to Venezuelan marsupials exclusively.

Methods based on quantitative analyses are the best to establish biogeographical associations, as they provide objectivity and produce consistent and repeatable results. Specifically, this kind of analyses applied to the presence-absence data of several taxa in

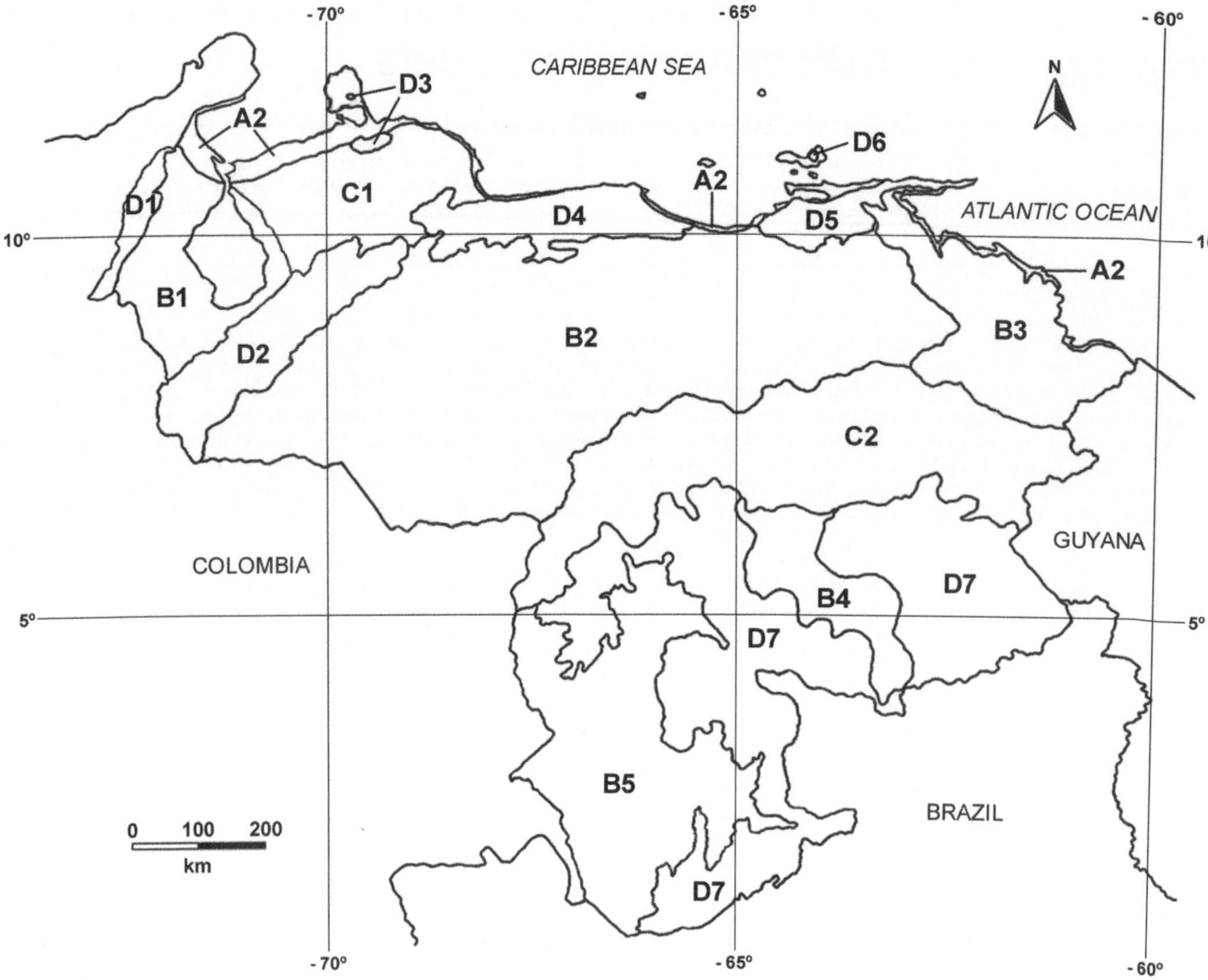

Figure 1. Physiographical sub-regions of Venezuela, according to Huber [5,6]. A2 Continental coast, B1 Maracaibo depression, B2 The Llanos, B3 Plain of the Orinoco River Delta and the boggy plain of the San Juan River, B4 Peneplain of Caura and Paragua Rivers, B5 Peneplain of the Casiquiare River, Upper Orinoco, C1 Lara-Falcón hill system, C2 Guayana Shield foothill system, D1 Serranía de Perijá, D2 The Andes, D3 Serranía de San Luis and Cerro Santa Ana, D4 Central Coastal Range, D5 Eastern Coastal Range, D6 Cerro Copey, D7 Guayana Massif. Huber's A1 sub-region (Insular coast) was not included in the analyses since there are no marsupials in these extremely dry and small islands.

a number of geographic areas [13] allows for the detection of repeatable biogeographical patterns within the data in the form of biotic elements (groups of taxa with similar distributions) and biotic regions (groups of regions with similar biotas). The quantitative classification method for defining boundaries between ordered locations outlined by McCoy et al. [14] is an extension of the probabilistic similarity techniques [15]. This procedure, complemented by a significance test of the resulting groups (see [16]), constitutes an objective method for measuring the statistical significance of groups obtained by numerical classification (e.g., [17–21].

Venezuela can be considered an excellent model to identify and characterize significant biotic regions and chorotypes regarding its marsupial fauna because of the afore mentioned high physiographical diversity and the availability of a relatively large amount of information on the geographical distribution of the marsupial species. Therefore, the main aim of this study is to determine by quantitative analyses the distribution patterns of the marsupial species in Venezuela, using the physiographical sub-regions defined by Huber [5,6] as Operational Geographical Units (OGUs; [22]). The specific goals are: (i) to determine significant biotic regions; (ii) to identify associations of species with similar distribution patterns (chorotypes); and (iii) to characterize the biotic regions and the chorotypes using limiting environmental variables for the geographic distribution of these species in Venezuela. To our knowledge, the present study constitutes the first attempt to address these issues for the marsupial fauna of northern South America.

Materials and Methods

Study Area and Distributional Data

Venezuela is located in northern South America, between 0°45′ N and 15°40′ N, and between 59°45′ W and 73°25′ W. It covers a total emerged area of 916,445 km^2. For analytical purposes this

Table 1. Presence (1) or absence (0) of marsupial species in each physiographical sub-region of Venezuela (for abbreviations see Fig. 1).

Species	A2	B1	B2	B3	B4	B5	C1	C2	D1	D2	D3	D4	D5	D6	D7
Caenolestes fuliginosus	0	0	0	0	0	0	0	0	0	1	0	0	0	0	0
Caluromys lanatus	0	1	0	0	0	1	0	0	1	1	0	0	0	0	0
Caluromys philander	0	0	0	0	1	1	0	1	0	0	0	0	0	0	1
Caluromys trinitatis	0	0	1	1	0	0	1	0	0	1	0	1	1	1	0
Chironectes minimus	0	0	0	1	1	0	0	0	1	1	0	1	1	0	1
Didelphis imperfect	0	0	0	0	0	1	0	1	0	0	0	0	0	0	1
Didelphis marsupialis	0	1	1	1	1	1	1	1	1	1	1	1	1	0	1
Didelphis pernigra	0	0	0	0	0	0	0	0	0	1	0	0	0	0	0
Gracilinanus dryas	0	0	0	0	0	0	0	0	0	1	0	0	0	0	0
Gracilinanus marica	0	0	0	0	0	0	0	0	1	1	1	1	1	0	0
Lutreolina crassicaudata	0	0	1	1	0	0	0	1	0	0	0	0	0	0	1
Marmosa demerarae	0	0	0	1	1	1	0	1	0	1	1	1	1	0	1
Marmosa lepida	0	0	0	0	1	0	0	0	0	0	0	0	0	0	0
Marmosa murina	0	1	0	1	1	1	0	1	1	1	0	1	1	0	1
Marmosa robinsoni	0	1	1	1	0	0	1	0	1	1	1	1	1	1	0
Marmosa tyleriana	0	0	0	0	0	0	0	0	0	0	0	0	0	0	1
Marmosa waterhousei	0	0	0	0	0	0	0	0	0	1	0	0	0	0	0
Marmosa xerophila	1	0	0	0	0	0	0	0	0	0	0	0	0	0	0
Marmosops cracens	0	0	0	0	0	0	1	0	0	0	0	0	0	0	0
Marmosops fuscatus	0	0	0	0	0	0	1	0	1	1	1	1	1	0	0
Marmosops impavidus	0	0	0	0	0	0	0	0	0	1	0	0	0	0	0
Marmosops neblina	0	0	0	0	0	0	0	0	0	0	0	0	0	0	1
Marmosops pakaraimae	0	0	0	0	0	0	0	0	0	0	0	0	0	0	1
Marmosops pinheiroi	0	0	0	0	1	0	0	0	0	0	0	0	0	0	1
Metachirus nudicaudatus	0	1	1	1	1	1	0	1	1	1	0	0	0	0	1
Monodelphis adusta	0	0	0	0	0	0	0	0	0	1	0	0	0	0	0
Monodelphis brevicaudata	0	0	0	0	1	1	0	1	0	0	0	0	0	0	1
Monodelphis palliolata	0	1	0	1	0	0	1	0	1	1	1	1	1	0	0
Monodelphis reigi	0	0	0	0	0	0	0	0	0	0	0	0	0	0	1
Monodelphis species A	0	0	1	0	0	0	0	0	0	0	0	0	0	0	0
Philander andersoni	0	0	0	0	1	1	0	0	0	0	0	0	0	0	0
Philander deltae	0	0	0	1	0	0	0	0	0	0	0	0	0	0	0
Philander mondolfii	0	1	1	0	0	0	0	1	1	1	0	0	0	0	1

study area was divided into 15 OGUs (see Fig. 1) that correspond to the physiographical sub-regions defined by Huber [5,6].

Information on the distribution of 33 marsupial species in Venezuela (32 belonging to Order Didelphimorphia, Family Didelphidae, and one to Order Paucituberculata, Family Caenolestidae; see Appendix S1 in File S1) was obtained from Pérez-Hernández [23] and Pérez-Hernández et al. [24], and was complemented with data reported in 30 articles published between 1987 and 2013 (see Appendix S2 in File S1). Two species, *Gracilinanus emiliae* and *Gracilinanus agilis*, were excluded from our analyses because distribution information is unreliable. *Gracilinanus emiliae* shows a relatively wide geographic distribution in northern South America (Colombia, Guyana, Surinam, French Guiana and northeastern Brazil), but in Venezuela it has only been found in the Orinoco River Delta, which constitutes a very marginal site within its distribution area [25]. Similarly, *G. agilis* has been reported in two locations that need to be fully corroborated (in forest areas of the Andes and the Central Coastal Range; see [9,25–27]). Additionally, regarding the taxonomic observations by Voss et al. [28,29], *Marmosops parvidens* has not been included in the analyses, since its presence in the peneplain of the Casiquiare River and Lara-Falcon hill system sub-regions must also be confirmed. Moreover, following Voss et al. [29], we have considered the specimens from the Guayana Massif subregion as *M. pakaraimae*, although initially assigned to *M. parvidens*.

The altitudinal range reported for each species was obtained from the information available in the following collections: Estación Biológica de Rancho Grande (EBRG), Museo de Biología de la Universidad Central de Venezuela (MBUCV), Museo de Historia Natural La Salle (MHNLS) and Colección de Vertebrados de la Universidad de Los Andes (CVULA).

Since several territories in Venezuela remain poorly explored due to their difficult accessibility, the information on the marsupial fauna in some physiographical sub-regions is incomplete. This is the case in certain areas of Serranía de Perijá, piedemonte andino (Andes), Guayana Massif, Orinoco River Delta, and the peneplains of the Caura, Paragua and Casiquiare Rivers. Additionally, it should be noted that our data sources do not allow us to determine the strength of the relationship between the absences of species and the trapping efforts in each sub-region. Nevertheless, despite these drawbacks, it can be assumed that the geographical distribution of the marsupial species in Venezuela is relatively well known (e.g., [8–10]). Consequently, the data used in this study are enough to detect reliable general distribution patterns of the marsupial species using physiographical sub-regions as biogeographical units.

Classification of Physiographical Regions and Species

A presence-absence matrix for the 33 species (Operational Taxonomic Units; OTUs) in the 15 OGUs was constructed (Table 1). The classification of physiographical sub-regions (Q-mode) and species (R-mode) was conducted by using R programming language [30] (see Code S1). Two similarity matrices were obtained (one for each mode) using the Baroni-Urbani and Buser's similarity index (see [31] for details) and implemented in the *simba* package [32]. This pairwise index, when applied to sites, takes into account the number of species unique to each OGU, and both the number of species that co-occur and the number of species that are absent in both OGUs of the same pair. Likewise, when it is applied to species, it takes into account the number of sites unique to each OTU, and both the number of sites where the two OTUs co-occur and the number of sites where the two OTUs are absent. According to Baroni-Urbani and Buser [31], considering common absences is valuable because they highlight differences that are biogeographically informative. Shared absences are, however, multiplied by shared presences in their index to emphasize shared presences and to prevent the possibility that two OGUs would show high similarity because of shared absences alone (see [21] for details).

The relationships between OGUs and OTUs were determined by applying function *hclust* and the unweighted pair-group method using arithmetic averages (UPGMA) from dissimilarity matrices (1-similarity matrices), and assembled into a dendrogram (see [17] for details). We obtained two matrices of significant similarities, one for the physiographical sub-regions and the other for the species, following the table of critical values in Baroni-Urbani and Buser [31]; these matrices reported the similarities that were higher, lower or equal to the similarity value expected at random. The determination of Operational Biogeographical Units (OBUs; see [33]) and chorotypes was accomplished by following the method by McCoy et al. [14] and the approach by Real et al. [16], that consists of testing for the presence of weak or strong significant boundaries between physiographical sub-regions or species. At each node of the corresponding dendrogram, a submatrix that only included the sub-regions or species involved in the node was extracted from the matrix of significant similarities. This submatrix was divided into three zones: A and B, corresponding to the groups separated by the node; and A*B corresponding to the intersection between the former two zones. The number of significant similarities (higher and lower than expected) in each zone was used to compute the parameters DW, DW(A*A), DW(B*B) and DS at each node. DW and DS measure the efficiency of a boundary to separate two groups of physiographical regions or species, whose fauna or geographical distribution, respectively, is similar within but not between each group. DW, which can be separated into DW(A*A) and DW(B*B), measures whether the similarities that are higher than expected tend to be located in zones A and B, but not in A*B. DS measures whether the similarities lower than expected tend to appear in A*B, but not in A or B (see [19] for details). In order to identify significant boundaries, a G-test of independence of the distribution of significant similarities in zones A, B and A*B (after Yates' correction) was conducted for each node of the dendrogram. This test gives the parameter GW for weak segregation and the parameter GS for strong segregation (see [14] for details).

Relationships between the Biogeographical Units and Chorotypes with Environmental Variables

To detect altitudinal and/or climatic trends in the biotic regions and chorotypes the following variables were considered: mean altitude (MA), mean precipitation of the driest month (MDP), temperature range (TR) and mean evapotranspiration (ME). Both MDP and TR values were obtained from Álvarez Bernal [34], ME values from Legórburu [35] and MA values from Hearn et al. [36]. For each variable, the average in each sub-region was calculated (Table 2). We used only four variables in order to avoid overfitting in the ordination analysis. We selected these variables because we considered them to be limiting for the geographic distribution of the marsupial species in Venezuela.

We conducted an ordination analysis of the species presence-absence table constrained by climatic and altitudinal variables. Since a detrended correspondence analysis (DCA) on the species table resulted in gradient lengths < 2 standard deviation, we used a linear constrained ordination (redundancy analysis, RDA). An ANOVA-like permutation test with 9999 permutations was performed to assess the statistical significance between species composition and environmental variables for each axis. The *vegan*

Table 2. Climatic and altitudinal variables for each physiographical sub-region of Venezuela used in the RDA analysis.

Sub-region	MA (m)	MDP (mm)	TR (°C)	ME (mm)
A2	50	8	8	500
B1	61	20	9	1000
B2	150	5	10	1050
B3	51	25	8	1100
B4	202	25	9	1200
B5	642	30	10	1250
C1	406	5	8	750
C2	348	25	9	1100
D1	1561	5	9	1300
D2	1791	25	8	900
D3	529	8	8	800
D4	1066	5	10	900
D5	598	10	8	800
D6	198	5	8	400
D7	855	65	11	1300

MA, mean altitude (in m); MDP, mean precipitation of the driest month (in mm); TR, temperature range, difference between the mean of maximum temperatures of every month and the mean of minimum temperatures of every month (in °C); ME, mean evapotranspiration (in mm).

R package [37] was used to perform an RDA analysis and a permutation test.

Results and Discussion

The test for the presence of significant boundaries between physiographical sub-regions revealed four OBUs in Venezuela (Table 3). From the species similarity matrix, both the matrix of significant similarities (see Table S1 in File S1) and the corresponding dendrogram were computed (Fig. 2). The test for significant boundaries (Table 3) revealed 11 chorotypes, separated either by weak or strong boundaries (see Table S2 in File S1, and Figs. 3 and 4).

In the ordination analysis of the species presence-absence table constrained by environmental variables, the four variables considered jointly explained 49.4% of the total variance (Fig. 5) and the permutation test revealed a significant association only for the first ($P < 0.001$) and second ($P = 0.002$) axes. The first axis explained 25.8% of the variance and had relatively high positive associations with ME, MDP and TR. The second axis explained 16.3% of the variance and was positively associated with MA and, to a lesser extent, with ME.

Biogeographical Regions

Our Q-mode analysis separated the Continental coast (OBU I) from the other sub-regions by a strong boundary ($DS > 0$ and GS $P < 0.001$; Table 3). The Continental coast corresponds entirely to a xerophytic zone, characterized by extremely dry lowlands with high temperatures throughout the year [38]. Redundancy analysis partially corroborated this fact, showing a high association of OBU I with low mean values of altitude and evapotranspiration (Fig. 5). This analysis did not reveal the low MDP as an important constraint, likely because other sub-regions not enclosed in OBU I had similar values for this variable (see Table 2). Only one marsupial species (*Marmosa xerophyla*) occurs in the Continental coast sub-region, because of its dependence on this dry habitat (see below).

A strong boundary ($DS > 0$ and GS $P < 0.01$; Table 3) separated OBU II from OBUs III and IV, which in turn were only separated by a very weak boundary ($DW > 0$ and GW $P < 0.05$; Table 3). In light of these results and excluding the particular xerophytic zone, Venezuela can be physiographically divided into two main areas that correspond to the territories south (OBU II) and north of the Orinoco River (OBUs III and IV; see Fig. 2). This pattern is concordant with the two main geological regions that are recognized, in general terms, for this country [3]: a southern region that includes Guayana, and a northern region that comprises other mountain ranges, hills and plains.

Specifically, OBU II, which includes the Amazon and Guayana Shield territories (Figure 2), was characterized by a wide TR and relatively high values of both MDP (see also [38]) and ME (Fig. 5). Operational Biogeographical Units III and IV jointly encompassed all the sub-regions located north of the Orinoco River, except A2. In particular, OBU III was composed of the Llanos and Cerro Copey sub-regions, which can be environmentally defined by a low mean altitude and evapotranspiration (Fig. 5). Nevertheless, these sub-regions are physiographically very different. Thus, whereas the Llanos is a vast grassland plain with a bimodal seasonal climate [38], Cerro Copey is a mountain of moderate elevation (960 m.a.s.l.) with a semiarid, or even arid climate, and surrounded by desert plains. In spite of the relatively low altitude of Cerro Copey, in this sub-region there is predominance of green forests and montane grasslands that receive the moisture provided by the trade winds; lower areas are dominated by dry and semideciduous forests. Therefore, the association between the Llanos and Cerro Copey obtained in our Q-mode analysis is not due to a true physiographical relationship, but likely because: i) the only two species found in Cerro Copey (*Caluromys trinitatis* and *Marmosa robinsoni*) also occur in the Llanos; ii) the Llanos shows a relatively low number of species; and iii) the similarity index used in our analysis takes into account shared absences and, although they are multiplied by shared presences, they may influence the association. From the other five species found in the Llanos, one (*Didelphis marsupialis*) has a wide distribution range in Venezuela,

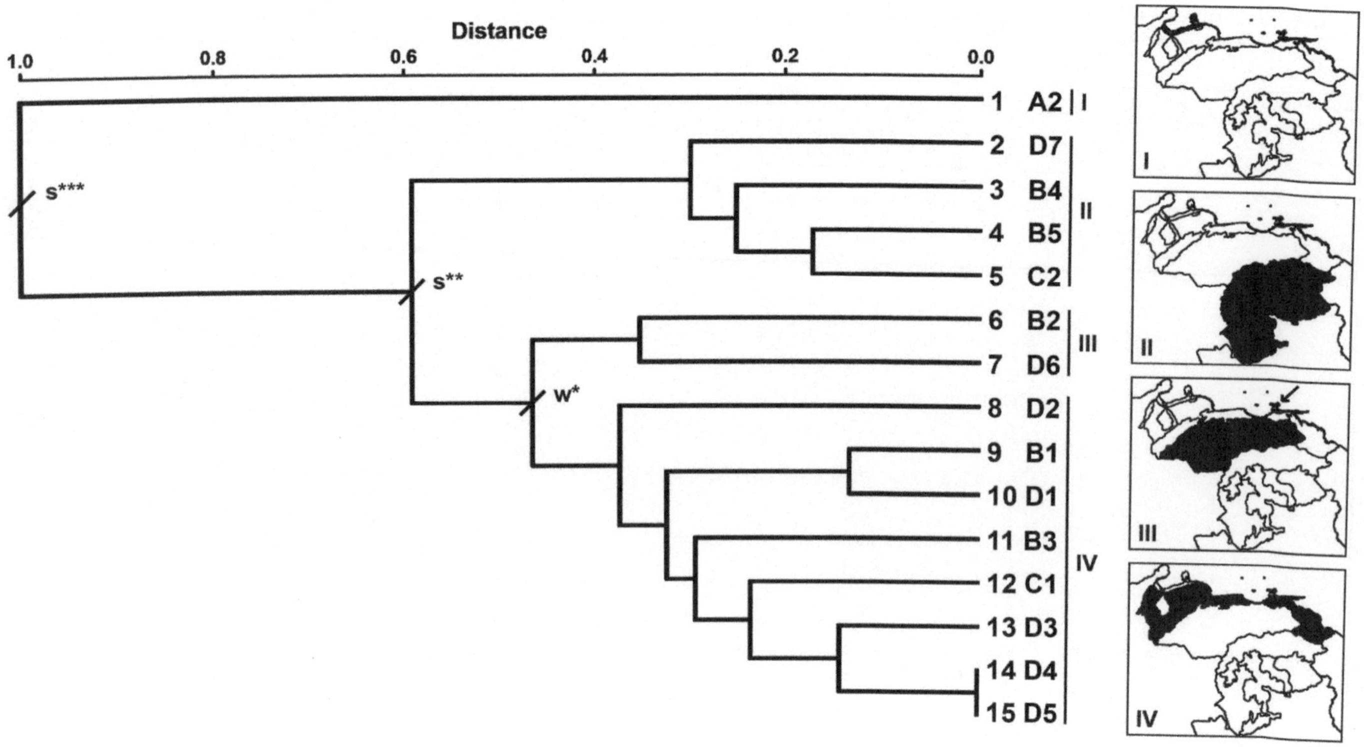

Figure 2. Classification dendrogram of the 15 physiographical sub-regions of Venezuela constructed according to the dissimilarity matrix of marsupial species presence-absence in each sub-region. I–IV, OBUs; w, weak boundary; s, strong boundary; ***$P < 0.001$; **$P < 0.01$; *$P < 0.05$. Maps show the physiographical sub-regions included in each OBU.

three (*Metachirus nudicaudatus*, *Philander mondolfii and Lutreolina crassicaudata*) have a limited and marginal distribution, and one (*Monodelphis* species A) is probably endemic of this sub-region (see below). The pattern obtained for the Llanos concerning the marsupial species composition is concordant with that reported for the non-volant eutherians in Venezuela, i.e. low specific richness, presence of species with a wide distribution and paucity of endemic species (see [10] for details).

Operational Biogeographical Unit IV also showed a very heterogeneous composition. It included the mountain sub-regions of northern Venezuela and two northern flood plains. None of the environmental variables considered was associated with this biogeographical unit. The inclusion in this OBU of such climatically and physiographically diverse sub-regions is likely the result of the wide distribution of many species throughout the north of Venezuela. In fact, most of these zones show a wide range of altitudes, with corresponding variability in temperature and rainfall. In particular, Serranía de Perijá, Andes (Cordillera de Mérida) and the Coastal Ranges show the highest altitudinal range (up to around 5000 m a.s.l.) in Venezuela. In concordance with our results, studies in other mammal groups support the biogeographical affinities between these sub-regions (e.g., [39–41]).

The Plain of the Orinoco River Delta, the boggy plain of the San Juan River and the Maracaibo depression are warm and wet areas that show high evapotranspiration mean values. In these sub-regions there are a relatively high number of species that are also present in other physiographical areas of OBU III. In particular, five of these taxa (*Didelphis marsupialis*, *Marmosa murina*, *Marmosa robinsoni*, *Metachirus nudicaudatus* and *Monodelphis palliolata*) occur in both the Maracaibo depression and the Orinoco River Delta in spite of their wide geographic separation. Nevertheless, previous studies revealed that this delta system and the sub-regions south of the Orinoco River show close biogeographical relationships when all mammalian species are considered; in fact, both areas share a relatively high number of species (see [9,10] for details). Further quantitative analyses using other kinds of OGUs and other environmental variables are needed to elucidate the causes of the particular pattern obtained here for this sub-region.

Chorotypes

The first four chorotypes (I–IV) comprised only one species and were separated from other groups of species by very strong boundaries (Figs. 3 and 4). None of the variables considered were clearly related these chorotypes (Fig. 5). In particular, chorotype I exclusively included *M. xerophila* and was significantly separated from the other chorotypes due to its very limited geographic distribution and strict habitat requirements (e.g., [42]). This species occurs only in xerophytic zones of the Continental coast sub-region (OBU I) that are characterized by meadows and thickets, and altitudes ranging from 0 to 100 m a.s.l.

Other species with a very restricted distribution is *Marmosops cracens*, the unique species of chorotype II. This taxon is an endemic marsupial of Venezuela that has only been recorded in one locality, in the hills of Falcón State (see [43]).

The single species of chorotype III, *Monodelphis* species A (*sensu* Pine and Handley [44]), occurs exclusively in savanna environments of the Llanos, especially in areas densely covered by grasses at elevations from 20 to 575 m a.s.l. [45]. Interestingly, the taxonomic status of the specimens from these environments attributed to genus *Monodelphis* is controversial (see [44] for details); they have been either assigned to a subspecies of *M. brevicaudata* (*M. brevicaudata orinoci*; e.g., [46]), reported as a full species (*M. orinoci*;

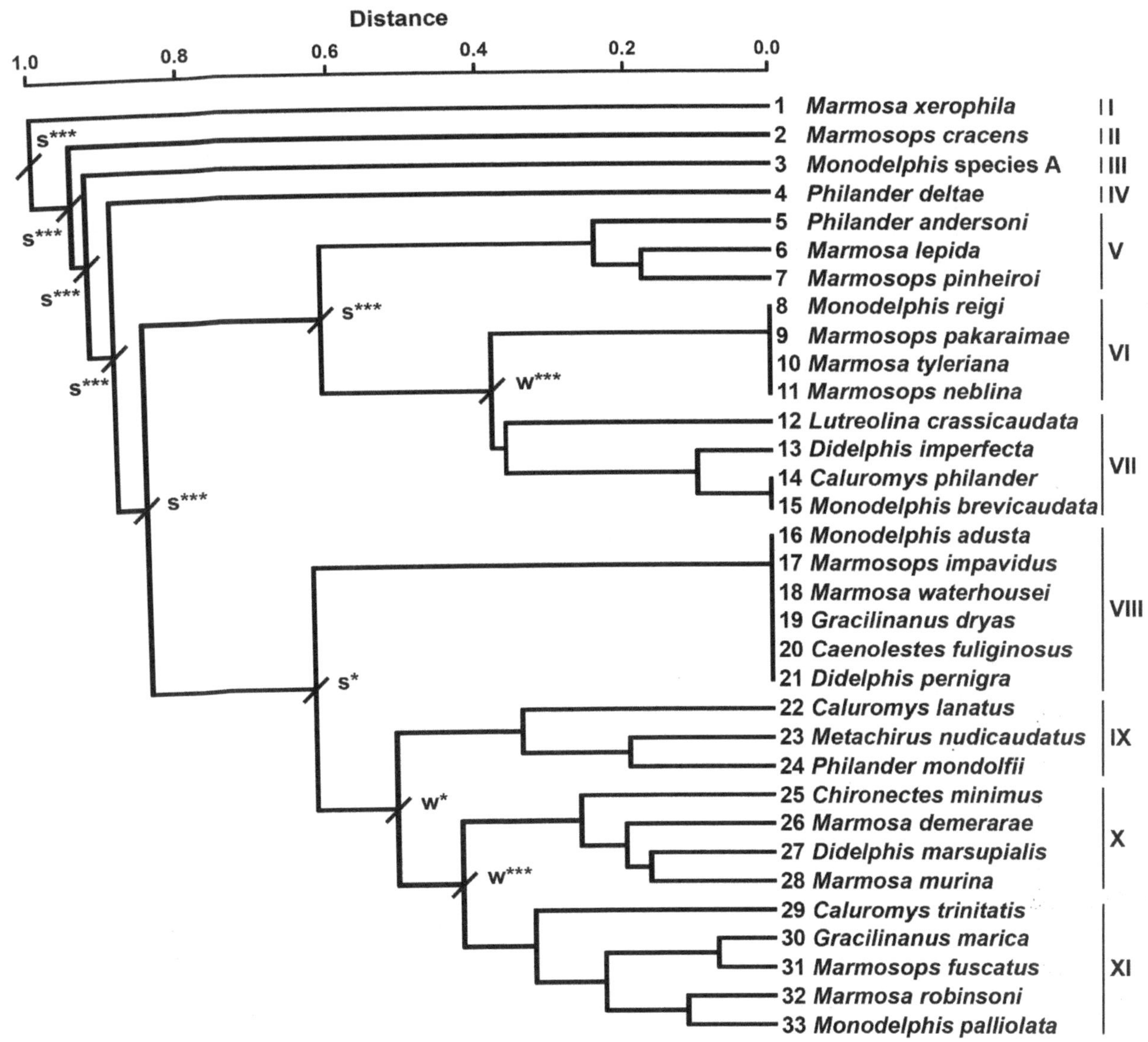

Figure 3. Classification dendrogram of the 33 marsupial species according to their dissimilarity matrix based on presence-absence in each of the 15 physiographical sub-regions of Venezuela. I–XI, chorotypes; w, weak boundary; s, strong boundary; ***$P < 0.001$; *$P < 0.05$.

e.g., [45,47]), or considered a taxonomic entity pending specific assignation (*Monodelphis* species A [44]). In this study we followed the latter option because it is based on the most recent and sound revision of this taxonomic problem. If this morphotype is finally recognized as a full species, it will constitute the only known endemic marsupial in the Llanos.

Philander deltae, the species of chorotype IV, has been found exclusively in the Orinoco River Delta and the boggy plain of the San Juan River sub-region. In light of the current information, it can be considered the only marsupial species endemic to this zone. The vegetation characteristics of this subregion are peculiar within the Guayana Shield, consisting basically of temporary or permanently flooded forests and palm groves. The forests are generally dense and highly mixed with palms [6]. Among non-volant vertebrates, besides *P. deltae*, only one amphibian (*Hyalinobatrachium mondolfii*; [48,49]), one reptile (*Anolis deltae*; [50]) and one mammal (*Dasyprocta guamara*; [9,51]) species have been reported as endemics of this sub-region. As for mammals in particular, the few surveys performed in the Orinoco River Delta determines the relatively poor knowledge of its biodiversity and consequently might be partially related to the scarce number of endemic and also of exclusive species found up to now in this zone (see [8–10]).

As for the chorotypes formed by more than one species, a strong boundary separated two main associations of species. In general terms, this boundary segregates the chorotypes formed by species from the Llanos and/or distributed south of the Orinoco River (chorotypes V–VII) from those composed of species that either occur exclusively north of this river or show a wide distribution throughout Venezuela (chorotypes VIII-XI; see Fig. 4).

A strong boundary separated chorotype V from chorotypes VI and VII (Fig. 3). The former comprised three species (*Marmosa lepida*, *Marmosops pinheiroi* and *Philander andersoni*), which occur south of the Guayana Shield foothill system. Although these species jointly cover a wide altitudinal range (*M. lepida*: 63 m a.s.l.; *M. pinheiroi*, 150–1374 m a.s.l.; *P. andersoni*, 0–180 m a.s.l.), the physiographic regions in which they occur are all characterized by a long rainy period throughout the year [38]. In fact, our RDA revealed that the relatively high MDP is associated with chorotype

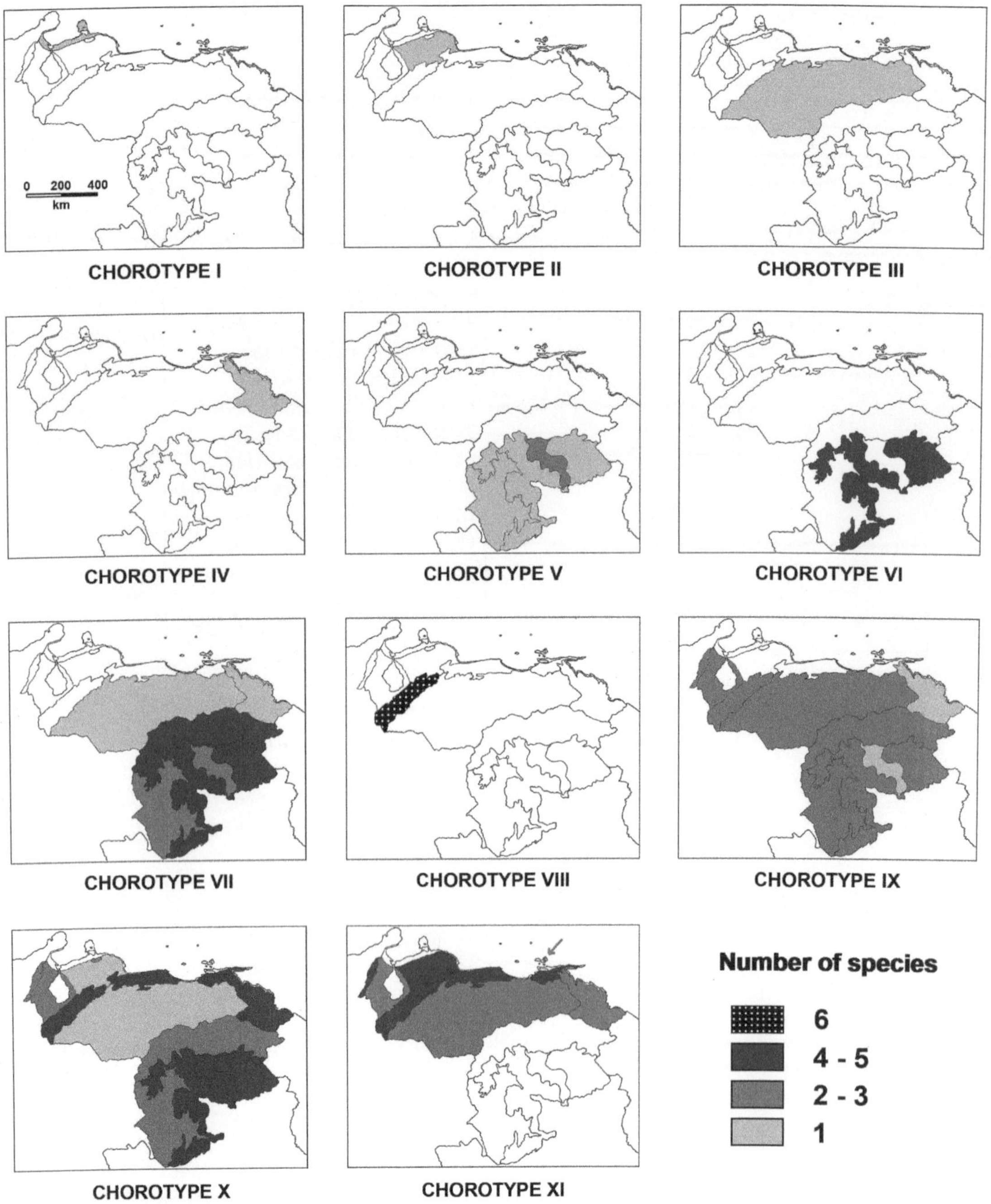

Figure 4. Distribution of the 11 chorotypes of marsupial species in Venezuela. The hatchings indicate the number of species from the chorotype in each sub-region.

V (Fig. 5). It is worth mentioning that the presence of *M. lepida* in Venezuela is based on a single and relatively recent record [52]; specifically, in the margin of the mouth of the Nichare River (north of the peneplain of the Caura and Paragua Rivers sub-region). In fact, the distribution of *M. lepida* in South America is known from a very low number of records, mainly corresponding to sites below 600 m a.s.l. around the western periphery of the Amazon basin [42].

A weak boundary accounted for the separation between chorotypes VI and VII (Fig. 3). The species included in chorotype VI are found in the Roraima formation, in the Guayana Massif sub-region. Most species of this chorotype occur in premontane and montane habitats in the tepuis (*Marmosops pakaraimae*, 800–1500 m a.s.l., [29]; *Marmosa tyleriana*, 1500–2200 m a.s.l.; *Monodelphis reigi*, 1300–1374 m a.s.l.), the characteristic flat-topped rock formations of this sub-region (Pantepui region [53]); only one species (*Marmosops neblina*) has also been found at lower altitudes in the Roraima formation (140–2000 m a.s.l.). High values of MDP, TR and ME accounted for the distribution of this chorotype (Fig. 5). In fact, the Guayana Massif sub-region is characterized by a very short dry season, a wide temperature range related to the

Table 3. Significant segregations between the physiographical sub-regions and the species on the dendrogram constructed using UPGMA and on the dissimilarity matrix of marsupial distributions.

	Groups set up by UPGMA			Weak boundary					Strong boundary		
	Group A	Group B	Distance	DW(A*A)	DW(B*B)	DW	GW	P	DS	GS	P
Regions	III	IV	0.464	−0.176	0.343	0.172	5.367	*	−0.116	0.089	n.s.
	II	III–IV	0.592	0.640	0.384	0.512	31.000	***	0.183	8.109	**
	I	II–IV	1.000	0.000	0.157	0.079	4.936	*	0.599	43.482	***
Chorotypes	VI	VII	0.385	0.707	0.416	0.562	16.228	***	0.000	0	n.s.
	X	XI	0.432	0.640	0.515	0.577	19.048	***	0.000	0	n.s.
	IX	X–XI	0.521	0.227	0.221	0.260	4.249	*	−0.071	0	n.s.
	V	VI–VII	0.615	0.252	−0.001	0.277	0.014	n.s.	0.085	17.030	***
	VIII	IX–XI	0.633	0.707	0.207	0.457	37.544	***	0.082	5.681	*
	V–VII	VIII–XI	0.857	0.130	0.147	0.139	64.568	***	0.488	160.889	***
	IV	V–XI	0.896	0.000	−0.220	−0.110	3.997	n.s.	0.178	12.738	***
	III	IV–XI	0.928	0.000	−0.247	−0.123	3.778	n.s.	0.235	21.328	***

GW and GS indicate weak segregation and strong segregation between the groups, respectively. ***$P < 0.001$; **$P < 0.01$; *$P < 0.05$. DW(A*A) and DW(B*B) quantify the internal homogeneity of each group. DW and DS measure the value of each boundary.

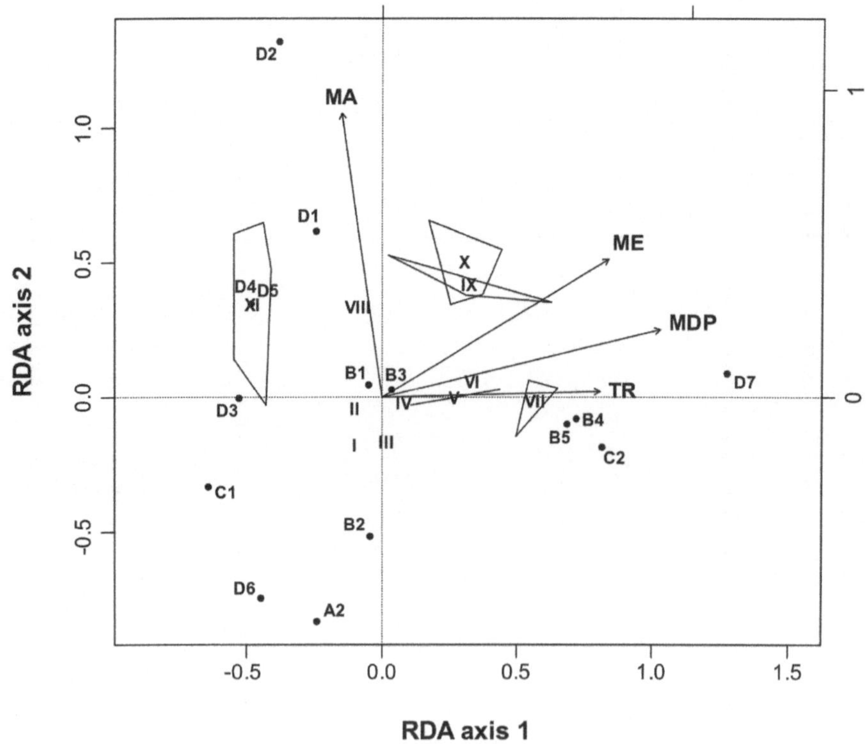

Figure 5. Triplot for the redundancy analysis (RDA) of presence-absence table and environmental variables. MA, mean altitude; MDP, mean precipitation of the driest month; TR, temperature range; ME, mean evapotranspiration. Chorotypes (I–XI) with convex hulls for their species and physiographical sub-regions (OBU I: A2; OBU II: B4, B5, C2, D7; OBU III: B2, B6; OBU IV: B1, B3, C1, D1, D2, D3, D4, D5) are plotted. Only the first and second RDA axes, which explain 25.8% and 16.3% of the total variation, respectively, are shown. Species and regions are scaled symmetrically by the square root of eigenvalues.

wide altitudinal range of the tepuis and abundant precipitation throughout the year.

The marsupial species richness of the Guayana Massif is clearly higher than that of the other sub-regions of Venezuela, with the exception of the Andes (see Table 1). Three out of the four species of chorotype VI (*M. reigi*, *M. tyleriana* and *M. pakaraimae*) are endemic to this southern zone. This is concordant with results obtained in other vertebrate groups, such as amphibians, reptiles, aves and eutherians, which reveal that the Guayana Massif constitutes an important centre of endemisms [10,48,49,54–57]. The fact that the three endemic marsupial species of this physiographical sub-region have been found exclusively at high altitudes in the tepuis seems congruent with the proposal by Señaris and Rojas-Runjaic [49] to recognize the Guayanan highlands as a separate bioregion ("Pantepui" *sensu* these authors) from the other areas of the Guayana Massif. High MDP and ME, and wide TR were related to this chorotype (Fig. 5). In fact, the tepuis region is climatically characterized by a very humid climate and a very short dry season.

Chorotype VII included *Caluromys philander*, *Monodelphis brevicaudata*, *Didelphis imperfect* and *L. crassicaudata* (Fig. 3). The same environmental variables reported for chorotype VI were also related to the distribution of chorotype VII. The mean altitude did not account for its distribution since their species show a relatively wide altitudinal range although, in general, most of them occur below 1000 m a.s.l. (*C. philander*, 0–600 m a.s.l.; *L. crassicaudata*, 0–940 m a.s.l.; *M. brevicaudata*, 82–1104 m a.s.l.); only *D. imperfecta* markedly surpasses this altitude (0–2550 m a.s.l.). The fact that high values of MDP and ME are associated with the distribution of this chorotype (Fig. 5) evidences the wet characteristics of the physiographical sub-regions where its species occur. *Caluromys philander*, *D.imperfecta* and *M. brevicaudata* have been found in the physiographical sub-regions south of the Orinoco River. *Lutreolina crassicaudata* has only been reported in four localities, each one belonging to a different physiographical sub-region (Llanos, Plain of the Orinoco River Delta and the boggy plain of the San Juan River, Guayana Shield foothill system, and Guayana Massif). In fact, *L. crassicaudata* is uncommon not only in Venezuela but also in the rest of northern South America, occurring exclusively in grasslands closely associated with aquatic habitats, from flooded areas to small streams [24,58].

As mentioned above, the other chorotypes included, in general terms, taxa found either exclusively north of the Orinoco River (chorotypes VIII and XI) or widely distributed throughout Venezuela (chorotypes IX and X; Fig. 4). Chorotype VIII was separated by a strong boundary from chorotypes IX–XI and exclusively included Andean species (Fig. 3). Redundancy analysis revealed MA as the only explanatory environmental variable for the distribution of this chorotype (Fig. 5). Specifically, all its species occur between 1000 and 4000 m a.s.l., the latter being the highest altitude reached by a marsupial species (*D. pernigra*) in Venezuela. In this wide range of altitudes, there is high physiographical and climatic heterogeneity, from very warm xerophytic zones (for example, the middle basin of the Chama River [59]) to extremely cold páramo areas at the highest altitudes. This heterogeneity leads, in turn, to diversity in vegetation types (see [60] for details) and to local differences in flora, kinds of soil and hydrology [61].

Regarding the species of chorotype VIII, both *Caenolestes fuliginosus* (2200–2460 m a.s.l.) and *M. impavidus* (2250–2460 m a.s.l.) have been found exclusively in the Páramo Tamá (Estado

Táchira), West of the Táchira depression and, therefore, not reaching the Cordillera de Mérida. As for the other species of this chorotype, *Didelphis pernigra* (2150–4000 m a.s.l.), *Gracilinanus dryas* (2210–2410 m a.s.l.) and *Marmosa waterhousei*, which was recently reported in Venezuela, (1000–1200 m a.s.l. [62]), have only been detected in the Cordillera de Mérida, whereas *Monodelphis adusta* (1000–1100 m a.s.l.) occurs in this mountain range and in the west to the Táchira depression. Of all the representatives of this chorotype, only *G. dryas* is endemic to the Venezuelan Andes.

Interestingly, 18 out of the 33 marsupial species studied here were found in the Andes sub-region. When other non-marsupial vertebrate groups are considered, such as amphibians, reptiles [50,54,63] or eutherians [10,40], it is remarkable that this sub-region not only shows a relatively high number of species but also of endemics. The wide altitudinal range and high ecological complexity of the Venezuelan Andes are likely related to this fact [10].

A weak boundary separated chorotypes IX from X and XI (Fig. 3). The former gathered species that globally occur in all physiographical sub-regions, except the Continental coast, Lara-Falcón hill system, Cerro Copey, and the Central and Eastern Coastal Ranges (Fig. 4). High ME and MA environmentally define this group (Fig. 5). The exclusive lack of all species of this chorotype in the xerophytic Continental coast and the surrounding areas of Cerro Copey, two particularly dry sub-regions, can probably explain the implication of ME in the climatic characterization of this chorotype. Species of chorotype IX occur in wide altitudinal ranges (*Caluromys lanatus*: 100–1130 m a.s.l.; *M. nudicaudatus*: 25–2225 m a.s.l.; *P. mondolfii*: 0–1550 m a.s.l.) and are mainly associated with forest environments.

Chorotype X was separated from chorotype XI by a weak boundary (Fig. 3) and its distribution was positively associated with all environmental variables considered (Fig. 5). The joint distribution of the species of chorotype X covers all physiographical sub-regions except the Continental coast and Cerro Copey (Fig. 4). Since a large number of species of this chorotype occur in mountain areas, with high evapotranspiration mean values, this variable shows a high weight in the redundancy analysis (Fig. 5). These species have been found in a wide range of altitudes (*Chironectes minimus*: 0–2100 m a.s.l.; *D. marsupialis*: 0–2550 m a.s.l.; *Marmosa demerarae*: 40–1500 m a.s.l.; *M. murina*: 0–1347 m a.s.l.). Although *D. marsupialis* is a generalist species, this is not the case of the other three taxa. Thus, *M. demerarae* and *M. murina* live in tropical humid forests and are also frequent in second growth and disturbed habitats, and *Chironectes minimus* is a semiaquatic species with a distribution closely tied to tropical forest streams and lakes [64].

Chorotype XI included species with a wide distribution throughout northern and central Venezuela, extending as a whole from the Serranía de Perijá sub-region in the west to the Orinoco River Delta in the east (Fig. 4). Low MDP, TR and ME characterize climatically the distribution of this chorotype (Fig. 5). As shown in the RDA analyses, the variability in mean altitude for this chorotype is very large. Moreover, its species cover a wide altitudinal range, from the lowlands of the Llanos to the high altitudes in the Andes, coast mountains and Serranía de San Luis. Specifically, *M. robinsoni* (0–1160 m a.s.l.) shows a wide distribution north to the Orinoco River, from the Llanos, on the left shore of the Orinoco River, to the northernmost sub-regions of Central and Eastern Coastal Ranges, Cerro Santa Ana, and Cerro Copey. In our dendrogram (Fig. 3), *M. robinsoni* and *C. trinitatis* appear in the same chorotype due to their similar general distribution and, particularly, to their presence on Margarita Island, where they are the only marsupial species reported up to now. *Gracilinanus marica*, *Marmosops fuscatus* and *M. palliolata* show a very similar distribution in Venezuela. The former is a montane species (0–1750 m a.s.l.) found in the Cordillera de Mérida, Coastal Ranges, Serranía de San Luis and the Llanos of Monagas state. It has been detected in savanna edge habitats and/or in deciduous, humid or cloud forests [65,66]. The other two species occur in a relatively wide range of altitudes in mountain areas or within their limits (*M. fuscatus*, 0–2232 m a.s.l.; and *M. palliolata*, 0–1500 m a.s.l.).

Concluding Remarks

This study constitutes the first approach to detect common distribution patterns of marsupials in Venezuela by means of quantitative methods. Specifically, we identified significant biotic regions and chorotypes using physiographical regions as OGUs. In addition, the use of limiting macroclimatic and altitudinal variables allowed us to define OBUs and chorotypes environmentally. It is worth bearing in mind that since the configuration and limits of OBUs and the composition of chorotypes depend on the type of OGUs considered (e.g., [67,68]), the method applied in the present study constitutes only a specific way to identify biogeographical patterns. Consequently, results have been interpreted exclusively from a physiographical perspective and only very general considerations can be outlined under a historical point of view. Specifically, information on the paleogeographic evolution of the proto-Orinoco and Orinoco rivers from the Paleogene to the present [1], the uplift of the Andes [2], and the recent results based on a time-calibrated molecular phylogeny of the opossums in South America [69] clearly indicate that the biogeographical pattern presented here is posterior to the Late Miocene. Additionally, molecular analyses [69] also suggest that the diversification of opossums in South America (about 17 Ma BP) occurred entirely within moist-forest biomes, with a subsequent invasion of dry-forest areas occurring during or after the Late Miocene (10 Ma, approximately). Undetermined processes during the later Tertiary and Quaternary history of marsupials led to the current distribution of the species of this mammalian group in South America.

In order to understand the current geographic distribution of the marsupial species in this part of South America, additional chorological information and analyses using other OGUs are needed. In particular, future surveys and taxonomic studies should define biotic boundaries, chorotypes and their characterization more accurately. Likewise, further biogeographical analyses based on historical and ecological variables are essential to complement the distribution patterns presented here.

Supporting Information

Code S1 R code used to perform the classification of physiographical regions and species from the presence-absence matrix. (ZIP-archived directory containing an R script calling functions and data).

File S1 Appendix S1, Nomenclature authorities for all species names mentioned in this paper. Appendix S2, Sources of information on the distribution of the marsupial species in Venezuela. Table S1, Matrix of significant similarities for the physiographical regions in Venezuela. Table S2, Matrix of significant similarities for the marsupial species from Venezuela.

Acknowledgments

We are grateful to the following curators or collection managers for allowing access to information on the marsupial species in their collections: F. Bisbal and J. Sánchez (Estación Biológica de Rancho Grande), M. Salazar and C. Ferreira (Museo de Biología de la Universidad Central de Venezuela), B. Rivas and D. Lew (Museo de Historia Natural La Salle), and P. Soriano (Colección de Vertebrados de la Universidad de Los Andes).

Author Contributions

Conceived and designed the experiments: JV RPH. Analyzed the data: GB MSF. Contributed reagents/materials/analysis tools: JV RPH GB MSF. Wrote the paper: JV RPH GB.

References

1. Díaz de Gamero ML (1996) The changing course of the Orinoco River during the Neogene: a review. Palaeogeogr Palaeoclimatol Palaeoecol 123: 385–402.
2. Hoorn C, Wesselingh P, ter Steege H, Bermudez MA, Mora A, et al. (2010) Amazonia through time: Andean uplift, climate change, landscape evolution, and biodiveristy. Science 330: 927–931.
3. Urbani F (2011) Un esbozo de la geología de Venezuela. BioLlania Edición Esp 10: 27–32.
4. Anderson RP, Gutiérrez EE, Ochoa J, García FJ, Aguilera M (2012) Faunal nestedness and species-area relationship for small non-volant mammals in "sky island" of northern Venezuela. Stud Neotrop Fauna Environ 47: 157–170.
5. Huber O (1997) Ambientes fisiográficos y vegetales de Venezuela. In: La Marca E, editor. Vertebrados actuales y fósiles de VenezuelaMérida Serie Catálogo Zoológico de Venezuela pp. 279–291.
6. Huber O (2007) Los grandes paisajes vegetales. In: Cunill Grau P, editor. GeoVenezuela 2. Caracas: Fundación Polar. pp. 538–575.
7. Mittermeier R, Goetsch-Mittermeier C, Robles-Gil P (1997) Megadiversidad: los países biológicamente más ricos del mundo. México: CEMEX. 501 p.
8. Sánchez J, Lew D (2012) Lista actualizada y comentada de los mamíferos de Venezuela. Mem Fund La Salle Cien Nat 173–174: 173–238.
9. Linares OJ (1998) Mamíferos de Venezuela. Caracas: Sociedad Conservacionista Audubon de Venezuela. 691 p.
10. Ochoa J, Aguilera M (2003) Mamíferos. In: Aguilera M, Azócar A, González Jiménez E, editors. Biodiversidad en Venezuela. Caracas: Fundación Polar, Ministerio de Ciencia y Tecnología, Fondo Nacional de Ciencia, Tecnología e Innovación (FONACIT)pp. 650–672.
11. Aguilera Meneses M, Pérez-Hernández R (2007) Distribución geográfica de la fauna venezolana. In: Cunill Grau P, editors. GeoVenezuela 2Caracas: Fundación Polar 576–615.
12. Aguilera Meneses M, Pérez-Hernández R (2011) Mapa sobre las Unidades Zoogeográficas de la República Bolivariana de Venezuela. Caracas: Fundación Polar. Map A.MN013.
13. Birks HJB (1987) Recent methodological developments in quantitative descriptive biogeography. Ann Zool Fennici 24: 165–178.
14. McCoy ED, Bell SS, Walters K (1986) Identifying biotic boundaries along environmental gradients. Ecology 67: 749–759.
15. Crick RE (1980) Integration of paleobiogeography and paleogeography: evidence from Arenigian nautiloid biogeography. J Paleontol 54: 1218–1236.
16. Real R, Guerrero JC, Ramírez JM (1992) Identificación de fronteras bióticas significativas para los anfibios en la cuenca hidrográfica del Sur de España. Doñana Acta Vert 19: 53–70.
17. Márquez AL, Real R, Vargas JM, Salvo AE (1997) On identifying common distribution patterns and their causal factors: a probabilistic method applied to pteridophytes in the Iberian Peninsula. J Biogeogr 24: 613–631.
18. Márquez AL, Real R, Vargas JM (2001) Methods for comparison of biotic regionalizations: the case of pteridophytes in the Iberian Peninsula. Ecography 24: 659–670.
19. Sans-Fuentes MA, Ventura J (2000) Distribution patterns of the small mammals (Insectivora and Rodentia) in a transitional zone between the Eurosiberian and the Mediterranean regions. J Biogeogr 27: 755–764.
20. Baez JC, Real R, Vargas JM, Flores-Moya A (2004) A biogeographical analysis of the genera *Audouinella* (Rhodophyta), *Cystoseira* (Phaeophyceae) and *Cladophora* (Chlorophyta) in the western Mediterranean Sea and Adriatic Sea. Phycologia 43: 404–415.
21. Olivero J, Márquez AL, Real R (2013) Integrating fuzzy logistic and statistics to improve the reliable delimitation of biogeographic regions and transition zones. Syst Biol 62: 1–21.
22. Crovello TJ (1981) Quantitative biogeography: an overview. Taxon 30: 563–575.
23. Pérez-Hernández R (1989) Distribution of the family Didelphidae (Mammalia-Marsupialia) in Venezuela. In: Redford KH, Eisenberg JF, editors. Advances in Neotropical mammalogyGainesville: The SandHill Crane Press, Inc pp. 363–410.
24. Pérez-Hernández R, Soriano P, Lew D (1994) Marsupiales de Venezuela. Caracas: Lagoven. 76 p.
25. Creighton GK, Gardner AL (2007a) Genus *Gracilinanus*. In: Gardner AL, editor. Mammals of South America.Chicago and LondonThe University of Chicago Press pp. 43–50.
26. Linares OJ (1997) New locality records of mouse opossums from Venezuela (Marsupialia: Didelphidae). Mammalia 61: 255–259
27. Bisbal FJ, Naveda Rodríguez A (2010) Mamíferos de la cuenca del río Guárico, Estados Aragua, Carabobo y Guárico, Venezuela. Mem Fund La Salle Cien Nat 172: 69–89.
28. Voss RS, Lunde DP, Simmons NB (2001) Mammals of Paracou, Frenc Guiana: A Neotropical lowland rainforest. Part 2: nonvolant species. Bull Am Mus Nat Hist 265: 1–236.
29. Voss RS, Lim BK, Díaz-Nieto JF, Jansa S (2013) A new species of *Marmosops* (Marsupialia: Didelphidae) from the Pakaraima highlands of Guyana, with remarks on the origin of the endemic Pantepui mammals fauna. Am Mus Novit 3778: 1–27.
30. R Development Core Team (2012) R: A language and environment for statistical computing. Vienna: R Foundation for Statistical Computing.
31. Baroni-Urbani C, Buser MW (1976) Similarity of binary data. Syst Zool 25: 251–259.
32. Jurasinski G (2012) simba: A Collection of functions for similarity analysis of vegetation data. R package version 0.3–4. Available: http://CRAN.R-project.org/package = simba. Accessed 2013 Jan 09.
33. Doadrio I (1988) Delimitation of areas in the Iberian Peninsula on the basis of freshwater fishes. Bonn Zool Beitr 39: 113–128.
34. Álvarez Bernal F (1983) Atlas climatológico de Venezuela. Caracas: Imprenta Universitaria de la Universidad Central de Venezuela. 132 p.
35. Legórburu MJ (1995) Balance ambiental de Venezuela 1994–1995. Caracas: Ministerio del Ambiente y de los Recursos Naturales Renovables. 77 p.
36. Hearn Jr P, Hare T, Schruben P, Sherrill D, LaMar C, et al. (2000) Global GIS Database: Digital Atlas of Central and South America, version 5.6. U.S.Geological Survey and Environmental Systems Research Institute, Inc
37. Oksanen J, Blanchet FG, Kindt R, Legendre P, Minchin PR, et al. (2012) vegan: Community Ecology Package. R package version 2.0–3. Available: http://CRAN.R-project.org/package = vegan. Accessed 2013 Jan 29.
38. Walter H, Medina E (1971) Caracterización climática de Venezuela sobre la base de climadiagramas de estaciones particulares. Bol Soc Venez Cien Nat 29: 211–240.
39. Duarte MA, Viloria AL (1992) Nuevo hallazgo de *Cryptotis thomasi* (Merriam, 1897) (Mammalia: Insectivora) en la Sierra de Perijá, Noreste de Colombia. Acta Cient Venez 43: 240–242.
40. Soriano PJ, Díaz de Pascual A, Ochoa J, Aguilera M (1999) Biogeographic analysis of the mammals communities in the Venezuelan Andes. Interciencia 24: 17–25.
41. Woodman N, Péfaur JE (2007) Magnorder Epitheria. In: Gardner AL, editor. Mammals of South AmericaChicago and LondonThe University Chicago Press 177187
42. Creighton GK, Gardner AL (2007b) Genus *Marmosa*. In: Gardner AL, editor. Mammals of South AmericaChicago and LondonThe University of Chicago Press pp. 51–61.
43. Handley Jr CO, Gordon LK (1979) New species of mammals from northern South America: Mouse opossums, genus *Marmosa* Gray. In: Eisenberg JF, editor. Vertebrate Ecology in Northern NeotropicsWashington DCSmithsonian Institution Presspp. 65–72.
44. Pine RH, Handley Jr CO (2007) Genus *Monodelphis*. In: Gardner AL, editor. Mammals of South AmericaChicago and LondonThe University Chicago Press 82107
45. Ventura J, Pérez-Hernández R, López-Fuster MJ (1998) A morphometric assessment of the systematics of *Monodelphis brevicaudata* (Didelphimorphia: Didelphidae) in Venezuela. J Mammal 79: 104–117.
46. Gardner AL (1993) Order Didelphimorphia. In: Wilson DE, Reeder DM, editors. Mammal species of the WorldWashington DCThe Smithsonian Institution Presspp. 15–24.
47. Reig OA, Gardner AL, Bianchi NO, Patton JL (1977) The chromosomes of the Didelphidae (Marsupialia) and their evolutionary significance. Biol J Linn Soc Lond 9: 191–216.
48. Barrio Amorós CL (2004) Amphibians of Venezuela systematic list, distribution and references, an update. Rev Ecol Latinoam 9: 1–48.
49. Señaris JC, Rojas-Runjaic F (2009) Distribución y biogeografía. In: Molina C, Señaris JC, Lampo M, Rial A, editors. Anfibios de Venezuela. Estado del conocimiento y recomendaciones para su conservación.CaracasConservation International, Instituto de Zoología y Ecología Tropical, Fundación La Salle de Ciencias Naturales, IVIC, Gold Reserve Inc.pp. 41–47.
50. Rivas GA, Molina CR, Ugueto GN, Barros TR, Barrio-Amorós CL, et al. (2012) Reptiles of Venezuela: an updated and commented checklist. Zootaxa 3211: 1–64.

51. Ojasti J (1972) Revisión preliminar de los picures o agutíes de Venezuela (Rodentia, Dasyproctidae). Mem Soc Cien Nat La Salle 93: 159–204.
52. Ochoa J, García JF, Caura S, Sánchez J (2008) Mamíferos de la cuenca del río Caura, Venezuela: listado taxonómico y distribución conocida. Mem Fund La Salle Cien Nat 170: 5–80.
53. Mayr E, Phelps Jr WH (1967) The origin of the bird fauna of the south Venezuelan highlands. Bull Am Mus Nat Hist 136: 269–328.
54. Péfaur JE, Rivero JA (2000) Distribution, species-richness, endemism, and conservation of Venezuelan amphibians and reptiles. Amphib Reptile Conserv 2: 42–70.
55. La Marca E (2003) Anfibios. In: Aguilera M, Azócar A, González Jiménez E, editors. Biodiversidad en VenezuelaCaracas: Fundación Polar, Ministerio de Ciencia y Tecnología, Fondo Nacional de Ciencia, Tecnología e Innovación (FONACIT)pp. 582–595.
56. Lentino M (2003) Aves. In: Aguilera M, Azócar A, González Jiménez E, editors. Biodiversidad en VenezuelaCaracas Fundación Polar, Ministerio de Ciencia y Tecnología, Fondo Nacional de Ciencia, Tecnología e Innovación (FONACIT)pp. 610–648.
57. Gardner AL, Creighton K (2007) Genus *Marmosops*. In: Gardner AL, editor. Mammals of South America. Chicago and LondonThe University of Chicago Press. pp. 61–74.
58. Emmons LH, Feer F (1997) Neotropical rainforest mammals: A field guide, 2nd edition. Chicago: The University Chicago Press. 307 p.
59. Soriano PJ, Ruiz A (2003) Arbustales xerófilos. In: Aguilera M, Azócar A, González Jiménez E, editors. Biodiversidad en VenezuelaCaracasFundación Polar, Ministerio de Ciencia y Tecnología, Fondo Nacional de Ciencia, Tecnología e Innovación (FONACITpp. 696–715.
60. Azocar A, Fariñas M (2003) Páramos. In: Aguilera M, Azócar A, González Jiménez E, editors. Biodiversidad en VenezuelaCaracasFundación Polar, Ministerio de Ciencia y Tecnología, Fondo Nacional de Ciencia, Tecnología e Innovación (FONACIT)pp. 716–733.
61. van der Hammen T, Cleef AM (1986) Development of the high Andean páramo flora and vegetation. In: Vuilleumier F, Monasterio M, editors. High altitude tropical biogeographyOxfordOxford University Presspp. 153–201.
62. Gutiérrez EE, Soriano P, Rossi RV, Murillo JJ, Ochoa J, et al. (2011) Occurrence of *Marmosa waterhousei* in the Venezuelan Andes, with comments on its biogeographic significance. Mammalia 75: 381–386.
63. Barrio Amorós CL (1998) Sistemática y biogeografía de los anfibios (Amphibia) de Venezuela. Acta Biol Venez 18: 1–93.
64. Stein BR, Patton JL (2007) Genus *Chironectes*. In: Gardner AL, editor. Mammals of South AmericaChicago and LondonThe University of Chicago Presspp 14–17.
65. Handley Jr CO (1976) Mammals of the Smithsonian Venezuelan Project. Brigham Young University Science Bulletin 20: 1–89.
66. Hershkovitz P (1992) The South American Gracile Mouse Opossum, genus *Gracilinanus* Gardner and Creighton, 1989 (Marmosidae, Marsupialia): A taxonomic review with notes on general morphology and relationships. Fieldiana Zool 70: 1–56.
67. Palomo LJ, Vargas JM, Jiménez-Gómez MP (1994) Distribution patterns in Iberian Peninsula Rodents. Pol Ecol Stud 20: 497–502.
68. Real R, Márquez AL, Guerrero JC, Vargas JM, Palomo LJ (1996) Modelos de distribución de los insectívoros de la Península Ibérica. Doñana Acta Vert 23: 123–142.
69. Jansa SA, Barker FK, Voss RS (2013) The early diversification history of didelphid marsupials: a window into South America "Splendid Isolation". Evolution. doi: 10.1111/evo.12290.

High Temperature and Salinity Enhance Soil Nitrogen Mineralization in a Tidal Freshwater Marsh

Haifeng Gao[1], Junhong Bai[1]*, Xinhua He[2], Qingqing Zhao[1], Qiongqiong Lu[1], Junjing Wang[1]

1 State Key Laboratory of Water Environment Simulation, School of Environment, Beijing Normal University, Beijing, China, 2 NSW Department of Primary Industries, West Pennant Hills, New South Wales, Australia

Abstract

Soil nitrogen (N) mineralization in wetlands is sensitive to various environmental factors. To compare the effects of salinity and temperature on N mineralization, wetland soils from a tidal freshwater marsh locating in the Yellow River Delta was incubated over a 48-d anaerobic incubation period under four salinity concentrations (0, 10, 20 and 35‰) and four temperature levels (10, 20, 30 and 40°C). The results suggested that accumulated ammonium nitrogen (NH_4^+-N) increased with increasing incubation time under all salinity concentrations. Higher temperatures and salinities significantly enhanced soil N mineralization except for a short-term (≈10 days) inhibiting effect found under 35‰ salinity. The incubation time, temperature, salinity and their interactions exhibited significant effects on N mineralization ($P<0.001$) except the interactive effect of salinity and temperature ($P>0.05$), while temperature exhibited the greatest effect ($P<0.001$). Meanwhile, N mineralization processes were simulated using both an effective accumulated temperature model and a one-pool model. Both models fit well with the simulation of soil N mineralization process in the coastal freshwater wetlands under a range of 30 to 40°C ($R^2=0.88$–0.99, $P<0.01$). Our results indicated that an enhanced NH_4^+-N release with increasing temperature and salinity deriving from the projected global warming could have profound effects on nutrient cycling in coastal wetland ecosystems.

Editor: Fei-Hai Yu, Beijing Forestry University, China

Funding: This work was financially supported by the National Science Foundation for Innovative Research Group (51121003), the National Basic Research Program Cover Letter (2013CB430406), National Natural Science Foundation of China (51179006), the Program for New Century Excellent Talents in University (NECT-10-0235) and the Fok Ying Tung Foundation (132009). The funders had no role in study design, data collection and analysis, decision to publish, or preparation of the manuscript.

Competing Interests: The authors have declared that no competing interests exist.

* E-mail: junhongbai@163.com

Introduction

Nitrogen (N) is the most limiting nutrient in flooded soils [1–3], especially in coastal wetland ecosystems [4,5]. Studies have shown that N availability greatly affects plant growth, diversity and possible successional changes [6–8]. Organic N transformed into inorganic N via N mineralization primarily determines the soil N availability [2,9], which is also closely related to greenhouse gas emissions (i.e., N_2O, NO, and other forms). Therefore, a better understanding of N mineralization processes is important for improving the sustainability of N limiting ecosystems while reducing the risk of atmospheric pollution [10–12].

Nitrogen mineralization in wetlands is sensitive to various environmental factors, including air temperature, water tables, flooding periods and soil properties [3,6,13–15]. Numerous studies have demonstrated that soil temperature is a primary environmental factor that influences N mineralization processes in wetland soils [2,3,16]. For instance, Noe et al. [17] reported that N mineralization fluxes and turnover rates in continuously freshwater tidal forests to salt impacted tidal forest to oligohaline marsh along the Waccamaw river and Savannah river in the US were generally lower in March and December compared to June and September due to differences in microbial activity as affected by seasonal variation of soil temperature. Gilliam et al. [18] suggested that mild freezing might inhibit N mineralizing microbes and stimulate N immobilizing groups. However, Zhang et al. [19] demonstrated that mild freezing had few effects on N mineralization, while Gao et al. [2] observed higher N mineralization rates in winter than in summer in a seasonal flooding wetland. Therefore, a better understanding of how variation of air temperature could affect N mineralization is urgently needed since the potential increase of the earth's surface temperature could be 1.8–4.0°C at the end of this century [20].

Salinity is another important factor that influences N mineralization in coastal wetlands [2,6,21]. Specifically, alluvial soils are typically subjected to seepage or flooding during high tides and coastal storms. Moreover, the poor drainage from high groundwater tables contributes to salt accumulation in surface soils [2,21]. However, studies have shown conflicting effects (i.e., inhibiting or improving) of salinity on nitrogen mineralization in wetlands [6,17,21,22]. For example, Pathak and Rao [23] suggested that N mineralization decreased with increasing salinity and was biologically mediated at higher salinity levels. Gao et al. [2] found a slight increase in the net N mineralization rate in tidal salt marshes after freshwater input. However, Chen and Twilley [22] reported that relative N mineralization of mangrove soils decreased exponentially with increasing distance from the mouth of the Shark River estuary in the USA or decreasing salinity. More recently, Noe et al. [17] demonstrated that salinification increased net N mineralization fluxes in tidal freshwater wetland soils along

Table 1. Physico-chemical properties of the tested soil.

	Average	SD
Moisture (%)	21.50	0.43
Bulk density (g cm^{-3})	1.79	0.05
Silt+Sand (%)	98.19	1.10
Clay (%)	1.79	1.10
Salinity (‰)	0.53	0.21
pH value	8.38	0.16
SOM (g kg^{-1})	6.47	1.92
Total N (g kg^{-1})	0.38	0.05
Total P (g kg^{-1})	0.65	0.02
C:N ratio	9.88	1.48
NH_4^+-N (mg kg^{-1})	1.46	0.57
NO_3^--N (mg kg^{-1})	1.82	0.63

SD: Standard deviation.

the floodplains of both the Waccamaw and Savannah rivers in the USA. However, these studies have primarily relied on sites positioned along salinity gradients. This configuration may not be sufficient to identify the effects of salinity on N mineralization due to the different environmental conditions at these sites.

Studies on the salinification effects on nutrient biogeochemistry in tidal freshwater wetlands are limited even though these wetlands are sensitive to sea level rise and increased salinity due to their location between non-tidal freshwater rivers and tidal saline estuaries [17,24]. As most tidal freshwater marshes may be substantially disturbed by the conversion of freshwater marshes to brackish marshes caused by accelerated sea-level rise effects [25,26], changing salinity may become a dominant factor influencing ecological processes in tidal freshwater marshes in coastal regions. However, few studies have simultaneously examined the effects of temperature plus salinity caused by global warming in N-limiting tidal freshwater wetlands. Therefore, the primary objectives of this study were to (1) simulate N mineralization processes for different temperatures and soil salinity levels in a tidal freshwater wetland of the Yellow River Delta and (2) investigate the effects of temperature, salinity and their interactions on N mineralization rates.

Materials and Methods

Ethics statement

Our study area is located in the Yellow River Delta wetland nature preserve, which is owned by the Chinese government. Specific permits are required for conducting research in the preserve. However, our sampling sites were not located in any strictly protected areas containing endangered or protected species.

Site description

The study area (37°35′–38°12′ N,118°33′–119°20′ E) is located in the estuary of the Yellow River Delta Wetland Nature Preserve, Shandong, China. This area has a typical monsoon (warm temperate) climate with an annual mean temperature of 11.9°C; the highest and lowest temperatures (39.7°C and −23.3°C, respectively) occur in July and January, respectively. Moreover, the annual mean precipitation is 640 mm (70% of the precipitation falls between June and August). Typical freshwater wetlands dominated by *P. australis* were selected. These sites were 150 m away from the southern riverbank along the lower reach of the Yellow River, where the soil is Eutric Fluvisol [27] with >85% of silt. The study site has a relatively higher topography compared with the floodplains near the river and can be flooded by the Yellow River after the annual flow-sediment regulation in late June. However, this area is likely to be flooded with seawater in the future due to potential increases in sea level.

Soil sampling

Plant root system mainly distributed in upper soils, though the rhizome of *P. australis* could reach to more than 70cm soil depth. In general, top soils exhibited higher N contents and are more sensitive to environmental changes (e.g. temperature, hydrological changes) compared to deeper soils [4,28]. Moreover, top soils (0–20 cm) in five sampling sites with five replicates were randomly collected from the freshwater marsh (*P. australis*) in September 2008 and transported to the laboratory at 4°C with an ice cooler. The soils were sieved through a 2 mm nylon sieve to remove recognizable plant litters, coarse roots, and stones and subsequently divided into two portions. One portion was stored at 4 °C for incubation, soil moisture, and initial NH_4^+-N and NO_3^--N analyses. The other portion was air-dried for three weeks and subsequently ground and sieved (0.15 mm) for physical and chemical analyses (Table 1). Five additional soil cores (100 cm^3) were collected to determine the soil bulk density.

Incubation and chemical analysis

Every 20-g soil sample (oven-dried adjusted) was submerged with 15 ml of adjusted sea salt solution at the salinities 0‰, 10‰, 20‰ or 35‰ [29,30] in 50 ml glass tubes for the salinity treatment experiments. The sea salt contained 55.03% Cl^-, 30.59% Na^+, 7.68% SO_4^{2-}, 3.68% Mg^{2+}, 1.18% Ca^{2+} and 1.11% K^+ [31]. The highest salinity (35‰), which is similar to seawater at study site, was used simulate a seawater flooding intrusion caused by future sea level rise. The 10 and 20‰ salinity gradients represented mixed river and seawater flooding situations due to the location between non-tidal freshwater rivers and tidal saline estuaries [17,24]. Deionized water was used for the 0‰ salinity, which was used as a control scenario. Alcohol-wetted burning cotton was stretched into the tube for approximately two or three seconds.

Then, the cotton was slowly pulled out while the flame shrank until extinction. The oxygen above the water surface in the tube was completely consumed through combustion. Immediately thereafter, the tubes were tightly capped with rubber bungs to prevent aeration. Nitrification could be inhibited in high moisture environments, especially in submerged conditions [28,30–32]. Therefore, ammonium ought to reflect N mineralization [33,34].

Each salinity treatment had 84 replicates, which were divided into four incubation groups. The groups remained in the dark for 48 days in the constant temperature incubators at 10, 20, 30 or 40°C. A pilot experiment observed the balance of the amount of mineralized N over 48-d incubation under higher temperature ranges (20°C, 30°C and 40°C) based on the potential mineralized N of the first-order reaction kinetics model. No obvious changes in N mineralization rates at 10°C were observed over a 48-d incubation period. We designated 48 days as our incubation period. Generally, these temperatures represented the mean seasonal air temperatures at the study site. Three tubes from each salt treatment and for each temperature were randomly chosen at 2, 5, 9, 14, 21, 35, and 48 days for analysis.

Soil NH_4^+-N was extracted with 2 M KCl (soil:solution = 1:2.5). Extracts were mixed for 1 hour on a longitudinal shaker and filtered through #42 Whatman filter paper and stored at 4°C to determine NH_4^+-N with an Astoria Analyzer 300 system (Astoria Pacific International, Taiwan). Soil organic matter (SOM) was determined according to Walkley and Black [35]. Total carbon (TC) and total nitrogen (TN) were measured using an elemental analyzer (Vario El, Elementar Co., Germany). The initial soil nitrate nitrogen (NO_3^--N) and ammonium nitrogen (NH_4^+-N) were determined in the 2 M KCl extracts with an Astoria Analyzer 300 system (Astoria Pacific International, Taiwan). Soil pH (H_2O) was measured using a pH meter (soil: water = 1:5). Water content was evaluated by drying the soil to a constant weight at 105°C for 24 h in an oven.

Nitrogen mineralization models

Nitrogen mineralization was estimated using an effective accumulated temperature model (temperature model) [33,36] and a first-order reaction kinetic model (one-pool model) [37]. The temperature model is described by

$$N = k[(T - T_0)D]^n, \qquad (1)$$

where N is the mineralized soil organic N (mg kg^{-1}), k is the constant mineralization rate, T_0 is the base point temperature (i.e., 15°C, below which microorganism activity is weak and N mineralization may be trivial), D is the incubation time at temperature T (day), and n is a mineralization constant.

The one-pool model is given by

$$N = N_0\left(1 - e^{-kt}\right), \qquad (2)$$

where k is the instantaneous release rate, N_o is the potential mineralized N as $t \to \infty$ (mg kg^{-1}), and t is the incubation time (day).

The active N fraction (ANF) in the SOM was calculated according to the following equation [38]:

$$\mathrm{ANF} = N_0/\mathrm{TN}, \qquad (3)$$

where N_o is the potential mineralized N as $t \to \infty$ (mg kg^{-1}), and TN is initial total nitrogen.

Statistical analysis

Nitrogen mineralization data were transformed using a square root transformation to ensure normality and homogeneity of the variance. Three-way ANOVA was used to test the effects of incubation time, temperature, salinity and their interactions on nitrogen mineralization. Non-linear regression analysis was implemented to determine model parameters. Statistical analyses were conducted using the SPSS 13.0 software package.

Results

Nitrogen mineralization process

Figure 1 shows the accumulated soil NH_4^+-N over a 48-d incubation period for four salinity levels and temperature. In general, the accumulated NH_4^+-N ranged from 0.05 to 33.74 mg kg^{-1} and was generally significantly increased with increasing incubation time for temperatures of 20°C, 30°C and 40°C ($r^2 = 0.77$, $P < 0.001$, Table 2 and Figure 1). ANOVA analysis showed that the incubation time significantly affected N mineralization rates ($P < 0.001$). However, a decreasing trend was found at 10°C, irrespective of the salinity concentrations (Figure 1). The incubation temperature also exhibited a significant effect on N mineralization in this study ($P < 0.001$, Table 2). The accumulated NH_4^+-N under incubation temperatures was ranked as follows: 40°C>30°C>20°C>10°C, regardless of the chosen salinity. Significantly higher mineralized N was observed at higher temperatures. It is noted that a much lower level of the accumulated NH_4^+-N was maintained over the incubation time under four salinity levels at 10°C. By the end of the 48-d incubation period, the accumulated NH_4^+-N at 40°C was two and five times higher than at 30°C and 20°C, respectively. Meanwhile, the accumulated NH_4^+-N for a given temperature remained nearly the same during the entire incubation period under all salinity treatments. One exception was found for a temperature of 40°C after 10 days.

The accumulated NH_4^+-N was lower before 10 days into the incubation period at 20°C, 30°C and 40°C under 35‰ salinity concentration compared to the other salinity treatments. As shown in Table 2, salinity also showed significant effects on the mineralized N ($P < 0.001$). The accumulated NH_4^+-N was generally higher at salinities of 20‰ and 35‰ compared with salinities 0‰ and 10‰ (Figure 1). In particular, the accumulated NH_4^+-N under 0‰ salinity treatment showed significantly lower values after 10 days of incubation at 30°C and 40°C than other salinity treatment ($P < 0.05$). Incubation time, temperature, salinity and their interaction effects on N mineralization were all significant ($P < 0.001$, Table 2), except for the salinity and temperature interaction ($P > 0.05$, Table 2). Generally, temperature exhibited the greatest effect on N mineralization according to the F values, followed by the incubation time, whereas salinity exhibited the least effects on N mineralization (Table 2).

Nitrogen mineralization modeling

The accumulated NH_4^+-N at 10°C was not simulated using the temperature and one-pool models because it remained nearly stable over the 48-d incubation period (Figure 2a). The parameters (k, n) for both models were identified based on the experiment data under other incubation temperatures, i.e., 20, 30 and 40°C (Tables 3 and 4). Generally, k ranged from 0.004 to 2.552 and increased with increasing incubation temperature and n ranged from 0.346 to 1.299 and decreased with increasing temperature for the temperature model (Table 3). Moreover, k was highest at 0‰ salinity and 40°C and lowest at 35‰ salinity and 20°C; n was highest at 35‰ salinity and 20°C and lowest at 10‰ salinity and

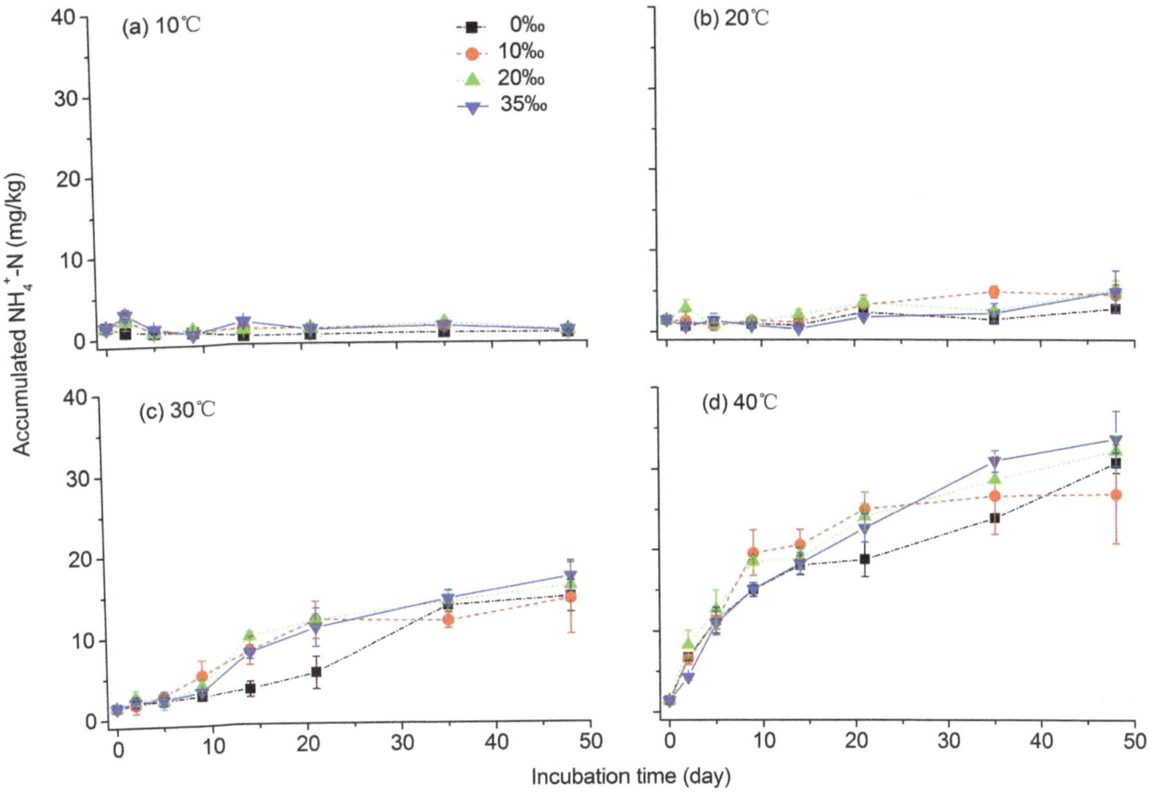

Figure 1. Soil accumulated ammonium nitrogen over 48 days of laboratory incubation under four temperature and salinity levels.

40°C. In addition, no pronounced changing trends were observed between salinity concentrations at the same incubation temperature for both k and n. Both the effective accumulated temperature and one-pool models fit well with the predicted N mineralization under higher temperatures of 30 and 40°C (Figure 2). The regression analyses (Tables 3 and 4) showed that these parameters were ideal for both models with a better fitness ($R^2 = 0.88$–0.99, $P<0.01$) at both 30 and 40°C than at 20°C ($R^2 = 0.30$–0.85, $P< 0.05$).

In general, N_0 and k increased with increasing incubation temperature for the one-pool model (Table 4). Moreover, N_0 increased and k decreased with increasing salinity (i.e., from 10 to 35‰). At 20°C and 40°C, N_0 was lower than the observed accumulated NH_4^+-N at the end of the incubation period; the opposite was true at 30°C (Figure 2; Table 4). The ANF increased with increasing incubation temperature (Table 5). The potential mineralized N from SOM ranged from 0.65% to 9.36% of total N (0.38 g/kg, Table 1) under the four salinity treatments. The incubation times required to mineralize half of the N_0 ($t_{1/2}$) were generally less than 2 weeks and longest for 4 weeks. Additionally, a rapid decrease in $t_{1/2}$ was observed from 30°C to 40°C under 35‰ salinity compared with the other salinity treatments. Compared with the temperature model over the entire 48-d incubation period, the one-pool model better represented the observed accumulated NH_4^+-N values at 30°C for all the four salinity concentrations.

Table 2. Results of three-way ANOVA on the soil accumulated N mineralization.

Source	df	F	P
Salinity	3	19.2	< 0.001
Temperature	3	2686.3	< 0.001
Time	6	282.5	< 0.001
Salinity × Temperature	9	1.9	> 0.05
Salinity × Time	18	3.6	< 0.001
Temperature × Time	18	90.9	< 0.001
Salinity × Temperature × Time	54	2.1	< 0.001
Error	178		

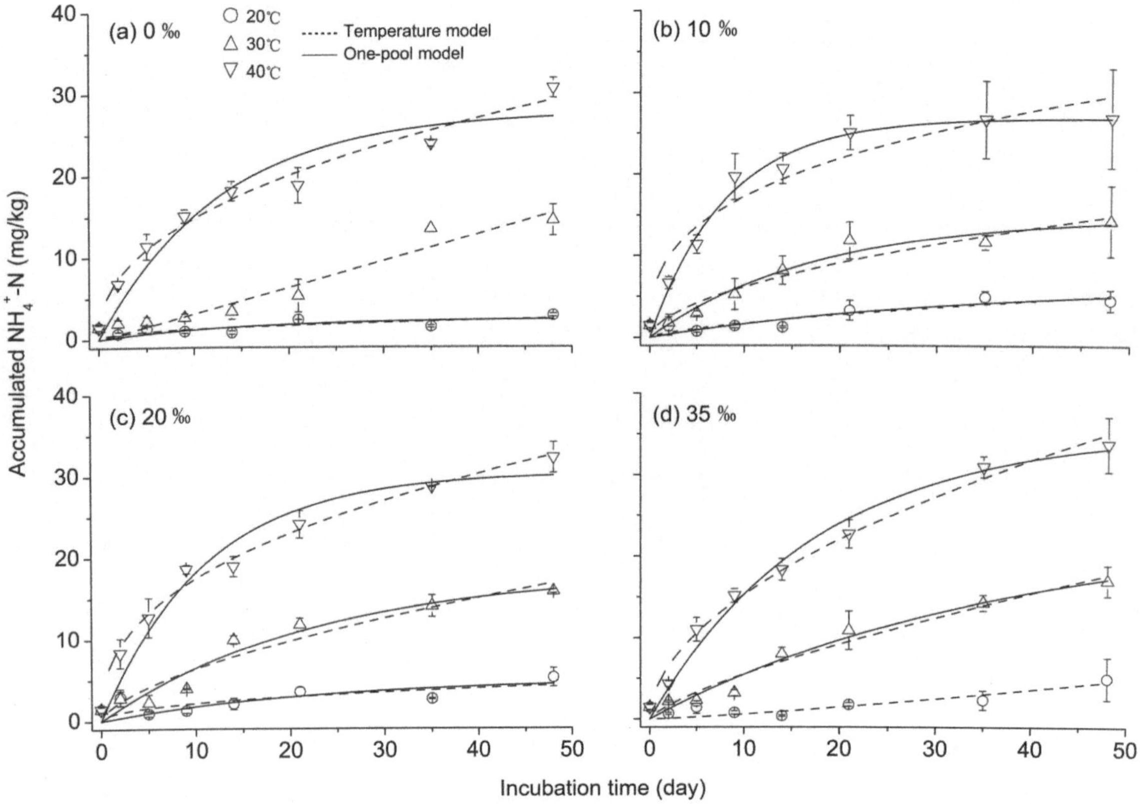

Figure 2. Comparisons of predicted (solid lines) and observed (symbols with dashed lines) cumulative mineral-N over 48 days of incubation under four temperature and salinity levels.

Discussion

Effects of temperature and salinity on N mineralization

The accumulated NH_4^+-N increased with increasing incubation time under almost all temperature treatments except at 10°C (Figures 1 and 2). This finding is consistent with results of previous studies in wetland and paddy soils [17,33,34,37]. The reason for the significant effect of incubation temperature on N mineralization in this study ($P<0.01$, Table 2) was that because temperature could greatly affect soil microbial processes [6,10,16,29,39] and higher temperature favored microbial and enzyme activity [2,40] and thus enhanced N mineralization processes. In contrast, lower temperatures (e.g. 10°C) exhibited inhibiting effects (Figure 1) due to lower microbial and enzyme activities [19,41]. Grenon et al. [42] also showed that most microbial metabolic rates are positively related to temperature. However, at extremely lower temperature (e.g. freezing) in non-growing seasons, higher N mineralization rates might occur due to lower N immobilization of microbes [2,19]. Root mortality at freezing temperature might enhance N immobilization sine more available C would be released from detritus and enhance microbial activity [19]. Moreover, lower microbial biomass was likely correlated with faster net N mineralization [18,19]. Therefore, microbial activities can be used to predict the N mineralization rate because soil microorganisms are primarily responsible for the transformation of organic N to mineral N [10,34].

Studies have reported that N mineralization decreased with increasing salinity and was biologically mediated at higher salinities [12,20,23] because microbial activity can be inhibited at high salinities (>500 mS m^{-1}) [6,29,43]. Muhammad et al. [44] and Tripathi et al. [45] also presented that both soil respiration and soil microbial biomass were negatively correlated with soil salinity. Additionally, Wong et al. [46] addressed that high salinity (30 dS m^{-1}) increased electrolyte concentrations and caused the soil to flocculate, preventing the organic N release. However, obvious inhibiting effects of high salinity on N mineralization processes were not observed in this study. A higher accumulated NH_4^+-N corresponded with higher salinity treatments, which was supported by Neo et al. [17], who suggested that increasing salinity from 0.1 to 3.5‰ contributes to soil N mineralization in tidal forested wetlands. Chen and Twilley [22] also observed higher N mineralization with increasing salinity in mangrove soils. The ANOVA analysis further showed significant effects of salinity on the accumulated NH_4^+-N ($P<0.001$, Table 2). However, the lower accumulated NH_4^+-N in the first 10 incubation days before its increase under 35‰ salinity compared to the other salinity treatments might imply a short-term inhibiting effect before enhancing NH_4^+-N release. Khoi et al. [47] also observed that adverse effects of salinity on N mineralization were short-lived, whereas the rate of N mineralization recovered in later stages. Further studies are needed to testify this effect in different wetland soils under different vegetation.

Generally, the effects of increasing salinity on N mineralization can be explained as follows. Firstly, under high salinity conditions, a shift in microbial population structure occurs [46]. Muhammad et al. [44] found higher proportions of fungal biomass in more saline soils compared with less saline soils. Therefore, the soil ammonification process can be enhanced as an effect of salinity on specific microbial function groups responsible for ammonification, which results in increased accumulated NH_4^+-N at higher salinities

Table 3. Parameter values used for the effective accumulated temperature model over 48 days of waterlogged incubation experiments.

Salt added, ‰	20°C			30°C			40°C		
	k	*n*	R^2	*k*	*n*	R^2	*k*	*n*	R^2
0	0.029±0.187	0.449±0.183	0.604	0.017±0.019	1.013±0.174	0.936	1.344±0.326	0.434±0.037	0.974
10	0.106±0.095	0.705±0.176	0.839	0.366±0.204	0.563±0.092	0.927	2.552±1.061	0.346±0.065	0.889
20	0.371±0.402	0.447±0.221	0.449	0.287±0.196	0.619±0.112	0.914	1.931±0.307	0.398±0.025	0.986
35	0.004±0.008	1.299±0.392	0.808	0.130±0.072	0.746±0.090	0.963	0.817±0.185	0.530±0.035	0.987

Table 4. Parameter values used for the one-pool model over 48 days of waterlogged incubation experiments.

Salt added, ‰	20°C			30°C			40°C		
	N_0 (mg kg^{-1} soil)	*k* (day^{-1})	R^2	N_0 (mg kg^{-1} soil)	*k* (day^{-1})	R^2	N_0 (mg kg^{-1} soil)	*k* (day^{-1})	R^2
0	2.483±0.776	0.061±0.045	0.497	–	–	–	27.826±2.870	0.078±0.021	0.880
10	6.906±3.277	0.027±0.020	0.846	15.033±1.378	0.055±0.011	0.965	26.908±0.792	0.123±0.012	0.983
20	5.181±2.875	0.041±0.044	0.302	18.201±2.795	0.044±0.014	0.944	30.417±2.17	0.091±0.018	0.923
35	–	–	–	24.332±5.079	0.025±0.008	0.974	35.569±2.117	0.056±0.008	0.982

– no fitting parameter.

Table 5. Estimates of active N fraction (ANF) and time required to mineralize half of the N_0 ($t_{1/2}$) using the one-pool model.

Salt added, ‰	20°C		30°C		40°C	
	ANF (%)	$t_{1/2}$ (d)	ANF (%)	$t_{1/2}$ (d)	ANF (%)	$t_{1/2}$ (d)
0	0.65	11.4	-	-	7.32	8.9
10	1.82	25.7	3.96	12.6	7.08	5.6
20	1.36	16.9	4.79	15.8	8.00	7.6
35	-	-	6.40	27.7	9.36	12.4

(Figure 1). Secondly, Na^+ is an important and dominant seawater component. High electrical conductivity solutions, particularly those high in Na^+, could rapidly alter the composition of exchange sites on clays, causing NH_4^+ adsorbed on to clay surfaces to be desorbed [46], leading to the increase in mineralized N. Thirdly, the inhibiting effects of salinity on N mineralization processes might be mitigated by high temperatures [29], because N mineralization was largely improved at higher temperatures (i.e., 30°C and 40°C, Figure 2). Meanwhile, the ANOVA analysis also showed a significant effect of temperature (larger than the salinity effect) on the accumulated NH_4^+-N ($P<0.001$, Table 2). Although no significantly interactive effect of temperature and salinity was observed, ($P>0.05$, Table 2), the significant interactive effect was displayed ($P<0.001$, Table 2) if incubation time was further took into account. This indicated that incubation time was also an important factor influencing the accumulated NH_4^+-N [4]. Therefore, long-term incubation experiments are required to better understand the soil N mineralization process.

Nitrogen mineralization models

It is well known that mineralization is a process in which soil organic compounds are decomposed by microorganisms into inorganic compounds. Moreover, microbial mineralization activity depends on enzyme kinetics, which follows a first-order dynamical model [48]. Models and their corresponding parameters are essential to mathematically describe soil N mineralization kinetics [49]. The first-order reaction kinetic model was firstly used by Stanford and Smith to simulate aerobic long-term mineralization dynamics [37], while other exponential models have been used for N mineralization in recent studies [6,48,50].In this study, the ideal parameters for both the effective accumulated temperature and one-pool models indicated that both models could simulate N mineralization in tidal freshwater wetlands at higher temperatures of 30 and 40°C (Figure 2). However, a better simulation was observed at 40°C for the temperature model, whereas a similar performance was obtained between the temperature and one-pool models at less than 40°C, which is consistent with results of previous studies [6,38,50].

The effective accumulated temperature model, which is regarded as the empirical formula derived from the first-order reaction kinetic model, explains the enzyme kinetics. In general, the mineralization constant (n) for the temperature model should be between 0 and 1 based on the theory of enzyme kinetics [33,38]. Moreover, n decreased with increasing temperature, suggesting that mineralized N per effective accumulated temperature decreased with increasing effective accumulated temperature due to higher enzyme activity [38]. However, a higher n was observed for either lower or higher salinities in this study (Table 2). The k value reflects the mineral extent of mineralization [47]. Compared with lower temperature (20°C), the larger k values at higher temperature exhibited higher mineral extent of mineralization, indicating temperature is a vital factor of N mineralization. Bai et al. [33] and Li et al. [48] also demonstrated that the effective accumulated temperature model (temperature model) was a N mineralization model with temperature as the dominant factor. Bai et al. [33] also testified that a temperature model can efficiently simulate N mineralization in inland salt marshes. Therefore, the temperature model can be used to describe the N mineralization process when the temperature exhibits substantial effects on N mineralization.

For the one-pool model, k (day^{-1}) is considered as the percentage of the remaining mineral N that is mineralized per day [37], which decreased with increasing salinity. In contrast, N_0 generally increased with increasing salinity, implying that a higher N mineralization potential existed under higher salinity conditions. Additionally, the ratio of potential mineralized N (N_0) to total N represents the active N fraction (ANF) in the SOM. The general increasing of ANF and decreasing of incubation time to mineralize half of N_0 ($t_{1/2}$) corresponded with increasing incubation temperature (Table 5), which again indicated that increasing temperature could substantially accelerate N mineralization. The time of $t_{1/2}$ for all salinity treatments ranged from > 10 days to < 10 days from 20°C to 40°C, indicating a quick mineralization at higher temperatures. Meanwhile, smaller $t_{1/2}$ reflected large value of the mineralization rate constant (k) (Table 5) and indicated that the small ANF of a soil was easily mineralized and readily used by microorganisms [38]. However, compared with other studies [38,48], N_0 and ANF were much smaller at 20°C and 30°C (Table 5). Although the incubation time was not as long (12 to 40 weeks) as in other studies [33,49–51], all mineralization half times under all treatments were within a 48-d incubation period (Table 5). This time frame reflected the N mineralization process in the N-limiting tidal freshwater wetland soils. Furthermore, the rapid decrease in $t_{1/2}$ from 30°C to 40°C was observed under 35‰ salinity than under other salinity treatments (Table 5), indicating that a higher salinity might contribute to an enhanced N mineralization. However, the short-term inhibiting effects of 35‰ salinity on the N mineralization rate might also occur at higher temperature of 40°C in the first 10 incubation days (Figures 1 and 2) due to lower enzyme activity [2] and higher N immobilization of microbes [2,19].

Theoretically, N_0 should be greater than the observed cumulative mineral N over the incubation period [48,49] because N_0 is defined as the entire pool of potentially mineralizable N. However, the potential mineralized N at 20°C and 40°C from the one-pool model were very close to or even smaller than the observed mineralized N (Table 4). Nevertheless, the one-pool model had a better fit to the observed data than the double first-order exponential model (two-pool model) [52], which was used to simulate the accumulated NH_4^+-N over the incubation period.

Moreover, the two-pool model did not fit the observations well and contained large standard errors (data not shown) This finding contradicts the results of Li et al. [48], who concluded that a two-pool model was better for determining accumulated mineralizable N in paddy soils. However, Cordovil et al. [38] reported that the two-pool model had no advantage to fit the experimental data over one-pool model as it could not identify materials with different mineralization rate kinetics. Stanford and Smith [37] reported that the pool containing the compounds that contribute to the potentially mineralizable N was similar in most soils. Therefore, only one pool is often required [38].

Conclusions

NH_4^+-N release was significantly enhanced with increasing temperature and salinity over a 48-d incubation period. Moreover, accumulated NH_4^+-N was significantly higher under higher salinity than under lower salinity levels at higher temperatures. Except for the interaction of temperature and salinity, incubation time, temperature and salinity and their interactive effects exhibited significant effects on N mineralization; and temperature dominated the effects. Accumulated NH_4^+-N increased with increasing incubation time, which fit well with the simulation by both the effective accumulated temperature and one-pool models at temperatures up to 40°C. Although we do not compare N mineralization among different soil types overlaying vegetation in this study, our results indicated that NH_4^+-N release was increased in tidal freshwater wetland soils after the conversion to brackish wetlands due to sea-level rise, which might occur as a result of projected global warming. Therefore, the soil N mineralization response to increasing temperature and salinity in tidal freshwater wetlands should be integrated with studies on the structure and functions of coastal ecosystems in response to global warming and sea-level rise. Additionally, further investigation is needed to reveal the effects of salt anions (e.g. Cl^-, SO_4^{2-}) and on N mineralization in wetland soils. The combined study on N mineralization though lab and *in situ* incubation experiments should be focused on in the future.

Acknowledgments

The authors are grateful to the Editor and two anonymous reviewers for their valuable comments and suggestions. We also express our thanks to the college of Life Science and Engineering, Huanggang Normal University and Miss Lijuan Wang for their experimental assistance.

Author Contributions

Conceived and designed the experiments: HG JB. Performed the experiments: HG QL QZ JW. Analyzed the data: HG JB XH. Contributed reagents/materials/analysis tools: JB. Wrote the paper: HG JB XH. Revised for the final version: JB XH.

References

1. Downing JA, Mcclain M, Twilley R, Melack JM, Elser J, et al. (1999) The impact of accelerating land-use change on the N-Cycle of tropical aquatic ecosystems: Current conditions and projected changes. Biogeochemistry 46: 109–148.
2. Gao HF, Bai JH, Xiao R, Yan DH, Huang LB, et al. (2012) Soil net nitrogen mineralization in salt marshes with different flooding periods in the Yellow River Delta, China. CLEAN–Soil, Air, Water 40: 1111–1117.
3. Reddy KR, DeLaune RD (2008) Biogeochemistry of Wetlands: Science and Applications. Boca Raton: CRC Press.
4. Deng W, Bai JH, Li AN, Wang GX, Hu JM, et al. (2012) Typical Wetland System Pattern Changes and Ecohydrological Processes: a Case Study of Huang-Huai-Hai Regions. Beijing: Science Press.
5. Mitsch WJ, Gosselink JG (2007) Wetlands (Fourth Edition). Hoboken: John Wiley & Sons, Inc..
6. Bai JH, Gao HF, Xiao R, Wang JJ, Huang C (2012) A review of soil nitrogen mineralization as affected by water and salt in coastal wetlands: issues and methods. CLEAN–Soil, Air, Water 40: 1099–1105.
7. Zaman M, Saggar S, Blennerhassett JD, Singh J (2009) Effect of urease and nitrification inhibitors on N transformation, gaseous emissions of ammonia and nitrous oxide, pasture yield and N uptake in grazed pasture system. Soil Biol Biochem 41: 1270–1280.
8. Prusty BAK, Chandra R, Azeez PA (2009) Distribution of carbon, nitrogen, phosphorus, and sulfur in the soil in a multiple habitat system in India. Soil Res 47: 177–189.
9. Wang CH, Wan SQ, Xing XR, Zhang L, Han XG (2006) Temperature and soil moisture interactively affected soil net N mineralization in temperate grassland in Northern China. Soil Biol Biochem 38: 1101–1110.
10. Bagherzadeh A, Brumme R, Beese F (2008) Temperature dependence of nitrogen mineralization and microbial status in O-H horizon of a temperate forest ecosystem. J Forest Res 19: 37–43.
11. Koch O, Tscherko D, Kandeler E (2007) Temperature sensitivity of microbial respiration, nitrogen mineralization, and potential soil enzyme activities in organic alpine soils. Global Biogeochem Cycles 21: B4017.
12. Li S, Zhao K, Lu H (2013) Nitrogen mineralization characteristics of disturbed and undisturbed soil samples for four main soil types on the Loess Plateau. Commun Soil Sci Plant Anal 44: 1659–1673.
13. Chen YT, Borken W, Stange CF, Matzner E (2012) Dynamics of nitrogen and carbon mineralization in a fen soil following water table fluctuations. Wetlands 32: 579–587.
14. Ehrenfeld JG, Yu S (2012) Patterns of nitrogen mineralization in wetlands of the New Jersey pinelands along a shallow water table gradient. Am Midl Nat 167: 322–335.
15. Fujita Y, van Bodegom PM, Olde Venterink H, Runhaar H, Witte JM (2013) Towards a proper integration of hydrology in predicting soil nitrogen mineralization rates along natural moisture gradients. Soil Biol Biochem 58: 302–312.
16. Gao JQ, Ouyang H, Xu XL, Zhou CP, Zhang F (2009) Effects of temperature and water saturation on CO_2 production and nitrogen mineralization in alpine wetland soils. Pedosphere 19: 71–77.
17. Noe G, Krauss K, Lockaby BG, Conner W, Hupp C (2013) The effect of increasing salinity and forest mortality on soil nitrogen and phosphorus mineralization in tidal freshwater forested wetlands. Biogeochemistry 114: 225–244.
18. Gilliam FS, Cook A, Lyter S (2010) Effects of experimental freezing on soil nitrogen (N) dynamics in soils of a net nitrification gradient in an N-saturated hardwood forest ecosystem. Can J For Res 40: 436–444.
19. Zhang X, Bai W, Gilliam FS, Wang Q, Han X, et al. (2011) Effects of *in situ* freezing on soil net nitrogen mineralization and net nitrification in fertilized grassland of northern China. Grass Forage Sci 66: 391–401.
20. Solomon S, Qin D, Manning M, Chen Z, Marquis M, et al. (2007) IPCC,2007: Climate Change 2007: The Physical Science Basis. Contribution of Working Group I to the Fourth Assessment Report of the Intergovernmental Panel on Climate Change, Cambridge University Press, Cambridge, UK. And New York, USA.
21. Fang H, Liu G, Kearney M (2005) Georelational analysis of soil type, soil salt content, landform, and land use in the Yellow River Delta, China. Environ Manage 35: 72–83.
22. Chen R, Twilley RR (1999) Patterns of mangrove forest structure and soil nutrient dynamics along the Shark River estuary, Florida. Estuaries 22:955–970.
23. Pathak H, Rao DL (1998) Carbon and nitrogen mineralization from added organic matter in saline and alkali soils. Soil Biol Biochem 30: 695–702.
24. Whigham D, Baldwin A, Barendregt A (2009) Tidal freshwater wetlands. In: Perillo G, Wolanski E, Cahoon D, Brinson M, Coastal Wetlands: An Integrated Ecosystem Approach. Amsterdam, Elsevier. pp. 515–534.
25. Carlson AE, LeGrande AN, Oppo DW, Came RE, Schmidt GA, et al. (2008) Rapid early Holocene deglaciation of the Laurentide ice sheet. Nat Geosci 1: 620–624.
26. Craft C, Clough J, Ehman J, Joye S, Park R, et al. (2008) Forecasting the effects of accelerated sea-level rise on tidal marsh ecosystem services. Front Ecol Environ 7: 73–78.
27. IUSS-ISRIC-FAO (2006) World reference base for soil resources-a framework for international classification, correlation and communication. Rome: FAO.
28. Bai JH, Gao HF, Deng W, Yang ZF, Cui BS, et al. (2010) Nitrification potential of marsh soils from two natural saline –alkaline wetlands. Biol Fertil Soils 46: 525–529.
29. Rysgaard S, Thastum P, Dalsgaard T, Christensen PB, Sloth NP (1999). Effects of salinity on NH_4^+ adsorption capacity, nitrification, and denitrification in Danish estuarine sediments. Estuaries 22: 21–30.
30. Lodhi A, Arshad M, Azam F, Sajjad MH (2009) Changes in mineral and mineralizable N of soil incubated at varying salinity, moisture and temperature regimes. Pak J Bot 41: 967–980.
31. Pidwirny M (2006) Seawater. http://www.eoearth.org/article/Seawater

32. Waring SA, Bremner JM (1964) Production in soil under waterlogged conditions as an index of nitrogen availability. Nature 201: 951–952.
33. Bai JH, Ouyang H, Deng W, Wang QG, Chen H, et al. (2005) Nitrogen mineralization processes of soils from natural saline-alkalined wetlands, Xianghai National Nature Reserve, China. Can J Soil Sci 85: 359–367.
34. Haddad SA, Tabatabai MA, Loynachan TE (2013) Biochemical processes controlling soil nitrogen mineralization under waterlogged conditions. Soil Sci Soc Am J 77: 809–816.
35. Walkley A, Black IA (1934) An examination of the Degtjareff method for determining soil organic matter, and a proposed modification of the chromic acid titration method. Soil Science 37: 29–38.
36. Zhu ZL (1982) Nitrogen mineralization and prediction of nitrogen availability of Taihu region. Acta PedologSin 21: 29–36 (in Chinese with English abstract).
37. Stanford G, Smith SJ (1972) Nitrogen mineralization potentials of soil. Soil Sci Soc Am J 36: 465–472.
38. Cordovil CMDS, Coutinho J, Goss M, Cabral F (2005) Potentially mineralizable nitrogen from organic materials applied to a sandy soil: fitting the one-pool exponential model. Soil Use Manage 21: 65–72.
39. Ma L-N, Lü X-T, Liu Y, Guo J-X, Zhang N-Y, et al. (2011) The effects of warming and nitrogen addition on soil nitrogen cycling in a temperate grassland, Northeastern China. PLoS ONE 6: e27645.
40. Zhang XL, Wang QB, Li LH, Han XG (2008) Seasonal variations in nitrogen mineralization under three land use types in a grassland landscape. Acta Oecol 34: 322–330.
41. Schimel JP, Bilbrough C, Welker JM (2004) Increased snow depth affects microbial activity and nitrogen mineralization in two Arctic tundra communities. Soil Biol Biochem 36: 217–227.
42. Grenon F, Bradley RL, Titus BD (2004) Temperature sensitivity of mineral N transformation rates, and heterotrophic nitrification: possible factors controlling the post-disturbance mineral N flush in forest floors. Soil Biol Biochem 36: 1465–1474.
43. Rietz DN, Haynes RJ (2003) Effects of irrigation-induced salinity and sodicity on soil microbial activity. Soil Biol Biochem 35: 845–854.
44. Muhammad S, Müller T, Joergensen R (2006) Decomposition of pea and maize straw in Pakistani soils along a gradient in salinity. Biol Fertil Soils 43: 93–101.
45. Tripathi S, Kumari S, Chakraborty A, Gupta A, Chakrabarti K, et al. (2006) Microbial biomass and its activities in salt-affected coastal soils. Biol Fertil Soils 42: 273–277.
46. Wong V, Dalal R, Greene R (2008) Salinity and sodicity effects on respiration and microbial biomass of soil. Biol Fertil Soils 44: 943–953.
47. Khoi CM, Guong VT, Merckx R (2006) Predicting the Release of Mineral Nitrogen from Hypersaline Pond Sediments Used for Brine Shrimp Artemia franciscana Production in the Mekong Delta. Aquaculture 257: 221–231.
48. Li SX (2008) Soil and Plant Nitrogen in Dryland Areas of China(in Chinese). Beijing: Science Press. 83 p.
49. Dou ZX, Toth JD, Jabro JD, Fox RH, Fritton DD (1996) Soil nitrogen mineralization during laboratory incubation: Dynamics and model fitting. Soil Biol Biochem 28: 625–632.
50. Li HL, Han Y, Cai ZC (2003) Nitrogen mineralization in paddy soils of the Taihu Region of China under anaerobic conditions: dynamics and model fitting. Geoderma 115: 161–175.
51. Wang WJ, Smith CJ, Chen D (2003) Towards a standardised procedure for determining the potentially mineralisable nitrogen of soil. Biol Fertil Soils 37: 362–374.
52. Richter J, Nuske A, Habenicht W, Bauer J (1982) Optimized N-mineralization parameters of loess soils from incubation experiments. Plant Soil 68: 379–388.

The Prevalence of Hepatitis E Virus Infections among Swine, Swine Farmers and the General Population in Guangdong Province, China

Huanbin Liang[§], **Shuo Su**[§], **Shengchao Deng**[§], **Honglang Gu, Fangxiao Ji, Lifang Wang, Chumin Liang, Heng Wang, Guihong Zhang***

MOA Key Laboratory of Animal Vaccine Development, College of Veterinary Medicine, South China Agricultural University, Guangzhou, China

Abstract

Hepatitis E virus (HEV) infection is widespread in China, but few studies have been carried out in Guangdong Province. This study aimed to characterize the prevalence of HEV infections among swine, swine farmers and the general population in Guangdong Province. We conducted an epidemiological study that included swine, swine farmers and health examination attendees in Guangdong from 2011 to 2013.. The overall seroprevalence of anti-HEV antibodies in swine was 64.7%. The results revealed that growing pigs, sows and boars (OR ranges from 3.5 to 21.5) have a higher risk than nursery pigs. HEV RNA in swine bile showed that HEV is epidemic in swine in the Pearl River Delta, with the highest prevalence of 22.73% in Foshan. Some genomes of HEV strains from each district were sequenced. Phylogenetic analysis of partial open reading frame 2 (ORF2) shows that they belong to genotype IV and are most closely related to isolates from China. In total, 307 participants were enrolled in the study, including 114 swine farmers and 193 attendees from hospitals. IgG anti-HEV was detected in 48.25% of swine farmers and in 38.34% of the general population. Seroprevalence rates were almost stratified by age, with a higher positive rate for males compared to females across all age groups. Women on swine farms appeared to have a lower risk of infection compared to the general population, revealing that the risk factors for HEV infection are not unique. The results suggested that there were other risk factors for HEV infection. HEV infection is prevalent in Guangdong, but due to the small sample sizes, more investigations are needed to assess the potential impact of HEV infection, and many additional risk factors should be considered.

Editor: Xiang-Jin Meng, Virginia Polytechnic Institute and State University, United States of America

Funding: This work was supported by the Industry technology system of modern agriculture construction fund (No.CARS-36) and the Ph.D. Programs Foundation of Ministry of Education of China (No.20104404110009 and No.20114404120010). The funders had no role in study design, data collection and analysis, decision to publish, or preparation of the manuscript.

Competing Interests: The authors have declared that no competing interests exist.

* E-mail: guihongzh@scau.edu.cn

§ These authors contributed equally to this work.

Introduction

The hepatitis E virus (HEV) is a small non-enveloped RNA virus that belongs to the genus Hepevirus in the family Hepeviridae [1]. There are 4 HEV genotypes but only 1 serotype. Genotypes 1 and 2 infect only humans and are mainly endemic to developing countries such as those in Asia, Africa and South America. Genotypes 3 and 4 infect humans, pigs and other animal species in America, Europe and Asia. Obviously, the genotypes differ with respect to epidemiological distribution and host species [2,3].

Hepatitis E virus (HEV) infection is widespread in China [4,5]. However, few studies have been conducted in Guangdong, and such studies were only for human infections [6,7]. Those studies also only dealt with prevalence and not risk factors, so consequently, it is necessary to identify risk factors for HEV infection. Consumption of fecally contaminated water has played an important role in hepatitis E epidemics in China [8,9]. Although evidence gathered suggested that eating raw or inadequately cooked meat and offal from pigs could cause HEV infection [10,11], this transmission route has not been reported in China. On the other hand, recent studies in China have shown the seroprevalence of HEV in swine, swine farmers and the general population [4,12–15]. Nevertheless, there is a lack of such research in Guangdong Province, China. Meanwhile, an increasing number of HEV infections have been detected in nearby districts, including Hunan [5], Hong Kong [16], and Taiwan [17]. Consequently, additional surveys should be carried out, and we aim to determine the relationship between human and swine HEV infections. Recently, the first hepatitis E vaccine was approved by the State Food and Drug Administration in China [18]. Prevention and control programs that include vaccination would need to specifically target people living in regions with relatively higher prevalence. Thus, it is important to understand the local epidemiology of Guangdong Province.

Materials and Methods

Study Area and Recruitment of the Study Population

Swine bile samples and serum samples were collected from 2011 to 2013 from Guangdong Province, China. Two hundred eighty-eight swine bile samples were collected in the Pearl River Delta. Bile samples were collected from nursery pigs (<4 wks), growing pigs (4 wks-6 mos), sows (>7mos) and boars (>7mos). Five hundred sixty-one samples of swine serum and 114 samples of farmers' serum samples were collected from swine farms located in Guangdong. Thirty-four different farms were sampled, and the information of farms and samples was listed in TABLE 1. Health examination attendees were enrolled at the Third Affiliated Hospital of Sun Yat-sen University. One hundred ninety-three samples of human serum were collected from the attendees at that hospital, which is located in Canton. A questionnaire was also administered to collect relevant information from both the swine farmers and the urban attendees.

Ethical Considerations

This study protocol was reviewed and approved by the Institutional Review Board at the Guangdong Center for Disease Control and Prevention. An informed consent form was provided to and signed by each participant. Human sampling procedures were also approved by the Guangdong Center for Disease Control and Prevention.

Pig sampling procedures were approved by the Animal Care and Use Committee of Guangdong Province, China. Our sampling processes were assisted by local authorities and veterinarians. All animal research was conducted under the guidance of the CDC's Institutional Animal Care and Use Committee, and all animal research was conducted in a facility accredited by the Association for Assessment and Accreditation of Laboratory Animal Care International. The animal research in our study was approved by the Guangdong Province Animal Disease Control Center. The contract-numbers of the approval documents from the ethic committees is 2013-02.

RNA extraction and reverse transcription-nested PCR

Total viral RNA was extracted using Trizol (Invitrogen, Carlsbad, CA, USA) according to the manufacturer's instructions. We designed two sets of degenerate HEV primers to detect genetically divergent strains (partial ORF2 gene) of HEV, using nest PCR after reverse transcription (RT). The primers HEV-EXF (5'- GAR GCY TCT AAT TAT GCY CAG TA -3') and HEV-EXR (5'- AAA GCC ARA GCA CAT CAT TAG C -3') were used as external primers, and HEV-INF (5'- TTG GCG CTC RGT TGA GAC CTC -3') and HEV-INR (5'- TRG CTA TAC CCT TRT CCT GCT G -3') were used as internal primers. The PCR profile for both rounds was as follows: 94°C for 5 min, 30 cycles at 94°C for 30 s, 56°C for 30 s, and 72°C for 45 s, and then a final extension at 72°C for 10 min. A negative control (water) and a positive control (HEV RNA) were included in each RT and PCR run. Furthermore, Primer Premier 5.0 was used to design primers to amplify the genes of HEV.

Table 1. The information of swine farms and samples.

Farm	Bile(positive/total)				Farm	Serum(positive/total)			
	N	G	S	B		N	G	S	B
Dongguan1	1/5	0/7	0/1	1/1	A	4/12	0	26/30	6/7
Dongguan2	0	1/6	1/1	0/1	B	6/13	0	25/33	7/7
Foshan1	0/3	1/5	1/1	1/1	C	1/7	0	18/28	6/6
Foshan2	2/8	1/6	1/2	0	D	4/14	5/12	14/20	2/3
Foshan3	1/7	2/10	0/1	0	E	6/12	3/6	16/18	3/4
Guangzhou1	0/2	1/9	0/1	0	F	2/12	8/18	11/20	4/4
Guangzhou2	0	1/7	0/1	0	G	3/8	9/11	17/24	5/5
Huizhou1	0/2	2/11	0/2	0	H	3/9	7/10	16/22	5/5
Huizhou2	1/5	1/7	0	0	I	5/16	7/8	19/25	6/6
Huizhou3	0	2/13	0	0	J	3/9	5/5	19/19	5/5
Jiangmen1	0/3	1/10	1/1	1/1	K	4/11	4/4	16/22	4/5
Jiangmen2	1/9	2/12	1/2	0	L	5/10	4/6	11/25	4/5
Jiangmen3	1/3	0/5	2/2	0					
Shenzhen1	0/2	0/4	0/1	0					
Shenzhen2	0/2	1/6	1/1	0					
Zhaoqing1	0/6	1/8	0	0					
Zhaoqing2	0/7	1/10	0/1	0					
Zhongshan1	2/6	1/7	0/1	0					
Zhongshan2	1/4	1/5	0/1	0					
Zhongshan3	1/6	2/9	0/1	0					
Zhuhai1	1/6	0/8	1/2	0					
Zhuhai2	1/4	1/4	0/2	0					
Total	13/90	23/169	9/25	3/4		46/133	52/80	208/286	57/62

Bile samples were collected from 22 different swine farms including large-scale farms and family-scale farms in 9 districts of Pearl River Delta. Swine sera were sampled from 12 large-scale swine farm in Guangdong. N, nursery pig; G, growing pig; S, sow; B, boar.

Detection of antibodies to HEV in serum

To detect immunoglobulin G (IgG) against HEV (anti-HEV), a commercial enzyme-linked immunosorbent assay (ELISA) from Wantai Biopharmaceutical, Inc. (Beijing, China) was used, and laboratory analysis was carried out according to the manufacturer's instructions [17,19-20]. The kits for detection of antibodies to HEV in serum differ for swine and human. For swine, a sandwich antigen based ELISA was used to detect the total antibodies; but for human, an indirect method was used to detect only the IgG-class antibodies.

Sequencing and phylogenetic analysis

The PCR products of the second round were purified (Invitrogen) and then ligated into pMD18-T vectors (Takara Dalian Co. Ltd., China). The plasmids were extracted, and the inserts were sequenced at Life technologies (Shanghai, China). Ten sequences isolated in this study were deposited in GenBank under accession number KJ001824-KJ001832 and JX855794. Percent identity was calculated with Lasergene (DNAstar Version 7.1.0). Genetic distances between the known strains and the virus isolates were calculated with the MEGA software (Version 5), and phylogenetic trees were constructed by the neighbor-joining method. Twenty-five related HEV strains were used as references in Fig 1.

Statistical Analysis

The chi-square test was used to analyze categorical variables when appropriate. The magnitude of the association between variables and seropositivity is expressed as an odds ratio (OR) with 95% confidence intervals (95% CI). Univariate and multivariate logistic regression analysis was carried out to identify which variables were significantly associated with anti-HEV seropositivity. Logistic regression was performed using R (R version 2.15.3 for windows x64). The results of the OR and its 95% CI were prepared using R with a Fourfold plot [21]. A P value of less than 0.05 was considered significant. The best fit trendline for the positive rate data was selected based on the R-squared value of the curves drawn using the Microsoft Office Excel 2007 program (Microsoft, Redmond, WA). Questionnaire data collected from farmers and urban attendees were digitized and verified.

Results

The prevalence of antibodies against HEV among pigs

Among the 561 swine serum samples collected from swine farms, the overall seroprevalence was 64.7% (363/561). Swine sera were divided into 4 groups according to the pig herds on the farms. The positive rate of HEV-specific IgG in all groups ranged from 34.59% to 91.93% (TABLE 2). The data show some possible risk factors for HEV infection and revealed that boars had an approximately 21.5-fold higher risk (odds ratio [OR], 21.56; 95% confidence interval [CI], 9.47–49.10; $p<0.0001$), sows had a nearly 5-fold higher risk (OR, 5.04; 95%CI, 3.27–7.77; $p<0.0001$) and growing pigs had a 3.5-fold higher risk (OR, 3.51; 95%CI, 1.95–6.33; $p<0.0001$) than nursery pigs. Sows were divided into two groups to determine if the risk factor for HEV infection was due to multiparity. The results showed that multiparous sows had an approximately 2.5-fold higher risk (OR, 2.46; 95%CI, 1.45–4.18; $p<0.01$) than pre-farrowing sows, and there were statistically significant differences between them (Fig 2).

HEV RNA detection in pig bile samples and phylogenetic analysis

The 228 pig bile samples that were collected from 9 districts of the Pearl River Delta had been stored in Guangdong since 2011. As shown in TABLE 3, the detection rates of HEV RNA varied among districts of Guangdong Province. The highest prevalence was 22.73% (95%CI, 2.5–33.86) in Foshan, and the lowest was 6.25% (95%CI, 0–14.64) in Zhaoqing. The prevalence of HEV RNA in Guangzhou and Shenzhen ranged from 10% to 12.5%, while those in Huizhou, Zhuhai, and Dongguan ranged from 15% to 18.18%. Like Foshan, Jiangmen and Zhongshan also had a high prevalence of over 20% (Fig 3).

Phylogenetic tree constructed by the neighbor-joining method based on the partial nucleotide sequence of the ORF2 region (509 nucleotides) of 34 HEV isolates. Ten isolates were compared with 24 HEV reference sequences of different genotypes to analyze the genetic characteristics and evolutionary regularity. The nucleotide sequence identity among the 10 swine HEV isolates obtained from pigs from different farms in three years ranged between 94.3 and 99.8%. Based on the phylogenetic and homology analysis, the isolates we obtained from the 9 different districts shared high nucleotide similarities, and they were all belonged to genotype IV HEV. The isolates also shared the same phylogenetic branch with other Chinese isolates. The sequence similarity was approximately 87% between the isolates and the other genotype IV strains that had been isolated from HEV patients in China and Japan (Fig 1).

HEV-specific IgG detection in humans

The seroprevalence of IgG anti-HEV in serum samples collected from swine farmers was almost as high than that of the general population. However, it was surprising to find that the positive rates in urban people (38.34%, 74/193) were very high and almost as high as the total prevalence found in swine farm workers (48.25%, 55/114).

Univariate analysis was used to compare HEV seroprevalence rates between swine farmers and urban attendees. As seen in TABLE 4, the overall positive rates for gender ($p>0.23$) and exposure to pigs ($p>0.05$) appeared to have weak associations with HEV seropositivity. However, when age was stratified into groups, differences were visualized using the best fit trendline. Samples of ages ≤29, 30–39, 40–49, 50–59, and ≥60 years were examined (TABLE 4). There were significant differences between the age groups, and anti-HEV was distributed among all age groups. Seroprevalence rates stratified by age are shown in TABLE 4. Without any age group showing a plateau, the best fit curves showed that the positive rate for HEV IgG increased steadily with age (Fig 4.1–4.3). The trendlines suggested a higher positive rate for males than females across all age groups, especially among the swine farmers (Fig 4.1–4.3). The trendline of swine farmers in the 40–49 age group had a steeper slope (OR, 4.14; 95%CI, 2.15–7.96) (Fig 4.4–4.6). Nevertheless, it should be noted that women from swine farms appeared to have a lower risk of infection compared to that of the general population (Fig 4.5).

Discussion

Seroprevalence and risk factor of swine HEV in Guangdong

Some previous studies have been given a preliminary overview of the seroprevalence situation of swine HEV in China. A previous serological investigation found that the seroprevalence of anti-HEV IgG in pig herds was 26.8% (22/82) [22]. Another survey revealed that the prevalence of swine HEV was as high as 91.7% (77/84) [23]. The present study showed that a great HEV infection pressure exists in Guangdong pig herds, which had an overall seroprevalence of 64.7% (363/561), with an even higher positive rate in sows (72.72%, 208/286) and boars (91.93%, 57/62). The higher infection pressure on farms may have increased

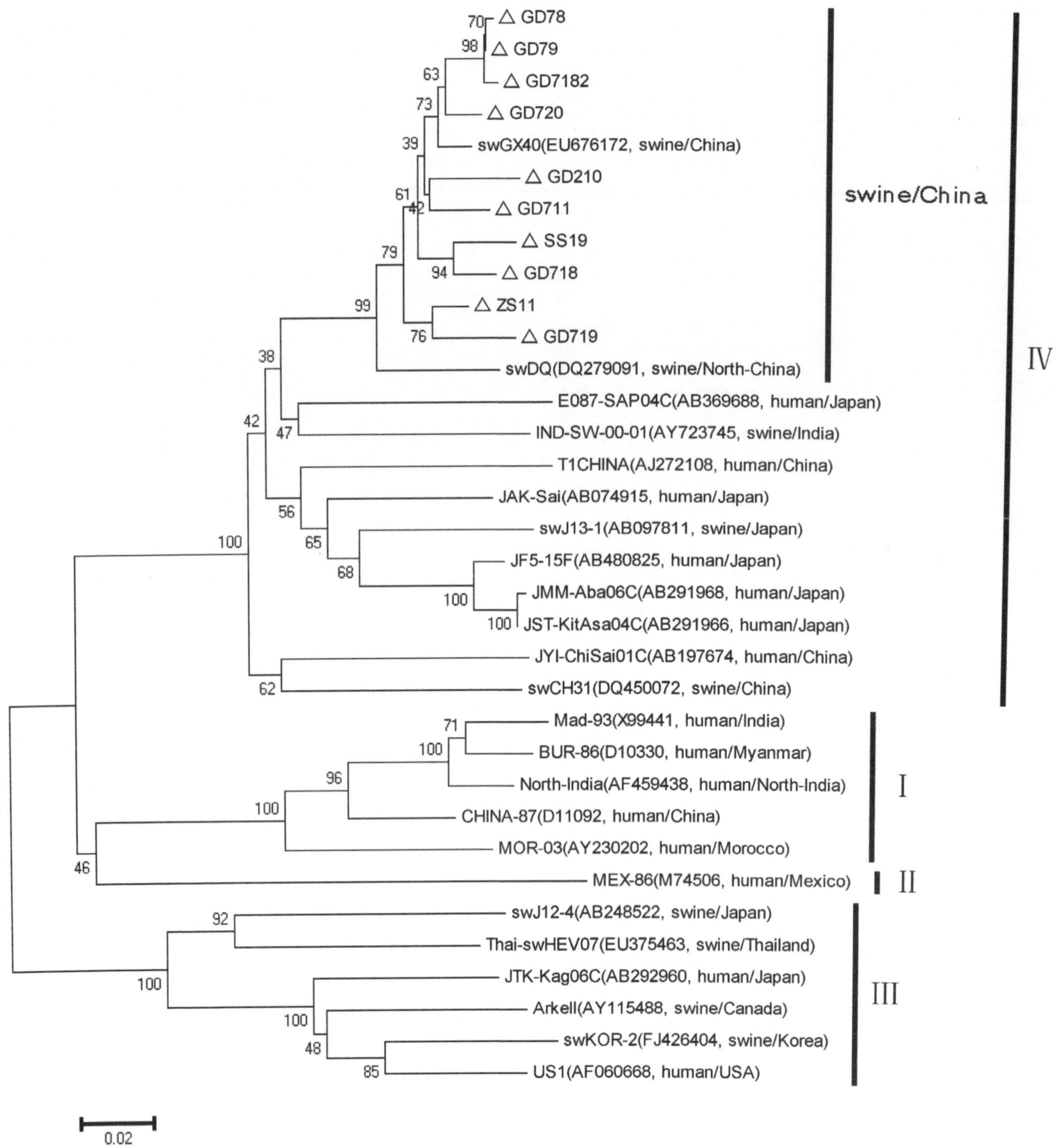

Figure. 1. A phylogenetic tree based on the full nucleotide sequence of HEV. Total RNA positive bile samples are 48, from 9 different districts in Guangdong. We have chosen at least one sample from each district by random sampling. The amplification products of ORF2 (509 nucleotides, primer sequences were HEV-INF and HEV-INR as described above) from 10 positive bile samples were sequenced and compared. The nucleotide sequence identity among the 10 swine HEV isolates obtained from pigs from different farms in three years ranged between 94.3 and 99.8%. Phylogenetic trees were constructed by the neighbor-joining method based on the partial nucleotide sequence of the ORF2 region (509 nucleotides). The bootstrap values (expressed as percentages) were determined on 1000 re-samplings of the data sets. △were the isolates in this study.

the probability of infection and consequently reduced the period between exposure and infection as described by Bouwknegt et al. [24]. With long-term exposure, the seroprevalence of swine HEV obviously accumulated with age. As the most senior animals on the farms, boars had the highest risk compared with piglets (approximately 21.5-fold) (Fig 2). In the present study, the HEV

Table 2. The prevalence of antibodies against HEV among pigs in 2011–2013, Guangdong Province, China.

	Serum sample		**Univariate analysis**		
	Positive/Total (n)	**Positive rate (%)**	**OR[a] (95% CI [b])**	**X^2 [c]**	**P value**
Different swine farm					
Nursery pig (<4 wks)	46/133	34.59	ref[d]		
Growing pig (>4 wk-6 mos)	52/80	65.00	3.51(1.95–6.33)	17.40	$<10^4$
Sow (>7 mos)	208/286	72.72	5.04(3.27–7.77)	53.74	$<10^4$
Boar (>7 mos)	57/62	91.93	21.56(9.47–49.10)	53.53	$<10^4$
Sow					
Pre-farrowing sow	58/96	60.42	ref		
Multiparous sow	150/190	78.95	2.46(1.45–4.18)	10.13	<0.01

a: OR, odds ratio;
b: 95%CI, 95%confidence interval;
c: X^2, Chi-Square Test;
d: ref, reference.

seroprevalence among multiparous sows and pre-farrowing sows was found to be 60.47% and 78.95%, respectively. According to a previous study conducted by Casas et al. [25], pre-farrowing sows had a relatively low seropositive (17%). These discrepancies might be caused by the different sensitivities of ELISA kits and variations. This study indicated that multiparous sows had an approximately 2.5-fold higher risk (OR, 2.46) than pre-farrowing sows. The reasons are still not clear. Recent studies revealed that stress from farrowing or high levels of steroid hormones from the physiological changes during pregnancy might reactivate HEV replication

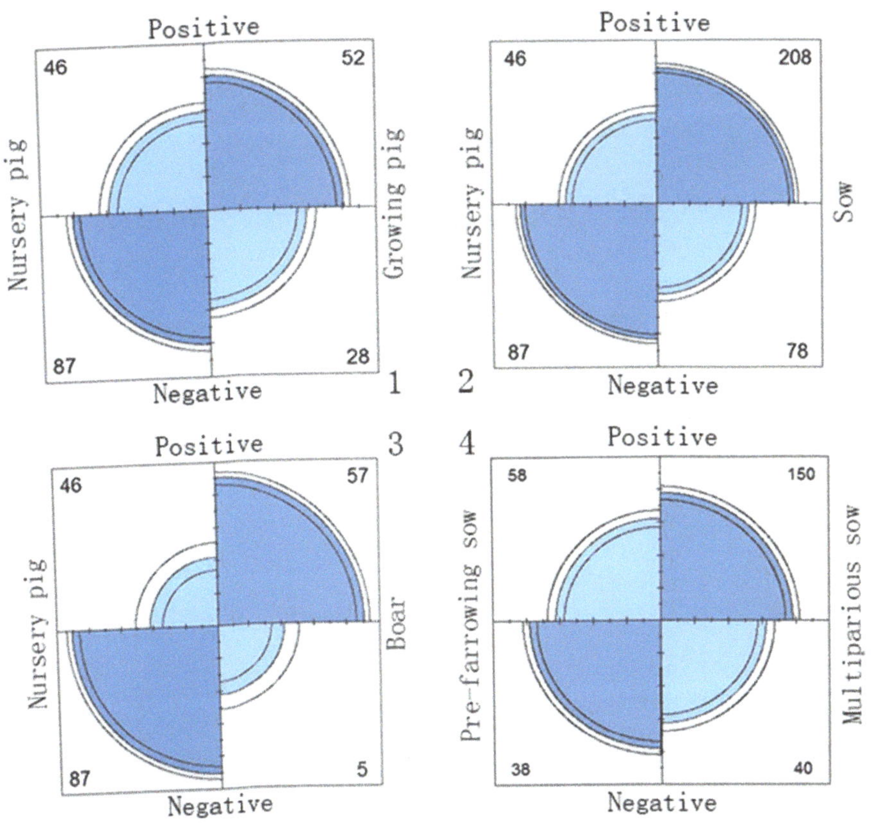

Figure. 2. Fourfold plot for odds ratio (OR). Univariate analysis results are presented in the Fourfold plot diagram. Fourfold display for seropositive data of sows: Evidence for pig herds bias or multiparity bias? The area of eachshaded quadrant shows the frequency, standardized to equate the margins for different herds and seropositive. Circular arcs show the limits of a 95% confidence interval for the odds ratio. 1. Compared with nursery pigs, the OR of growing pigs was 3.51(95%CI, 1.95–6.33; p<0.0001); 2. Compared with nursery pigs, the OR of sows was 5.04(95%CI, 3.27–7.77; p<0.0001); 3. Compared with nursery pigs, the OR of boars was 21.56(95%CI, 9.47–49.10; p<0.0001); 4. Compared with pre-farrowing sows, the OR of multiparous sows was 2.46(95%CI, 1.45–4.18; p<0.01).

Table 3. The prevalence of HEV RNA in pig bile samples collected from the Pearl River Delta. [b]

District	No. of bile samples	No. of positive HEV RNA samples	Analysis	
			positive rate (%)	95% CI
Dongguan	22	4	18.18	2.5–33.86
Foshan	44	10	22.73	10.35–35.11
Guangzhou	20	2	10.00	0–23.14
Huizhou	40	6	15.00	3.93–26.07
Jiangmen	48	10	20.83	9.34–32.32
Shenzhen	16	2	12.50	0–28.70
Zhaoqing	32	2	6.25	0–14.64
Zhongshan	40	8	20.00	8.24–31.76
Zhuhai	26	4	15.38	1.51–29.25

[b]Pig bile samples have been collected from 9 districts of the Pearl River Delta since 2011.

[26,27]. However, risk factors for HEV infection could be diverse. Not only pregnancy but also age could be a possible risk factor. Much more study should be progressed to figure out.

Comparison of human HEV prevalence between Guangdong and other provinces in China

In China, many surveys have shown an irregular seroprevalence of HEV in several provinces. Only approximately 10–20% of people have had an HEV infection in the Jilin province of China

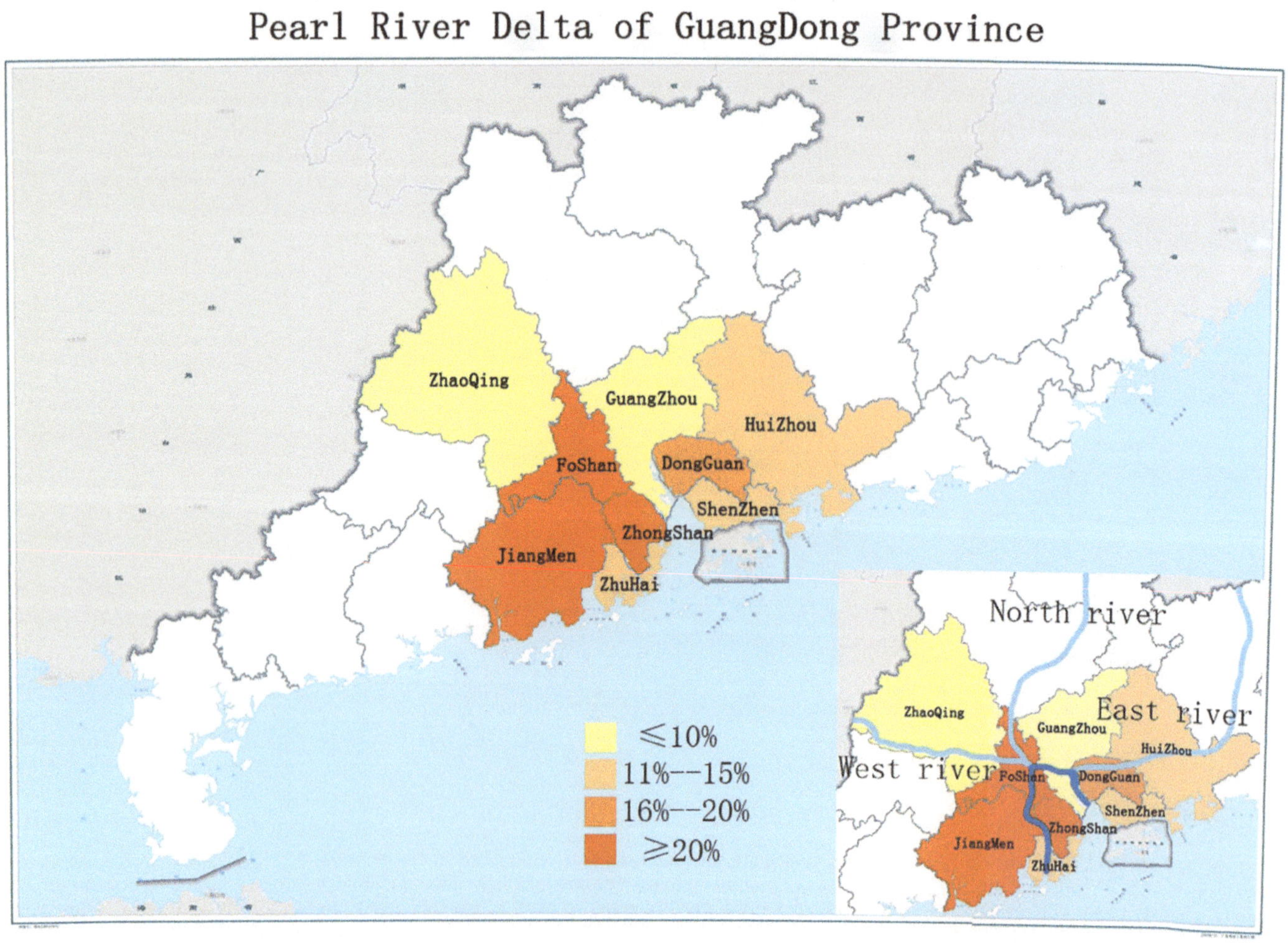

Figure. 3. A map indicating the prevalence of HEV RNA in the Pearl River Delta of Guangdong Province.

Table 4. The anti-HEV rates of the population in Guangdong province, China, according to age, gender, and pig-exposure status.

	Sample		Univariate analysis			
	Positive/Total (n)	Positive rate (%)	OR[a] (95% CI[b])	X^2 [c]	P value	
Age						
≤29	22/88	25.00	ref[d]			
30–39	24/76	31.58	1.38(0.60–3.21)	0.58	0.45	
40–49	51/88	57.95	4.14(2.15–7.96)	18.35	$<10^4$	**
50–59	14/29	48.28	2.80(1.09–7.22)	4.51	<0.05	*
≥60	18/26	69.23	6.75(2.60–17.54)	15.35	$<10^4$	**
Gender						
UF	30/77	38.96	ref			
UM	44/116	37.93	0.96(0.60–1.54)	0.03	≈1	
FF	19/47	40.43	ref			
FM	36/67	53.73	1.71(0.72–4.08)	1.46	0.23	
Pig-exposure						
UF	30/77	38.96	ref			
FF	19/47	40.43	1.06(0.54–2.08)	0.03	≈1	
UM	44/116	37.93	ref			
FM	36/67	53.73	1.90(0.99–3.66)	3.69	0.05	

a: OR, odds ratio; b: 95%CI, 95%confidence interval; c: X^2, Chi-Square Test; d: ref, reference; UF, urban female; UM, urban male; FF, swine farm female; FM, swine farm male; **, significant difference; *, different.

[15], while prevalence in northeastern China (47.7%–50%) has been reported to be very high [28]. However, such differences could have arisen because of differences in the study sample sizes or might reflect true regional differences. In the current survey, compared with the seroprevalence of HEV in the adjacent districts of Hunan (22.4%, 488/2181) and Hong Kong (28.7%, 129/450), Guangdong (38.3%, 74/193) had a significantly higher positive rate [5,16]. Guangdong rates were comparable to those in Taiwan (12%) and Korea (23.1%) but less than that in the highly endemic southwest France (52.5%), where the same diagnostic assay was used [17,29–30]. Nevertheless, direct comparisons of studies were limited by the use of different diagnostic assays [20]. The assay kit (Wantai) used in the current study, which had been used for a survey in Taiwan, is more sensitive than the HEV diagnostic serology kits previously available [17]. As Guangdong is a developed and populous area in China, its high population density, frequent public events, and abundant water resources in some regions of this province may contribute to its higher endemicity.

Risk factors for human HEV infection among genders and ages

The present study indicates the same conclusion as has been reported in several previous surveys [5,16–17], which is that the

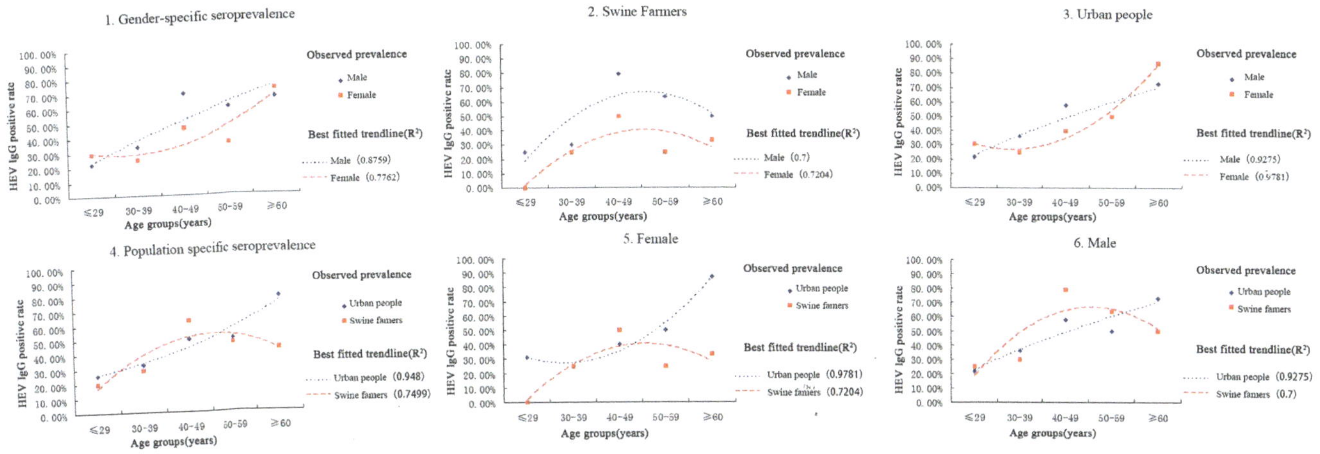

Figure. 4. The estimated seroprevalence of the hepatitis E virus in Guangdong Province. The best fit trendline for the positive rate data was selected based on the R-squared value of the curves drawn using the Microsoft Office Excel 2007 program (Microsoft, Redmond, WA). 1. Overall gender-specific seroprevalence; 2. Gender-specific seroprevalence of swine farm; 3. Gender-specific seroprevalence of the general population; 4. Overall population-specific seroprevalence; 5. Population-specific seroprevalence of females; 6. Population-specific seroprevalence of males.

positive rate for HEV IgG increased with age (TABLE 4) and that males commonly had higher rates than females, as shown in Fig 4.1–4.3. The significantly higher seroprevalence of HEV on swine farms compared to the general population had also been reported in several countries [31]. However, the trend is the opposite in Denmark [32] and northern Thailand [33]. According to Fig 4.4–4.6, the trendline for swine farmers showed a steeper slope in the 40–49 age group, revealing that occupational exposure could be a strong risk factor for HEV infection. Occupational exposure to pigs may be a relevant risk factor, but it is not the only risk factor for HEV infection. The higher prevalence of HEV infection in males than in females is even more obvious on swine farms. In the present study, the positive rates of urban people (38.34%, 74/193) were very high and very close to the total prevalence of swine farm workers (48.25%, 55/114). Although in contrast to urban people, swine farm workers came in contact with infected pigs in closer intervals. In addition, an increase in the HEV seroprevalence was noted in swine farmers of advanced age, while that of the general population remained on an upward trend across all ages (Fig 4.4–4.6). This is not surprising because several risk factors contributed to the seroprevalence of HEV in the general population. Validation of this trend using a larger cohort is required.

These results may be associated with the complicated exposure factors in cities and the single exposure factor on swine farms. The positive rate for urban females is higher than the positive rate for farm females, which confirms this conjecture (Fig 4.5). This may induce a similar seroprevalence between different genders and may be responsible in part for the high seroprevalence of the general population in both genders. Due to low cultural awareness and a lack of personal hygiene knowledge, workers on the swine farms in Guangdong are ill-equipped to protect themselves from HEV infection. The seroprevalence of females on swine farms is obviously lower than that of males and even lower than the overall seroprevalence of the general population (Fig 4.2 and Fig 4.5). Altogether, these findings may explain why the seroprevalence of urban females is slightly higher than those working on farms.

Elementary molecular epidemiology survey of swine HEV in Guangdong

In the current study, a total of 16.67% of pigs had HEV RNA in their bile. Nonetheless, this prevalence was significantly higher than that found in the Netherlands [34], the United States [35], Spain [25], and especially Japan [36]. The contact between workers and infected pig livers could be a possible route of HEV transmission [37,38]. Water sources contaminated with swine feces may have facilitated HEV transmission between farms [8,25]. As shown in Fig 3, the Pearl River consists of 3 tributaries, which are named the North River, the East River, and the West River (Fig. 3). The three branches intersect in Foshan (22.73%, 10/44) from 3 different directions and then end up in the Southern Sea of China through the Jiangmen (20.83%, 10/48) and the Zhongshan (20%, 8/40) districts. This reveals that the prevalence of swine HEV in each district accumulated along the Pearl River. These results strongly imply that swine HEV epidemic in Guangdong might mainly be water borne. However, some water samples collected from different districts in Guangdong have been tested for HEV RNA, nothing was found by this time (data not shown). Many more studies should be carried out to determine if a virus could transmit through contaminated river water.

The identity and relationship of HEV isolates in Guangdong with other isolates were shown in this study. In the U.S., HEV isolates from the same geographic region showed higher percentages of similar identities to each other than with those from different geographic regions [39]. In the present study, we found that ten new isolates belong to genotype IV and share the same phylogenetic branch as other Chinese isolates (Fig 1). They all displayed high homology with the swine HEV strains swDQ and swGX40 and exhibited a high identity with the human strains E087-SAP04C that were originally isolated from Japan. In other words, the results revealed that the genotype IV HEV strain currently exists in the pig herds of Guangdong Province. Pork safety should be of great concern for the public health in Guangdong Province.

In summary, this survey shows the possible relationship among swine, swine farmers and the general population in Guangdong Province, China. Exhaustive reports and epidemiological investigations in Guangdong were carried out, and the results highlighted pork safety concerns. Other risk factors for HEV infection were implied in this study. Meanwhile, the prevalence of HEV in Guangdong could provide a reference for vaccinations. However, there were some limitations to this study. The sample size was relatively small and may not be fully representative of Guangdong. Many more surveys should be carried out to confirm these findings. Thus, an extensive public health surveillance system needs to be established and maintained in Guangdong Province.

Supporting Information

Fig. S1 Analysis of nucleotide identity of partial ORF2 among 10 HEV isolates and with represent HEV isolate.

Acknowledgments

We thank Dr.Shuo Su of the South China Agricultural University for for his contributions in the major research design and technical editing assistance.

Author Contributions

Conceived and designed the experiments: GZ HL. Performed the experiments: HG FJ LW CL. Analyzed the data: HW. Wrote the paper: HL. Planned the study and revised the manuscript: SS. Contributed with sample collection: SD. Read and approved the final manuscript: HL SS SD HG FJ LW CL HW GZ.

References

1. Kamar N, Bendall R, Legrand-Abravanel F, Xia NS, Ijaz S, et al. (2012) Hepatitis E. Lancet 379: 2477–2488.
2. Purcell RH, Emerson SU (2008) Hepatitis E: an emerging awareness of an old disease. J Hepatol 48: 494–503.
3. Okamoto H (2007) Genetic variability and evolution of hepatitis E virus. Virus Res 127: 216–228.
4. Dong C, Dai X, Shao JS, Hu K, Meng JH (2007) Identification of genetic diversity of hepatitis E virus (HEV) and determination of the seroprevalence of HEV in eastern China. Arch Virol 152: 739–746.
5. Dong C, Dai X, Liang J, Dong M, Meng J (2012) Seroprevalence of hepatitis e virus varies considerably among chinese provinces. Hepat Mon 12: 386–390.
6. Zhen Y, Liang M, Wang J (2000) [Detection of IgG antibody to hepatitis G virus by ELISA]. Zhonghua Shi Yan He Lin Chuang Bing Du Xue Za Zhi 14: 151–153.
7. Zhou H, Jiang CW, Li LP, Zhao CY, Wang YC, et al. (2008) [Comparison of the reliability of two ELISA kits for detecting IgM antibody against hepatitis E virus]. Zhonghua Yu Fang Yi Xue Za Zhi 42: 667–671.
8. Zhuang H, Cao XY, Liu CB, Wang GM (1991) Epidemiology of hepatitis E in China. Gastroenterol Jpn 26 Suppl 3: 135–138.
9. Dalton HR, Bendall R, Ijaz S, Banks M (2008) Hepatitis E: an emerging infection in developed countries. Lancet Infect Dis 8: 698–709.

10. Matsuda H, Okada K, Takahashi K, Mishiro S (2003) Severe hepatitis E virus infection after ingestion of uncooked liver from a wild boar. J Infect Dis 188: 944.
11. Tamada Y, Yano K, Yatsuhashi H, Inoue O, Mawatari F, et al. (2004) Consumption of wild boar linked to cases of hepatitis E. J Hepatol 40: 869–870.
12. Cheng PN, Wang RH, Wu IC, Wu JC, Tseng KC, et al. (2007) Seroprevalence of hepatitis E virus infection among institutionalized psychiatric patients in Taiwan. J Clin Virol 38: 44–48.
13. Lu J, Zhou Y, Lin X, Jiang Y, Tian R, et al. (2009) General epidemiological parameters of viral hepatitis A, B, C, and E in six regions of China: a cross-sectional study in 2007. PLoS One 4: e8467.
14. Yu Y, Sun J, Liu M, Xia L, Zhao C, et al. (2009) Seroepidemiology and genetic characterization of hepatitis E virus in the northeast of China. Infect Genet Evol 9: 554–561.
15. Zhu G, Qu Y, Jin N, Sun Z, Liu T, et al. (2008) Seroepidemiology and molecular characterization of hepatitis E virus in Jilin, China. Infection 36: 140–146.
16. Chiu DM, Chan MC, Yeung AC, Ngai KL, Chan PK (2013) Seroprevalence of hepatitis E virus in Hong Kong, 2008–2009. J Med Virol 85: 459–461.
17. Lee JT, Shao PL, Chang LY, Xia NS, Chen PJ, et al. (2013) Seroprevalence of Hepatitis E Virus Infection among Swine Farmers and the General Population in Rural Taiwan. PLoS One 8: e67180.
18. Zhu FC, Zhang J, Zhang XF, Zhou C, Wang ZZ, et al. (2010) Efficacy and safety of a recombinant hepatitis E vaccine in healthy adults: a large-scale, randomised, double-blind placebo-controlled, phase 3 trial. Lancet 376: 895–902.
19. Tang X, Yang C, Gu Y, Song C, Zhang X, et al. (2011) Structural basis for the neutralization and genotype specificity of hepatitis E virus. Proc Natl Acad Sci U S A 108: 10266–10271.
20. Bendall R, Ellis V, Ijaz S, Ali R, Dalton H (2010) A comparison of two commercially available anti-HEV IgG kits and a re-evaluation of anti-HEV IgG seroprevalence data in developed countries. J Med Virol 82: 799–805.
21. Friendly M. (1994) A fourfold display for 2 by 2 by k tables. Technical Report 217, York University, Psychology Department.
22. Meng XJ, Dea S, Engle RE, Friendship R, Lyoo YS, et al. (1999) Prevalence of antibodies to the hepatitis E virus in pigs from countries where hepatitis E is common or is rare in the human population. J Med Virol 59: 297–302.
23. Wang YC, Zhang HY, Xia NS, Peng G, Lan HY, et al. (2002) Prevalence, isolation, and partial sequence analysis of hepatitis E virus from domestic animals in China. J Med Virol 67: 516–521.
24. Bouwknegt M, Frankena K, Rutjes SA, Wellenberg GJ, de Roda HA, et al. (2008) Estimation of hepatitis E virus transmission among pigs due to contact-exposure. Vet Res 39: 40.
25. Casas M, Cortes R, Pina S, Peralta B, Allepuz A, et al. (2011) Longitudinal study of hepatitis E virus infection in Spanish farrow-to-finish swine herds. Vet Microbiol 148: 27–34.
26. Fernandez-Barredo S, Galiana C, Garcia A, Vega S, Gomez MT, et al. (2006) Detection of hepatitis E virus shedding in feces of pigs at different stages of production using reverse transcription-polymerase chain reaction. J Vet Diagn Invest 18: 462–465.
27. Navaneethan U, Al MM, Shata MT (2008) Hepatitis E and pregnancy: understanding the pathogenesis. Liver Int 28: 1190–1199.
28. Taniguchi M, Kim SR, Mishiro S, Takahashi K, Shin MH, et al. (2009) Epidemiology of hepatitis E in Northeastern China, South Korea and Japan. J Infect 58: 232–237.
29. Park HK, Jeong SH, Kim JW, Woo BH, Lee DH, et al. (2012) Seroprevalence of anti-hepatitis E virus (HEV) in a Korean population: comparison of two commercial anti-HEV assays. BMC Infect Dis 12: 142.
30. Mansuy JM, Bendall R, Legrand-Abravanel F, Saune K, Miedouge M, et al. (2011) Hepatitis E virus antibodies in blood donors, France. Emerg Infect Dis 17: 2309–2312.
31. Drobeniuc J, Favorov MO, Shapiro CN, Bell BP, Mast EE, et al. (2001) Hepatitis E virus antibody prevalence among persons who work with swine. J Infect Dis 184: 1594–1597.
32. Christensen PB, Engle RE, Hjort C, Homburg KM, Vach W, et al. (2008) Time trend of the prevalence of hepatitis E antibodies among farmers and blood donors: a potential zoonosis in Denmark. Clin Infect Dis 47: 1026–1031.
33. Hinjoy S, Nelson KE, Gibbons RV, Jarman RG, Mongkolsirichaikul D, et al. (2012) A Cross-Sectional Study of Hepatitis E Virus Infection in Healthy People Directly Exposed and Unexposed to Pigs in a Rural Community in Northern Thailand. Zoonoses Public Health.
34. Bouwknegt M, Lodder-Verschoor F, van der Poel WH, Rutjes SA, de Roda HA (2007) Hepatitis E virus RNA in commercial porcine livers in The Netherlands. J Food Prot 70: 2889–2895.
35. Feagins AR, Opriessnig T, Guenette DK, Halbur PG, Meng XJ (2007) Detection and characterization of infectious Hepatitis E virus from commercial pig livers sold in local grocery stores in the USA. J Gen Virol 88: 912–917.
36. Yazaki Y, Mizuo H, Takahashi M, Nishizawa T, Sasaki N, et al. (2003) Sporadic acute or fulminant hepatitis E in Hokkaido, Japan, may be food-borne, as suggested by the presence of hepatitis E virus in pig liver as food. J Gen Virol 84: 2351–2357.
37. Jary C (2005) Hepatitis E and meat carcasses. Br J Gen Pract 55: 557–558.
38. Perez-Gracia MT, Mateos ML, Galiana C, Fernandez-Barredo S, Garcia A, et al. (2007) Autochthonous hepatitis E infection in a slaughterhouse worker. Am J Trop Med Hyg 77: 893–896.
39. Huang FF, Haqshenas G, Guenette DK, Halbur PG, Schommer SK, et al. (2002) Detection by reverse transcription-PCR and genetic characterization of field isolates of swine hepatitis E virus from pigs in different geographic regions of the United States. J Clin Microbiol 40: 1326–1332.

Permissions

All chapters in this book were first published in PLOS ONE, by The Public Library of Science; hereby published with permission under the Creative Commons Attribution License or equivalent. Every chapter published in this book has been scrutinized by our experts. Their significance has been extensively debated. The topics covered herein carry significant findings which will fuel the growth of the discipline. They may even be implemented as practical applications or may be referred to as a beginning point for another development.

The contributors of this book come from diverse backgrounds, making this book a truly international effort. This book will bring forth new frontiers with its revolutionizing research information and detailed analysis of the nascent developments around the world.

We would like to thank all the contributing authors for lending their expertise to make the book truly unique. They have played a crucial role in the development of this book. Without their invaluable contributions this book wouldn't have been possible. They have made vital efforts to compile up to date information on the varied aspects of this subject to make this book a valuable addition to the collection of many professionals and students.

This book was conceptualized with the vision of imparting up-to-date information and advanced data in this field. To ensure the same, a matchless editorial board was set up. Every individual on the board went through rigorous rounds of assessment to prove their worth. After which they invested a large part of their time researching and compiling the most relevant data for our readers.

The editorial board has been involved in producing this book since its inception. They have spent rigorous hours researching and exploring the diverse topics which have resulted in the successful publishing of this book. They have passed on their knowledge of decades through this book. To expedite this challenging task, the publisher supported the team at every step. A small team of assistant editors was also appointed to further simplify the editing procedure and attain best results for the readers.

Apart from the editorial board, the designing team has also invested a significant amount of their time in understanding the subject and creating the most relevant covers. They scrutinized every image to scout for the most suitable representation of the subject and create an appropriate cover for the book.

The publishing team has been an ardent support to the editorial, designing and production team. Their endless efforts to recruit the best for this project, has resulted in the accomplishment of this book. They are a veteran in the field of academics and their pool of knowledge is as vast as their experience in printing. Their expertise and guidance has proved useful at every step. Their uncompromising quality standards have made this book an exceptional effort. Their encouragement from time to time has been an inspiration for everyone.

The publisher and the editorial board hope that this book will prove to be a valuable piece of knowledge for researchers, students, practitioners and scholars across the globe.

List of Contributors

Michael J. Osland and Richard H. Day
U.S. Geological Survey, National Wetlands Research Center, Lafayette, Louisiana, United States of America

Jack C. Larriviere and Andrew S. From
Five Rivers Services, LLC, U.S. Geological Survey, National Wetlands Research Center, Lafayette, Louisiana, United States of America

Qiang He and Baoshan Cui
State Key Laboratory of Water Environmental Simulation, School of Environment, Beijing Normal University, Beijing, China

Yuan An
School of Agriculture and Biology, MOA Key Laboratory of Urban Agriculture (South), Shanghai Jiao Tong University, Shanghai, China

Gao Hu, Fang Lu and Bao-Ping Zhai
Key Laboratory of Integrated Management of Crop Diseases and Insect Pests (Ministry of Education), College of Plant Protection, Nanjing Agricultural University, Nanjing, China

Ming-Hong Lu and Wan-Cai Liu
Division of Pest Forecasting, China National Agro-Tec Extension and Service Center, Beijing, China

Wei-Gen Xu and Xue-Hui Jiang
Plant Protection Station of Zhejiang Province, Hangzhou, China

Mathilde C. Paul
Centre de coopération internationale en recherche agronomique pour le développement (CIRAD), UR AGIRs, Montpellier, France
Université de Toulouse, INP-ENVT, UMR ENVT INRA 1225 IHAP, Toulouse, France

Marius Gilbert
Biological Control and Spatial Ecology, Université Libre de Bruxelles, Brussels, Belgium
Fonds National de la Recherche Scientifique, Brussels, Belgium

Nguyen Viet Khong
National Institute of Veterinary Research, Hanoi, Vietnam

Stéphanie Desvaux
Centre de coopération internationale en recherche agronomique pour le développement (CIRAD), UR AGIRs, Montpellier, France
Direction Régionale de l'Alimentation, de l'Agriculture et de la Forêt de Languedoc- Roussillon, Montpellier, France

Harena Rasamoelina Andriamanivo
FOFIFA-DRZV, Antananarivo, Madagascar
Faculté de Médecine d'Antananarivo - Département Vétérinaire, Antananarivo, Madagascar

Marisa Peyre and Véronique Chevalier
Centre de coopération internationale en recherche agronomique pour le développement (CIRAD), UR AGIRs, Montpellier, France

Weerapong Thanapongtharm
Department of Livestock Development, Bangkok, Thailand

Jessica L. O'Connell and Maggi Kelly
Department of Environmental Sciences, Policy and Management, University of California, Berkeley, Berkeley, California, United States of America

Kristin B. Byrd
United States Geological Survey, Menlo Park, California, United States of America

Liqun Zhu, Naijuan Hu and Zhengwen Zhang
College of Agriculture, Nanjing Agricultural University, Nanjing, China

Minfang Yang and Xinhua Zhan
College of Resources and Environmental Science, Nanjing Agricultural University, Nanjing, China

Jianbo Wang and Hongwei Ni
Key Laboratory for Vegetation Ecology, Ministry of Education, Institute of Grassland Science, Northeast Normal University, Changchun, Jilin Province, P. R. China

Tingcheng Zhu
Key Laboratory for Vegetation Ecology, Ministry of Education, Institute of Grassland Science, Northeast Normal University, Changchun, Jilin Province, P. R. China

Haixiu Zhong, Xiaoling Fu and Jifeng Wang
Institute of Natural Resources and Ecology, Heilongjiang Academy of Sciences, Harbin, Heilongjiang Province, P.R. China

Paula Meli
Natura y Ecosistemas Mexicanos A.C., México DF, México
Departamento de Ciencias de la Vida, Universidad de Alcalá, Alcaláde Henares, Madrid, Spain

José María Rey Benayas
Departamento de Ciencias de la Vida, Universidad de Alcalá, Alcaláde Henares, Madrid, Spain

Patricia Balvanera and Miguel Martínez Ramos
Centro de Investigaciones en Ecosistemas, Universidad Nacional Autónoma de México, Morelia, Michoacán, México

Gao Hu, Fang Lu, Bao-Ping Zhai and Xiao-Xi Zhang
Key Laboratory of Integrated Management of Crop Diseases and Pests (Ministry of Education), College of Plant Protection, Nanjing Agricultural University, Nanjing, China

Ming-Hong Lu and Wan-Cai Liu
Division of Pest Forecasting, China National Agro-Tec Extension and Service Center, Beijing, China

Feng Zhu
Plant Protection Station of Jiangsu Province, Nanjing, China

Xiang-Wen Wu
Plant Protection Station of Shanghai City, Shanghai, China

Gui-Hua Chen
Plant Protection Station of Jinhua City, Jinhua, China

Kun Ma and Mingxiang Zhang
School of Nature Conservation, Beijing Forestry University, Beijing, People's Republic of China

Liangzhi You
School of Nature Conservation, Beijing Forestry University, Beijing, People's Republic of China
Environment and Production Technology, International Food Policy Research Institute, Washington, D.C., United States of America

Junguo Liu
School of Nature Conservation, Beijing Forestry University, Beijing, People's Republic of China
Ecosystems Services & Management Program, International Institute for Applied Systems Analysis, Laxenburg, Austria

Michael Grace
Water Studies Centre and School of Chemistry, Monash University, Clayton, Australia

Wenzhong Tang, Baoqing Shan, Wenqiang Zhang, Hong Zhang and Yuekui Ding
State Key Laboratory on Environmental Aquatic Chemistry, Research Center for Eco-Environmental Sciences, Chinese Academy of Sciences, Beijing, China

Lishuo Wang
State Key Laboratory on Environmental Aquatic Chemistry, Research Center for Eco-Environmental Sciences, Chinese Academy of Sciences, Beijing, China
China University of Mining & Technology (Beijing), Beijing, China

Junhong Bai
State Key Laboratory of Water Environment Stimulation, School of Environment, Beijing Normal University, Beijing, P. R. China
Wetland Biogeochemistry Laboratory, Soil and Water Science Department, University of Florida, Gainesville, Florida, United States of America

Rong Xiao, Qingqing Zhao, Qiongqiong Lu and Junjing Wang
State Key Laboratory of Water Environment Stimulation, School of Environment, Beijing Normal University, Beijing, P. R. China

K. Ramesh Reddy
Wetland Biogeochemistry Laboratory, Soil and Water Science Department, University of Florida, Gainesville, Florida, United States of America

Yuanchun Zou, Guoping Wang, Xiaonan Lou, Xiaofei Yu and Xianguo Lu
Key Lab of Wetland Ecology and Environment, Northeast Institute of Geography and Agroecology, Chinese Academy of Sciences, Changchun, China

Huihua Zhang, Junjian Chen, Guoyi Yang and Dingqiang Li
Guangdong Institute of Eco-environmental and Soil Sciences, Guangzhou, China

Li Zhu
Management School, Jinan University, Guangzhou, China

Steven C. Zeug and Bradley J. Cavallo
Cramer Fish Sciences, Auburn, California, United States of America

Nicolas Gaidet, Julien Cappelle, Alexandre Caron, Gilles Balança and Vladimir Grosbois
CIRAD-ES, UPR AGIRS, Montpellier, France

Ahmed B. Ould El Mamy
Centre National d'Elevage et de Recherches Vétérinaires, Nouakchott, Islamic Republic of Mauritania

Graeme S. Cumming and Mduduzi Ndlovu
Percy FitzPatrick Institute, DST/NRF Center of Excellence, University of Cape Town, Cape Town, South Africa

Patricia Gil, Saliha Hammoumi and Renata Servan de Almeida
CIRAD-BIOS, UMR CIRAD/INRA CMAEE, Montpellier, France

Sasan R. Fereidouni
Friedrich- Loeffler-Institut, Insel Riems, Germany

Giovanni Cattoli
OIE/FAO Reference Laboratory for Avian Influenza and Newcastle Disease, Istituto Zooprofilattico Sper. le delle Venezie, Legnaro, Italy

Celia Abolnik
ARC-Onderstepoort Veterinary Institute, and Department of Production Animal Studies, Faculty of Veterinary Science, University of Pretoria, Onderstepoort, South Africa

Josphine Mundava
National University of Science and Technology, Bulawayo, Zimbabwe

Bouba Fofana
Direction Nationale des Eaux et Forêts du Mali, Bamako, Mali

Yelli Diawara
Parc National du Banc d'Arguin, Nouakchott, Islamic Republic of Mauritania

Renata Hurtado
Department of Microbiology, Institute of Biomedical Science, University of São Paulo, São Paulo, Brazil
Department of Preventive Medicine and Animal Health, Faculty of Veterinary Medicine University of São Paulo, São Paulo, Brazil

Scott H. Newman
Food & Agriculture Organisation of the United Nations, Wildlife Health & Ecology Unit, EMPRES Animal Health, Rome, Italy

Tim Dodman
Wetlands International, Wageningen, The Netherlands

Rami S. Mangoubi, Nathan C. Lowry and Mukund Desai
Algorithms and Software, Charles Stark Draper Laboratory, Inc., Cambridge, Massachusetts, United States of America

Matteo Convertino
Risk and Decision Science Team, Environmental Laboratory, Engineering Research and Development Center, United States Army Corps of Engineers, Concord, Massachusetts, United States of America
Department of Agricultural and Biological Engineering - Institute of Food and Agricultural Sciences, University of Florida, Gainesville, Florida, United States of America
Florida Climate Institute, University of Florida-Florida State University, Gainesville, Florida, United States of America

Igor Linkov
Risk and Decision Science Team, Environmental Laboratory, Engineering Research and Development Center, United States Army Corps of Engineers, Concord, Massachusetts, United States of America
Department of Engineering and Public Policy, Carnegie Mellon University, Pittsburgh, Pennsylvania, United States of America

Hang Wang, HongYi Li, ZhiJian Zhang and Xin Hua Xu
College of Natural Resource and Environmental Sciences, China Academy of West Development, Zhejiang University, Hangzhou, Zhejiang Province, China

Jeffrey D. Muehlbauer
Curriculum for the Environment & Ecology, University of North Carolina at Chapel Hill, Chapel Hill, North Carolina, United States of America

Qiang He
Department of Civil & Environmental Engineering, University of Tennessee, Knoxville, Tennessee, United States of America

ChunLei Yue
Institute of Ecology, Zhejiang Forestry Academy, Hangzhou, China

DaQian Jiang
Department of Earth and Environmental Engineering, Henry Krumb School of Mines, Columbia University, New York, New York, United States of America

Roland I. Hall, Andrea J. Farwell, D.George Dixon and Johan A. Wiklund
Department of Biology, University of Waterloo, Waterloo, Ontario, Canada

Brent B. Wolfe
Department of Geography and Environmental Studies, Wilfrid Laurier University, Waterloo, Ontario, Canada

Thomas W. D. Edwards
Department of Earth & Environmental Sciences, University of Waterloo, Waterloo, Ontario, Canada

Jacint Ventura
Departament de Biologia Animal, Biologia Vegetal i Ecologia, Facultat de Biocie`ncies, Universitat Autònoma de Barcelona, Cerdanyola del Vallès, Spain

Guillem Bagaria
Departament de Biologia Animal, Biologia Vegetal i Ecologia, Facultat de Biociències, Universitat Autònoma de Barcelona, Cerdanyola del Vallès, Spain
CREAF, Cerdanyola del Vallès, Spain

Maria Assumpció Sans-Fuentes
School of Geography and Development, University of Arizona, Tucson, Arizona, United States of America
Departament de Fisiologia Animal, Facultat de Farmàcia, Universitat de Barcelona, Barcelona, Spain

Roger Pérez-Hernández
Instituto de Zoología y Ecología Tropical, Facultad de Ciencias, Universidad Central de Venezuela, Caracas, Venezuelaé

Haifeng Gao, Junhong Bai, Qingqing Zhao, Qiongqiong Lu and Junjing Wang
State Key Laboratory of Water Environment Simulation, School of Environment, Beijing Normal University, Beijing, China

Xinhua He
NSW Department of Primary Industries, West Pennant Hills, New South Wales, Australia

Huanbin Liang, Shuo Su, Shengchao Deng, Honglang Gu, Fangxiao Ji, Lifang Wang, Chumin Liang, Heng Wang and Guihong Zhang
MOA Key Laboratory of Animal Vaccine Development, College of Veterinary Medicine, South China Agricultural University, Guangzhou, China

Index

$150
I